COMPUTER PRINCIPLES AND DESIGN IN VERILOG HDL

COMPUTER PRINCIPLES AND DESIGN IN VERILOG HDL

Yamin Li

Hosei University, Japan

清华大学出版社
TSINGHUA UNIVERSITY PRESS

Registered office:
John Wiley & Sons Singapore Pte. Ltd., I Fusionopolis Walk, #07-01 Solaris South Tower, Singapore 138628.

For details of our global editorial offices, for customer services and for information about how to apply for permission to reuse the copyright material in this book please see our website at www.wiley.com.

Library of Congress Cataloging-in-Publication Data applied for.

ISBN: 9781118841099

Typeset in 9/11pt TimesLTStd by SPi Global, Chennai, India

1 2015

Brief Contents

Contents

List of Figures

List of Tables

Preface

Welcome to read this book. The book describes computer principles, computer designs, and how to use Verilog HDL (hardware description language) to implement the designs. Besides the source codes of Verilog HDL, the simulation waveforms are also included in the book for easy understanding of the designs. Briefly, the contents of the book include

1. Binary number system and basic logic circuit design.
2. Computer arithmetic algorithms and implementations.
3. Single-cycle, multiple-cycle, and pipelined CPU designs.
4. Interrupt and exceptions, and precise interrupt in pipelined CPUs.
5. Floating-point algorithms and design of a pipelined CPU with FPU (floating-point unit).
6. Memory hierarchy and designs of instruction and data caches and TLBs (translation lookaside buffers).
7. Design of a CPU with FPU, caches, TLBs, and interrupt/exceptions.
8. Multithreading CPU and multicore CPU designs.
9. Input/output interface controllers, and I2C and PCI (peripheral component interconnect) buses.
10. High-performance supercomputers and interconnection networks.

For each design, we describe the principle, illustrate the schematic circuit in detail, give the implementation in Verilog HDL from scratch, and show the simulation waveforms. All the Verilog HDL codes were compiled and simulated with ModelSim, and some of them were implemented on Altera FPGA (field programmable gate array) boards. If readers use Xilinx FPGA boards, only the memory codes that use the Altera LPM (library of parameterized modules) need to be revised with Xilinx Block Memory Generator. Also, some readers may use VHDL. There would not be a tremendous amount of work in producing a VHDL version. There are also some C codes in the book that were compiled with gcc for x86 on CentOS.

Via Wiley's web page (www.wiley.com\go\li\verilog), readers can download all the source codes listed in the book, including the Verilog HDL codes, the memory initialization files in both `mif` and `hex` formats, the C codes, and the MIPS assembly codes. The test benches that are not shown in the book, and the `asmsim.jar`, a MIPS graphical assembler and simulator, are also available for download.

All the figures in the book can be downloaded by instructors who teach courses related to Digital Circuit Design, Verilog HDL, Computer Principles, Computer Organization, Microprocessor Design, or Computer Architecture. Full-color, well-sized PDFs are available for each figure and can be used to prepare lecture notes.

The contents of the book cover a wide range of levels. Chapter 1, Computer Fundamentals and Performance Evaluation, provides an overview of the book. The contents of Chapter 2, A Brief Introduction to Logic Circuit Design and Verilog HDL, Chapter 3, Computer Arithmetic Algorithms and Implementations, and Chapter 4, Instruction Set Architecture and ALU Design, are suitable for second year students. Chapter 5, Single-Cycle CPU Design in Verilog HDL, Chapter 6, Exceptions and Interrupts Handling and Design in Verilog HDL, Chapter 7, Multiple-Cycle CPU Design in Verilog HDL, Chapter 8, Design of Pipelined CPU with Precise Interrupt in Verilog HDL, and some parts of Chapter 14, are for third year students. The contents of the next four chapters, Chapter 9, Floating-Point Algorithms and FPU Design in Verilog HDL, Chapter 10, Design of Pipelined CPU with FPU in Verilog HDL, Chapter 11, Memory Hierarchy and Virtual Memory Management, and Chapter 12, Design of Pipelined CPU with Caches and TLBs in Verilog HDL, can be studied in the fourth year. And the rest, Chapter 13, Multithreading CPU and Multicore CPU Design in Verilog HDL, Chapter 14, Input/Output Interface Controller Design in Verilog HDL, and Chapter 15, High-Performance Computers and Interconnection Networks, are intended for graduate students.

Although no prerequisite course or knowledge is required for this book, knowing Digital Logic Design, C programming language, and Verilog HDL or VHDL in advance will help students to understand the contents of the book quickly. We hope readers can understand the computer principles deeply and enhance their CPU design skills by using CAD/CAE tools and HDLs. As the saying goes, "Doing is better than seeing." The author encourages readers to simulate the circuits given in the book and implement/simulate your own ideas while reading the book. One more important issue besides understanding the contents of the book is to prepare readers with the skills for research. Therefore, we have listed some exercise problems to encourage readers to investigate some techniques that are not explained in the book.

All source codes included in the book can be used and modified freely; the author is not liable for any resulting penalty through the use of these codes for CPU or circuit design. Any comments are welcome, errata and bug reports are especially welcome. Email to: yamin@ieee.org or yamin@computer.org.

Yamin Li

1

Computer Fundamentals and Performance Evaluation

Welcome to read "Computer Principles and Design in Verilog HDL". This book starts from the very beginning – introducing the basic logic operations. You will learn that any digital circuit can be designed by using three kinds of logic gates: the AND gate, the OR gate, and the NOT gate. Then the methods of designing the basic components of the CPU (central processing unit), such as the adder, the subtracter, the shifter, the ALU (arithmetic logic unit), and register file, by using these gates, will be presented. After introducing the designs of various simple CPUs, this book describes how to design a multicore CPU with each core containing an IU (integer unit), an FPU (floating-point unit), an instruction cache, a data cache, an instruction TLB (translation lookaside buffer), a data TLB, and the mechanism of interrupt/exceptions.

The design of popular I/O (input/output) interfaces, such as the UART (universal asynchronous receiver transmitter), PS/2 keyboard/mouse and VGA (video graphics array) interfaces, the I2C (inter-integrated circuit) serial bus controller, and the PCI (peripheral component interconnect) parallel bus controller, will be also described. Except for the PS/2 mouse interface controller, all of the circuits of CPUs, I/O interfaces, and bus controllers were designed in Verilog HDL (hardware description language) and simulated with ModelSim. The Verilog HDL source codes and simulation waveforms are also included in the book. Finally, the book describes how to design an interconnection network for connecting multiple CPU and memory modules together to construct a high-performance supercomputer.

This chapter introduces some basic concepts, the organization of modern computers, and the evaluation method of computer performance.

1.1 Overview of Computer Systems

Suppose that you have a personal computer, which has 8 GB memory and the clock frequency is 4 GHz. Question: Are the "G" in 8 GB and "G" in 4 GHz same? The answer will be given in the last part of this section. Before that, we introduce the terminologies of "computer" and "computer system", a brief history of the computer, instruction set architectures (ISAs), and the differences between RISC (reduced instruction set computer) and CISC (complex instruction set computer).

1.1.1 Organization of Computer Systems

You may know what a "single-chip computer" is, or have heard about it (it does not matter even if you have never). It is an IC (integrated circuit) chip, in which there is a CPU (or a microprocessor), a small amount of memory, and some I/O interface controllers. Generally, any IC chip or printed circuit board that contains these three kinds of components is called a computer.

Computer Principles and Design in Verilog HDL, First Edition. Yamin Li.
© 2015 Tsinghua University Press. All rights reserved. Published 2015 by John Wiley & Sons Singapore Pte Ltd.
Companion Website: www.wiley.com/go/li/verilog

Then, one more question: is it possible for an end user to use this computer directly? The answer is "no". The computer we often say is actually a "computer system". A computer system consists of not only a computer, but also the software, I/O devices, and power supply.

The essential software is an operating system (OS). It manages all the resources of the computer, I/O devices, and other software, and provides an interface for users to use the computer system. The MS-DOS (Microsoft disk operating system) was the early operating system for IBM PC. The various versions of Windows were then provided with graphical user interfaces (GUIs). Another OS is the open-source Linux kernel. The various versions of the operating systems that support GUI, such as Fedora, Ubuntu, OpenSUSE, CentOS, and Scientific Linux, were developed on the top of the Linux kernel. All of the above are called "system software".

Software is a collection of programs and related data. A program provides the instructions for telling a computer what to do and how to do it. The instructions are represented in binary format that a computer can understand. An executable program is usually generated by a compiler, based on the source codes of the program which are prepared by the programmers. Programmers develop source codes in a high-level programming language, C for instance, with a text editor. A debugger is often used by programmers to debug the program. The compiler, editor, debugger, and other libraries are also programs, sometimes we call them "utilities". All programs other than the system software and utilities are called "applications".

I/O devices, also known as peripheral devices, are like the wheels of a car. The keyboard, mouse, display, hard disk, and printer are typical examples of I/O devices. A network can also be considered as an I/O device. Other I/O devices include the scanner, video camera, microphone, speaker, CD/DVD drive, and so on.

As mentioned above, a computer consists of a CPU, memory, and I/O interfaces. An I/O interface is a hardware controller that makes the communication between an I/O device and the CPU (and memory) possible. Memory provides places where the programs and data can be stored. The CPU reads instructions and data from memory and executes the instructions that perform operations on data.

Inside a CPU, there are multiplexers, ALUs, FPUs, register files, and a control unit which decodes instructions and controls the operations of all other components. Because the speed of the memory is much smaller than that of the CPU, in modern CPUs there is an instruction cache and a data cache. For the purpose of the fast virtual address translation, an instruction TLB and a data TLB are also fabricated in the CPUs. This book focuses on the computer and computer design in Verilog HDL.

As a summary, Figure 1.1 illustrates the organization of a computer system. You can see that your personal computer is not a computer, but is a computer system. A high-end mobile phone can also be considered as a computer system.

The computer and I/O devices belong to computer hardware. Therefore, we can say that a computer system consists of the computer hardware and the computer software (and a power supply).

1.1.2 A Brief History of the Computer

One of the earliest machines designed to assist people in calculations was said to be the Abacus, which was invented about 4400 years ago and is still being used by merchants, traders, and clerks in Asia, Africa, and elsewhere. Since then, various mechanical and electrical analog computers were invented. From 1940, computers entered the electronic era. Electronic computers can be classified into four generations. Each generation used a new technology to build computers.

The first-generation computers (1946–1955) were distinguished primarily for their use of vacuum tubes as the main electronic components. Several special-purpose electronic computers were developed between 1940 and 1945. The electronic numerical integrator and computer (ENIAC) built in 1946 was said to be the first general-purpose digital computer. However, ENIAC had an architecture that required rewiring a plug-board to change its programming. Computers of this generation could only perform a single task, and they had no operating system.

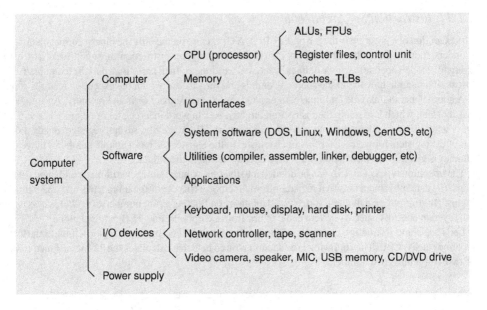

Figure 1.1 Computer system organization

The second-generation computers (1956–1963) used transistors as the processing elements and magnetic cores as their memory. Transistors were invented in 1947 as an alternative to vacuum tubes for use as electronic switches. The transistor was far superior to the vacuum tube, allowing computers to become smaller, faster, cheaper, more energy-efficient and more reliable than their first-generation predecessors. One of the most successful second-generation computers was the IBM 1401, which was introduced in 1959. By 1964, IBM had installed more than 100,000 units, capturing about one-third of the world market. Assembly programming languages became the major tool for software development, instead of the machine languages used in the first-generation computers. Operating systems and high-level programming languages were also being developed at this time.

The third-generation computers (1964–1970) were characterized by the transition from transistors to IC chips. The IC chip was invented in 1958 by Jack Kilby of Texas Instruments and Robert Noyce of Fairchild Semiconductor Corporation. In 1964, IBM announced System/360, one of the first computers to use ICs as its main processing technology. System/360 was designed to cover the complete range of applications, from small to large, both commercial and scientific. The design made a clear distinction between architecture and implementation, allowing IBM to release a suite of compatible designs at different prices. The number of transistors on an IC chip has been doubling approximately every 2 years, and the rate has held strong for more than half a century. The nature of this trend was first proposed by the Intel cofounder, Gordon Moore, in 1965.

The fourth-generation computers (1971 to the present) were distinguished primarily by their use of microprocessors and semiconductor memory. These were made possible by the improvements in IC design and manufacturing methods, which allowed engineers to create IC chips with tens of thousands of transistors (later hundreds of thousands, then millions, and now billions), a process now known as very large scale integration (VLSI). The personal computers first appeared during this time. In 1980, the MS-DOS was born, and in 1981 IBM introduced the personal computer for home and office use. Three years later, Apple gave us the Macintosh computers. The IBM PC used Intel 8086 as its microprocessor. The ubiquity of the PC platform has resulted in the Intel x86 becoming one of the most popular CPU architectures.

1.1.3   Instruction Set Architecture

The execution of a program is the job of the CPUs. A CPU can execute only the binary-format instructions it understands. A particular CPU has its own ISA. Generally, an instruction must consist of at least an operation code (opcode), which defines what will be done. Other parts that may be contained in an instruction include how to get the source operands and the place at which the execution result is stored.

A source operand may be an immediate or a data word in a register or in the memory. An immediate is a constant, which is given in the instruction directly. A data word in a register or memory is a variable. Of course, it is also allowed to put constants in memory. There may be several registers inside a CPU. All of the registers form a register file. Each register in the register file has a unique number. This register number is given in the instruction if a source operand is the register data.

The operation types of an ISA can be divided into the following: (i) integer arithmetic and logic calculations; (ii) data movement between register file and memory; (iii) conditional branches and unconditional jumps; (iv) subroutine call and return; (v) calculations on floating-point numbers; (vi) I/O accesses; and (vii) system controls, such as system calls, return from exceptions, and TLB manipulations.

Let's see some instruction examples of the Intel x86 and MIPS (microprocessor without interlocked pipeline stages) CPU. The following is a function written in C that calculates the 32-bit unsigned product of two 16-bit unsigned numbers.

```
unsigned int mul16 (unsigned int x, unsigned int y) { // mul by shift
       unsigned int a, b, c;                          // c = a * b
       unsigned int i;                                // counter
       a = x;                                         // multiplicand
       b = y;                                         // multiplier
       c = 0;                                         // product
       for (i = 0; i < 16; i++) {                     // for 16 bits
           if ((b & 1) == 1) {                        // LSB of b is 1
               c += a;                                // c = c + a
           }
           a = a << 1;                                // shift a 1-bit left
           b = b >> 1;                                // shift b 1-bit right
       }
       return(c);                                     // return product
}
```

Of course, you can use c = a * b to get the product, but we show here how to use the addition and shift operations to perform the multiplication. The following x86 assembly codes were generated by running the gcc -O4 -s command on an x86 machine under CentOS.

```
 1:    mul16:
 2:             pushl    %ebp
 3:             movl     %esp, %ebp
 4:             movl     8(%ebp), %ecx
 5:             pushl    %ebx
 6:             movl     12(%ebp), %edx
 7:             xorl     %ebx, %ebx
 8:             movl     $15, %eax
 9:             .p2align 2,,3
10:    .L6:
```

```
11:             testb   $1, %dl
12:             je      .L5
13:             addl    %ecx, %ebx
14:     .L5:
15:             sall    %ecx
16:             shrl    %edx
17:             decl    %eax
18:             jns     .L6
19:             movl    %ebx, %eax
20:             popl    %ebx
21:             leave
22:             ret
```

We do not explain all the instructions in the assembly codes above. The only instruction we examine is addl %ecx, %ebx in line 13. It performs the operation of c += a in the C codes. The addl (add long) points out that it is a 32-bit addition instruction; %ecx and %ebx are the registers of the two source operands; and %ebx is also the destination register. That is, addl %ecx, %ebx instruction adds the contents of the ecx and ebx registers, and places the sum in the ebx register. We can see that the x86 ISA has a two-operand format.

If we use the gcc -O4 -S command to compile the same C codes on a MIPS machine, we will get the following assembly codes:

```
 1:     mul16:
 2:             move    $6, $0
 3:             li      $3, 15
 4:     $L6:
 5:             andi    $2, $5, 0x1
 6:             addiu   $3, $3, -1
 7:             beq     $2, $0, $L5
 8:             srl     $5, $5, 1        # delay slot
 9:             addu    $6, $6, $4
10:     $L5:
11:             bgez    $3, $L6
12:             sll     $4, $4, 1        # delay slot
13:             j       $31
14:             move    $2, $6           # delay slot
```

The instruction in the ninth line, addu $6, $6, $4, performs the c += a. The addu (add unsigned) points out that it is an unsigned addition instruction; the first $6 is the destination register and the others are the two source registers. That is, this instruction adds the contents of the $6 and $4 registers, and places the sum in the $6 register. We can see that the MIPS ISA has a three-operand format. Note the instruction order in the MIPS assembly codes: The instructions srl, sll, and move, following-up the beq, bgez, and j, respectively, are always executed before the control is transferred to the target address. This feature is named delayed branch, which we will describe in detail in Chapter 8.

There are ISAs that have a one-operand format. In the CPUs that implement the one-operand ISAs, a special register, called an accumulator, acts as a default source register and the default destination register. It doesn't appear in the instruction. The instruction needs only to give a register name or a memory address for the second source operand. Z80 and 6502 are examples of the one-operand ISAs.

Table 1.1 Category of instruction set architecture

Register/memory-oriented		Accumulator-oriented	Stack-oriented
Three-operand	Two-operand	One-operand	Zero-operand
add x, y, z	add x, y	add x	add

Are there ISAs that have no (zero) operand? The answer is "yes". Such ISAs are stack-oriented. The two source operands are popped from the top of the stack and the result is pushed onto the stack. The stack-top pointer is adjusted automatically according to the operation of the instruction. The Bytecode, an ISA of JVM (Java virtual machine), is a typical example of the zero-operand ISAs.

Table 1.1 summarizes the four add formats of ISAs where x, y, and z are the register numbers, or some of them are memory addresses.

All the instructions are represented in binary numbers. In x86 ISA, there are eight registers: eax, ebx, ecx, edx, ebp, esp, esi, and edi. Therefore, the encoding of a register has three bits ($\log_2 8 = 3$ or $2^3 = 8$). To shorten program codes, x86 uses a short opcode to encode the very commonly used instructions. The encodings of the x86 assembly program are shown below. From this we can see that the length of the instruction encodings is not fixed. If we do not decode the current instruction, we cannot know from where the next instruction starts. This feature makes the design of a pipelined x86 CPU difficult.

```
 1:    mul16:
 2:            pushl   %ebp            ; 01010101
 3:            movl    %esp, %ebp      ; 1000100111100101
 4:            movl    8(%ebp), %ecx   ; 100010000100110100001000
 5:            pushl   %ebx            ; 01010011
 6:            movl    12(%ebp), %edx  ; 100010110101010100001100
 7:            xorl    %ebx, %ebx      ; 0011000111011011
 8:            movl    $15, %eax       ; 1011100000001111
 9:            .p2align 2,,3           ; 00000000000000000000000000
                                       ; 10001101011101100000000000
10:    .L6:
11:            testb   $1, %dl         ; 11110110110000100000000001
12:            je      .L5             ; 0111010000000010
13:            addl    %ecx, %ebx      ; 0000000111001011
14:    .L5:
15:            sall    %ecx            ; 1101000111100001
16:            shrl    %edx            ; 1101000111101010
17:            decl    %eax            ; 01001000
18:            jns     .L6             ; 0111100111110010
19:            movl    %ebx, %eax      ; 1000100111011000
20:            popl    %ebx            ; 01011011
21:            leave                   ; 11001001
22:            ret                     ; 11000011
```

The following shows the encodings of the MIPS assembly codes. MIPS32 ISA has a general-purpose register file that contains thirty-two 32-bit registers and hence a register number has five bits. We can see that the length of the MIPS instructions is fixed: All the instructions are represented with 32 bits. This feature makes the design of a pipelined MIPS CPU easy.

```
 1:    mul16:
 2:            move    $6, $0        # 0000000000000000011000000100001
 3:            li      $3, 15        # 0010010000000011000000000001111
 4:    $L6:
 5:            andi    $2, $5, 0x1   # 0011000010100010000000000000001
 6:            addiu   $3, $3, -1    # 0010010001100011111111111111111
 7:            beq     $2, $0, $L5   # 0001000010000000000000000000010
 8:            srl     $5, $5, 1     # 0000000000001010010100001000010
 9:            addu    $6, $6, $4    # 0000000110001000011000000100001
10:    $L5:
11:            bgez    $3, $L6       # 0000010001100001111111111111010
12:            sll     $4, $4, 1     # 0000000000001000010000001000000
13:            j       $31           # 0000001111000000000000000001000
14:            move    $2, $6        # 0000000110000000001000000100001
```

Although the length of each MIPS instruction is longer than that of some x86 instructions, the operation of each MIPS instruction is very simple. We say that the MIPS belongs to RISC and the x86 belongs to CISC.

1.1.4   CISC and RISC

CISC is the general name for CPUs that have a complex instruction set. An instruction set is said to be complex if there are some instructions that perform complex operations or the instruction formats are not uniform. The Intel x86 family, Motorola 68000 series, PDP-11, and VAX are examples of CISC.

The CISC instruction set tries to enhance the code density so that a computer system can use a small amount of memory, including cache, to store as many instructions as possible for reducing the cost and improving the performance.

CISC adopts two measures to reduce the code size – it lets an instruction perform as many operations as possible and makes the encoding of each instruction as short as possible. The first measure results in using microcode to implement the complex instructions, and the second measure results in a variable length of the instruction formats. As a consequence, it becomes difficult to design and implement a pipelined CISC CPU to obtain substantial performance improvements.

Is every instruction in a CISC complex? No. There are some very simple instructions in a CISC. The analysis of the instruction mix generated by CISC compilers shows that about 80% of executed instructions in a typical program uses only 20% of an instruction set and these instructions perform the simple operations and use only the simple addressing modes. We can see that almost all instructions in the x86 codes listed in Section 1.1.3 are simple instructions.

RISC is the general name for CPUs that have a small number of simple instructions. In the 1980s, the team headed by David Patterson of the University of California at Berkeley investigated the existing ISAs, proposed the term of RISC, and made two CPU prototypes: RISC-I and RISC-II. This concept was adopted in the designs of Sun Microsystems' SPARC microprocessors. Meanwhile, the team headed by John Hennessy of Stanford University did similar research and created MIPS CPUs. Actually, John Cocke of IBM Research originated the RISC concept in the project IBM 801, initiated from 1974. The first computer to benefit from this project was IBM PC/RT (RISC technology), the ancestor of the RS/6000 series.

There are two main features in a RISC CPU. One is the fixed length of the instruction formats. This feature makes fetching an instruction in one clock cycle possible. The other feature is the so-called load/store architecture. It means that only the load and store instructions transfer data between the register file and memory, and other instructions perform operations on the register operands. This feature makes the operations of the RISC instructions simple. Both features make the design of the pipelined RISC CPUs easier than that of the CISC CPUs.

Table 1.2 Some base units

Powers of 2			Powers of 10					
Memory capacity			Clock frequency			Cycle length		
K	kilo	2^{10}	1,024	K	kilo	10^3	m	milli	10^{-3}
M	mega	2^{20}	1,048,576	M	mega	10^6	µ	micro	10^{-6}
G	giga	2^{30}	1,073,741,824	G	giga	10^9	n	nano	10^{-9}
T	tera	2^{40}	1,099,511,627,776	T	tera	10^{12}	p	pico	10^{-12}
P	peta	2^{50}	1,125,899,906,842,624	P	peta	10^{15}	f	femto	10^{-15}
E	exa	2^{60}	1,152,921,504,606,846,976	E	exa	10^{18}	a	atto	10^{-18}
Z	zetta	2^{70}	1,180,591,620,717,411,303,424	Z	zetta	10^{21}	z	zepto	10^{-21}
Y	yotta	2^{80}	1,208,925,819,614,629,174,706,176	Y	yotta	10^{24}	y	yocto	10^{-24}

The SUN Microsystems SPARC, AMD 29000 family, SGI MIPS, IBM PowerPC, HP PA-RISC, and ARM are examples of RISC.

Was the CISC replaced by RISC? No. The reason is simple – the market. The Intel x86 ISA is widely used in the IBM-compatible PC, which is the most common computer system in the world. There is a huge amount of software resources that we cannot throw away.

Today, most CISC CPUs use a decoder to convert CISC instructions into RISC instructions (micro-operations) and then use RISC cores to execute these instructions. Meanwhile, many RISC CPUs add more new instructions to support the complex operations, the multimedia operations for instance.

1.1.5 Definitions of Some Units

Now, we answer the question raised in the beginning of this section. The G in 8 GB and G in 4 GHz are different. The first G equals $2^{30} = 1,073,741,824$; the second G equals $10^9 = 1,000,000,000$. A power of 10 is used for denoting the clock frequency; a power of 2 is used for denoting the memory capacity. Table 1.2 lists some of the base units.

The unit of byte is often used for counting the digital information, which commonly consists of 8 bits. Other units include half word (16 bits), word (32 bits), and long word (64 bits). Note that the x86 defines a word as 16 bits, a long word as 32 bits, and a quad word as 64 bits (only supported by x86-64).

1.2 Basic Structure of Computers

As we described in the previous section, a computer consists of a CPU, memory, and I/O interfaces. This section describes these three components briefly.

1.2.1 Basic Structure of RISC CPU

The job of a CPU is to execute instructions. Figure 1.2 shows a simplified structure of a RISC CPU. The instruction is fetched from the instruction memory. The content of PC (program counter) is used as the address of the instruction memory. In the register file, there are a certain number of registers which can store data. The ALU is responsible for calculations. Each of the two input operands of ALU can be either a register datum or an immediate provided in the instruction. The multiplexers (mux in the figure) are used for selecting an input from two inputs. The result calculated by ALU is saved into the register file. If the instruction is a load instruction, the data read from the data memory will be saved into the register file. If the instruction is a store instruction, the data read from the register file will be saved into the data memory. In both cases, the ALU output is used as the address of the data memory.

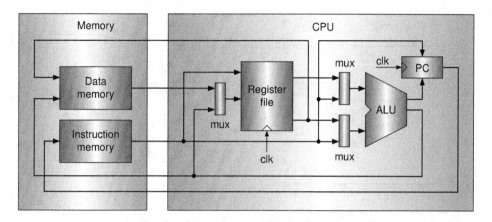

Figure 1.2 Simplified structure of RISC CPU

If the instruction is a conditional branch instruction, the flags, outputs of ALU, are used to determine whether jumping to the target address or not. The target address of the branch can be calculated by adding an immediate to the current PC. If the branch is not taken, the PC + 4 is saved into PC for fetching the next instruction (a 32-bit instruction has four bytes and the PC holds the byte address of the instruction memory). Figure 1.2 is actually a single-cycle CPU that executes an instruction in a single clock cycle.

In the single-cycle CPU, the execution of an instruction can be started only after the execution of the prior instruction has been completed. The pipelined CPU divides the execution of an instruction into several stages and allows overlapping execution of multiple instructions.

Figure 1.3 shows a simplified RISC pipelined CPU. There are five stages: (i) instruction fetch (IF), (ii) instruction decode (ID), (iii) execution (EXE), (iv) memory access (MEM), and (v) write back (WB). The pipeline registers are inserted in between the stages for storing the temporary results. Under ideal conditions, a new instruction can enter the pipeline during every clock cycle and five instructions are overlapped in execution. Therefore, the CPU throughput can be improved greatly.

From early 1980s, CPUs have been rapidly increasing in speed, much faster than the main memory. Referring to Figure 1.4, to hide the performance gap between the CPU and memory, instruction cache and data cache are fabricated on the CPU chips. A cache is a small amount of fast memory that is located in between the CPU and main memory and stores copies of the data or instructions from the most frequently used main memory locations. On a cache hit, the CPU can get data or instructions immediately without accessing the external main memory. Therefore, the average memory accessing time can be shortened.

In modern computer systems, the main memory is commonly implemented with DRAM (dynamic random access memory). DRAMs have large capacity and are less expensive, but the memory control circuit

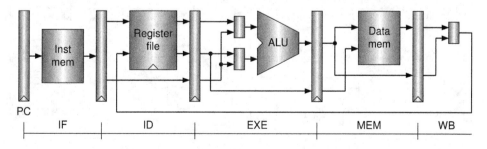

Figure 1.3 Simplified structure of pipelined CPU

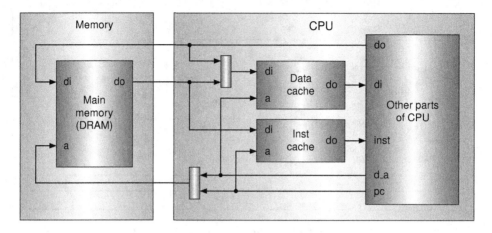

Figure 1.4   On-chip dedicated instruction cache and data cache

is complex. Some high-performance computer systems use the SRAM (static random access memory) as the main memory. The SRAM is faster but more expensive than DRAM.

1.2.2   Multithreading and Multicore CPUs

The pipelined CPU tries to produce a result on every clock cycle. The superscalar CPU tries to produce multiple results on every cycle by means of fetching and executing multiple instructions in a clock cycle. However, due to the control and data dependencies between instructions, the average number of instructions executed by a superscalar CPU per cycle is about 1.2. In order to achieve this 20% performance improvement, a superscalar must be designed with heavy extra circuits, such as register renaming, reservation stations, and reorder buffers, to support the parallel executions of multiple instructions. The superscalar CPUs cannot improve the performance further because of the lack of ILP (instruction level parallelism). Although the compilers can adopt the techniques of the loop-unrolling and static instruction scheduling, the improvement in performance is still limited.

Multithreading CPUs try to execute multiple threads in parallel. A thread is a sequential execution stream of instructions. Figure 1.5 shows the structure of a simplified multithreading CPU which can execute four threads simultaneously.

Each thread has a dedicated program counter and a register file, but the instruction cache, data cache, and functional units (FUs) are shared by all the threads. Sharing FUs and caches will increase

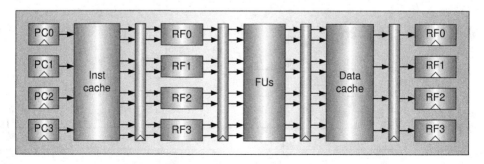

Figure 1.5   Simplified structure of a multithreading CPU

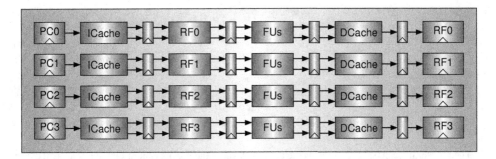

Figure 1.6 Simplified structure of a multicore CPU

the utilization of these components but make the control complex. Multithreading CPUs improve performance by exploiting the TLP (thread level parallelism).

A multicore CPU is an IC chip in which multiple ordinary CPUs (cores) are fabricated. An ordinary CPU may be a pipelined CPU, a superscalar CPU, or a multithreading CPU. Figure 1.6 shows the structure of a simplified multicore CPU in which there are four pipelined CPUs.

In a multicore CPU, each core has a dedicated level 1 (L1) instruction cache and a data cache. All the cores may share a level 2 (L2) cache, or each core has a dedicated L2 cache. Compared to the multithreading CPU, in the multicore CPU the FUs and L1 caches cannot be shared by all the cores. This decreases the utilization of the components but makes the design and implementation simple.

A multicore CPU can be considered as a small-scale multiprocessor system. The name "multicore" was given by the industry; it has almost the same meaning as the chip-multiprocessor, a name proposed by academia prior to the multicore.

1.2.3 Memory Hierarchy and Virtual Memory Management

Memory is a place in which the programs that are executing currently are stored temporarily. A program consists of instructions and data. The CPU reads instructions from memory and calculates on the data. As we have mentioned before, the speed of the memory is much smaller than that of the CPU. Therefore, high-speed caches are inserted in between the CPU and memory. Figure 1.7 shows a typical memory hierarchy. There are three levels of caches (L1, L2, and L3). An on-chip cache is one that is fabricated in the CPU chip; an off-chip cache is one that is implemented with the dedicated SRAM chip(s) outside the CPU. Commonly, the L1 on-chip cache consists of separated instruction cache and data cache, but the L2 on-chip cache is shared by instructions and data.

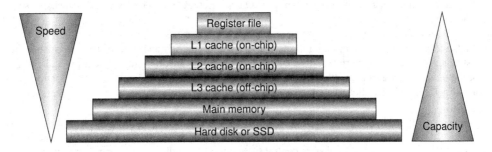

Figure 1.7 Memory hierarchy

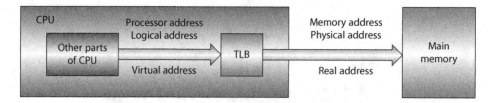

Figure 1.8   TLB maps virtual address to physical address

The register file can be considered as the topmost level of the memory hierarchy. It has a small capacity but the highest speed. The hard disk is at the bottom position of the memory hierarchy. In addition to storing files, the hard disk provides a large space for implementing the virtual memory. There is a trend that the SSD (solid-state drive) will replace the hard disk drive.

The virtual memory provides a large virtual address space for running programs (processes) to exist in. Each process uses its own virtual address space, usually starting from 0. The size of virtual address space is determined by the bits of the program counter. For example, a 32-bit program counter can access 4 GB of virtual memory. The virtual addresses cannot be used to access the main memory directly because multiple programs are allowed to be executed simultaneously. The virtual addresses must be mapped to physical memory locations. In a paging management mechanism, the virtual address space is divided into pages – blocks of contiguous virtual memory addresses. Thus, a virtual address consists of a virtual page number and an offset within a page. Only the virtual page number is needed to be mapped to a physical page number.

This mapping is handled by the operating system and the CPU. The operating system maintains a page table describing the mapping from virtual to physical pages for each process. Every virtual address issued by a running program must be mapped to the corresponding physical memory page containing the data for that location. If the operating system had to intervene in every memory access, the execution performance would be slow. Therefore, in the modern CPUs, the TLBs are fabricated for speeding up the mapping, as shown as in Figure 1.8.

The organization of the TLB is very similar to that of the cache. The TLB stores copies of the most frequently used page table entries. On a TLB hit, the physical page number can be obtained immediately without disturbing the operating system, so the actual mapping takes place very fast. Generally, the virtual address is also known as the logical address or processor address. And the physical address is also known as the real address or memory address. Note that for a particular architecture, these names may have different meanings.

1.2.4   Input/Output Interfaces and Buses

In a computer system, CPU communicates with I/O devices via I/O interfaces. There are multiple I/O devices in a computer system, and the multiple I/O interfaces are connected by a common bus, as shown as in Figure 1.9.

An I/O interface may have several I/O ports, and each I/O port has a unique I/O address. The CPU can use the I/O address to write control information to a control register and to read state information from a state register in an I/O interface. Although these two registers are different, but their I/O addresses can be the same. This is a significantly different feature from the memory, where the read information from a memory location is definitely the same as the information that was written into the same location before reading.

When an I/O device wants to communicate with the CPU, by pressing a key of the keyboard for instance, it sends an interrupt request to the CPU via the I/O interface. Once the CPU receives the request, it stops the current work and transfers control to an interrupt handler – a pre-prepared program. In the interrupt handler, the CPU can use I/O address to read or write I/O data. The Intel x86 has two special

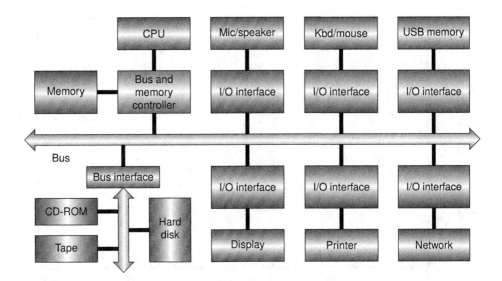

Figure 1.9 I/O interfaces in a computer system

instructions, in and out, for these operations. We say that the x86 has a dedicated I/O address space. But most of RISC CPUs use memory access instructions, load and store for example, to read and write I/O data. We say that these kinds of CPUs adopt a memory-mapped I/O address space.

1.3 Improving Computer Performance

Computer performance has improved incredibly since the first electronic computer was created. This rapid improvement came from the advances in IC technology used to build computers and the innovation in computer design. This section describes computer performance evaluation, trace-driven simulation, and high-performance parallel computers.

1.3.1 Computer Performance Evaluation

If we focus only on the execution time of real programs, then we can say that the shorter the execution time, the higher the performance. Therefore, we simply define the performance as the reciprocal of the time required. To calculate the execution time of a program, we have the following equation:

$$\text{Time} = I \times \text{CPI} \times \text{TPC} = \frac{I \times \text{CPI}}{F}$$

where I is the number of executed instructions of a program, CPI is the average clock cycles per instruction, and TPC is the time per clock cycle which is the reciprocal of the clock frequency (F).

Many researchers around the world are trying to improve computer performance by reducing the value of each of the three terms in the expression above. Architecture designers and compiler developers are trying to reduce the number of required instructions (I) of a program. Architecture and CPU designers are trying to decrease the CPI. And CPU designers and IC engineers are trying to increase the clock frequency (F).

Note that these three parameters are not independent. For example, CISC CPUs may reduce the instruction count I by providing complex instructions, but it may result in an increase of CPI; RISC CPUs may decrease CPI and TPC, but it may cause the increase of I.

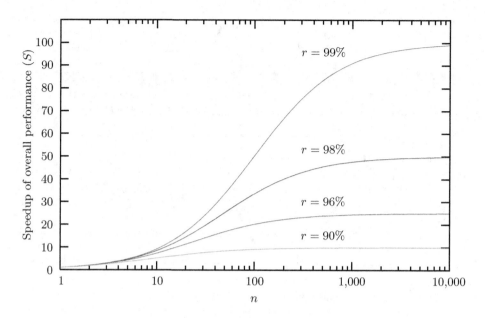

Figure 1.10   Amdahl's Law examples

Also, the clock frequency F cannot get higher unlimitedly. We know that electronic signals travel in a circuit at about two-thirds of the speed of light; this means that in a nanosecond (ns) the signal travels about 20 cm. If F = 20 GHz, in one clock cycle, the signal can travels only 1 cm. This will result in some parts of the circuit being in the "current" cycle and other parts in the "previous" cycle.

When we calculate the expected improvement to an overall system when only part of the system is improved, we often use Amdahl's Law. Amdahl's Law states that the performance improvement to be gained from using some faster mode of execution is limited by the fraction of the time the faster mode can be used. Let P_n be the performance with enhancement, P_o be the performance without enhancement, T_n be the execution time with enhancement, and T_o be the execution time without enhancement. Then the speedup is given by

$$S = \frac{P_n}{P_o} = \frac{T_o}{T_n} = \frac{T_o}{T_o \times r/n + T_o \times (1-r)} = \frac{1}{r/n + (1-r)}$$

where r is the fraction enhanced and n is the speedup of the enhanced section. Let $n \to \infty$; then we get the upper bound of the overall speedup (S), which is $1/(1-r)$. For example, if $r = 50\%$, no matter how big n is, the overall speedup S cannot be larger than 2. Figure 1.10 shows some examples of Amdahl's Law, from which we can see the upper bound of the speedup.

1.3.2   Trace-Driven Simulation and Execution-Driven Simulation

Trace-driven simulation is a method for estimating the performance of potential computer architectures by simulating their behavior in response to the instruction and data references contained in an input trace.

As shown in Figure 1.11, a real machine is used to execute a benchmark program and write the executed instruction information, such as the instruction address, instruction opcode, and data reference address, to a trace file. This trace is then fed into an architecture simulator for the performance study, such as cache performance, ILP, accuracy of branch prediction, TLB performance, and the utilization of FUs.

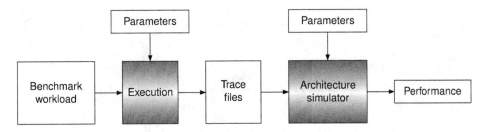

Figure 1.11   Trace-driven simulation

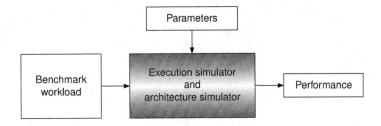

Figure 1.12   Execution-driven simulation

Through these simulations, we can find the performance bottleneck and change the architecture configurations to eliminate the bottleneck. The architecture simulator can run any machine as long as the trace format is known, but the trace files are very large in general for real benchmark programs. Execution-driven simulation is another method that does not require storing traces, as shown in Figure 1.12.

The execution-driven simulation requires the execution of the benchmark program: as the program is being executed, the performance study is also carried out at the same time. Execution-driven simulation must run the benchmark and simulator on the same machine and is much slower than trace-driven simulation. If a benchmark program will be used many times, it is better to generate the trace file once, and use the method of the trace-driven simulation.

1.3.3   High-Performance Computers and Interconnection Networks

A high-performance computer commonly means a computer system in which there are multiple CPUs or multiple computers. High-performance computers can be categorized into multiprocessors and multicomputers. In a multiprocessor system, there are multiple CPUs that communicate with each other through the shared memory. We also call it a parallel system. The shared memory can be centralized or distributed. Supercomputers are commonly parallel systems with the distributed shared memory (DSM). In a multicomputer system, there are multiple computers that communicate with each other via message-passing, over TCP/IP for example, and the memory in a computer is not shared by other computers. We call it a distributed system. Grid and cloud computer systems are usually distributed systems but in which there may be several supercomputers. Note that in some parallel systems, the distributed memories are shared by all CPUs by means of message-passing. An industry standard interface for this is the MPI (message-passing interface), which is installed in many popular parallel computing platforms.

Figure 1.13 illustrates the organization of a supercomputer in which the interconnection network provides the communication paths between CPUs and remote memories. A memory is said to be remote if

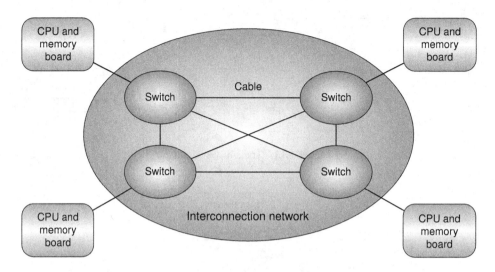

Figure 1.13   Supercomputer and interconnection network

the memory is not located on the same board with the CPU. If the memory is located on the same board with the CPU, we call it local memory. A CPU can access the local memory directly without disturbing the interconnection network. Access to the local memory is faster than access to the remote memory; we call this feature nonuniform memory access (NUMA).

The supercomputer shown in Figure 1.13 is a DSM parallel system. A DSM system can be very large. When we want to build a small parallel system, a server for instance, we can use the architecture of symmetric multiprocessors (SMPs). In an SMP system, there is a common bus connecting CPUs and memory modules. The memory accesses in an SMP have uniform memory access (UMA).

The www.top500.org supercomputer site announces the rank of supercomputers in the world in June and November every year. Currently, there are millions of CPU cores in top supercomputers. The number of cores in top supercomputers will increase year by year.

Amdahl's Law is also suitable for calculating the speedup of a program using a supercomputer. If the sequential fraction of a program is 1%, the theoretical speedup using a supercomputer cannot exceed 100 no matter how many cores are used, as plotted for $r = 99\%$ in Figure 1.10.

1.4   Hardware Description Languages

HDLs are languages used for the hardware design, just like C and Java are languages for software development. HDLs are easier to use than the schematic capture, especially for designing large-scale hardware circuits. The most popular HDLs are Verilog HDL and VHDL (very high speed integrated circuit HDL). Both are covered by IEEE standards. Other high-level HDLs include SystemVerilog and SystemC.

Below is the Verilog HDL code that implements a 4-bit counter. The file name is `time_counter_verilog`, the same as the module name; the extension name is `.v`. There are two input signals: `enable` and `clk` (clock). The output signal is a 4-bit `my_counter` of reg (register) type. If `enable` is a 1, the counter is increased at the rising edge of the `clk`. `always` is a keyword. `posedge` (positive edge) is also a keyword, standing for the rising edge. `4'h1` is a 4-bit constant 1, denoted with the hexadecimal number. The symbol of "`<=`" can be also a "`=`". You see that, even if you do not understand the low level of the circuit design, you can design hardware, just as you can develop software in C or Java although you do not know the assembly programming languages.

```
module time_counter_verilog (enable, clk, my_counter);    // a counter example
    input        enable;                                   // input,  1 bit
    input        clk;                                      // input,  1 bit
    output [3:0] my_counter;                               // output, 4 bits
    reg    [3:0] my_counter = 0;                           // register type
    always @ (posedge clk) begin                           // positive edge
        if (enable)                                        // if (enable == 1)
            my_counter <= my_counter + 4'h1;               //     my_counter++
    end
endmodule
```

Figure 1.14 shows the simulation waveform of the counter using the ModelSim, a powerful HDL simulator. The output signal, my_counter, is denoted in hexadecimal numbers. Counting happens on the clock rising edges when enable is a 1.

The following is the VHDL code (time_counter_vhdl.vhdl) that does the same work above. The Verilog HDL is said to be like the programming language C and VHDL is like C++, but actually it is more like Ada.

```
LIBRARY IEEE;                                             – a counter example
USE IEEE.STD_LOGIC_1164.ALL;
USE IEEE.STD_LOGIC_ARITH.ALL;
USE IEEE.STD_LOGIC_UNSIGNED.ALL;
ENTITY time_counter_vhdl IS
  PORT (clk        : IN  STD_LOGIC;                        – input,  1 bit
        enable     : IN  STD_LOGIC;                        – input,  1 bit
        my_counter : OUT STD_LOGIC_VECTOR (3 DOWNTO 0)     – output, 4 bits
  );
END time_counter_vhdl;
ARCHITECTURE a_counter OF time_counter_vhdl IS
  SIGNAL cnt : STD_LOGIC_VECTOR (3 DOWNTO 0) := "0000";    – initialize cnt
BEGIN
  my_counter <= cnt;                                       – assign to output
  PROCESS (clk) BEGIN
    IF (clk'EVENT AND clk = '1') THEN                      – positive edge
      IF (enable = '1') THEN                               – if (enable == 1)
        cnt <= cnt + '1';                                  –     cnt++
      END IF;
    END IF;
  END PROCESS;
END a_counter;
```

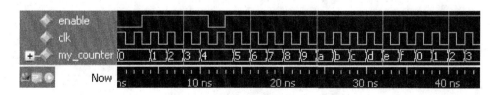

Figure 1.14 Waveform of time counter

Both examples use the high-level style to implement the counter. There are also low-level styles that we can use to design the circuits. This book uses only Verilog HDL for the designs of CPUs and I/O interfaces/buses.

Exercises

1.1 What are the main differences between RISC and CISC?

1.2 Investigate the meaning of each instruction of x86 and MIPS in the code examples of mul16 given in this chapter.

1.3 Why is the microcode difficult to be pipelined?

1.4 Suppose that we have two machines: Machine A has a clock rate of 1 GHz and machine B has a clock rate of 2 GHz. We have made the measurements for these two machines as listed in the following table. Calculate the execution time and the MIPS (million instructions per second) of each machine.

Machine	Clock frequency	CPI				Executed instructions
		1	2	3	4	
A	1 GHz	50%	35%	10%	5%	20,200,000
B	2 GHz	10%	10%	30%	50%	22,000,000

1.5 Let $n = 10$ and $r = 75\%$. Calculate the overall speedup S by using Amdahl's Law and give the upper bound of S.

1.6 Let $n = 1,000,000$ and $S = 500,000$. Calculate the fractions r and $1 - r$ by using Amdahl's Law. What are the meanings of these numbers when we compare the performance of a supercomputer to that of a uniprocessor computer?

2

A Brief Introduction to Logic Circuits and Verilog HDL

Logic circuit design is the foundation of computer design. This chapter briefly introduces the basic concept of the logic circuits and Verilog HDL (hardware description language), a language for implementing the circuits.

There are two types of the logic circuits: combinational circuits and sequential circuits. A combinational circuit can be defined as one whose outputs are dependent only on the present inputs. Full adder and multiplexer are two typical examples of combinational circuits. A sequential circuit can be defined as one whose outputs depend not only on the present inputs but also on the past history of inputs. A sequential circuit is used to construct a finite state machine; it needs flip-plops to record the current state. The counter and the control circuit of vending machines are two examples of the sequential circuits.

2.1 Logic Gates

Logic circuits consist of inputs, logic gates, and outputs. There are three basic logic gates: the AND gate, the OR gate, and the NOT gate. The following four gates are not basic gates but they are commonly used in the logic circuit designs: the NAND (NOT-AND) gate, the NOR (NOT-OR) gate, the XOR (Exclusive OR) gate, and the XNOR (Exclusive NOR) gate.

Figure 2.1 shows a logic circuit that consists of two inputs, seven gates as mentioned above, and seven outputs, one for each gate.

Basically, any unique names, except for the keywords of Verilog HDL, can be assigned to the inputs and outputs. The logic expressions of the outputs are given below. Note that the symbol "+" stands for an OR operation.

$$
\begin{aligned}
\text{AND Gate \ (AND):} \quad & f\_and &&= a \cdot b = ab \\
\text{OR Gate \ (OR):} \quad & f\_or &&= a + b \\
\text{NOT Gate \ (NOT):} \quad & f\_not &&= \bar{a} \\
\text{NOT-AND Gate \ (NAND):} \quad & f\_nand &&= \overline{ab} \\
\text{NOT-OR Gate \ (NOR):} \quad & f\_nor &&= \overline{a+b} \\
\text{Exclusive OR Gate \ (XOR):} \quad & f\_xor &&= a \oplus b = \bar{a}b + a\bar{b} \\
\text{Exclusive NOR Gate \ (XNOR):} \quad & f\_xnor &&= a \odot b = \bar{a}\bar{b} + ab = \overline{a \oplus b}
\end{aligned}
$$

Table 2.1 lists the values of the outputs under all the combinations of the input values. These are the results we expect. Such table is called a truth table.

Computer Principles and Design in Verilog HDL, First Edition. Yamin Li.
© 2015 Tsinghua University Press. All rights reserved. Published 2015 by John Wiley & Sons Singapore Pte Ltd.
Companion Website: www.wiley.com/go/li/verilog

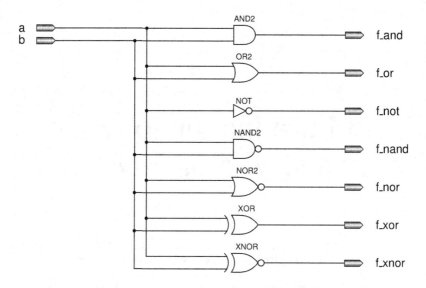

Figure 2.1   Three basic gates and four common gates

Table 2.1   Outputs of seven gates (truth table)

Input		Output						
a	b	f_and	f_or	f_not	f_nand	f_nor	f_xor	f_xnor
0	0	0	0	1	1	1	0	1
1	0	0	1	0	1	0	1	0
0	1	0	1	1	1	0	1	0
1	1	1	1	0	0	0	0	1

The following are the laws of Boolean algebra. Boolean algebra is a logical algebra of the truth values of 0 and 1, introduced by George Boole (1815–1864). It is ideal for expressing the behavior of logic circuit. The last two equations are called De Morgan's Law, introduced by Augustus De Morgan (1806–1871).

1. $A \cdot A = A$ (Idempotent law)
2. $A + A = A$ (Idempotent law)
3. $A \cdot B = B \cdot A$ (Commutative law)
4. $A + B = B + A$ (Commutative law)
5. $(A \cdot B) \cdot C = A \cdot (B \cdot C)$ (Associative law)
6. $(A + B) + C = A + (B + C)$ (Associative law)
7. $A \cdot (B + C) = (A \cdot B) + (A \cdot C)$ (Distributive law)
8. $A + (B \cdot C) = (A + B) \cdot (A + C)$ (Distributive law)
9. $A \cdot (A + B) = A$ (Absorption law)
10. $A + (A \cdot B) = A$ (Absorption law)
11. $A \cdot 0 = 0$ (Annulment law)
12. $A \cdot 1 = A$ (Identity law)
13. $A + 0 = A$ (Identity law)

14. $A + 1 = 1$ (Annulment law)
15. $A \cdot \overline{A} = 0$ (Complement law)
16. $A + \overline{A} = 1$ (Complement law)
17. $\overline{\overline{A}} = A$ (Double negation law)
18. $\overline{A \cdot B} = \overline{A} + \overline{B}$ (De Morgan's Law)
19. $\overline{A + B} = \overline{A} \cdot \overline{B}$ (De Morgan's Law)

Let's take an example to show how to design a combinational circuit. Suppose that we want to design a circuit with three inputs ($a0$, $a1$, and s) and an output (y). The output y will be the same as the input $a0$ if the input s is 0; y is the same as $a1$ otherwise. We call this circuit a 1-bit 2-to-1 multiplexer, denoted as mux2x1.

The general steps of designing a combinational circuit are as follows:

1. Write the truth table for each output.
2. Use the Karnaugh map to get a simplified expression for each output.
3. Design the circuit based on these expressions.
4. Verify the correctness of the circuit through logic simulation.

The truth table of the mux2x1 is shown in Table 2.2. Figure 2.2 shows the Karnaugh map of the output y, from where we get its expression as follows. The expression gives the conditions at which the output y becomes a 1.

$$y = \overline{s} \, a0 + s \, a1$$

Table 2.2 Truth table of a 1-bit 2-to-1 multiplexer

	Input		Output	Comment
s	$a1$	$a0$	y	
0	0	0	0	y is the same as $a0$
0	0	1	1	
0	1	0	0	
0	1	1	1	
1	0	0	0	y is the same as $a1$
1	0	1	0	
1	1	0	1	
1	1	1	1	

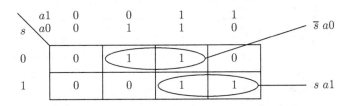

Figure 2.2 Karnaugh map for a 2-to-1 multiplexer

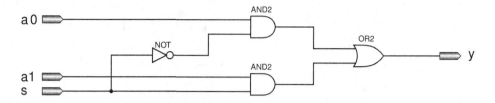

Figure 2.3   Schematic diagram of a 2-to-1 multiplexer

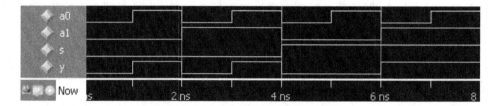

Figure 2.4   Waveform of a 2-to-1 multiplexer

Figure 2.3 shows the circuit schematic diagram and Figure 2.4 is the simulation waveform where the high voltage stands for logic 1 and the low voltage stands for logic 0. From Figure 2.4, we can see that, when s is 0, y is the same as $a0$, and when s is 1, y is the same as $a1$. This is what we wanted.

The logic expression of y can be written as a sum of products directly from the truth table and simplified with the laws of Boolean algebra:

$$y = \bar{s}\,\overline{a1}\,a0 + \bar{s}\,a1\,a0 + s\,a1\,\overline{a0} + s\,a1\,a0$$

$$= \bar{s}\,(\overline{a1} + a1)\,a0 + s\,a1\,(\overline{a0} + a0)$$

$$= \bar{s}\,a0 + s\,a1$$

2.2   Logic Circuit Design in Verilog HDL

Verilog HDL is a language for digital circuit design. It allows designers to design at various levels of abstraction. The Verilog HDL code listed below implements the circuit in Figure 2.1.

```
module gates7_structural (a,b,f_and,f_or,f_not,f_nand,f_nor,f_xor,f_xnor);
    input   a, b;                                           // inputs
    output f_and, f_or, f_not, f_nand, f_nor, f_xor, f_xnor;    // outputs
    and    i1 (f_and,  a, b);                    // and   (out, in1, in2)
    or     i2 (f_or,   a, b);                    // or    (out, in2, in2)
    not    i3 (f_not,  a);                       // not   (out, in)
    nand   i4 (f_nand, a, b);                    // nand  (out, in1, in2)
    nor    i5 (f_nor,  a, b);                     // nor   (out, in1, in2)
    xor    i6 (f_xor,  a, b);                    // xor   (out, in1, in2)
    xnor   i7 (f_xnor, a, b);                    // xnor  (out, in1, in2)
endmodule
```

The code starts with `module` and ends with `endmodule`; both are reserved keywords. `gates7_` `structural` is the name of the circuit (module name). The text in the parentheses is the list of inputs and outputs. The `input` and `output` are also keywords that define the input and output signals, respectively. The text starting with `//` is treated as a comment. The rest of the code is the main body that defines the operations of the circuit.

This example uses a low level of abstraction (gate level). The `and`, `or`, `not`, `nand`, `nor`, `xor`, and `xnor` are gate names. Following a gate name is an instance name. The first signal in the parentheses is the output of the gate and the rest of the signals are the inputs. The code is saved as a file. The file name should be same as the module name, and the extension name is `.v`.

To check the correctness of the logic operations, we must simulate the code. Therefore, a test code (called test bench) is required. Below is the code of the test bench. Following `` `timescale``, the first `1ns` defines time unit and the second `1ns` defines the minimum time precision we can use in the simulation. The module name is `gates7_structural_tb`, and no inputs/outputs are required.

```verilog
`timescale 1ns/1ns
module gates7_structural_tb;
    reg a,b;
    wire f_and,f_or,f_not,f_nand,f_nor,f_xor,f_xnor;
    gates7_structural g (a,b,f_and,f_or,f_not,f_nand,f_nor,f_xor,f_xnor);
    initial begin
            a = 0; b = 0;
        #1 $display (" a=%b",a," b=%b",b," f_and=%b",f_and,
                    " f_or=%b",f_or," f_not=%b",f_not,
                    " f_nand=%b",f_nand," f_nor=%b",f_nor,
                    " f_xor=%b",f_xor," f_xnor=%b",f_xnor);
        #1 a = 1; b = 0;
        #1 $display (" a=%b",a," b=%b",b," f_and=%b",f_and,
                    " f_or=%b",f_or," f_not=%b",f_not,
                    " f_nand=%b",f_nand," f_nor=%b",f_nor,
                    " f_xor=%b",f_xor," f_xnor=%b",f_xnor);
        #1 a = 0; b = 1;
        #1 $display (" a=%b",a," b=%b",b," f_and=%b",f_and,
                    " f_or=%b",f_or," f_not=%b",f_not,
                    " f_nand=%b",f_nand," f_nor=%b",f_nor,
                    " f_xor=%b",f_xor," f_xnor=%b",f_xnor);
        #1 a = 1; b = 1;
        #1 $display (" a=%b",a," b=%b",b," f_and=%b",f_and,
                    " f_or=%b",f_or," f_not=%b",f_not,
                    " f_nand=%b",f_nand," f_nor=%b",f_nor,
                    " f_xor=%b",f_xor," f_xnor=%b",f_xnor);
        #1 a = 0; b = 0;
        #1 $finish;
    end
    initial begin
        $dumpfile("gates7_structural.vcd");
        $dumpvars;
    end
endmodule
```

It consists three parts. The first part invokes the `gates7_structural` module. The signals of a and b are declared as `reg` (register) type because they will be assigned with different values several times. The second part is an `initial` block that describes the behaviors of signals of a and b. # describes the time delay in the unit of nanoseconds (ns), which was defined in the beginning of the code. The `$display` statements show messages on the display. `%b` indicates the display of a signal in binary format. The last part describes the output file. The format of the output file is `vcd` (value change dump). This `vcd` file can be read by some applications to show the outputs graphically, GTKWave for example. If you do not use the `vcd` file further, this part can be deleted.

A tool (software) that integrates a compiler and a simulator is required. There are many such tools developed by famous companies, such as Cadence, Synopsys, Mentor Graphics, LogicVision, Xilinx, and Altera. There are also some open-source free tools; `Icarus Verilog` is an example. The `iverilog` command in the following text checks the correctness of the grammar and generates an output file `a.out`. Rather than `a.out`, you can designate an output file name with the `-o` option in the `iverilog` command line. The `vvp` command executes `a.out`, which shows the execution results of the test bench.

```
[cpu_verilog]$ iverilog gates7_structural_tb.v gates7_structural.v
[cpu_verilog]$ vvp a.out
VCD info: dumpfile gates7_structural.vcd opened for output.
 a=0 b=0 f_and=0 f_or=0 f_not=1 f_nand=1 f_nor=1 f_xor=0 f_xnor=1
 a=1 b=0 f_and=0 f_or=1 f_not=0 f_nand=1 f_nor=0 f_xor=1 f_xnor=0
 a=0 b=1 f_and=0 f_or=1 f_not=1 f_nand=1 f_nor=0 f_xor=1 f_xnor=0
 a=1 b=1 f_and=1 f_or=1 f_not=0 f_nand=0 f_nor=0 f_xor=0 f_xnor=1
```

The following code also implements the circuit in Figure 2.1. It uses the dataflow style. `assign` is a keyword that assigns a logic operation to an output. The parentheses are required because the NOT operation has the highest priority.

```
module gates7_dataflow (a,b,f_and,f_or,f_not,f_nand,f_nor,f_xor,f_xnor);
    input   a, b;                                       // inputs
    output f_and, f_or, f_not, f_nand, f_nor, f_xor, f_xnor;    // outputs
    assign f_and  =   a & b;                    // and
    assign f_or   =   a | b;                    // or
    assign f_not  = ~ a;                        // not
    assign f_nand = ~(a & b);                   // nand = not(a and b)
    assign f_nor  = ~(a | b);                   // nor = not(a or  b)
    assign f_xor  =   a ^ b;                    // xor
    assign f_xnor = ~(a ^ b);                   // xnor = not(a xor b)
endmodule
```

The test bench code is listed below. Instead of using `$display` on every change of the time, we used `$monitor` once to display the output messages. Combining with the variable `$time`, `$monitor` makes the test bench code shorter. The two `always` statements give the behaviors of signals of a and b.

```
`timescale 1ns/1ns
module gates7_dataflow_tb;
    reg a,b;
    wire f_and,f_or,f_not,f_nand,f_nor,f_xor,f_xnor;
    gates7_dataflow g7 (a,b,f_and,f_or,f_not,f_nand,f_nor,f_xor,f_xnor);
    initial begin
```

```
        $display("time\ta\tb\tand\tor\tnot\tnand\tnor\txor\txnor");
        a = 0;
        b = 0;
        #5 $finish;
    end
    initial begin
        $monitor ("%2d:\t%b\t%b\t%b\t%b\t%b\t%b\t%b\t%b",
                  $time,a,b,f_and,f_or,f_not,f_nand,f_nor,f_xor,f_xnor);
    end
    always #1 a = !a;
    always #2 b = !b;
endmodule
```

The following text shows the compilation command and the execution results of the test bench code.

```
[cpu_verilog]$ iverilog gates7_dataflow_tb.v gates7_dataflow.v
[cpu_verilog]$ vvp a.out
time    a       b       and     or      not     nand    nor     xor     xnor
0:      0       0       0       0       1       1       1       0       1
1:      1       0       0       1       0       1       0       1       0
2:      0       1       0       1       1       1       0       1       0
3:      1       1       1       1       0       0       0       0       1
4:      0       0       0       0       1       1       1       0       1
```

2.3   CMOS Logic Gates

This section describes how to use low-level CMOS (complementary metal–oxide–semiconductor) transistors to design logic gates and gives the gate implementations in Verilog HDL.

2.3.1   CMOS Inverter

In CMOS technology, both PMOS (p-type or p-channel metal–oxide–semiconductor) and NMOS (n-type or n-channel metal–oxide–semiconductor) transistors are used. For PMOS transistors, as shown in Figure 2.5(a), if the gate input is a 0, the switch is on, and the voltage of the drain is the same as that of the source; otherwise it is off. On the other hand, for the NMOS, as shown in Figure 2.5(b), if the input gate is 1, the transistor is on, and the voltage of drain is the same as that of source; otherwise the transistor is off. Figure 2.5(c) shows a CMOS inverter (a NOT gate).

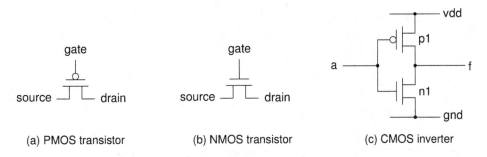

(a) PMOS transistor   (b) NMOS transistor   (c) CMOS inverter

Figure 2.5   Schematic diagram of a CMOS inverter

The following Verilog HDL code implements the CMOS inverter. `supply0`, `supply1`, `pmos`, and `nmos` are keywords that stand for ground, power supply, PMOS transistor, and NMOS transistor, respectively.

```
module cmosnot (f, a);                      // cmos inverter
    input    a;                             // input    a
    output   f;                             // output f = ~a
    supply1 vdd;                            // logic 1 (power)
    supply0 gnd;                            // logic 0 (ground)
    // pmos (drain, source, gate);
    pmos p1 (f,      vdd,    a);
    // nmos (drain, source, gate);
    nmos n1 (f,      gnd,    a);
endmodule
```

Below is the code of the test bench.

```
`timescale 1ns/1ns
module cmosnot_tb;
    reg  a;
    wire f;
    cmosnot cmos_not (f,a);
    initial begin
            a = 0;
        #1 a = 1;
        #1 a = 0;
        #1 $finish;
    end
    initial begin
        $monitor("%2d:\ta = %b\tf = %b",$time,a,f);
    end
endmodule
```

The execution results of the test bench code are shown below, from where we can see that the output f is the reverse of the input a.

```
[cpu_verilog]$ iverilog cmosnot_tb.v cmosnot.v
[cpu_verilog]$ vvp a.out
 0:     a = 0    f = 1
 1:     a = 1    f = 0
 2:     a = 0    f = 1
```

2.3.2   CMOS NAND and NOR Gates

For a NAND gate, the output is 1 if an input is a 0. Figure 2.6(a) shows the CMOS implementation of a 2-input NAND gate. When both a and b are 1, p1 and p2 are off, and n1 and n2 are on, and the output is 0. Otherwise, the output is 1.

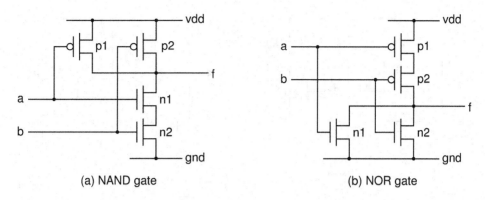

(a) NAND gate (b) NOR gate

Figure 2.6 Schematic diagram of NAND and NOR gates

Below is the transistor-level Verilog HDL code for the NAND gate. `wire w_n` defines a wire that is used to connect n1 and n2 transistors. Basically, every wire in a circuit requires a name in its Verilog HDL implementation code.

```
module cmosnand (f, a, b);            // cmos nand
    input    a, b;                    // inputs: a, b
    output   f;                       // output: f = ~(a & b)
    supply1 vdd;                      // logic 1 (power)
    supply0 gnd;                      // logic 0 (ground)
    wire     w_n;                     // wire: connects 2 nmos transistors
    // pmos (drain, source, gate);
    pmos p1 (f,      vdd,    a);
    pmos p2 (f,      vdd,    b);
    // nmos (drain, source, gate);
    nmos n1 (f,      w_n,    a);
    nmos n2 (w_n,    gnd,    b);
endmodule
```

For a NOR gate, the output is 0 if an input is a 1. Figure 2.6(b) shows the CMOS implementation of a 2-input NOR gate. When both a and b are 0, p1 and p2 are on, and n1 and n2 are off, the output is 1. Otherwise, the output is 0. Below is the transistor-level Verilog HDL code for the NOR gate.

```
module cmosnor (f, a, b);             // cmos nor
    input    a, b;                    // inputs: a, b
    output   f;                       // output: f = ~(a | b)
    supply1 vdd;                      // logic 1 (power)
    supply0 gnd;                      // logic 0 (ground)
    wire     w_p;                     // wire: connects 2 pmos transistors
    // nmos (drain, source, gate);
    nmos n1 (f,      gnd,    a);
    nmos n2 (f,      gnd,    b);
    // pmos (drain, source, gate);
    pmos p1 (w_p,    vdd,    a);
    pmos p2 (f,      w_p,    b);
endmodule
```

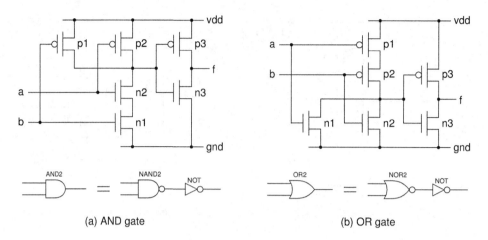

(a) AND gate   (b) OR gate

Figure 2.7   Schematic diagram of AND and OR gates

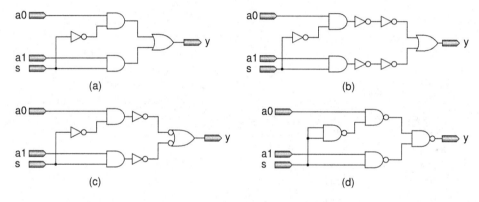

Figure 2.8   Implementing a multiplexer using only NAND gates

Figure 2.7 shows the CMOS implementations of the AND and OR gates. An AND gate is constructed by appending a NOT gate to the output of a NAND gate (Figure 2.7(a)). Similarly, an OR gate is constructed by appending a NOT gate to the output of a NOR gate (Figure 2.7(b)).

From Figure 2.7, we know that an AND gate uses two more transistors than a NAND gate. Therefore, if we can design a circuit using only NAND gates, the cost of the circuit will become lower. Figure 2.8 illustrates how to use only NAND gates to design circuits.

Figure 2.8(a) shows a 2-to-1 multiplexer designed with AND, OR, and NOT gates. Figure 2.8(b) adds two NOT gates to each output of the AND gates. Figure 2.8(c) combines a NOT gate and an OR gate. By applying De Morgan's Law, we get the final circuit of the multiplexer that uses only NAND gates, as shown as in Figure 2.8(d). Some gate array chips contain an array of NAND-only gates.

2.4   Four Levels/Styles of Verilog HDL

Verilog HDL supports a variety of description levels. This section takes the 2-to-1 multiplexer as an example to show the implementations in the following four levels (styles): (i) transistor switch level, (ii) gate level or structural style, (iii) dataflow style, and (iv) behavioral style.

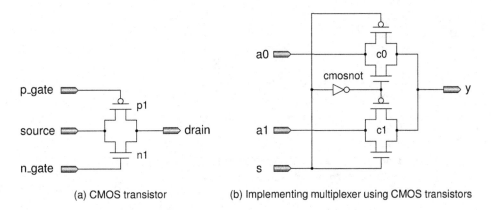

(a) CMOS transistor   (b) Implementing multiplexer using CMOS transistors

Figure 2.9   Implementing a multiplexer using CMOS transistors

2.4.1   Transistor Switch Level

Figure 2.9 shows the schematic circuit of a 2-to-1 multiplexer designed with CMOS transistors. Figure 2.9(a) is a CMOS transistor that contains a PMOS transistor and an NMOS transistor. If the input p_gate is 0 and the input n_gate is 1, both transistors of p1 and n1 are on and the output drain is the same as the input source.

The multiplexer shown in Figure 2.9(b) uses two CMOS transistors and a CMOS NOT gate that we already described. When the input s is a 0, c0 is on and c1 is off; thus the output y is the same as the input a0. Otherwise, c0 is off and c1 is on, and the output y is the same as the input a1. The Verilog HDL code of the 2-to-1 multiplexer implemented with CMOS transistors is listed below.

```
module mux2x1_cmos (a0, a1, s, y);      // multiplexer using cmos transistors
    input   s, a0, a1;                  // inputs
    output y;                           // output
    wire    sn;                         // internal wire, output of cmosnot
    // cmosnot   (f,   a);              // a cmos invert: (out, in)
    cmosnot inv (sn, s);
    // cmoscmos (drain, source, n_gate, p_gate);
    cmoscmos c0 (y,      a0,      sn,      s);
    cmoscmos c1 (y,      a1,      s,      sn);
endmodule
```

The code listed above invokes cmosnot (CMOS NOT gate) and cmoscmos (CMOS transistor) modules. The cmosnot module was given in the previous section. The cmoscmos module and the multiplexer test bench are listed below.

```
module cmoscmos (drain, source, n_gate, p_gate);                      // cmos gate
    input   source, n_gate, p_gate;
    output drain;
    pmos p1 (drain, source, p_gate);      // pmos name (drain, source, gate);
    nmos n1 (drain, source, n_gate);      // nmos name (drain, source, gate);
endmodule
```

```
`timescale 1ns/1ns
module mux2x1_cmos_tb;
    reg   s,a0,a1;
    wire y;
    mux2x1_cmos mux2x1 (a0,a1,s,y);
    initial begin
            a0 = 0;
            a1 = 0;
            s  = 0;
            $display("time\ts\ta1\ta0\ty");
            $monitor("%2d:\t%b\t%b\t%b\t%b",$time,s,a1,a0,y);
        #8 $finish;
    end
    always #1 a0 = !a0;
    always #2 a1 = !a1;
    always #4 s  = !s;
endmodule
```

The compilation and execution results are shown below, from where we can see that the multiplexer works correctly.

```
[cpu_verilog]$ iverilog mux2x1_cmos_tb.v mux2x1_cmos.v cmosnot.v cmoscmos.v
[cpu_verilog]$ vvp a.out
time      s          a1         a0          y
 0:       0          0          0           0
 1:       0          0          1           1
 2:       0          1          0           0
 3:       0          1          1           1
 4:       1          0          0           0
 5:       1          0          1           0
 6:       1          1          0           1
 7:       1          1          1           1
 8:       0          0          0           0
```

2.4.2   Logic Gate Level

We give two circuits of the multiplexer implemented with logic gates. Figure 2.10(a) shows the circuit of the multiplexer implemented with two tri-state gates. A tri-state gate is a logic gate with three states:

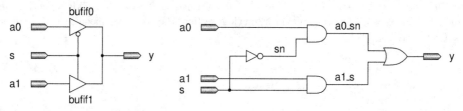

(a) A multiplexer built with tri-state gates   (b) A multiplexer built with AND-OR-NOT gates

Figure 2.10   Schematic diagram of multiplexer using tri-state and ordinary gates

logic 0, logic 1, or high impedance. If the input s is a 0, the output of the tri-state buffer bufif1 is in high impedance and the output y is the same as the input a0. Otherwise, the output of the tri-state buffer bufif0 is in high impedance and the output y is the same as the input a1.

The Verilog HDL code implementing the multiplexer with tri-state gates is listed below. bufif0 and bufif1 are two tri-state gates supported by Verilog HDL.

```
module mux2x1_3s (a0, a1, s, y);        // multiplexer using tri-state gates
    input  s, a0, a1;                   // inputs
    output y;                           // output
    // bufif0 (out, in,  ctl);          // tri-state buffer: ctl==0: out=in;
    bufif0 b0 (y,   a0,  s);            //    ctl == 1: out = high-impedance;
    // bufif1 (out, in,  ctl);          // tri-state buffer: ctl==1: out=in;
    bufif1 b1 (y,   a1,  s);            //    ctl == 0: out = high-impedance;
endmodule
```

Figure 2.10(b) shows the circuit of the multiplexer implemented with ordinary AND, OR, and NOT gates. Its Verilog HDL code is listed below.

```
module mux2x1_gate (a0, a1, s, y);      // multiplexer using ordinary gates
    input  s, a0, a1;                   // inputs
    output y;                           // output
    wire   sn, a0_sn, a1_s;             // internal wires
    not i0 (sn,    s);                  // not (out, in);
    and i1 (a0_sn, a0,    sn);          // and (out, in1, in2);
    and i2 (a1_s,  a1,    s );          // and (out, in1, in2);
    or  i3 (y,     a0_sn, a1_s);        // or  (out, in1, in2);
endmodule
```

2.4.3   Dataflow Style

Different from transistor and gate levels, the Verilog HDL code in dataflow style does not use any particular component (transistor or gate). It uses the keyword assign and logic expression to assign a value to a net or an output. A net or an output can be assigned only once in the code. Below is the dataflow style Verilog HDL code that implements the multiplexer.

```
module mux2x1_dataflow1 (a0, a1, s, y);    // multiplexer, dataflow style
    input  s, a0, a1;                      // inputs
    output y;                              // output
    assign y = ~s & a0 | s & a1;           // logic expression
endmodule
```

Below is another implementation which also uses assign but does not use logic expression. The input s is queried. If it is true (logic 1), the a1 input is assigned to y; otherwise a0 is assigned to y. This code is somewhat in behavioral style.

```
module mux2x1_dataflow2 (a0, a1, s, y);    // multiplexer, dataflow style
    input  s, a0, a1;                      // inputs
    output y;                              // output
    assign y = s ? a1 : a0;                // if (s==1) y=a1; else y=a0;
endmodule
```

2.4.4   Behavioral Style

Behavioral style is the highest level of abstraction. The main feature of this style is that the function-ality of a circuit is expressed by an algorithmic description. By using the behavioral style, a designer who does not know the details of low-level logic circuit design can design the circuits, just as a pro-grammer can develop software in C or Java although he or she does not know the assembly language programming at all.

The behavioral Verilog HDL code must be inside procedural blocks. There are two types of procedural blocks: `always` and `function`. Inside a block, we can use control statements similar to C program-ming language, such as `if-else`, `case`, and `for` loop. A signal can be assigned multiple times with different values inside a block. Such signals must be declared as `reg` type outside the procedural blocks. Although `reg` stands for register, we can design the combinational circuits with signals of `reg` type. In a combinational circuit, there is no register.

The following example code of the 2-to-1 multiplexer uses `always` block and the `if-else` statement. As mentioned above, the output y is declared as `reg` type (it cannot use `wire` type). All the combinational values of s are checked inside the block. This ensures that a combinational circuit will be generated although y is declared as a register type.

```
module mux2x1_behavioral_if_else (a0,a1,s,y);    // multiplexer, if else
    input   s, a0, a1;                           // inputs
    output y;                                    // output
    reg    y;                                    // y cannot be a wire
    always @ (s or a0 or a1) begin               // always block
        if (s) begin                             // if (s == 1)
            y = a1;                              //     y = a1;
        end else begin                           // if (s == 0)
            y = a0;                              //     y = a0;
        end
    end
endmodule
```

The following code uses only the `if` statement but assigns a default value to y first. This also ensures that a combinational circuit will be generated.

```
module mux2x1_behavioral_if (a0,a1,s,y);         // multiplexer, default first
    input   s, a0, a1;                           // inputs
    output y;                                    // output
    reg    y;                                    // y cannot be a wire
    always @ (s or a0 or a1) begin               // always block
        y = a0;                                  // y = a0;
        if (s) begin                             // if (s == 1)
            y = a1;                              //     y = a1;
        end
    end
endmodule
```

The following code uses the `always` block and the `case` statement. All the combinational cases of s are checked inside the block. This ensures that a combinational circuit will be generated.

```
module mux2x1_behavioral_case_all (a0,a1,s,y);      // multiplexer, all cases
    input   s, a0, a1;                              // inputs
    output y;                                       // output
    reg    y;                                       // y cannot be a wire
    always @ (s or a0 or a1) begin                  // always block
        case (s)                                    // cases:
            1'b0: y = a0;                           // if (s == 0) y = a0;
            1'b1: y = a1;                           // if (s == 1) y = a1;
        endcase
    end
endmodule
```

The following code shows how to use default in the case statement. All the cases that are not specified obviously go to the case of default.

```
module mux2x1_behavioral_case_default (a0,a1,s,y);   // multiplexer, default
    input   s, a0, a1;                              // inputs
    output y;                                       // output
    reg    y;                                       // y cannot be a wire
    always @ (s or a0 or a1) begin                  // always block
        case (s)                                    // cases:
            1'b1:    y = a1;                        // if (s == 1) y = a1;
            default: y = a0;                        // other cases y = a0;
        endcase
    end
endmodule
```

The four examples given above use the always block. The next code uses a function block. A function has a name. The name is also the output of the function. This output can be assigned to a net or an output with assign outside the function. The input arguments to the function are specified as input inside the function. We can use local variable names for these arguments. The following example uses case statement inside the function.

```
module mux2x1_behavioral_function_case_all (a0,a1,s,y);      // multiplexer,
    input   s, a0, a1;           // inputs                    //      function
    output y;                    // output                    //      all cases
    assign y = sel (a0,a1,s);    // call a function with parameters
    function sel;                // function name (= return value)
        input a,b,c;             // notice the order of the input arguments
        case (c)                 // cases:
            1'b0: sel = a;       // if (c==0) return value = a
            1'b1: sel = b;       // if (c==1) return value = b
        endcase
    endfunction
endmodule
```

The following code also uses the function block. Inside the function, the if-else statement is used.

```
module mux2x1_behavioral_function_if_else (a0,a1,s,y);      // multiplexer,
    input  s, a0, a1;            // inputs                   //       function
    output y;                    // output                   //       if else
    assign y = sel (a0,a1,s);    // call a function with parameters
    function sel;                // function name (= return value)
        input a,b,c;             // notice the order of the input arguments
        if (c) sel = b;          // if (c==1) return value = b
        else   sel = a;          // if (c==0) return value = a
    endfunction
endmodule
```

In addition to `always` and `function`, there is also another procedural block: `initial`, which is executed once the simulation starts. This is useful in writing the test benches, which were described in the previous section.

2.5   Combinational Circuit Design

The outputs of combinational circuits are dependent only on the present inputs. This section introduces some common combinational circuits, including 32-bit multiplexers, demultiplexer, decoder, priority encoder, and barrel shifter, which will be used in the design of CPUs. The combinational arithmetic circuits, such as full adder and parallel multiplier, will be given in Chapter 3.

2.5.1   Multiplexer

A multiplexer selects an input from multiple inputs. We already introduced a 1-bit 2-to-1 multiplexer. The following code implements a 32-bit 2-to-1 multiplexer. The [31:0] denotes a bus that has 32 bits.

```
module mux2x32 (a0,a1,s,y);              // multiplexer, 32 bits
    input   [31:0] a0, a1;               // inputs, 32 bits
    input          s;                    // input,   1 bit
    output  [31:0] y;                    // output, 32 bits
    assign         y = s ? a1 : a0;      // if (s==1) y=a1; else y=a0;
endmodule
```

The following code implements a 32-bit 4-to-1 multiplexer. A 4-to-1 multiplexer selects one input from four inputs and needs a 2-bit selection signal. The `assign` statement can also be put in above the function block.

```
module mux4x32 (a0,a1,a2,a3,s,y); // 4-to-1 multiplexer, 32-bit
    input   [31:0] a0, a1, a2, a3; // inputs, 32 bits
    input   [1:0] s;               // input,   2 bits
    output  [31:0] y;              // output, 32 bits
    function  [31:0] select;       // function name (= return value, 32 bits)
        input [31:0] a0,a1,a2,a3;  // notice the order of the input arguments
        input  [1:0] s;            // notice the order of the input arguments
        case (s)                   // cases:
            2'b00: select = a0;    // if (s==0) return value = a0
            2'b01: select = a1;    // if (s==1) return value = a1
```

```
            2'b10: select = a2;   // if (s==2) return value = a2
            2'b11: select = a3;   // if (s==3) return value = a3
        endcase
    endfunction
    assign y = select(a0,a1,a2,a3,s);   // call the function with parameters
endmodule
```

2.5.2 Decoder

An m-2^m decoder works as follows: suppose there is a 1-bit input *ena*, m-bit input $n[m-1:0]$, and 2^m-bit output $d[2^m-1:0]$. If *ena* = 1, then 1-bit $d[n] = 1$, other bits are 0; if *ena* = 0, all the bits of output $d[2^m-1:0]$ are 0. Table 2.3 is the truth table of a 3-8 decoder.

We can write the logic expression for each output directly from the truth table; for example, $d[3] = ena\ \overline{n[2]}\ n[1]\ n[0]$. Figure 2.11 shows the circuit of the 3-8 decoder. The inputs and outputs of the gate signals can be labeled with wire names so that there is no need to connect them to the input or output pins. Based on the logic expressions, we can write the Verilog HDL code in structural or dataflow style easily (we do not show it here).

Below is a Verilog HDL implementation of the 3-8 decoder in behavioral style. Inside the always block, every bit of d is assigned with a default value 0. Then a 1 is assigned to the nth bit of d if *ena* is a 1.

Table 2.3 Truth table of a 3-8 decoder

Input				Output							
ena	$n[2]$	$n[1]$	$n[0]$	$d[7]$	$d[6]$	$d[5]$	$d[4]$	$d[3]$	$d[2]$	$d[1]$	$d[0]$
1	0	0	0	0	0	0	0	0	0	0	1
1	0	0	1	0	0	0	0	0	0	1	0
1	0	1	0	0	0	0	0	0	1	0	0
1	0	1	1	0	0	0	0	1	0	0	0
1	1	0	0	0	0	0	1	0	0	0	0
1	1	0	1	0	0	1	0	0	0	0	0
1	1	1	0	0	1	0	0	0	0	0	0
1	1	1	1	1	0	0	0	0	0	0	0
0	x	x	x	0	0	0	0	0	0	0	0

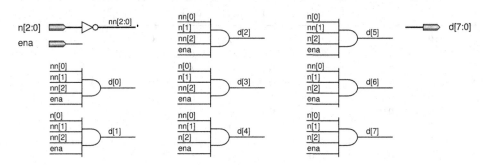

Figure 2.11 Schematic diagram of decoder with enable control

```
module decoder3e (n,ena,d);         // 3-8 decoder with enable
    input   [2:0] n;                // inputs:  n, 3 bits
    input         ena;              // input:   enable
    output [7:0] d;                 // outputs: d, 2^3 = 8 bits
    reg    [7:0] d;                 // d cannot be a wire
    always @ (ena or n) begin       // always block
        d    = 8'b0;                // let d = 00000000 first
        d[n] = ena;                 // then let n-th bit of d = ena (0 or 1)
    end
endmodule
```

2.5.3   Encoder

An encoder performs the reverse operation of the decoder. It has a maximum of 2^m inputs and m outputs. The symbol and the truth table of an 8-3 encoder are shown in Figure 2.12(b) and Table 2.4, respectively. Figure 2.12(a) is a 3-8 decoder for the comparison with the 8-3 encoder.

As shown in Figure 2.12(b), the input $d[7:0]$ has 8 bits ($m = 3$) and the output $n[2:0]$ has 3 bits. Another output, g, indicates whether there is a 1 in the $d[7:0]$ or not. If we do not add g, we cannot

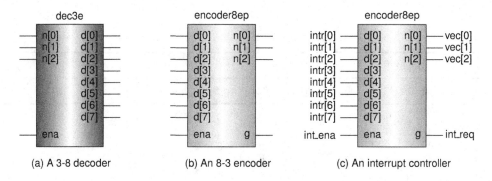

(a) A 3-8 decoder   (b) An 8-3 encoder   (c) An interrupt controller

Figure 2.12   Decoder and encoder

Table 2.4   Truth table of an 8-3 encoder

				Input							Output		
ena	$d[7]$	$d[6]$	$d[5]$	$d[4]$	$d[3]$	$d[2]$	$d[1]$	$d[0]$	$n[2]$	$n[1]$	$n[0]$	g
1	0	0	0	0	0	0	0	0	0	0	0	0
1	0	0	0	0	0	0	0	1	0	0	0	1
1	0	0	0	0	0	0	1	0	0	0	1	1
1	0	0	0	0	0	1	0	0	0	1	0	1
1	0	0	0	0	1	0	0	0	0	1	1	1
1	0	0	0	1	0	0	0	0	1	0	0	1
1	0	0	1	0	0	0	0	0	1	0	1	1
1	0	1	0	0	0	0	0	0	1	1	0	1
1	1	0	0	0	0	0	0	0	1	1	1	1
0	x	x	x	x	x	x	x	x	0	0	0	0

Table 2.5 Truth table of an 8-3 priority encoder

Input									Output			
ena	d[7]	d[6]	d[5]	d[4]	d[3]	d[2]	d[1]	d[0]	n[2]	n[1]	n[0]	g
1	0	0	0	0	0	0	0	0	0	0	0	0
1	0	0	0	0	0	0	0	1	0	0	0	1
1	0	0	0	0	0	0	1	x	0	0	1	1
1	0	0	0	0	0	1	x	x	0	1	0	1
1	0	0	0	0	1	x	x	x	0	1	1	1
1	0	0	0	1	x	x	x	x	1	0	0	1
1	0	0	1	x	x	x	x	x	1	0	1	1
1	0	1	x	x	x	x	x	x	1	1	0	1
1	1	x	x	x	x	x	x	x	1	1	1	1
0	x	x	x	x	x	x	x	x	0	0	0	0

distinguish whether the value of the input $d[7:0]$ is 00000001 or 00000000. The input *ena* is an enable signal. g will output a 0 if the *ena* is a 0, irrespective of the $d[7:0]$.

Table 2.4 does not list all combinations of the input $d[7:0]$. This encoder will generate the wrong output code when there is more than one input present at logic 1. For example, if both $d[1]$ and $d[2]$ are 1 at the same time, the resulting output is neither 001 nor 010, but will be 011, indicating $d[3]$ at a logic 1.

To solve this problem, priority encoders are commonly used. Table 2.5 shows the truth table of an 8-3 priority encoder. Each input has an assigned priority. In our priority encoder, the input $d[7]$ has the highest priority and $d[0]$ has the lowest priority. The value x stands for "don't care", meaning that it can be either a 1 or a 0, and has no effect on the output. Table 2.5 lists all combinations of the inputs $d[7:0]$ and *ena*.

From the truth table in Table 2.5, we can express the outputs as follows:

$$n[2] = ena \ (d[7] + \overline{d[7]}\,d[6] + \overline{d[7]}\,\overline{d[6]}\,d[5] + \overline{d[7]}\,\overline{d[6]}\,\overline{d[5]}\,d[4])$$

$$= ena \ (d[7] + d[6] + \overline{d[7]}\,\overline{d[6]}\,(d[5] + d[4]))$$

$$= ena \ (d[7] + d[6] + \overline{d[7]} + \overline{d[6]}\,(d[5] + d[4]))$$

$$= ena \ (d[7] + d[6] + d[5] + d[4])$$

$$n[1] = ena \ (d[7] + d[6] + \overline{d[5]}\,\overline{d[4]}\,d[3] + \overline{d[5]}\,\overline{d[4]}\,d[2])$$

$$n[0] = ena \ (d[7] + \overline{d[6]}\,d[5] + \overline{d[6]}\,\overline{d[4]}\,d[3] + \overline{d[6]}\,\overline{d[4]}\,\overline{d[2]}\,d[1])$$

$$g = ena \ (d[7] + d[6] + d[5] + d[4] + d[3] + d[2] + d[1] + d[0])$$

We can draw the logic diagram or write the dataflow style Verilog HDL codes based on the logic expressions of the outputs. Here we give a behavioral style version of the 8-3 priority encoder Verilog HDL codes, as shown below.

```
module encoder8ep (d,ena,n,g);        // 8-3 priority encoder with enable
    input  [7:0] d;                   // input:   d, 8 bits
    input        ena;                 // input:   enable
    output [2:0] n;                   // outputs: n, log_2 8 = 3 bits
    output       g;                   // output:  g = 1 if d is not 0
    assign       g = ena & |d;        // if there is at least a 1 in d
```

```
    assign       n = enc(ena, d);        // call a function enc
    function [2:0] enc;                   // the function enc
        input       e;                    // input of the function
        input [7:0] d;                    // input of the function
        casex ({e,d})                     // cases, x: don't care
            9'b1_1xxxxxxx: enc = 3'd7;    // d[7] has the highest priority
            9'b1_01xxxxxx: enc = 3'd6;    // d[6] is active, ignore d[5:0]
            9'b1_001xxxxx: enc = 3'd5;    // d[5] is active, ignore d[4:0]
            9'b1_0001xxxx: enc = 3'd4;    // d[4] is active, ignore d[3:0]
            9'b1_00001xxx: enc = 3'd3;    // d[3] is active, ignore d[2:0]
            9'b1_000001xx: enc = 3'd2;    // d[2] is active, ignore d[1:0]
            9'b1_0000001x: enc = 3'd1;    // d[1] is active, ignore d[0]
            default:       enc = 3'd0;    // d[0] has the lowest priority
        endcase
    endfunction
endmodule
```

The $|d$ operation in the expression of g is the reduction OR on 8-bit $d[7:0]$:

$$|d = d[7] + d[6] + d[5] + d[4] + d[3] + d[2] + d[1] + d[0]$$

The don't care x can be used with casex within the function. The underbar separates e and d for easy reading; it will be ignored by Verilog HDL compiler. Figure 2.13 shows the simulation waveform of the Verilog HDL codes given above for the 8-3 priority encoder.

In the design of interrupt controllers, we often use the priority encoder to indicate whether there is an interrupt and which input causes the interrupt request. For example, referring to Figure 2.12(c), eight interrupt requests $intr[7:0]$ can be connected to the inputs $d[7:0]$ of the 8-3 priority encoder; the outputs $n[2:0]$ can be used as the vector $vec[2:0]$ of the interrupt controller; the output g serves as the interrupt request int_req to inform the CPU; and the input ena can severe as the interrupt enable.

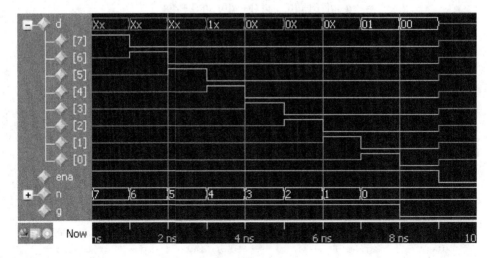

Figure 2.13   Waveform of a 8-3 priority encoder

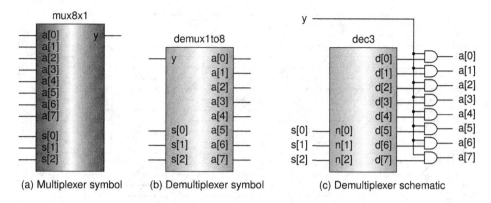

(a) Multiplexer symbol   (b) Demultiplexer symbol   (c) Demultiplexer schematic

Figure 2.14   A 1-to-8 demultiplexer

2.5.4   Demultiplexer

A demultiplexer performs the reverse operation of the multiplexer. It has only a 1-bit data input and an m-bit select signal. The output has 2^m bits. Depending on the select signal, one bit of the output is selected to take the state of the data input.

Figure 2.14(b) shows the symbol of a 1-to-8 demultiplexer. Figure 2.14(a) shows an 8-to-1 multiplexer for comparison with the 1-to-8 demultiplexer. Figure 2.14(c) is the logic schematic diagram of the 1-to-8 demultiplexer. It uses a 3-8 decoder and eight AND gates.

Below is the Verilog HDL code of the 1-to-8 demultiplexer. The shift operation (1 << s) implements the decoder. $\{8\{y\}\}$ replicates y to eight bits.

```
module demux1to8 (s,y,a);          // 1-to-8 demultiplexer
    input   [2:0] s;               // inputs:  s, dispatch control
    input         y;               // input:   y, to be dispatched
    output [7:0] a;                // outputs: only 1 bit's value = y
    assign a = (1 << s) & {8{y}};  // the value of (2^s)th bit = y
endmodule
```

2.5.5   Barrel Shifter

A 32-bit barrel shifter shifts a 32-bit input to the left or right by 0 to 31 bits based on a right input and a 5-bit sa (shift amount) input. There is also an arith (arithmetic) input that indicates whether to perform a logical shift or to perform an arithmetic shift when right is a 1. A logical shift right inserts zeroes in the emptied bit positions, and an arithmetic shift right replicates the sign bit in the emptied bit positions. The following examples show the three kinds of shift by 8 bits on a 32-bit input.

Original data (d):	11111111_00000000_00000000_11111111
Shift d to the left by 8 bits:	00000000_00000000_11111111_00000000
Logical shift d to the right by 8 bits:	00000000_11111111_00000000_00000000
Arithmetic shift d to the right by 8 bits:	11111111_11111111_00000000_00000000

Figure 2.15 shows a traditional implementation of a 32-bit left shifter that uses multiplexers. There are five stages in each of which a 32-bit 2-to-1 multiplexer is used. Each stage performs a shift to the left by a power of 2 (16, 8, 4, 2, 1) bits controlled by a bit of sa.

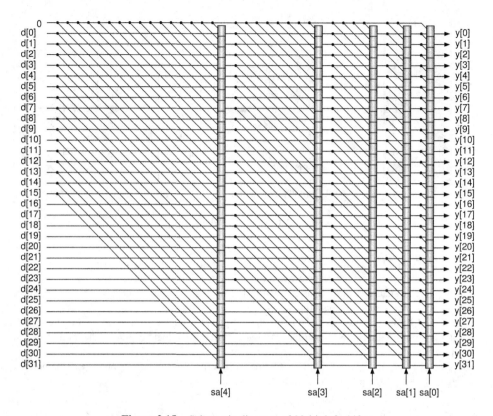

Figure 2.15   Schematic diagram of 32-bit left shifter

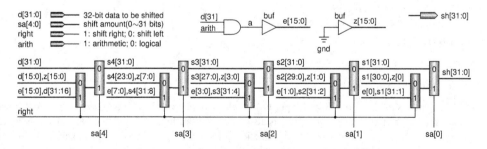

Figure 2.16   Schematic diagram of 32-bit barrel shifter

Figure 2.16 shows an implementation of a 32-bit barrel shifter. Other five multiplexers are used for selecting the result of the left shift or right shift. An AND gate generates the bits e, which will be inserted in the emptied bit positions for the right shift. e will be zero if arith is a 0 (logical shift); otherwise, all the bits of e are the same as the sign bit of the input d (arithmetic shift). The component named gnd outputs a 0, which is used for the left shift. A buf component replicates an input bit to multiple output bits, each of which is equal to the input bit.

The Verilog HDL code for the barrel shifter is listed below. It is almost identical to the schematic circuit shown in Figure 2.16.

```
module shift_mux (d,sa,right,arith,sh);          // a barrel shift using muxs
    input   [31:0] d;                            // input: 32-bit data
    input   [4:0] sa;                            // input: shift amount
    input          right;                        // 1: shift right; 0: left
    input          arith;                        // 1: arithmetic; 0: logical
    output  [31:0] sh;                           // output: shifted result
    wire    [31:0] t0, t1, t2, t3, t4;           // wires: outputs of muxs
    wire    [31:0] s1, s2, s3, s4;               // wires: outputs of muxs
    wire           a = d[31] & arith;            // a: filling bit
    wire    [15:0] e = {16{a}};                  // replicate a to 16 bits
    parameter      z = 16'b0;                    // a 16 bits zero
    wire    [31:0] sdl4,sdl3,sdl2,sdl1,sdl0;     // left  shifted data
    wire    [31:0] sdr4,sdr3,sdr2,sdr1,sdr0;     // right shifted data
    assign  sdl4  = {d[15:0],  z};               // shift to left  by 16 bits
    assign  sdl3  = {s4[23:0], z[7:0]};          // shift to left  by  8 bits
    assign  sdl2  = {s3[27:0], z[3:0]};          // shift to left  by  4 bits
    assign  sdl1  = {s2[29:0], z[1:0]};          // shift to left  by  2 bits
    assign  sdl0  = {s1[30:0], z[0]};            // shift to left  by  1 bit
    assign  sdr4  = {e,        d[31:16]};        // shift to right by 16 bits
    assign  sdr3  = {e[7:0],   s4[31:8]};        // shift to right by  8 bits
    assign  sdr2  = {e[3:0],   s3[31:4]};        // shift to right by  4 bits
    assign  sdr1  = {e[1:0],   s2[31:2]};        // shift to right by  2 bits
    assign  sdr0  = {e[0],     s1[31:1]};        // shift to right by  1 bit
    mux2x32 m_right4 (sdl4, sdr4, right, t4);    // select left or right
    mux2x32 m_right3 (sdl3, sdr3, right, t3);    // select left or right
    mux2x32 m_right2 (sdl2, sdr2, right, t2);    // select left or right
    mux2x32 m_right1 (sdl1, sdr1, right, t1);    // select left or right
    mux2x32 m_right0 (sdl0, sdr0, right, t0);    // select left or right
    mux2x32 m_shift4 (d,    t4, sa[4], s4);      // select not_shift or shift
    mux2x32 m_shift3 (s4,   t3, sa[3], s3);      // select not_shift or shift
    mux2x32 m_shift2 (s3,   t2, sa[2], s2);      // select not_shift or shift
    mux2x32 m_shift1 (s2,   t1, sa[1], s1);      // select not_shift or shift
    mux2x32 m_shift0 (s1,   t0, sa[0], sh);      // select not_shift or shift
endmodule
```

Verilog HDL supports the operators of shift left ($<<$), logical shift right ($>>$), and arithmetic shift right ($>>>$). $>>>$ must be used together with $signed()$; see the following behavioral style code that implements the barrel shifter. Note that the condition of the always block can be simply @*, which is added to the IEEE Verilog HDL standard of the 2001 version.

```
module shift (d,sa,right,arith,sh);          // barrel shift, behavioral style
    input   [31:0] d;                        // input: 32-bit data to be shifted
    input   [4:0] sa;                        // input: shift amount, 5 bits
    input          right;                    // 1: shift right; 0: shift left
    input          arith;                    // 1: arithmetic shift; 0: logical
    output  [31:0] sh;                       // output: shifted result
    reg     [31:0] sh;                       // will be combinational
    always @* begin                          // always block
```

```
        if (!right) begin            // if shift left
            sh = d << sa;            //     shift left sa bits
        end else if (!arith) begin   // if shift right logical
            sh = d >> sa;            //     shift right logical sa bits
        end else begin               // if shift right arithmetic
            sh = $signed(d) >>> sa;  //     shift right arithmetic sa bits
        end
    end
endmodule
```

Through the two implementation versions of the barrel shifter, you may find that the Verilog HDL code in behavioral style is more like a C program, and the Verilog HDL code in structural or dataflow style is like an assembly program.

All the Verilog HDL code examples given above are for designing combinational circuits. There is another type of the logic circuits, sequential circuits, which we will introduce in the next section.

2.6   Sequential Circuit Design

The outputs of sequential circuits are dependent not only on the present inputs but also on the past history of inputs. The components that can store data must be used to record the current state in sequential circuits. The D latch, D flip-flop (DFF), JK flip-flop (JKFF), and T flip-flop (TFF) are examples of such components.

2.6.1   D Latch and D Flip-Flop

A D latch can store one bit information. It can be implemented with an RS (reset-set) latch. Figure 2.17(a) shows the schematic circuit of an RS latch. When s (set, active-low) is a 0 (active) and r (reset, active-low) is a 1 (inactive), the output q is 1 (set); when s is a 1 (inactive) and r is a 0 (active), q is 0 (reset); when s and r are both 1 (inactive), q does not change (hold).

The schematic circuit of a D latch is shown in Figure 2.17(b). Three gates (two NAND gates and a NOT gate) are added to an RS latch. These additional gates generate the r and s signals. When the input c (control) is a 0, s and r are both 1, so the state q does not change; when c is a 1, the output q will be equal to the input d. Thus we say that a D latch is level-triggered. The Verilog HDL code of the D latch is shown below.

```
module d_latch (c,d,q,qn);       // d latch
    input   c, d;                // inputs:  c, d
    output q, qn;                // outputs: q, qn
```

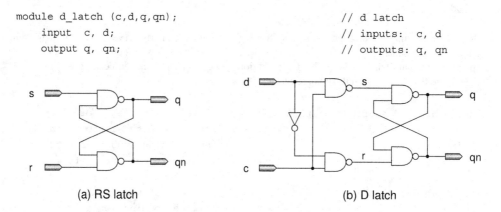

(a) RS latch   (b) D latch

Figure 2.17   Schematic diagram of a D latch

```
    wire    r, s;                        // internal wires
    nand nand1 (s,   d, c);              // nand (out, in1, in2);
    nand nand2 (r, ~d, c);               // nand (out, in1, in2);
    rs_latch rs (s, r, q, qn);           // use rs_latch module
endmodule
```

The following Verilog HDL code implements the RS latch that is used in the Verilog HDL code of the D latch.

```
module rs_latch (s,r,q,qn);             // rs latch
    input   s, r;                        // inputs:  set, reset
    output q, qn;                        // outputs: q, qn
    nand nand1 (q,   s, qn);             // nand (out, in1, in2);
    nand nand2 (qn, r, q);               // nand (out, in1, in2);
endmodule
```

Figure 2.18 shows the simulation waveform of the D latch from which we can see that, when the input c is a 1, the output q follows the input d, and q does not change when the input c is a 0. In the beginning (0–5 ns), the state of the D latch is unknown.

Different from a D latch, a DFF is edge-triggered. Figure 2.19 shows an academic version of the DFF circuit. It consists of two D latches and two NOT gates. The left D latch is called a master latch, and the right one is called a slave latch. When clk (clock) signal is a 0, the output of the master latch (q0) is equal to the input d, and the output of the slave latch does not change. When clk goes to 1 from 0, the output of the slave latch follows q0 but q0 does not change. Therefore, the state of a DFF is altered only when the clock changes from 0 to 1 (rising edge or positive edge). This is the meaning of "edge-triggered."

Below is the Verilog HDL code of the DFF. This module invokes the module of D latch which was already given above.

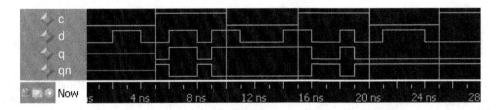

Figure 2.18   Waveform of a D latch

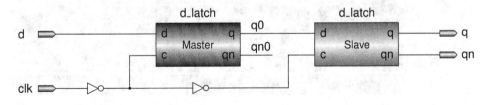

Figure 2.19   Schematic diagram of an academic D flip-flop

```
module d_flip_flop (clk,d,q,qn);              // dff using 2 d latches
   input   clk, d;                            // inputs:  clk, d
   output  q,  qn;                            // outputs: q, qn
   wire    q0, qn0;                           // internal wires
   wire    clkn, clknn;                       // internal wires
   not inv1 (clkn,  clk);                     // inverse of clk
   not inv2 (clknn, clkn);                    // inverse of clkn
   d_latch dlatch1 (clkn,  d,  q0, qn0);      // master d latch
   d_latch dlatch2 (clknn, q0, q,  qn);       // slave  d latch
endmodule
```

Figure 2.20 shows the simulation waveform of the DFF. We can see that the input d is stored to the DFF on the positive edge of clk.

An industry version of the DFF circuit is given in Figure 2.21, where prn and clrn are the preset and clear inputs, respectively. Both the signals are active-low.

The following Verilog HDL code was generated from the schematic circuit of Figure 2.21 by Quartus II Web Edition. Some wire names were simplified.

```
module d_ff (prn, clk, d, clrn, q, qn);      // dff generated from schematic
   input   prn, clk, d, clrn;                // prn: preset; clrn: clear (active low)
   output  q, qn;                            // outputs q, qn
   wire    wire_0, wire_1;                   // internal wire, see figure ch02_fig21
   wire    wire_2, wire_3;                   // internal wire, see figure ch02_fig21
   assign wire_0 = ~(wire_1 & prn   & wire_2);      // 3-input nand
   assign wire_1 = ~(clk    & clrn  & wire_0);      // 3-input nand
   assign wire_2 = ~(wire_3 & clrn  & d);           // 3-input nand
   assign wire_3 = ~(wire_1 & clk   & wire_2);      // 3-input nand
   assign q      = ~(prn    & wire_1 & qn);         // 3-input nand
   assign qn     = ~(q      & wire_3 & clrn);       // 3-input nand
endmodule
```

The simulation waveform of the industry version of the DFF is shown in Figure 2.22. The state of the DFF is 0 in the beginning because clrn is active. Then the state becomes 1 when prn is active. The rest shows that the value of d is stored into DFF in the positive edge of clk.

The DFF stores data on every positive edge of the clock if both the clear and preset inputs are inactive. Figure 2.23(a) shows a DFFE by adding an enable control signal e to the DFF: if e is a 0, the DFFE does not change even if a new clock edge arrives and/or the d input changes; otherwise, DFFE acts as a DFF. This is implemented with a 2-to-1 multiplexer: the DFFE is updated on every clock rising edge but the current state will be written again (no change) if the enable signal is inactive. Figure 2.23(b) shows a bad design of DFFE that uses an AND gate to prohibit the clock.

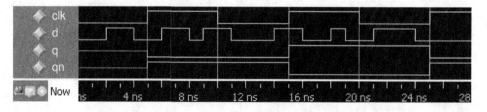

Figure 2.20   Waveform of an academic D flip-flop

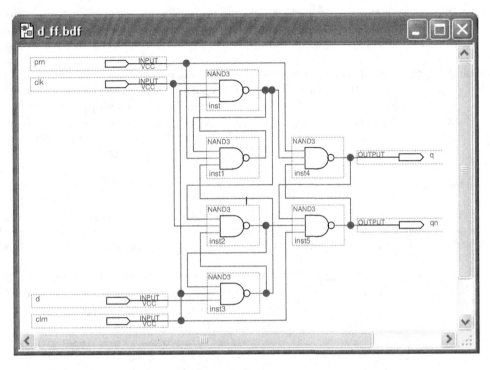

Figure 2.21   Schematic diagram of an industry D flip-flop

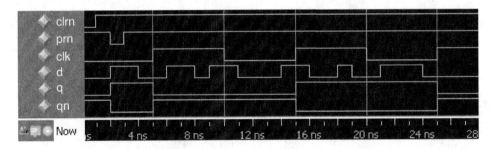

Figure 2.22   Waveform of an industry D flip-flop

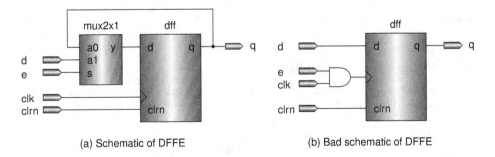

(a) Schematic of DFFE   (b) Bad schematic of DFFE

Figure 2.23   Schematic diagram of a D flip-flop with enable control

A DFF with a synchronous clear can be implemented by the following behavioral-style Verilog HDL code. `posedge` is a keyword that means a positive edge. The active-low clear input `clrn` is checked on the positive edge of `clk`: if `clrn` is active, the DFF will be reset to 0; otherwise the input `d` is stored.

```
module d_ff_sync (d,clk,clrn,q);     // dff with synchronous reset (to clk)
    input       d, clk, clrn;        // inputs d, clk, clrn (active low)
    output reg q;                    // output q, register type
    always @ (posedge clk) begin     // always block, posedge: rising edge
        if (!clrn) q <= 0;           // if clrn is asserted, reset dff
        else       q <= d;           // else store d to dff
    end
endmodule
```

The following code implements a DFF with an asynchronous clear. `negedge` is a keyword that means negative edge (falling edge). `clrn` is checked independently of `clk`. In fact, `clrn` is a level-triggered clear, not an edge-triggered clear, because if `clrn` is held active, the output q will continue to be held low.

```
module dff (d,clk,clrn,q);           // dff with asynchronous reset
    input       d, clk, clrn;        // inputs d, clk, clrn (active low)
    output reg q;                    // output q, register type
    always @ (posedge clk or negedge clrn) begin // always block, "or"
        if (!clrn) q <= 0;           // if clrn is asserted, reset dff
        else       q <= d;           // else store d to dff
    end
endmodule
```

A DFFE that has an enable control signal can be implemented by the following Verilog HDL code: an `if` statement controls the update of the DFFE.

```
module dffe (d,clk,clrn,e,q);        // dff (async) with write enable
    input       d, clk, clrn;        // inputs d, clk, clrn (active low)
    input       e;                   // enable
    output reg q;                    // output q, register type
    always @ (posedge clk or negedge clrn) begin // always block, "or"
        if (!clrn)  q <= 0;          // if clrn is asserted, reset dff
        else if (e) q <= d;          // else if enabled store d to dff
    end
endmodule
```

2.6.2 JK Latch and JK Flip-Flop

The DFF is used most commonly in digital circuit designs. There are two other flip-flops that are also used commonly: JKFF and TFF. Figure 2.24 shows the circuit of a JK latch where an RS latch is used for storing the state. The truth table is also given in the figure. The output expression is $q_n = c(j\,\overline{q} + \overline{k}\,q) + \overline{c}\,q$, where q_n is the next state of q.

The JK latch is rarely used because it toggles without speed control if all the three inputs are 1. Figure 2.25 shows the circuit of a JKFF with a preset and a clear. It consists of a JK latch (master latch) and a D latch (slave latch). This JKFF toggles on the positive edge of clock because the outputs of the D

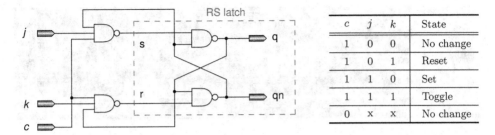

c	j	k	State
1	0	0	No change
1	0	1	Reset
1	1	0	Set
1	1	1	Toggle
0	x	x	No change

Figure 2.24   Schematic diagram of the JK latch

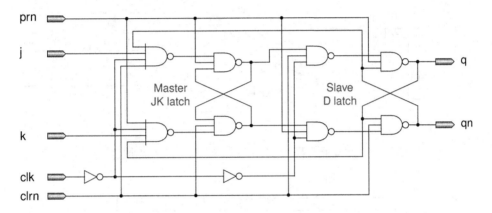

Figure 2.25   Schematic diagram of the JK flip-flop

latch, not the outputs of the JK latch, are sent back to the first-level gates. Note that at least a clear input is required; otherwise, the initial states of the JK and D latches cannot be determined.

The following Verilog HDL code uses $q_n = j\overline{q} + \overline{k}q$ to implement the JKFF.

```
module jkff (j,k,clk,clrn,q);              // jkff with asynchronous reset
    input      j, k, clk, clrn;            // inputs j, k, clock, reset
    output reg q;                          // output q, register type
    always @ (posedge clk or negedge clrn) begin // always block, "or"
        if (!clrn) q <= 0;                 // if clrn is active, reset jkff
        else       q <= j & ~q | ~k & q;   // else update jkff
    end
endmodule
```

Figure 2.26 shows the simulation waveform of the JKFF code.

2.6.3   T Latch and T Flip-Flop

Figure 2.27 shows the circuit of a T latch where an RS latch is used for storing the state. When the t and c inputs are both high, the output q toggles. The truth table is also given in the figure. The output expression is $q_n = c(t\overline{q} + \overline{t}q) + \overline{c}q$, where q_n is the next state of q. Similar to the JK latch, the T latch is rarely used.

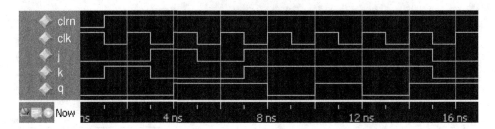

Figure 2.26   Waveform of the JK flip-flop

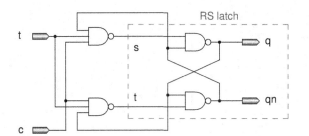

c	t	State
1	0	No change
1	1	Toggle
0	x	No change

Figure 2.27   Schematic diagram of the T latch

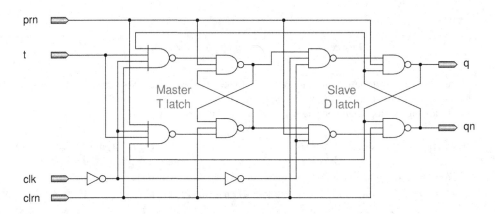

Figure 2.28   Schematic diagram of the T flip-flop

Figure 2.28 shows the circuit of a TFF. It consists of a T latch (master latch) and a D latch (slave latch). The output does not change until the next positive edge of the clock (clk). The input t determines whether to toggle the output: if t is a 1, the state is toggled in the clock edge; otherwise, the state does not change. prn and clrn are the preset and clear inputs, respectively.

The following Verilog HDL code uses $q_n = t\overline{q} + \overline{t}q$ to implement the TFF.

```
module tff (t,clk,clrn,q);              // tff with asynchronous reset
    input      t, clk, clrn;            // inputs t, clock, reset
    output reg q;                       // output q, register type
```

```
    always @ (posedge clk or negedge clrn) begin // always block, "or"
        if (!clrn) q <= 0;                       // if clrn is active, reset tff
        else       q <= t & ~q | ~t & q;         // else update tff
    end
endmodule
```

Figure 2.29 shows the simulation waveform of the TFF code.

2.6.4 Shift Register

Using multiple DFFs, we can design a shift register, as shown in Figure 2.30. It can convert a serial signal di to a parallel signal q[2:0]. It can also convert a parallel signal d[2:0] to a serial signal do. The load signal is used to load the parallel data d[2:0] to the DFFs (load = 1). mux2x1 is a 1-bit 2-to-1 multiplexer. When shift is performed, the load signal should be 0.

2.6.5 FIFO Buffer

FIFO (first-in first-out) is a special buffer for queuing data. Figure 2.31 shows a schematic diagram of a FIFO of depth 4. R1, R2, R3, and R4 are four registers (DFFs). The data in the input port enter the R1 register when the write signal is asserted. Then the data are passed to the rightmost empty register. When the read signal is asserted, the data in the R4 register will be read out and other data go to their right registers automatically.

There are four RS latches: F1, F2, F3, and F4. The *i*th latch indicates whether the *i*th register contains data or not. The inputs of the RS latches are active high: if S (set) is a 1, the latch is set; if R (reset) is a 1, the latch is cleared. Such a latch can be designed with NOR gates. The RS latches are cleared initially,

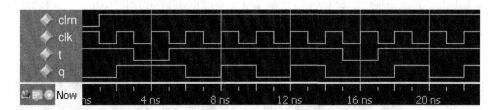

Figure 2.29 Waveform of the T flip-flop

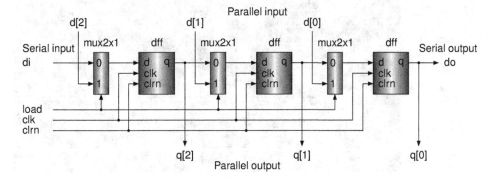

Figure 2.30 Schematic diagram of a shift register

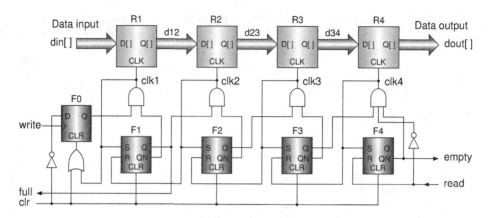

Figure 2.31   Schematic diagram of a FIFO of depth 4

therefore the outputs of QNs are high. These latches and the AND gates are used to generate clock pulse signals for the edge-triggered registers R1, R2, R3, and R4. If Q of the $(i-1)$th latch becomes 1 and Q of the ith latch is 0 (QN is 1), then clk$i$ becomes 1 (a rising edge) that sets the ith latch and resets the $(i-1)$th latch. clk$i$ will go back to 0 once clk$i+1$ becomes 1. Thus, a pulse of clk$i$ is generated that is used to store the data of the $(i-1)$th register into the ith register.

The DFF F0 is used to generate a write pulse for clk1. The NOT gate in the right side prevents the data in the R4 register from being overwritten when the read signal is asserted. A 1 of the empty signal indicates that the FIFO is empty. A 1 of the full signal indicates that the FIFO is full.

Figure 2.32 shows the simulation waveforms of the FIFO circuit. We set the delay time of the AND gates to 1 ns in the simulation. We can see that the data of 0xe1 (0x indicates the data are represented in hexadecimal format) are passed from R1 to R4, and at 86 ns the FIFO is full. The registers R1, R2, R3, and R4 hold the data of 0xe1, 0xe2, 0xe3, and 0xe4, respectively. These data are read out sequentially from the output port, resulting in the FIFO becoming empty.

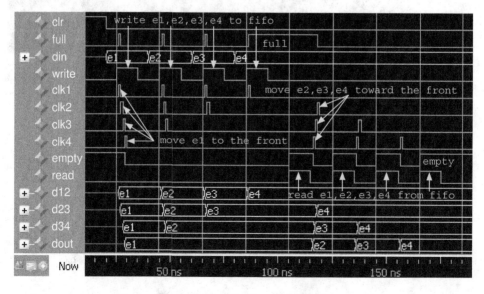

Figure 2.32   Waveforms of FIFO4

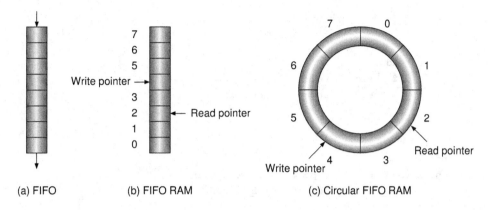

Figure 2.33   A circular FIFO implemented with RAM

There is no address in a FIFO buffer, meaning that a register in the FIFO cannot be accessed randomly. FIFOs are used in many places, such as cache block replacement circuits, the keyboard scan code buffer, and network communication buffers. Figure 2.33 shows another FIFO implementation in which a traditional RAM (random access memory) is used for queuing the data.

We prepare two n-bit internal pointers, the write pointer and the read pointer, as the addresses of the RAM. Each write or read operation will result in an increment of the corresponding pointer. Therefore, the FIFO is actually circulated. The depth of the FIFO is 2^n and the next location to the location $2^n - 1$ is 0.

The following Verilog HDL code implements such a RAM-based FIFO. A 1 of the `ready` signal indicates that the FIFO is not empty. `overflow` means that the FIFO is full and the data written last are lost.

```
module fifo (clk,clrn,read,write,data_in,data_out,ready,overflow);   // fifo
    input        clk, clrn;                   // clock and reset
    input        read;                        // fifo read, active high
    input        write;                       // fifo write, active high
    input  [7:0] data_in;                     // fifo data input
    output [7:0] data_out;                    // fifo data output
    output       ready;                       // fifo has data
    output reg   overflow;                    // fifo overflow flag
    reg    [7:0] fifo_buff [7:0];             // fifo buffer of depth 8
    reg    [2:0] write_pointer;               // fifo write pointer
    reg    [2:0] read_pointer;                // fifo read pointer
    always @ (posedge clk or negedge clrn) begin
        if (!clrn) begin
            write_pointer <= 0;               // clear write pointer
            read_pointer  <= 0;               // clear read pointer
            overflow      <= 0;               // clear overflow flag
        end else begin
            if (write) begin
                if ((write_pointer + 3'b1) != read_pointer) begin
                    fifo_buff[write_pointer] <= data_in;     // push data
                    write_pointer <= write_pointer + 3'd1; // pointer++
```

```
            end else begin
                overflow <= 1;                      // overflow
            end
        end
        if (read && ready) begin
            read_pointer <= read_pointer + 3'd1;    // pointer++, pop
            overflow <= 0;                          // clear overflow
        end
    end
end
assign ready = (write_pointer != read_pointer);     // has data
assign data_out = fifo_buff[read_pointer];          // data output
endmodule
```

Figure 2.34 shows the simulation waveforms of the RAM-based FIFO. The data of 0xe1, 0xe2, ... , and 0xf3 are written to the FIFO, but the data read from the FIFO are 0xe1, 0xe2, ... , 0xe9, 0xf1, 0xf2, and 0xf3. 0xea, 0xeb, ... , 0xf0 are lost as a result of the FIFO overflow. Although there are eight locations in the FIFO, the code given above results in that the FIFO can hold only seven data at the same time.

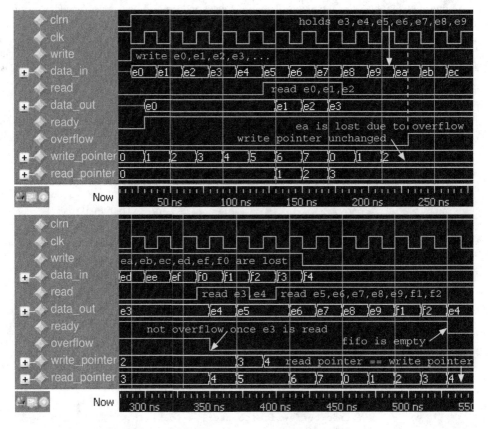

Figure 2.34   Waveforms of RAM-based FIFO

2.6.6 Finite State Machine and Counter Design

The flip-flops we have introduced can be used to design the circuit of finite state machines. A finite state machine is a way of modeling a system in which the system's outputs will depend on not only the current inputs but also the past history of inputs. It defines a finite set of states and behaviors, and how the system transits from one state to another when certain conditions are true.

A finite state machine can be implemented in software but here we focus on the hardware circuit design of the finite state machine. There are two models of the finite state machine: the Mealy model and the Moore model, as shown as in Figure 2.35.

There are a finite number of states that can be implemented with DFFs. Suppose that the number of states is N; then $n = \lceil \log_2 N \rceil$ flip-flops are needed. The module of next state is a combinational circuit that determines the next state based on the current state and inputs. The next state will be written into DFFs on the clock edge. The module of output function is also a combinational circuit that generates the outputs based on the current state (Moore model) or the combination of the current state and the current inputs (Mealy model). The general steps of designing a sequential circuit are described below.

1. Make a state transition diagram based on the problem statement.
2. Determine the number of flip-flops and assign binary codes to the states.
3. Fill a truth table for the next state signals.
4. Write the logic expression for each of the next state signals. Karnaugh maps may be used to get simplified expressions.
5. Fill a truth table for the output signals.
6. Write logic expressions of the output signals. Karnaugh maps may be used to get simplified expressions.
7. Build and simulate the circuit.

Let's take an example to show how to design a sequential circuit. Suppose we design a radix of six up/down counters with a seven-segment LED (light-emitting diode), as shown as in Figure 2.36. The state of the counter changes on the positive edge of the clock. If the input u is a 1, the counter value

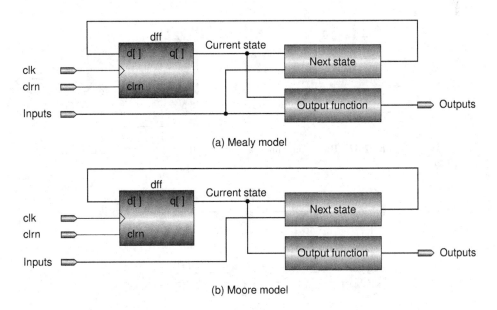

(a) Mealy model

(b) Moore model

Figure 2.35 Two models of the general finite state machine

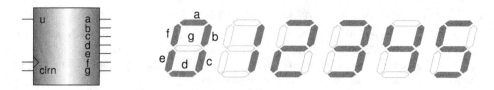

Figure 2.36   A counter with a seven-segment LED

will change in the sequence 0, 1, 2, 3, 4, 5, 0, 1, 2, ... If u is a 0, the counter value will change in the sequence 0, 5, 4, 3, 2, 1, 0, 5, 4, ... There are seven output signals, with each connecting to a segment of LED. A segment of the LED will be on if its control signal is a 0 (active-low).

Obviously, there are six states, thus three DFFs are needed. The circuit of the counter is shown in Figure 2.37. The module of dff3 contains three DFFs. The other module is a combinational circuit that generates signals of the next state (ns[2:0]) and LED control signals (a, b, c, d, e, f, and g). The current state (the outputs of dff3) is denoted with q[2:0].

Figure 2.38 shows the state transition diagram. The arrowed lines indicate the transitions of the states under the condition of the input. Figure 2.38 also shows that a 3-bit unique code is assigned to a state. Any code can be assigned to any state as long as all the codes are unique.

Table 2.6 is the truth table for the next state. Figure 2.39 shows the Karnaugh maps for each of the next state signals. From the Karnaugh maps, we can get the following expressions of the next state signals.

$$ns[0] = \overline{q[0]};$$

$$ns[1] = \overline{q[2]}\,\overline{q[1]}\,q[0]u + q[1]\,\overline{q[0]}u + q[1]\,q[0]\,\overline{u} + q[2]\,\overline{q[0]}\,\overline{u};$$

$$ns[2] = \overline{q[2]}\,\overline{q[1]}\,\overline{q[0]}\,\overline{u} + q[1]\,q[0]\,u + q[2]\,\overline{q[0]}\,u + q[2]\,q[0]\,\overline{u}.$$

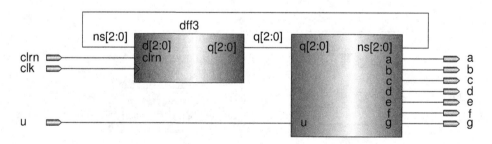

Figure 2.37   Block diagram of a counter with a seven-segment LED

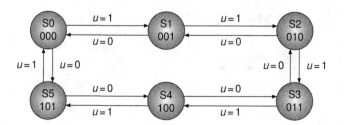

Figure 2.38   State transition diagram of the counter

Table 2.6 State transition table of the counter

	Current state			Input		Next state		
	$q[2]$	$q[1]$	$q[0]$	u		$ns[2]$	$ns[1]$	$ns[0]$
S0	0	0	0	1	S1	0	0	1
				0	S5	1	0	1
S1	0	0	1	1	S2	0	1	0
				0	S0	0	0	0
S2	0	1	0	1	S3	0	1	1
				0	S1	0	0	1
S3	0	1	1	1	S4	1	0	0
				0	S2	0	1	0
S4	1	0	0	1	S5	1	0	1
				0	S3	0	1	1
S5	1	0	1	1	S0	0	0	0
				0	S4	1	0	0

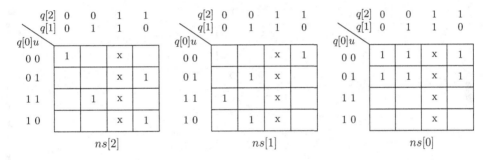

Figure 2.39 Karnaugh map for next state of the counter

Figure 2.40 shows the truth table and Karnaugh maps of the output signals. We get the following expressions of the output signals:

$$a = \overline{q[2]}\ \overline{q[1]}\ q[0] + q[2]\ \overline{q[0]};$$

$$b = q[2]\ \overline{q[0]};$$

$$c = q[1]\ \overline{q[0]};$$

$$d = \overline{q[2]}\ \overline{q[1]}\ q[0] + q[2]\ \overline{q[0]} = a;$$

$$e = q[2]\ \overline{q[0]} + q[0];$$

$$f = q[1]\ \overline{q[0]} + \overline{q[2]}\ q[0];$$

$$g = \overline{q[2]}\ \overline{q[1]}.$$

Now we can build the circuits of the up/down counter. Figure 2.41 shows the schematic diagram of the 3-bit DFFs. Figure 2.42 shows the schematic diagram of the next state for the counter. Figure 2.43 shows the schematic diagram of the output function for the counter. And Figure 2.44 shows the top schematic

q[2:0]	g f e d c b a
0 0 0	1 0 0 0 0 0 0
0 0 1	1 1 1 1 0 0 1
0 1 0	0 1 0 0 1 0 0
0 1 1	0 1 1 0 0 0 0
1 0 0	0 0 1 1 0 0 1
1 0 1	0 0 1 0 0 1 0

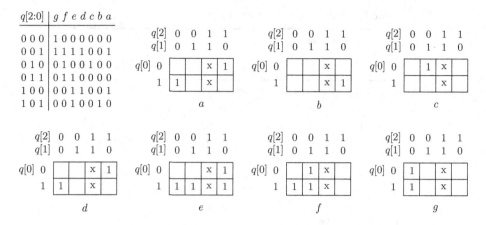

Figure 2.40   Karnaugh map for the output function of the counter

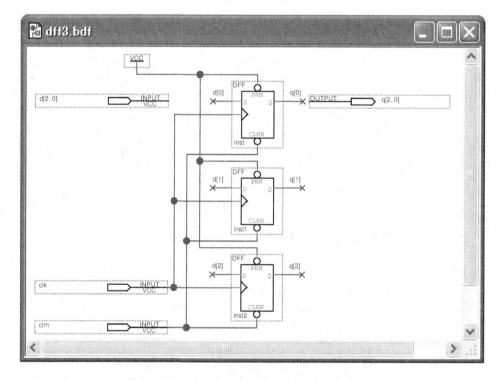

Figure 2.41   Schematic diagram of 3-bit D flip-flops

diagram of the counter with a seven-segment LED. It consists of three parts: (i) dff3, the 3-bit DFFs, (ii) next_state, the next state, and (iii) output_function, the output function.

Figure 2.45 shows the simulation waveform of the up/down counter. The first half shows the counting up, and the second half shows the counting down. The counter changes state on the clock rising edge. The LED control signals are also shown in the figure.

The purpose of describing the details of the counter design is not only for designing a counter but also for understanding the general procedure of sequential circuit designs. If we only want to design a radix of a six up/down counter with the seven-segment LED, we can implement it with the following behavioral-style Verilog HDL code where % is the operator of the modulo.

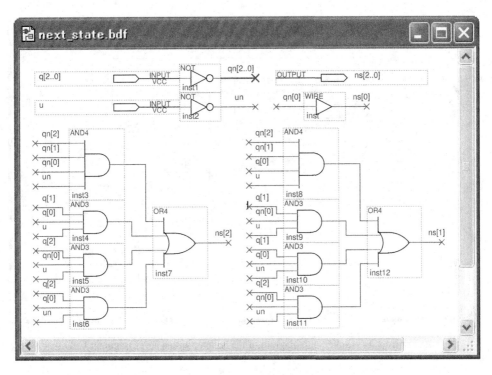

Figure 2.42 Schematic diagram of next state for the counter

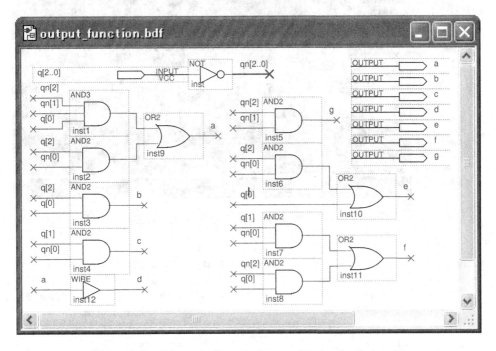

Figure 2.43 Schematic diagram of output function for the counter

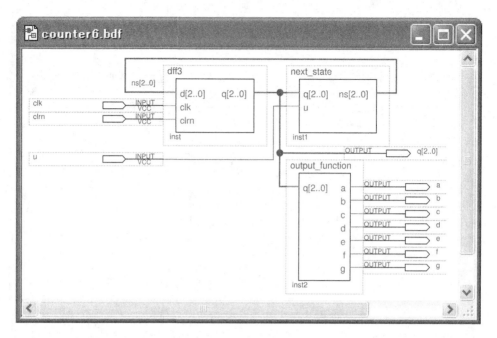

Figure 2.44   Schematic diagram of the counter with a seven-segment LED

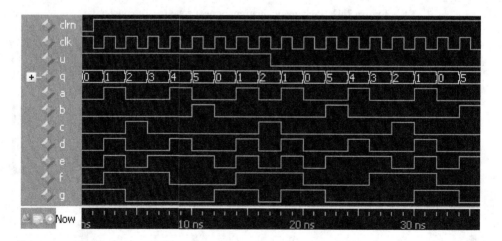

Figure 2.45   Waveform of the up/down counter

```
module counter_6 (u,clk,clrn,q,a,b,c,d,e,f,g);   // a counter with 7-seg LED
    input        clk, clrn;            // clk, clear (active low)
    input        u;                    // u==1: count up; u==0: count down
    output [2:0] q;                    // 3-bit counter output
    output       a, b, c, d, e, f, g;  // seven-segment LED control
    reg    [2:0] q;                    // register type
    always @ (posedge clk or negedge clrn) begin
```

```
        if (!clrn) q <= 0;                   // if clrn is asserted, counter=0
        else if (u) q <= (q + 1) % 6;        // if counter up, q++
        else if (q != 0) q <= q - 1;         // else            q-
            else        q <= 3'd5;
    end
    assign {g,f,e,d,c,b,a} = seg7(q);        // call function to get LED control
    function  [6:0] seg7;                    // the function, 7-bit return value
        input [2:0] q;                       // input argument
        case (q)                             // cases:
            3'd0 : seg7 = 7'b1000000;        // 0's LED control, 0: light on
            3'd1 : seg7 = 7'b1111001;        // 1's LED control, 1: light off
            3'd2 : seg7 = 7'b0100100;        // 2's LED control
            3'd3 : seg7 = 7'b0110000;        // 3's LED control
            3'd4 : seg7 = 7'b0011001;        // 4's LED control
            3'd5 : seg7 = 7'b0010010;        // 5's LED control
            default: seg7 = 7'b1111111;      // default: all segments light off
        endcase
    endfunction
endmodule
```

We will give another example of the sequential circuit in Chapter 7, where we describe how to design a multiple-cycle CPU.

Exercises

2.1 Try to execute the following Java program (`logic.java`) on your machine and explain the output results.

```
class logic {
    public static void main(String[] args) {
        int a = 0x5;
        int b = 0xc;
        int f_and  =   a & b;
        int f_or   =   a | b;
        int f_not  =  ~a;
        int f_nand = ~(a & b);
        int f_nor  = ~(a | b);
        int f_xor  =   a ^ b;
        int f_xnor = ~(a ^ b);
        System.out.format("a      = 0x%08x\n", a);
        System.out.format("b      = 0x%08x\n", b);
        System.out.format("f_and  = 0x%08x\n", f_and);
        System.out.format("f_or   = 0x%08x\n", f_or);
        System.out.format("f_not  = 0x%08x\n", f_not);
        System.out.format("f_nand = 0x%08x\n", f_nand);
        System.out.format("f_nor  = 0x%08x\n", f_nor);
```

```
        System.out.format("f_xor   = 0x%08x\n", f_xor);
        System.out.format("f_xnor = 0x%08x\n", f_xnor);
    }
}
```

Recommendation: use a command line mode to compile and execute the program:

```
$ javac logic.java
$ java logic
```

2.2   Suppose that we have a code fragment in C or Java as shown below.

```
if (((a == b) || (c >= d)) && (e != f)) {
    // action x :(
} else {
    // action y :)
}
```

Rewrite the `if` condition so that it becomes the following format (exchanged the `then` and `else` clauses). Hint: use De Morgan's Law.

```
if (_____) {
    // action y :)
} else {
    // action x :(
}
```

2.3   Download and install the Icarus Verilog or any other Verilog HDL simulators.

2.4   Write the test bench codes to simulate CMOS NAND and NOR gates.

2.5   Design the circuits of a three-input CMOS NAND gate and a three-input CMOS NOR gate, and simulate them with the downloaded simulators.

2.6   In shift operations, in addition to the logical shift left, logical shift right, and arithmetic shift right, there is also an arithmetic shift left that keeps the sign bit (bit 31) unchanged, as shown in the following example.

Original data (d):	11111111_00000000_00000000_11111111
Logical shift d to the left by 8 bits:	00000000_00000000_11111111_00000000
Arithmetic shift d to the left by 8 bits:	10000000_00000000_11111111_00000000
Logical shift d to the right by 8 bits:	00000000_11111111_00000000_00000000
Arithmetic shift d to the right by 8 bits:	11111111_11111111_00000000_00000000

Design a shifter that can perform the four types of shift operations described above. The input and output signals are the same as the barrel shifter given in this chapter.

2.7   Design an active-high RS latch with NOR gates.

2.8 Explain the reason why in Figure 2.19 two NOT gates were used to generate the control signal (c) for the slave D latch, instead of connecting the clk directly to the input c of the D latch.

2.9 Explain the reason why the DFFE shown in Figure 2.23(b) is a bad design.

2.10 Try to understand the following code and run it on your FPGA (field-programmable gate array) board.

```verilog
module minute_second (clk,m1,m0,s1,s0,dots);          // what's this?
    input        clk;                                 // 50MHz
    output [6:0] m1, m0;
    output [6:0] s1, s0;
    output [3:0] dots;


    reg          sec_clk = 1;
    reg    [24:0] clk_cnt = 0;
    always @ (posedge clk) begin
        if (clk_cnt == 25'd24999999) begin
            clk_cnt <= 0;
            sec_clk <= ~sec_clk;
        end else begin
            clk_cnt <= clk_cnt + 1;
        end
    end


    reg [2:0] min1 = 0, sec1 = 0;
    reg [3:0] min0 = 0, sec0 = 0;
    always @ (posedge sec_clk) begin
        if (sec0 == 4'd9) begin
            sec0 <= 0;
            if (sec1 == 3'd5) begin
                sec1 <= 0;
                if (min0 == 4'd9) begin
                    min0 <= 0;
                    if (min1 == 3'd5) begin
                        min1 <= 0;
                    end else begin
                        min1 <= min1 + 1;
                    end
                end else begin
                    min0 <= min0 + 1;
                end
            end else begin
                sec1 <= sec1 + 1;
            end
        end else begin
            sec0 <= sec0 + 1;
        end
```

```
    end

    assign m1 = seg7({1'b0,min1});
    assign s1 = seg7({1'b0,sec1});
    assign m0 = seg7(min0);
    assign s0 = seg7(sec0);
    assign dots = {1'b1,sec_clk,2'b11};

    //   0
    // 5   1
    //   6
    // 4   2
    //   3
    function [6:0] seg7;
        input [3:0] q;
        case (q)
            4'd0 : seg7 = 7'b1000000;
            4'd1 : seg7 = 7'b1111001;
            4'd2 : seg7 = 7'b0100100;
            4'd3 : seg7 = 7'b0110000;
            4'd4 : seg7 = 7'b0011001;
            4'd5 : seg7 = 7'b0010010;
            4'd6 : seg7 = 7'b0000010;
            4'd7 : seg7 = 7'b1111000;
            4'd8 : seg7 = 7'b0000000;
            4'd9 : seg7 = 7'b0010000;
            default: seg7 = 7'b1111111;
        endcase
    endfunction
endmodule
```

2.11   Still using eight registers, can you revise the `fifo.v` code given in this chapter so that the FIFO can hold eight data at the same time?

2.12   Design a control circuit for a vending machine of yours.

3

Computer Arithmetic Algorithms and Implementations

Computers can compute. This chapter describes binary numbers' arithmetic algorithms and their implementations in Verilog HDL (hardware description language). The algorithms include binary addition, subtraction, multiplication, division, and square root.

3.1 Binary Integers

This section introduces two types of the binary representations: unsigned number (absolute) and 2's complement representation for signed integers.

3.1.1 Binary and Hexadecimal Representations

We are familiar with the decimal system. However, all the information in computer systems, including instruction and data, is represented by the binary numbering system. A bit (a contraction for "binary digit") is the basic unit of the information that has only two values: either 0 or 1.

Question: What does stand for the following 32-bit binary code?

$$00110011110111100000000100000000$$

The correct answer is "don't know." The exact meaning of the code depends on where it will be used. If it is treated as an integer, its value is 870,187,264. If it is a floating-point number, its value is 0.0000001033786247717216610908508303007125. If it is an instruction and executed by a MIPS CPU (microprocessor without interlocked pipeline stages central processing unit), then it is `addi $30`, `$30, 256`, an immediate addition instruction. It may be an IP address, data of image or music, or something else.

Binary code is too long in representation and thus is hard to remember. Hexadecimal (hex) notation is much shorter and easier to remember. As shown in Table 3.1, 4-bit binary code is represented by one character. We need $2^4 = 16$ different characters. Because we are short of numbers, letters a, b, c, d, e, and f are used in the hex notation.

Then the binary code mentioned above can be represented in hex notation as

$$0011\ 0011\ 1101\ 1110\ 0000\ 0001\ 0000\ 0000_2 = 33de0100_{16}.$$

Computer Principles and Design in Verilog HDL, First Edition. Yamin Li.
© 2015 Tsinghua University Press. All rights reserved. Published 2015 by John Wiley & Sons Singapore Pte Ltd.
Companion Website: www.wiley.com/go/li/verilog

Table 3.1 Relationship between hexadecimal numbers and binary numbers

Binary number	0000	0001	0010	0011	0100	0101	0110	0111
Hexadecimal number	0	1	2	3	4	5	6	7
Binary number	1000	1001	1010	1011	1100	1101	1110	1111
Hexadecimal number	8	9	a	b	c	d	e	f

You can see that the conversion from binary to hex is easy. However, the conversion from binary to decimal is not so easy because (i) we must know the type of the binary number and (ii) the conversion itself takes longer time than that to hex.

3.1.2 Unsigned Binary Integers

Suppose that we use an n-bit binary number $b_{n-1}b_{n-2} \cdots b_1 b_0$, where b_i ($i = 0, 1, \cdots, n-2, n-1$) is a 0 or a 1, to denote an unsigned number: its value is

$$b_{n-1}b_{n-2} \cdots b_1 b_0 = b_{n-1} \times 2^{n-1} + b_{n-2} \times 2^{n-2} + \cdots + b_1 \times 2^1 + b_0 \times 2^0$$

Figure 3.1 shows an example of the unsigned binary number with $n = 16$. It is helpful to remember some values of the powers of 2, for example, $2^{16} = 65,536$.

An n-bit binary number can denote unsigned numbers from 0 (all the n bits are 0) to $2^n - 1$ (all the n bits are 1). The following lists some decimal values of 32-bit unsigned binary numbers.

```
0000 0000 0000 0000 0000 0000 0000 0000 = 0
0000 0000 0000 0000 0000 0000 0000 0001 = 1
0000 0000 0000 0000 0000 0000 0000 0010 = 2
...  ...
1111 1111 1111 1111 1111 1111 1111 1101 = 4,294,967,293
1111 1111 1111 1111 1111 1111 1111 1110 = 4,294,967,294
1111 1111 1111 1111 1111 1111 1111 1111 = 4,294,967,295
```

3.1.3 Signed Binary Integers (2's Complement Notation)

The 2's complement representation is widely used to denote signed integers. The value of an n-bit 2's complement binary number is

$$b_{n-1}b_{n-2} \cdots b_1 b_0 = -b_{n-1} \times 2^{n-1} + b_{n-2} \times 2^{n-2} + \cdots + b_1 \times 2^1 + b_0 \times 2^0$$

Notice the minus sign on the most significant bit (MSB) position. Figure 3.2 shows an example of the 2's complement binary number with $n = 16$.

2^{15}	2^{14}	2^{13}	2^{12}	2^{11}	2^{10}	2^9	2^8	2^7	2^6	2^5	2^4	2^3	2^2	2^1	2^0
b_{15}	b_{14}	b_{13}	b_{12}	b_{11}	b_{10}	b_9	b_8	b_7	b_6	b_5	b_4	b_3	b_2	b_1	b_0

Figure 3.1 Bit's significances of a 16-bit unsigned binary number

2^{15}	2^{14}	2^{13}	2^{12}	2^{11}	2^{10}	2^9	2^8	2^7	2^6	2^5	2^4	2^3	2^2	2^1	2^0
$-b_{15}$	b_{14}	b_{13}	b_{12}	b_{11}	b_{10}	b_9	b_8	b_7	b_6	b_5	b_4	b_3	b_2	b_1	b_0

Figure 3.2 Bit's significances of a 16-bit 2's complement signed number

An n-bit 2's complement binary number can denote signed numbers from -2^{n-1} (the MSB is 1 and others are 0) to $2^{n-1} - 1$ (the MSB is 0 and others are 1). The following lists some decimal values of 32-bit 2's complement binary numbers.

```
1000 0000 0000 0000 0000 0000 0000 0000 = -2,147,483,648
1000 0000 0000 0000 0000 0000 0000 0001 = -2,147,483,647
...  ...
1111 1111 1111 1111 1111 1111 1111 1110 = -2
1111 1111 1111 1111 1111 1111 1111 1111 = -1
0000 0000 0000 0000 0000 0000 0000 0000 = 0
0000 0000 0000 0000 0000 0000 0000 0001 = +1
...  ...
0111 1111 1111 1111 1111 1111 1111 1110 = +2,147,483,646
0111 1111 1111 1111 1111 1111 1111 1111 = +2,147,483,647
```

See the following two examples. The steps for calculating the 2's complement of an integer (calculating $-x$ from x) are to invert the binary equivalent of the number by changing all of the 1s to 0s and all of the 0s to 1s (also called 1's complement), and then to add a 1. That is, $-x = \bar{x} + 1$.

0111	(+7)	1001	(−7)
1000	(Invert)	0110	(Invert)
+ 0001	(Add 1)	+ 0001	(Add 1)
1001	(−7)	0111	(+7)

There are some other representations for signed integers such as sign-absolute representation (a sign bit followed by absolute) and biased representation (subtracting a bias from absolute). Table 3.2 lists the values of 4-bit binary numbers at different representations (the bias is 7 for biased representation). You can see that the same binary bit pattern can have different meanings.

3.2 Binary Addition and Subtraction

This section describes the binary addition and subtraction algorithms and their implementations in Verilog HDL. A carry-lookahead adder (CLA) is also described.

3.2.1 Ripple Adder and Subtracter Design

The circuit that adds two 1-bit numbers is called a half adder. The half adder performs $0 + 0 = 00$, $0 + 1 = 01$, $1 + 0 = 01$, and $1 + 1 = 10$. The left bit of the result is called a carry out and the right bit is called a sum.

Table 3.2   The values of 4-bit binary numbers at different representations

Binary number	Unsigned	2's complement	Sign-absolute	Biased
0000	0	0	+0	−7
0001	1	+1	+1	−6
0010	2	+2	+2	−5
0011	3	+3	+3	−4
0100	4	+4	+4	−3
0101	5	+5	+5	−2
0110	6	+6	+6	−1
0111	7	+7	+7	0
1000	8	−8	−0	+1
1001	9	−7	−1	+2
1010	10	−6	−2	+3
1011	11	−5	−3	+4
1100	12	−4	−4	+5
1101	13	−3	−5	+6
1110	14	−2	−6	+7
1111	15	−1	−7	+8

A full adder adds not only the two 1-bit numbers but also a carry in which is the carry out of the next bit to the right. Let's see the 4-bit addition example shown in Figure 3.3. The three inputs of a full adder are a, b, and ci (carry in), and the outputs are co (carry out) and s (sum).

Table 3.3 is the truth table for designing the circuit of the full adder. From the truth table or Karnaugh maps, we can get the following expressions:

$$s = \overline{a}\,\overline{b}\,ci + \overline{a}\ b\ \overline{ci} + a\ \overline{b}\ \overline{ci} + a\ b\ ci$$

$$co = a\ b + a\ ci + b\ ci$$

Then we can design the circuit of the full adder, as shown in Figure 3.4.

Because $s = \overline{a}\ \overline{b}\ ci + \overline{a}\ b\ \overline{ci} + a\ \overline{b}\ \overline{ci} + a\ b\ ci = a \oplus b \oplus ci$, we can also have the full adder circuit as shown in Figure 3.5.

The following is the structural style Verilog HDL code of the full adder. It is identical to the circuit shown in Figure 3.5.

```
module fa_structural (a,b,ci,s,co);      // full adder, structural style
    input   a, b, ci;                    // inputs:  a, b, carry_in
    output  s, co;                       // outputs: sum, carry_out
    wire    ab, bc, ca;                  // wires, outputs of and gates
    xor i1 (s,  a,  b,  ci);             // xor (out, in1, in2, in3);
    and i2 (ab, a,  b);                  // and (out, in1, in2);
    and i3 (bc, b,  ci);                 // and (out, in1, in2);
    and i4 (ca, ci, a);                  // and (out, in1, in2);
    or  i5 (co, ab, bc, ca);             // or  (out, in1, in2, in3);
endmodule
```

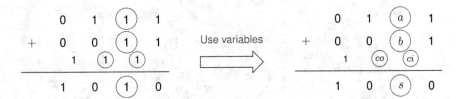

Figure 3.3 Addition of two 4-bit numbers

Table 3.3 Truth table of a full adder

Input signal			Output signal		Comment
a	b	ci	co	s	
0	0	0	0	0	$0 + 0 + 0 = 0\ 0$
0	0	1	0	1	$0 + 0 + 1 = 0\ 1$
0	1	0	0	1	$0 + 1 + 0 = 0\ 1$
0	1	1	1	0	$0 + 1 + 1 = 1\ 0$
1	0	0	0	1	$1 + 0 + 0 = 0\ 1$
1	0	1	1	0	$1 + 0 + 1 = 1\ 0$
1	1	0	1	0	$1 + 1 + 0 = 1\ 0$
1	1	1	1	1	$1 + 1 + 1 = 1\ 1$

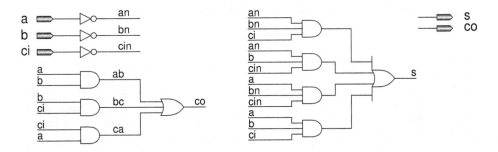

Figure 3.4 Schematic diagram of the full adder

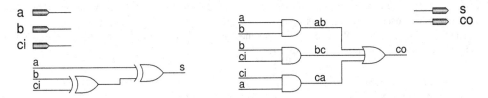

Figure 3.5 Schematic diagram of full adder (using XOR gates)

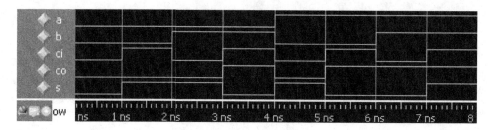

Figure 3.6   Waveform of the full adder

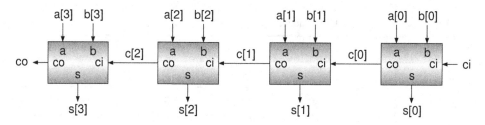

Figure 3.7   Schematic diagram of 4-bit ripple adder

The following is the dataflow style Verilog HDL code of the full adder.

```
module add1 (a,b,ci,s,co);              // full adder, dataflow style
    input   a, b, ci;                   // inputs:  a, b, carry_in
    output s, co;                       // outputs: sum, carry_out
    assign s  = a ^ b ^ ci;             // sum of inputs
    assign co = a & b | b & ci | ci & a;  // carry_out
endmodule
```

The following is the behavioral style Verilog HDL code of the full adder.

```
module fa_behavioral (a,b,ci,s,co);     // full adder, behavioral style
    input   a, b, ci;                   // inputs:  a, b, carry_in
    output s, co;                       // outputs: sum, carry_out
    assign {co,s} = a + b + ci;         // two-bit {co,s} output
endmodule
```

The simulation waveform of the full adder is shown in Figure 3.6.

The ripple adder is constructed with full adders connected in cascade, with the carry out of each full adder connected to the carry in of the next full adder in the chain. Figure 3.7 shows the circuit of a 4-bit ripple adder that uses four full adders.

The following Verilog HDL code implements the 4-bit ripple adder that invokes the add1 module. It is identical to the circuit shown in Figure 3.7.

```
module add4 (a,b,ci,s,co);              // 4-bit adder using four add1
    input   [3:0] a, b;                 // inputs: a and b
    input         ci;                   // input:  carry_in
    output [3:0] s;                     // output: sum
```

```
    output        co;              // output: carry_out
    wire    [2:0] c;              // internal carries
    // add1 (a,    b,    ci,    s,    co);
    add1 a0 (a[0], b[0], ci,    s[0], c[0]);  // ci: ci;   s: s[0]; co: c[0]
    add1 a1 (a[1], b[1], c[0],  s[1], c[1]);  // ci: c[0]; s: s[1]; co: c[1]
    add1 a2 (a[2], b[2], c[1],  s[2], c[2]);  // ci: c[1]; s: s[2]; co: c[2]
    add1 a3 (a[3], b[3], c[2],  s[3], co);    // ci: c[2]; s: s[3]; co: co
endmodule
```

The addition algorithm described above can be applied to both the unsigned representation and the 2's complement representation for signed numbers. Two 4-bit addition examples are shown below. The first (left) example shows the addition on unsigned numbers: $7 + 11$. The result should be 18 but it cannot be denoted with a 4-bit code. We say that an overflow has happened. The second (right) example shows the addition on signed numbers with the 2's complement representation: $(+7) + (-5)$. We get the correct result $+2$.

<table>
<tr><td colspan="2">Addition on unsigned numbers:</td><td colspan="2">Addition on signed numbers:</td></tr>
<tr><td>0111</td><td>(7)</td><td>0111</td><td>(+7)</td></tr>
<tr><td>+ 1011</td><td>(11)</td><td>+ 1011</td><td>(−5)</td></tr>
<tr><td>0010</td><td>(2)</td><td>0010</td><td>(+2)</td></tr>
</table>

Not surprisingly, full adders can be used to perform the subtraction operation. Because $-x = \bar{x} + 1$, $a - b = a + (-b) = a + \bar{b} + 1$. Figure 3.8 shows a circuit that can perform both addition and subtraction. If the sub input is a 0, $s = a + b + ci$; otherwise, $s = a + \bar{b} + \bar{ci} = a - b - ci$.

The following is the structural-style Verilog HDL code that implements the circuit shown in Figure 3.8.

```
module addsub4 (a,b,ci,sub,s,co);     // 4-bit adder/subtracter
    input   [3:0] a, b;              // inputs: a, b
    input         ci;                // input:   carry_in
    input         sub;              // input:   sub==1: s=a-b-ci
                                     // input:   sub==0: s=a+b+ci
    output  [3:0] s;                // output: sum
```

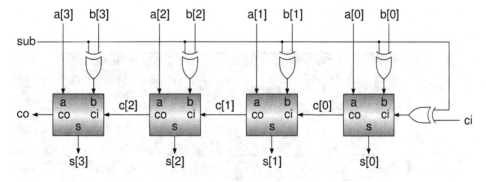

Figure 3.8 Schematic diagram of a 4-bit adder/subtracter

```
output          co;                          // output: carry_out
// sub==1, ci==0: a-b-ci = a+(-b)-0 = a+(~b+1)-0 = a+(b^sub)+(ci^sub)
// sub==1, ci==1: a-b-ci = a+(-b)-1 = a+(~b+1)-1 = a+(b^sub)+(ci^sub)
// sub==0, ci==0: a+b+ci = a+  b +0 = a+  b  +0 = a+(b^sub)+(ci^sub)
// sub==0, ci==1: a+b+ci = a+  b +1 = a+  b  +1 = a+(b^sub)+(ci^sub)
wire    [3:0] bx  = b  ^ {4{sub}};           // b  xor sub
wire          cix = ci ^ sub;                // ci xor sub
wire    [2:0] c;                             // internal carries
// add1 (a,     b,     ci,    s,     co);
add1 a0 (a[0], bx[0], cix,  s[0], c[0]); // b: bx[0], ci: cix;  co: c[0]
add1 a1 (a[1], bx[1], c[0], s[1], c[1]); // b: bx[1], ci: c[0]; co: c[1]
add1 a2 (a[2], bx[2], c[1], s[2], c[2]); // b: bx[2], ci: c[1]; co: c[2]
add1 a3 (a[3], bx[3], c[2], s[3], co);   // b: bx[3], ci: c[2]; co: co
endmodule
```

Figure 3.9 shows the simulation waveform of the 4-bit adder/subtracter. The numbers in the figure are represented in the signed decimal format.

3.2.2   Carry-Lookahead Adder Design

The ripple adder is area-efficient but is slow because the carry travels through the chain of the full adders. The CLA solves this problem by calculating the carry signals in advance. There are several implementations of the CLAs; here we just describe a tree-based CLA. Generally, the carry out of the $(i + 1)$th bit is

$$c_{i+1} = a_i \ b_i + a_i \ c_i + b_i \ c_i$$
$$= a_i \ b_i + (a_i + b_i) \ c_i$$
$$= g_i + p_i \ c_i$$

where $g_i = a_i \ b_i$ is a carry generator and $p_i = a_i + b_i$ is a carry propagator. Then

$$c_{i+1} = g_i + p_i \ g_{i-1} \ +$$
$$p_i \ p_{i-1} \ g_{i-2} \ +$$
$$\cdots \ +$$
$$p_i \ p_{i-1} \ \cdots \ p_1 \ g_0 \ +$$
$$p_i \ p_{i-1} \ \cdots \ p_1 \ p_0 \ c_0$$

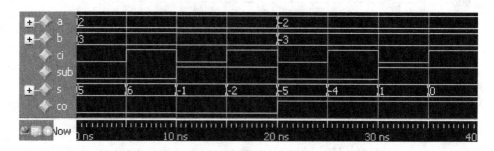

Figure 3.9   Waveform of a 4-bit adder/subtracter

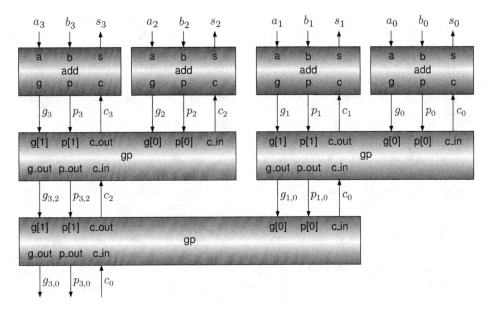

Figure 3.10 Four-bit carry-lookahead adder

Table 3.4 Eight-bit carry-lookahead adder

	Carry	Carry generator and propagator
1st group:	$c_0 = c\_in$	$g_{1,0} = g_1 + p_1\, g_0$
	$c_1 = g_0 + p_0\, c_0$	$p_{1,0} = p_1\, p_0$
2nd group:	$c_2 = g_{1,0} + p_{1,0}\, c_0$	$g_{3,2} = g_3 + p_3\, g_2$
	$c_3 = g_2 + p_2\, c_2$	$p_{3,2} = p_3\, p_2$
	1st & 2nd groups $\Rightarrow$	$g_{3,0} = g_{3,2} + p_{3,2}\, g_{1,0}$
		$p_{3,0} = p_{3,2}\, p_{1,0}$
3rd group:	$c_4 = g_{3,0} + p_{3,0}\, c_0$	$g_{5,4} = g_5 + p_5\, g_4$
	$c_5 = g_5 + p_5\, c_4$	$p_{5,4} = p_5\, p_4$
4th group:	$c_6 = g_{5,4} + p_{5,4}\, c_4$	$g_{7,6} = g_7 + p_7\, g_6$
	$c_7 = g_7 + p_7\, c_6$	$p_{7,6} = p_7\, p_6$
	3rd & 4th groups $\Rightarrow$	$g_{7,4} = g_{7,6} + p_{7,6}\, g_{5,4}$
		$p_{7,4} = p_{7,6}\, p_{5,4}$
	1st, 2nd, 3rd, 4th groups $\Rightarrow$	$g_{7,0} = g_{7,4} + p_{7,4}\, g_{3,0}$
		$p_{7,0} = p_{7,4}\, p_{3,0}$
	$c_8 = g_{7,0} + p_{7,0}\, c_0$	

Theoretically, all the carry bits can be generated in parallel by the expression given above, but it becomes difficult to implement in an IC chip. Figure 3.10 shows a 4-bit CLA in which a two-level binary tree is used for generating the carry bits.

Table 3.4 illustrates an 8-bit CLA in detail where a three-level tree is used.

We give the complete Verilog HDL codes for implementing a 32-bit CLA in the method described above. The following code generates the sum s, carry generator g, and carry propagator p, based on three inputs a, b, and c (carry in). The module name is add.

```
module add (a, b, c, g, p, s);          // adder and g, p
   input   a, b, c;                      // inputs:  a, b, c;
   output  g, p, s;                      // outputs: g, p, s;
   assign  s = a ^ b ^ c;                // output: sum of inputs
   assign  g = a & b;                    // output: carry generator
   assign  p = a | b;                    // output: carry propagator
endmodule
```

The following code generates carry generator g_out, carry propagator p_out, and carry out c_out based on 5-bit inputs: a 2-bit g, a 2-bit p, and a 1-bit c_in. We call this module a gp generator.

```
module gp (g,p,c_in,g_out,p_out,c_out); // carry generator, carry propagator
   input [1:0] g, p;                     // lower  level 2-set of g, p
   input       c_in;                     // lower  level carry_in
   output      g_out,p_out,c_out;        // higher level g, p, carry_out
   assign      g_out = g[1] | p[1] & g[0]; // higher level carry generator
   assign      p_out = p[1] & p[0];       // higher level carry propagator
   assign      c_out = g[0] | p[0] & c_in; // higher level carry_out
endmodule
```

By using two add modules and a gp generator, we can design a 2-bit CLA. Its Verilog HDL code is listed below.

```
module cla_2 (a, b, c_in, g_out, p_out, s);    // 2-bit carry lookahead adder
   input   [1:0] a, b;                             // inputs:  a, b
   input         c_in;                             // input:   carry_in
   output        g_out, p_out;                     // outputs: g, p
   output [1:0] s;                                 // output:  sum
   wire    [1:0] g, p;                             // internal wires
   wire          c_out;                            // internal wire
   // add (a,      b,    c,      g,     p,     s);  // generates g,p,s
   add a0 (a[0], b[0], c_in,  g[0], p[0], s[0]);   // add on bit 0
   add a1 (a[1], b[1], c_out, g[1], p[1], s[1]);   // add on bit 1
   // gp  (g, p, c_in, g_out, p_out, c_out);        // higher level g,p
   gp gp0 (g, p, c_in, g_out, p_out, c_out);        // higher level g,p
endmodule
```

By using two 2-bit CLA modules and a gp generator, we can design a 4-bit CLA. Its Verilog HDL code is listed below.

```
module cla_4 (a,b,c_in,g_out,p_out,s);        // 4-bit carry lookahead adder
   input   [3:0] a, b;                            // inputs:  a, b
   input         c_in;                            // input:   carry_in
   output        g_out, p_out;                    // outputs: g, p
   output [3:0] s;                                // output:  sum
   wire    [1:0] g, p;                            // internal wires
   wire          c_out;                           // internal wire
   cla_2 a0 (a[1:0],b[1:0],c_in, g[0],p[0],s[1:0]); // add on bits 0,1
```

```
    cla_2 a1 (a[3:2],b[3:2],c_out,g[1],p[1],s[3:2]);      // add on bits 2,3
    gp   gp0 (g,p,c_in, g_out,p_out,c_out);               // higher level g,p
endmodule
```

By using two 4-bit CLA modules and a gp generator, we can design an 8-bit CLA. Its Verilog HDL code is listed below.

```
module cla_8 (a,b,c_in,g_out,p_out,s);        // 8-bit carry lookahead adder
    input   [7:0] a, b;                        // inputs:  a, b
    input         c_in;                        // input:   carry_in
    output        g_out, p_out;                // outputs: g, p
    output  [7:0] s;                           // output:  sum
    wire    [1:0] g, p;                        // internal wires
    wire          c_out;                       // internal wire
    cla_4 a0 (a[3:0],b[3:0],c_in, g[0],p[0],s[3:0]);    // add on bits 0-3
    cla_4 a1 (a[7:4],b[7:4],c_out,g[1],p[1],s[7:4]);    // add on bits 4-7
    gp    gp0 (g,p,c_in, g_out,p_out,c_out);            // higher level g,p
endmodule
```

By using two 8-bit CLA modules and a gp generator, we can design a 16-bit CLA. Its Verilog HDL code is listed below.

```
module cla_16 (a,b,c_in,g_out,p_out,s);       // 16-bit carry lookahead adder
    input   [15:0] a, b;                       // inputs:  a, b
    input          c_in;                       // input:   carry_in
    output         g_out, p_out;               // outputs: g, p
    output  [15:0] s;                          // output:  sum
    wire    [1:0]  g, p;                       // internal wires
    wire           c_out;                      // internal wire
    cla_8 a0 (a[7:0], b[7:0], c_in, g[0],p[0],s[7:0]);  // add on bits 0-7
    cla_8 a1 (a[15:8],b[15:8],c_out,g[1],p[1],s[15:8]); // add on bits 8-15
    gp    gp0 (g,p,c_in, g_out,p_out,c_out);            // higher level g,p
endmodule
```

By using two 16-bit CLA modules and a gp generator, we can design a 32-bit CLA. Its Verilog HDL code is listed below.

```
module cla_32 (a,b,c_in,g_out,p_out,s);       // 32-bit carry lookahead adder
    input   [31:0] a, b;                       // inputs:  a, b
    input          c_in;                       // input:   carry_in
    output         g_out, p_out;               // outputs: g, p
    output  [31:0] s;                          // output:  sum
    wire    [1:0]  g, p;                       // internal wires
    wire           c_out;                      // internal wire
    cla_16 a0 (a[15:0], b[15:0], c_in, g[0],p[0],s[15:0]);   // + bits 0-15
    cla_16 a1 (a[31:16],b[31:16],c_out,g[1],p[1],s[31:16]);  // + bits 16-31
    gp     gp0 (g,p,c_in,g_out,p_out,c_out);
endmodule
```

The following is the final Verilog HDL code that implements a 32-bit CLA. What this module does is just to delete two output pins, g_out and g_out, from the module given above.

```
module cla32 (a,b,ci,s);       // 32-bit carry lookahead adder, no g, p outputs
    input   [31:0] a, b;                              // inputs: a, b
    input          ci;                                // input:  carry_in
    output  [31:0] s;                                 // output: sum
    wire           g_out, p_out;                      // internal wires
    cla_32 cla (a, b, ci, g_out, p_out, s);           // use cla_32 module
endmodule
```

Figure 3.11 shows the simulation waveform of the 32-bit CLA.

We will use the cla32 module in the designs of various CPUs, which will be described in later chapters.

3.3   Binary Multiplication Algorithms

This section describes unsigned and signed binary multiplication algorithms and their implementations in Verilog HDL. The Wallace tree, which is a high-speed multiplication circuit, is also introduced.

3.3.1   Unsigned Multiplier Design

Like the decimal multiplication, the unsigned binary multiplication can also be done with shifts and additions. For example,

$$
\begin{array}{rrrrrl}
  & 1 & 1 & 1 & 0 & (14)\\
\times & 1 & 0 & 1 & 0 & (10)\\
\hline
 0\ \ 0\ \ 0\ \ 0 & & & & &\\
 1\ \ 1\ \ 1\ \ 0 & & & & &\\
 0\ \ 0\ \ 0\ \ 0 & & & & &\\
+ 1\ \ 1\ \ 1\ \ 0 & & & & &\\
\end{array}
$$

The multiplication of two 1-bit numbers is the same as the AND operation. The additions can be done with an array of adders or with an adder iteratively (an accumulator or a register is required). The following C code implements the multiplication of two 16-bit unsigned numbers iteratively with shifts and additions.

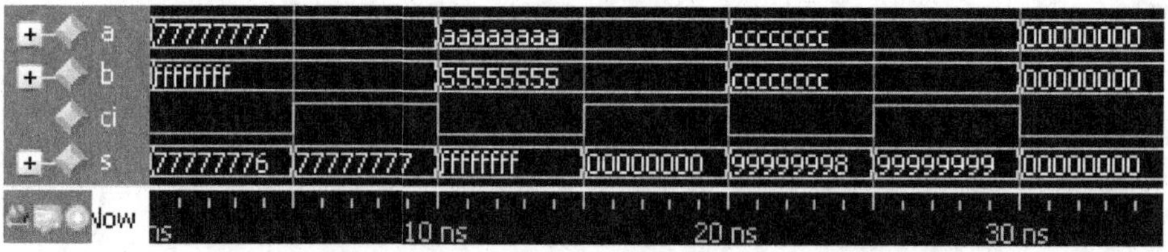

Figure 3.11   Waveform of the carry-lookahead adder

```c
#include <stdio.h>                              // mul_by_shift.c
unsigned int mul16 (unsigned int x, unsigned int y) {
    unsigned int a, b, c;
    unsigned int i;                             // counter
    a = x;                                      // multiplicand
    b = y;                                      // multiplier
    c = 0;                                      // product
    for (i = 0; i < 16; i++) {                  // for 16 bits
        if ((b & 1) == 1) {                     // LSB of b is 1
            c += a;                             // c = c + a
        }
        a = a << 1;                             // shift a 1-bit left
        b = b >> 1;                             // shift b 1-bit right
    }
    return(c);                                  // return product
}
main() {
    unsigned int x,y;
    fprintf(stderr,"input 1st 16-bit unsigned integer in hex: ");
    fscanf(stdin,"%x",&x);
    fprintf(stderr,"input 2nd 16-bit unsigned integer in hex: ");
    fscanf(stdin,"%x",&y);
    x &= 0xffff;                                // 16 bits
    y &= 0xffff;                                // 16 bits
    fprintf(stderr,"%04x * %04x = %08x\n", x, y, mul16(x, y));
}
```

The compilation command and the execution result are shown below.

```
[cpu_verilog]$ gcc mul_by_shift.c -o mul_by_shift
[cpu_verilog]$ ./mul_by_shift
input 1st 16-bit unsigned integer in hex: c9ae
input 2nd 16-bit unsigned integer in hex: f6e5
c9ae * f6e5 = c2819ca6
```

Note that the product of two 16-bit numbers has 32 bits. As an exercise, implement this iterative version of the multiplication in Verilog HDL.

3.3.2 Signed Multiplier Design

Suppose that we have two 8-bit signed numbers, A_8 and B_8, with 2's complement representation; try to calculate the product $Z_{16} = A_8 \times B_8$.

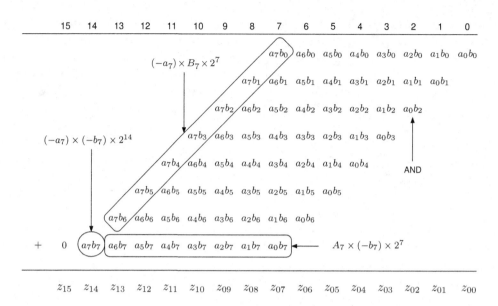

Figure 3.12   Multiplication of two 8-bit signed numbers

Each A_8 and B_8 may be a negative or nonnegative number. Because both A_8 and B_8 use 2's complement representation, we have

$$A_8 = a_7a_6a_5a_4a_3a_2a_1a_0 = -a_7 \times 2^7 + \sum_{i=0}^{6} a_i \times 2^i = -a_7 \times 2^7 + A_7$$

$$B_8 = b_7b_6b_5b_4b_3b_2b_1b_0 = -b_7 \times 2^7 + \sum_{i=0}^{6} b_i \times 2^i = -b_7 \times 2^7 + B_7$$

where A_7 and B_7 are 7-bit unsigned numbers. Because $(a + b)(x + y) = ax + ay + bx + by$, then

$$\begin{aligned}
Z_{16} &= A_8 \times B_8 \\
&= (-a_7 \times 2^7 + A_7) \times (-b_7 \times 2^7 + B_7) \\
&= a_7 \times b_7 \times 2^{14} + (-a_7 \times B_7) \times 2^7 + (-b_7 \times A_7) \times 2^7 + A_7 \times B_7
\end{aligned}$$

There are four terms in the expression. The first and last terms are positive, the same as the case of unsigned multiplication. The second and third terms are negative. These are shown in Figure 3.12.

The terms of $a_7 \times B_7$ and $b_7 \times A_7$ can be denoted as the following bit patterns where a bit $a_ib_j = a_i$ AND b_j.

$$a_7 \times B_7 = 0 \quad 0 \quad a_7b_6 \quad a_7b_5 \quad a_7b_4 \quad a_7b_3 \quad a_7b_2 \quad a_7b_1 \quad a_7b_0$$
$$b_7 \times A_7 = 0 \quad 0 \quad a_6b_7 \quad a_5b_7 \quad a_4b_7 \quad a_3b_7 \quad a_2b_7 \quad a_1b_7 \quad a_0b_7$$

We want to get $-a_7 \times B_7$ and $-b_7 \times A_7$ from these bit patterns. Because $-x = \bar{x} + 1$, we can calculate $(-a_7 \times B_7) + (-b_7 \times A_7)$ by inverting each pattern and adding a 1 to it, as shown below.

15	14	13	12	11	10	9	8	7
1	1	$\overline{a_7b_6}$	$\overline{a_7b_5}$	$\overline{a_7b_4}$	$\overline{a_7b_3}$	$\overline{a_7b_2}$	$\overline{a_7b_1}$	$\overline{a_7b_0}$
0	0	0	0	0	0	0	0	1
1	1	$\overline{a_6b_7}$	$\overline{a_5b_7}$	$\overline{a_4b_7}$	$\overline{a_3b_7}$	$\overline{a_2b_7}$	$\overline{a_1b_7}$	$\overline{a_0b_7}$
+ 0	0	0	0	0	0	0	0	1

This can be simplified as follows. Two 1s appear in the 8th and 15th bit positions, respectively.

15	14	13	12	11	10	9	8	7
0	0	$\overline{a_7b_6}$	$\overline{a_7b_5}$	$\overline{a_7b_4}$	$\overline{a_7b_3}$	$\overline{a_7b_2}$	$\overline{a_7b_1}$	$\overline{a_7b_0}$
0	0	$\overline{a_6b_7}$	$\overline{a_5b_7}$	$\overline{a_4b_7}$	$\overline{a_3b_7}$	$\overline{a_2b_7}$	$\overline{a_1b_7}$	$\overline{a_0b_7}$
+ 1	0	0	0	0	0	0	1	0

Then we can implement the signed binary multiplier by the method shown in Figure 3.13 where NAND gates are used in addition to AND gates. Also note that there are two 1s in the figure.

The following Verilog HDL code implements the signed binary multiplier. The first part generates partial product terms, and the second part adds the terms. Use of parentheses in the code ensures performing the additions in parallel.

```
module mul_signed (a,b,z);                          // 8x8 signed multiplier
    input    [7:0] a, b;                            // a, b
```

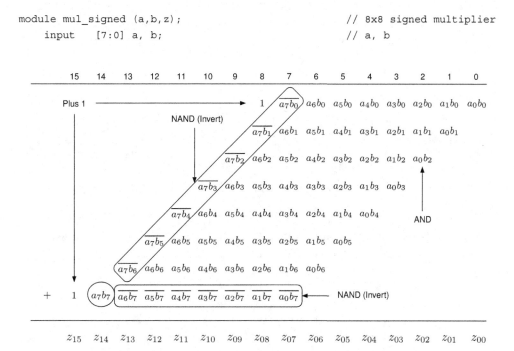

Figure 3.13 Implementing a signed multiplier using NAND gates

```
    output [15:0] z;                                    // z = a * b
    wire     [7:0] ab0 = a & {8{b[0]}};                 // a or 0 for b[0]
    wire     [7:0] ab1 = a & {8{b[1]}};                 // a or 0 for b[1]
    wire     [7:0] ab2 = a & {8{b[2]}};                 // a or 0 for b[2]
    wire     [7:0] ab3 = a & {8{b[3]}};                 // a or 0 for b[3]
    wire     [7:0] ab4 = a & {8{b[4]}};                 // a or 0 for b[4]
    wire     [7:0] ab5 = a & {8{b[5]}};                 // a or 0 for b[5]
    wire     [7:0] ab6 = a & {8{b[6]}};                 // a or 0 for b[6]
    wire     [7:0] ab7 = a & {8{b[7]}};                 // a or 0 for b[7]
    assign z = ((({8'b1,~ab0[7], ab0[6:0]}         +    // << 0, + 1 in bit 8
                  {7'b0,~ab1[7], ab1[6:0],1'b0})    +    // << 1
                 ({6'b0,~ab2[7], ab2[6:0],2'b0}     +    // << 2
                  {5'b0,~ab3[7], ab3[6:0],3'b0}))   +    // << 3
                (({4'b0,~ab4[7], ab4[6:0],4'b0}     +    // << 4
                  {3'b0,~ab5[7], ab5[6:0],5'b0})    +    // << 5
                 ({2'b0,~ab6[7], ab6[6:0],6'b0}     +    // << 6
                  {1'b1, ab7[7],~ab7[6:0],7'b0}))); //   << 7, + 1 in bit 15
endmodule
```

The following module is another implementation in which we use AND and NAND operations to generate the partial product terms in the first part.

```
module mul_signed_v2 (a,b,z);                           // 8x8 signed multiplier
    input    [7:0] a, b;                                // a, b
    output [15:0] z;                                    // z = a * b
    reg      [7:0] abi[7:0];                            // a[i] & b[j]
    integer        i, j;
    always @* begin
        for (i = 0; i < 7; i = i + 1)
            for (j = 0; j < 7; j = j + 1)
                abi[i][j] = a[i] & b[j];                //   a[i] & b[j]
        for (i = 0; i < 7; i = i + 1)
            abi[i][7] = ~(a[i] & b[7]);                 // ~(a[i] & b[7])
        for (j = 0; j < 7; j = j + 1)
            abi[7][j] = ~(a[7] & b[j]);                 // ~(a[7] & b[j])
        abi[7][7] = a[7] & b[7];
    end
    assign z = ((({8'b1,abi[0][7:0]}           +        // << 0, + 1 in bit 8
                  {7'b0,abi[1][7:0],1'b0})      +        // << 1
                 ({6'b0,abi[2][7:0],2'b0}       +        // << 2
                  {5'b0,abi[3][7:0],3'b0}))      +       // << 3
                (({4'b0,abi[4][7:0],4'b0}       +        // << 4
                  {3'b0,abi[5][7:0],5'b0})      +        // << 5
                 ({2'b0,abi[6][7:0],6'b0}       +        // << 6
                  {1'b1,abi[7][7:0],7'b0})));            // << 7, + 1 in bit 15
endmodule
```

Both versions perform the same calculation. The simulation waveform is shown in Figure 3.14. The first multiplication shows $(-1) \times (-1) = +1$.

3.3.3 Wallace Tree

Wallace tree is a high-speed implementation of a multiplier. A Wallace tree consists of three parts: (i) an AND gate array that multiplies each bit of one of the input number by each bit of the other; (ii) a carry-save adder (CSA) array that adds the outputs of the AND gates to generate two numbers; and (iii) A carry-propagate adder that adds the two numbers to generate the final product.

The key point of the Wallace tree is at the second part. This part has several levels. Each level consists of an array of CSAs that perform additions in parallel. There are no carry propagations among the adders; all the adders work independently. A CSA takes three inputs and generates two outputs (a carry bit and a sum bit). We use wires to denote these inputs and outputs. Therefore, each level of the tree reduces the number of wires by a factor of 3:2. A CSA will be used as long as there are any three input wires. If there are only two input wires left, we use a half adder. If there is just one wire left, we connect it to the next level.

Figure 3.15 shows the first level of an 8×8 Wallace tree. In the 7th bit position, there are eight input wires, so two CSAs and one half adder are used.

Figure 3.16 shows the circuit of the Wallace tree for the 7th bit. Four levels and seven adders are used. The carry output wires are sent to the 8th bit position; the sum output wires are used at the 7th bit position; and there are wires coming from the 6th bit position. In level 4, there are only two wires: one is the sum output wire, and the other is a carry output wire from the 6th bit position.

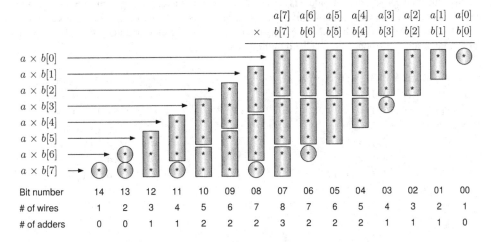

Figure 3.14   Waveform of a signed multiplier

$$\begin{array}{cccccccc} a[7] & a[6] & a[5] & a[4] & a[3] & a[2] & a[1] & a[0] \\ \times \quad b[7] & b[6] & b[5] & b[4] & b[3] & b[2] & b[1] & b[0] \end{array}$$

Bit number	14	13	12	11	10	09	08	07	06	05	04	03	02	01	00
# of wires	1	2	3	4	5	6	7	8	7	6	5	4	3	2	1
# of adders	0	0	1	1	2	2	2	3	2	2	2	1	1	1	0

Figure 3.15   First level of the 8-bit Wallace tree

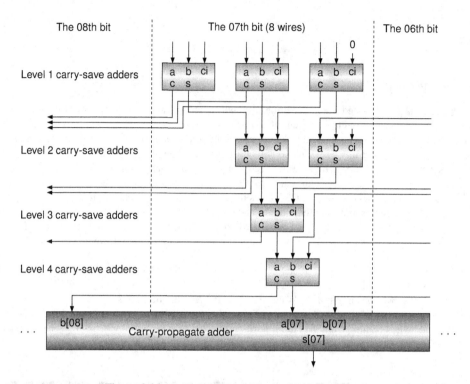

Figure 3.16 Adders for 7th bit in the 8-bit Wallace tree

Figure 3.17 shows the entire circuit of an 8×8 Wallace tree. The delay time of the circuit is $t = t_{and} + 4t_{csa} + t_{add}$, where t_{and} is the delay time of an AND gate, t_{csa} is the delay time of a CSA, and t_{add} is the delay time of a carry-propagate adder. A CLA can be used as the carry-propagate adder.

The following Verilog HDL code implements the 8×8 unsigned Wallace tree. It generates the partial product: the final carry-propagate adder is not included in this module. The inputs are two 8-bit unsigned numbers: a and b. The output z is the five least significant bits of the product. The outputs x and y are two 11-bit numbers that can be fed to the carry-propagate adder, or other modules, two 11-bit registers, for example, for implementing a pipelined multiplication.

```
module wallace_8x8 (a,b,x,y,z);              // 8*8 wallace tree
    input   [07:00] a;                       // 8-bit a
    input   [07:00] b;                       // 8-bit b
    output  [15:05] x;                       // sum high
    output  [15:05] y;                       // carry high
    output  [04:00] z;                       // sum low
    reg     [07:00] p [07:00];               // p[i][j]
    parameter zero = 1'b0;                   // constant 0
    integer i, j;
    always @ * begin
        for (i = 0; i < 8; i = i + 1)
            for (j = 0; j < 8; j = j + 1)
```

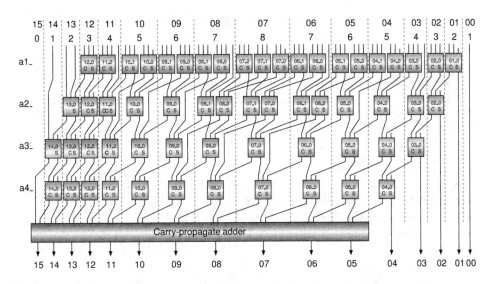

Figure 3.17   Schematic diagram of the 8-bit Wallace tree

```
                p[i][j] = a[i] & b[j];                      // p[i][j]=a[i]&b[j]
end
// level 1 ─────────────────────────────────────────────────────
wire  [2:0] s1 [12:1];
wire  [2:0] c1 [13:2];
//    15:
//    14:   p[07][07]
//    13:   p[06][07], p[07][06]
csa a1_12_0 (p[05][07], p[06][06], p[07][05], s1[12][0], c1[13][0]);
csa a1_11_0 (p[04][07], p[05][06], p[06][05], s1[11][0], c1[12][0]);
//    11:   p[07][04]
csa a1_10_1 (p[03][07], p[04][06], p[05][05], s1[10][1], c1[11][1]);
csa a1_10_0 (p[06][04], p[07][03],     zero, s1[10][0], c1[11][0]);
csa a1_09_1 (p[02][07], p[03][06], p[04][05], s1[09][1], c1[10][1]);
csa a1_09_0 (p[05][04], p[06][03], p[07][02], s1[09][0], c1[10][0]);
csa a1_08_1 (p[01][07], p[02][06], p[03][05], s1[08][1], c1[09][1]);
csa a1_08_0 (p[04][04], p[05][03], p[06][02], s1[08][0], c1[09][0]);
//    08:   p[07][01]
csa a1_07_2 (p[00][07], p[01][06], p[02][05], s1[07][2], c1[08][2]);
csa a1_07_1 (p[03][04], p[04][03], p[05][02], s1[07][1], c1[08][1]);
csa a1_07_0 (p[06][01], p[07][00],     zero, s1[07][0], c1[08][0]);
csa a1_06_1 (p[00][06], p[01][05], p[02][04], s1[06][1], c1[07][1]);
csa a1_06_0 (p[03][03], p[04][02], p[05][01], s1[06][0], c1[07][0]);
//    06:   p[06][00]
csa a1_05_1 (p[00][05], p[01][04], p[02][03], s1[05][1], c1[06][1]);
csa a1_05_0 (p[03][02], p[04][01], p[05][00], s1[05][0], c1[06][0]);
csa a1_04_1 (p[00][04], p[01][03], p[02][02], s1[04][1], c1[05][1]);
csa a1_04_0 (p[03][01], p[04][00],     zero, s1[04][0], c1[05][0]);
```

```
csa a1_03_0 (p[00][03], p[01][02], p[02][01], s1[03][0], c1[04][0]);
//    03:   p[03][00]
csa a1_02_0 (p[00][02], p[01][01], p[02][00], s1[02][0], c1[03][0]);
csa a1_01_0 (p[00][01], p[01][00],      zero, s1[01][0], c1[02][0]);
//    00:   p[00][00]
// level 2 ──────────────────────────────────────────────────────────
wire  [1:0] s2 [13:2];
wire  [1:0] c2 [14:3];
//    15:
//    14:   p[07][07]
csa a2_13_0 (p[06][07], p[07][06], c1[13][0], s2[13][0], c2[14][0]);
csa a2_12_0 (s1[12][0], c1[12][0],      zero, s2[12][0], c2[13][0]);
csa a2_11_0 (s1[11][0], p[07][04], c1[11][1], s2[11][0], c2[12][0]);
//    11:   c1[11][0]
csa a2_10_0 (s1[10][1], s1[10][0], c1[10][1], s2[10][0], c2[11][0]);
//    10:   c1[10][0]
csa a2_09_0 (s1[09][1], s1[09][0], c1[09][1], s2[09][0], c2[10][0]);
//    09:   c1[09][0]
csa a2_08_1 (s1[08][1], s1[08][0], p[07][01], s2[08][1], c2[09][1]);
csa a2_08_0 (c1[08][2], c1[08][1], c1[08][0], s2[08][0], c2[09][0]);
csa a2_07_1 (s1[07][2], s1[07][1], s1[07][0], s2[07][1], c2[08][1]);
csa a2_07_0 (c1[07][1], c1[07][0],      zero, s2[07][0], c2[08][0]);
csa a2_06_1 (s1[06][1], s1[06][0], p[06][00], s2[06][1], c2[07][1]);
csa a2_06_0 (c1[06][1], c1[06][0],      zero, s2[06][0], c2[07][0]);
csa a2_05_0 (s1[05][1], s1[05][0], c1[05][1], s2[05][0], c2[06][0]);
//    05:   c1[05][0]
csa a2_04_0 (s1[04][1], s1[04][0], c1[04][0], s2[04][0], c2[05][0]);
csa a2_03_0 (s1[03][0], p[03][00], c1[03][0], s2[03][0], c2[04][0]);
csa a2_02_0 (s1[02][0], c1[02][0],      zero, s2[02][0], c2[03][0]);
//    01:   s1[01][0]
//    00:   p[00][00]
// level 3 ──────────────────────────────────────────────────────────
wire  [0:0] s3 [14:3];
wire  [0:0] c3 [15:4];
//    15:
csa a3_14_0 (p[07][07], c2[14][0],      zero, s3[14][0], c3[15][0]);
csa a3_13_0 (s2[13][0], c2[13][0],      zero, s3[13][0], c3[14][0]);
csa a3_12_0 (s2[12][0], c2[12][0],      zero, s3[12][0], c3[13][0]);
csa a3_11_0 (s2[11][0], c1[11][0], c2[11][0], s3[11][0], c3[12][0]);
csa a3_10_0 (s2[10][0], c1[10][0], c2[10][0], s3[10][0], c3[11][0]);
csa a3_09_0 (s2[09][0], c1[09][0], c2[09][1], s3[09][0], c3[10][0]);
//    09:   c2[09][0]
csa a3_08_0 (s2[08][1], s2[08][0], c2[08][1], s3[08][0], c3[09][0]);
//    08:   c2[08][0]
csa a3_07_0 (s2[07][1], s2[07][0], c2[07][1], s3[07][0], c3[08][0]);
//    07:   c2[07][0]
csa a3_06_0 (s2[06][1], s2[06][0], c2[06][0], s3[06][0], c3[07][0]);
```

```
csa a3_05_0 (s2[05][0], c1[05][0], c2[05][0], s3[05][0], c3[06][0]);
csa a3_04_0 (s2[04][0], c2[04][0],      zero, s3[04][0], c3[05][0]);
csa a3_03_0 (s2[03][0], c2[03][0],      zero, s3[03][0], c3[04][0]);
//      02:    s2[02][0]
//      01:    s1[01][0]
//      00:    p[00][00]
// level 4 ─────────────────────────────────────────────────────────
wire  [0:0] s4 [14:4];
wire  [0:0] c4 [15:5];
//      15:    c3[15][0]
csa a4_14_0 (s3[14][0], c3[14][0],      zero, s4[14][0], c4[15][0]);
csa a4_13_0 (s3[13][0], c3[13][0],      zero, s4[13][0], c4[14][0]);
csa a4_12_0 (s3[12][0], c3[12][0],      zero, s4[12][0], c4[13][0]);
csa a4_11_0 (s3[11][0], c3[11][0],      zero, s4[11][0], c4[12][0]);
csa a4_10_0 (s3[10][0], c3[10][0],      zero, s4[10][0], c4[11][0]);
csa a4_09_0 (s3[09][0], c2[09][0], c3[09][0], s4[09][0], c4[10][0]);
csa a4_08_0 (s3[08][0], c2[08][0], c3[08][0], s4[08][0], c4[09][0]);
csa a4_07_0 (s3[07][0], c2[07][0], c3[07][0], s4[07][0], c4[08][0]);
csa a4_06_0 (s3[06][0], c3[06][0],      zero, s4[06][0], c4[07][0]);
csa a4_05_0 (s3[05][0], c3[05][0],      zero, s4[05][0], c4[06][0]);
csa a4_04_0 (s3[04][0], c3[04][0],      zero, s4[04][0], c4[05][0]);
//      03:    s3[03][0]
//      02:    s2[02][0]
//      01:    s1[01][0]
//      00:    p[00][00]
assign x[15] = c3[15][0];              assign y[15] = c4[15][0];
assign x[14] = s4[14][0];              assign y[14] = c4[14][0];
assign x[13] = s4[13][0];              assign y[13] = c4[13][0];
assign x[12] = s4[12][0];              assign y[12] = c4[12][0];
assign x[11] = s4[11][0];              assign y[11] = c4[11][0];
assign x[10] = s4[10][0];              assign y[10] = c4[10][0];
assign x[09] = s4[09][0];              assign y[09] = c4[09][0];
assign x[08] = s4[08][0];              assign y[08] = c4[08][0];
assign x[07] = s4[07][0];              assign y[07] = c4[07][0];
assign x[06] = s4[06][0];              assign y[06] = c4[06][0];
assign x[05] = s4[05][0];              assign y[05] = c4[05][0];
assign z[04] = s4[04][0];
assign z[03] = s3[03][0];
assign z[02] = s2[02][0];
assign z[01] = s1[01][0];
assign z[00] = p[00][00];
endmodule
```

Figure 3.18 shows the simulation waveform of the 8 × 8 unsigned Wallace tree with the partial product outputs x, y, and z (five LSBs of the product).

The following code generates the final 16-bit product z. It invokes the module of the Wallace tree listed above. We simply use the statement `assign z_high = x + y`. It can also be done by invoking a CLA.

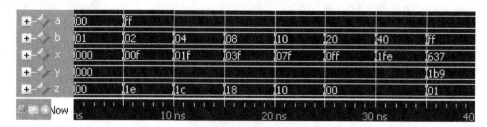

Figure 3.18   Waveform of the 8 × 8 Wallace tree (partial product)

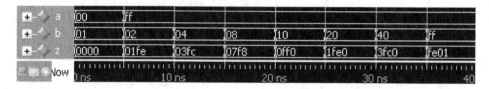

Figure 3.19   Waveform of 8 × 8 Wallace tree (product)

```
module wallace_8x8_product (a,b,z);              // 8*8 wt product
    input   [07:00] a;                           // 8 bits
    input   [07:00] b;                           // 8 bits
    output  [15:00] z;                           // product
    wire    [15:05] x;                           // sum high
    wire    [15:05] y;                           // carry high
    wire    [15:05] z_high;                       // product high
    wire    [04:00] z_low;                        // product low
    wallace_8x8 wt_partial (a, b, x, y, z_low);  // partial product
    assign z_high = x + y;
    assign z = {z_high,z_low};                    // product
endmodule
```

Figure 3.19 shows the simulation waveform of the 8 × 8 unsigned Wallace tree with the final product output z.

A register may be inserted in between the two modules given above to implement a two-stage pipelined multiplication, with each stage taking one clock cycle.

3.4   Binary Division Algorithms

This section introduces unsigned binary division algorithms, including the restoring algorithm, the non-restoring algorithm, the Goldschmidt algorithm, and the Newton–Raphson algorithm. The Verilog HDL codes that implement these algorithms and their simulation waveforms are also given.

3.4.1   Restoring Division Algorithm

Given a dividend a and a divisor b, the restoring division algorithm calculates the quotient q and the remainder r such that $a = b \times q + r$ and $r < b$, by subtracting b from the partial remainder (initially the

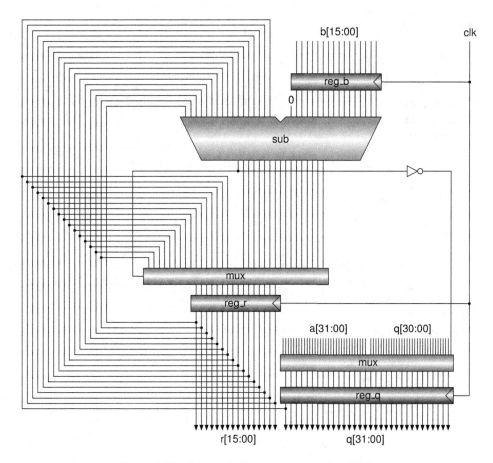

Figure 3.20 Schematic diagram of a restoring divider

MSB of a). If the result of the subtraction is not negative, we set the quotient bit to 1. Otherwise, b is added back to the result to restore the partial remainder. Then we shift the partial remainder with the remaining bits of a to the left by one bit for the calculation of the next quotient bit. This procedure is repeated until all the bits of a are shifted out.

Figure 3.20 shows the schematic diagram of a restoring divider. There are three registers: `reg_b`, `reg_r`, and `reg_q`, for storing the divisor b, remainder r, and quotient q, respectively. Initially, `reg_q` stores the dividend a. A subtracter is used to subtract b from the partial remainder. The MSB of the output of the subtracter is used to determine whether the result of the subtraction is negative or not. The multiplexer over `reg_q` is used to load a initially and to shift the content of `reg_q` (a and q) to the left later. The multiplexer over `reg_r` implements the restoring. If the result of the subtraction is negative, the multiplexer selects the original partial remainder. Otherwise, it selects the result of the subtraction. At each iteration, one bit of q is obtained from the sign bit of the subtracter result and written to the LSB of the `reg_q`.

The following Verilog HDL code implements the circuit shown in Figure 3.20. Some signals in the code are explained below. `start` means the start of the division; `busy` indicates that the divider is busy (cannot start a new division); `ready` indicates that the quotient and remainder are available; and `count` is the output of a counter that is used to control the iterations of the division.

```verilog
module div_restoring (a,b,start,clk,clrn,q,r,busy,ready,count);
    input    [31:0] a;                                        // dividend
    input    [15:0] b;                                        // divisor
    input           start;                                    // start
    input           clk, clrn;                                // clk,reset
    output   [31:0] q;                                        // quotient
    output   [15:0] r;                                        // remainder
    output reg      busy;                                     // busy
    output reg      ready;                                    // ready
    output   [4:0]  count;                                    // counter
    reg      [31:0] reg_q;
    reg      [15:0] reg_r;
    reg      [15:0] reg_b;
    reg      [4:0]  count;
    wire     [16:0] sub_out = {reg_r,reg_q[31]} - {1'b0,reg_b};  // sub
    wire     [15:0] mux_out = sub_out[16]?                    // restoring
                    {reg_r[14:0],reg_q[31]} : sub_out[15:0];  // or not
    assign q = reg_q;
    assign r = reg_r;
    always @ (posedge clk or negedge clrn) begin
        if (!clrn) begin
            busy  <= 0;
            ready <= 0;
        end else begin
            if (start) begin
                reg_q <= a;                                   // load a
                reg_b <= b;                                   // load b
                reg_r <= 0;
                busy  <= 1;
                ready <= 0;
                count <= 0;
            end else if (busy) begin
                reg_q <= {reg_q[30:0],~sub_out[16]};          // << 1
                reg_r <= mux_out;
                count <= count + 5'b1;                        // counter++
                if (count == 5'h1f) begin                     // finished
                    busy  <= 0;
                    ready <= 1;                               // q,r ready
                end
            end
        end
    end
endmodule
```

Figure 3.21 shows the part of the simulation waveform of the code listed above. It simulates 0x4c7f228a/0x6a0e, $q = $ 0xb8a6 and $r = $ 0x4d76 are available when ready is 1 at 330 ns; and then 0xffff00/4, q is 0x3fffc0, and r is 0.

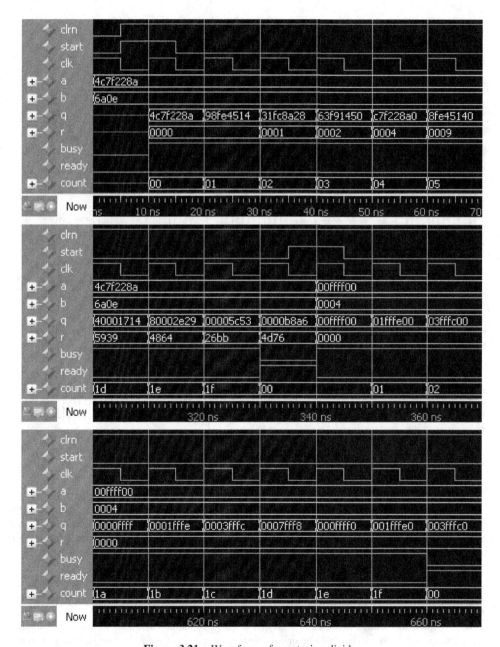

Figure 3.21 Waveform of a restoring divider

3.4.2 Nonrestoring Division Algorithm

In the restoring division algorithm described in the previous section, if the result of the subtraction r is negative, b is added back to r. That is, the remainder is restored by $r + b$, where r is the remainder in the current iteration. The restored remainder $r + b$ is then shifted to the left by one bit, that is, $2(r + b)$. Then b is subtracted from the shifted remainder, that is, $2(r + b) - b$.

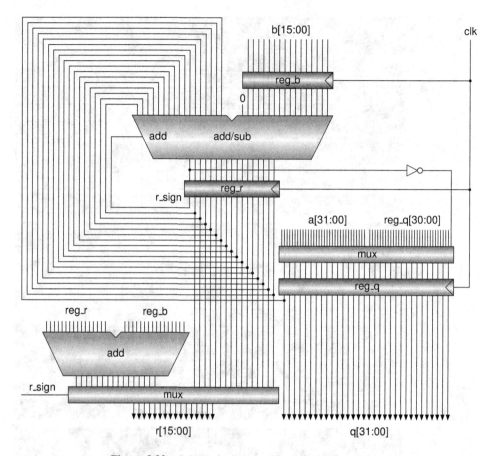

Figure 3.22   Schematic diagram of a nonrestoring divider

Because $2(r + b) - b = 2r + b$, we can use the negative remainder r directly for the calculation of the next iteration. This is the idea of the nonrestoring division algorithm. That is, if the partial remainder is negative, we shift it to the left directly and add b to the shifted partial remainder. If the partial remainder is not negative, we shift it to the left and subtract b from the shifted partial remainder (same as the restoring algorithm).

Figure 3.22 shows the schematic diagram of a nonrestoring divider from which we can see that there is no multiplexer over the register `reg_r`. However, instead of a subtracter in the restoring division algorithm, an adder/subtracter is required in the nonrestoring division algorithm.

Also, because the content in the `reg_r` may be negative in the last iteration, we must use an adder and a multiplexer to adjust the final remainder. If r is not negative, it is the final remainder. Otherwise, we must restore the remainder by adding b back to r, just as we did in the restoring division algorithm. If there is no need to have the remainder outputted (only outputting q), this part of the circuit can be deleted.

The following Verilog HDL code implements the circuit shown in Figure 3.22.

```
module div_nonrestoring (a,b,start,clk,clrn,q,r,busy,ready,count);
    input  [31:0] a;                              // dividend
    input  [15:0] b;                              // divisor
```

```
     input           start;                                    // start
     input           clk, clrn;                                // clk,reset
     output [31:0]  q;                                         // quotient
     output [15:0]  r;                                         // remainder
     output reg     busy;                                      // busy
     output reg     ready;                                     // ready
     output [4:0]   count;                                     // count
     reg    [31:0]  reg_q;
     reg    [15:0]  reg_r;
     reg    [15:0]  reg_b;
     reg    [4:0]   count;
     wire   [16:0]  sub_add = reg_r[15]?
                    {reg_r,reg_q[31]} + {1'b0,reg_b} :         // + b
                    {reg_r,reg_q[31]} - {1'b0,reg_b};          // - b
     assign q = reg_q;
     assign r = reg_r[15]? reg_r + reg_b : reg_r;              // adjust r
     always @ (posedge clk or negedge clrn) begin
         if (!clrn) begin
             busy  <= 0;
             ready <= 0;
         end else begin
             if (start) begin
                 reg_q <= a;                                   // load a
                 reg_b <= b;                                   // load b
                 reg_r <= 0;
                 busy  <= 1;
                 ready <= 0;
                 count <= 0;
             end else if (busy) begin
                 reg_q <= {reg_q[30:0],~sub_add[16]};          // << 1
                 reg_r <= sub_add[15:0];
                 count <= count + 5'b1;                        // count++
                 if (count == 5'h1f) begin                     // finish
                     busy  <= 0;
                     ready <= 1;                               // q,r ready
                 end
             end
         end
     end
endmodule
```

Figure 3.23 shows the part of the simulation waveform of the code listed above. It also simulates 0x4c7f228a/0x6a0e, the $q = $ 0xb8a6 and $r = $ 0x4d76 are available when `ready` is 1 at the 330 ns; and then 0xffff00/4, the q is 0x3fffc0 and r is 0.

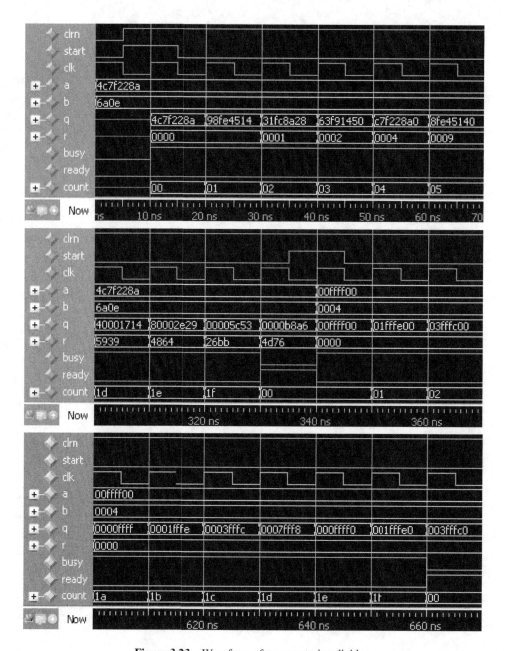

Figure 3.23   Waveform of a nonrestoring divider

3.4.3   Goldschmidt Division Algorithm

Given a and b in the format $0.1x \cdots x$, that is, $1/2 \leq a, b < 1$, the Goldschmidt division algorithm uses multiplications to get $q = a/b$ as shown below. Consider the fraction

$$\frac{a \times r_0 \times r_1 \times r_2 \times \cdots \times r_{n-1}}{b \times r_0 \times r_1 \times r_2 \times \cdots \times r_{n-1}}$$

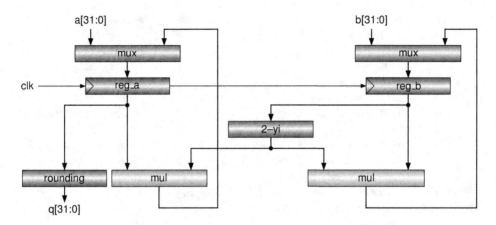

Figure 3.24   Schematic diagram of the Goldschmidt divider

If the denominator converges to 1, then the numerator converges to $a/b = q$.

The factor r_i can be calculated as follows: define $\delta = 1 - b$; then $0 < \delta \leq 1/2$ and $b = 1 - \delta$. Define $x_0 = a$, $y_0 = b = 1 - \delta$. Calculate

$$r_0 = 2 - y_0 = 1 + \delta$$
$$x_1 = x_0 \times r_0$$
$$y_1 = y_0 \times r_0 = (1 - \delta) \times (1 + \delta) = 1 - \delta^2$$
$$r_1 = 2 - y_1 = 1 + \delta^2$$
$$x_2 = x_1 \times r_1$$
$$y_2 = y_1 \times r_1 = (1 - \delta^2) \times (1 + \delta^2) = 1 - \delta^4$$
$$\cdots \quad \cdots$$
$$r_{i-1} = 2 - y_{i-1} = 1 + \delta^{2^{i-1}}$$
$$x_i = x_{i-1} \times r_{i-1}$$
$$y_i = y_{i-1} \times r_{i-1} = (1 - \delta^{2^{i-1}}) \times (1 + \delta^{2^{i-1}}) = 1 - \delta^{2^i}$$
$$\cdots \quad \cdots$$
$$r_{n-1} = 2 - y_{n-1} = 1 + \delta^{2^{n-1}}$$
$$x_n = x_{n-1} \times r_{n-1}$$
$$y_n = y_{n-1} \times r_{n-1} = (1 - \delta^{2^{n-1}}) \times (1 + \delta^{2^{n-1}}) = 1 - \delta^{2^n}$$

until y_n converges to 1. Then x_n converges to q. Why does $y_n \to 1$? This is because $y_n = 1 - \delta^{2^n}$ while $0 < \delta \leq 1/2$.

The schematic diagram of a 32-bit Goldschmidt divider is shown in Figure 3.24. Two registers `reg_a` and `reg_b` store a and b initially and then x_i and y_i, respectively. Both `reg_a` and `reg_b` are 64 bits, and the final quotient q is rounded to 32 bits.

The following Verilog HDL code implements the 32-bit Goldschmidt divider.

```
module goldschmidt (a,b,start,clk,clrn,q,busy,ready,count,yn);
    input   [31:0] a;                        // dividend: .1xxx...x
    input   [31:0] b;                        // divisor:  .1xxx...x
    input          start;                    // start
    input          clk, clrn;                // clock and reset
    output  [31:0] q;                        // quotient: x.xxx...x
    output reg     busy;                     // busy
```

```
    output reg      ready;                              // ready
    output   [2:0]  count;                              // counter
    output   [31:0] yn;                                 // .11111...1
    reg      [63:0] reg_a;                              // x.xxxx...x
    reg      [63:0] reg_b;                              // 0.xxxx...x
    reg      [2:0]  count;
    wire     [63:0] two_minus_yi = ~reg_b + 1'b1;       // 1.xxxx...x (2 - yi)
    wire     [127:0] xi = reg_a * two_minus_yi;         // 0x.xxx...x
    wire     [127:0] yi = reg_b * two_minus_yi;         // 0x.xxx...x
    assign          q = reg_a[63:32] + |reg_a[31:29];   // rounding up
    assign          yn = reg_b[62:31];
    always @ (posedge clk or negedge clrn) begin
        if (!clrn) begin
            busy  <= 0;
            ready <= 0;
        end else begin
            if (start) begin
                reg_a <= {1'b0,a,31'b0};                // 0.1x...x0...0
                reg_b <= {1'b0,b,31'b0};                // 0.1x...x0...0
                busy  <= 1;
                ready <= 0;
                count <= 0;
            end else begin
                reg_a <= xi[126:63];                    // x.xxx...x
                reg_b <= yi[126:63];                    // 0.xxx...x
                count <= count + 3'b1;                  // count++
                if (count == 3'h4) begin                // finish
                    busy  <= 0;
                    ready <= 1;                         // q is ready
                end
            end
        end
    end
endmodule
```

The speed of $y_n \to 1$ depends on δ. The slowest case is when $\delta = 1/2$. Figure 3.25 shows the simulation waveform of the Goldschmidt divider where $a = 0.75$ and $b = 0.5$ (the worst case), and we get $q = 1.5$. From the figure we know that the number of iterations must be at least 5. We can also see that $y_n \to 1$.

The multiplication can be done with a Wallace tree and takes two clock cycles. An addition takes one cycle. Including the rounding step, the 32-bit Goldschmidt divider takes $(2 + 1) \times 5 + 1 = 16$ cycles.

3.4.4 Newton–Raphson Division Algorithm

The Newton–Raphson division algorithm also uses multiplication to get the quotient. For calculating a/b, if we can calculate $1/b$ without using division, then $a/b = a \times (1/b)$.

Suppose that we have a function $f(x)$: how can we find an x_n so that $f(x_n) \approx 0$? First, guess an x_0. Using the tangential equation of $f(x)$ at x_0, $y - f(x_0) = f'(x_0)(x - x_0)$, and letting $y = 0$, we can find an x_1. x_1 is closer to x_n than x_0.

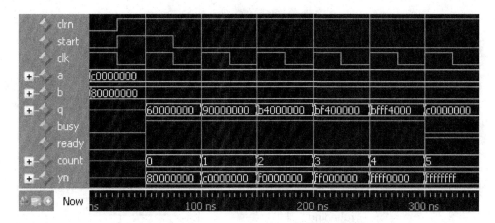

Figure 3.25   Waveform of the Goldschmidt divider

Generally, for $y - f(x_i) = f'(x_i)(x - x_i)$, let $y = 0$, so we get a new $x_{i+1} = x_i - f(x_i)/f'(x_i)$. Repeat it until x_n has sufficient accuracy. Let $f(x) = 1/x - b$; then $f(x) = 0$ at $x = 1/b$. We get

$$x_{i+1} = x_i(2 - x_i b)$$

Thus, we can calculate a/b with the following steps:

1. Apply a bit shift to b to scale it so that $0.5 \le b < 1$. That is, $b = 0.1xx \cdots xx$.
2. The same bit shift should be applied to a so that the quotient does not change.
3. Use some MSBs of b to get x_0 from the ROM (read only memory) table, or let $x_0 = 1.5$.
4. Repeat the calculation of $x_{i+1} = x_i(2 - x_i b)$, until it has enough accuracy.
5. Calculate $a \times x_n$ to get an approximation of the quotient q.

Suppose that x_i has p accurate bits. This means $|(x_i - 1/b)/(1/b)| \le 2^{-p}$. We can derive that $|(x_{i+1} - 1/b)/(1/b)| \le 2^{-2p}$. That is, the number of accurate bits will double after an iteration.

Figure 3.26 shows the schematic diagram of a 32-bit Newton–Raphson divider. It performs $x_{i+1} = x_i(2 - x_i b)$ and $a \times x_n$. The three registers reg_a, reg_b, and reg_x store the dividend a, the divisor b, and the iteration variable x_i, respectively.

The following Verilog HDL code implements the 32-bit Newton–Raphson divider.

```
module newton (a,b,start,clk,clrn,q,busy,ready,count);
    input   [31:0]  a;              // dividend:  .1xxx...x
    input   [31:0]  b;              // divisor:   .1xxx...x
    input           start;          // start
    input           clk, clrn;      // clock and reset
    output  [31:0]  q;              // quotient:  x.xxx...x
    output  reg     busy;           // busy
    output  reg     ready;          // ready
    output  [1:0]   count;          // counter
    reg     [33:0]  reg_x;          // xx.xxxxx...xx
    reg     [31:0]  reg_a;          //    .1xxxx...xx
    reg     [31:0]  reg_b;          //    .1xxxx...xx
    reg     [1:0]   count;
```

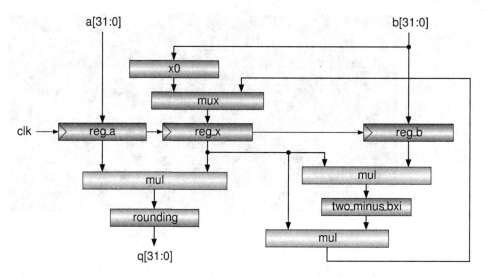

Figure 3.26   Schematic diagram of 32-bit Newton–Raphson divider

```
// x_{i+1} = x_i * (2 - x_i * b)
wire    [65:0] axi =  reg_x * reg_a;          //   xx.xxxxx...x
wire    [65:0] bxi =  reg_x * reg_b;          //   xx.xxxxx...x
wire    [33:0] b34 = ~bxi[64:31] + 1'b1;      //    x.xxxxx...x
wire    [67:0] x68 =  reg_x * b34;            //  xxx.xxxxx...x
wire    [7:0] x0   =  rom(b[30:27]);
assign         q   =  axi[64:33] + |axi[32:30]; // rounding up
always @ (posedge clk or negedge clrn) begin
    if (!clrn) begin
        busy  <= 0;
        ready <= 0;
    end else begin
        if (start) begin
            reg_a <= a;                       //    .1xxxx...x
            reg_b <= b;                       //    .1xxxx...x
            reg_x <= {2'b1,x0,24'b0}          // 01.xxxx0...0
            busy  <= 1;
            ready <= 0;
            count <= 0;
        end else begin
            reg_x <= x68[66:33];              //  xx.xxxxx...x
            count <= count + 2'b1;            //  count++
            if (count == 2'h2) begin          //  3 iterations
                busy  <= 0;
                ready <= 1;                   //  q is ready
            end
        end
    end
end
```

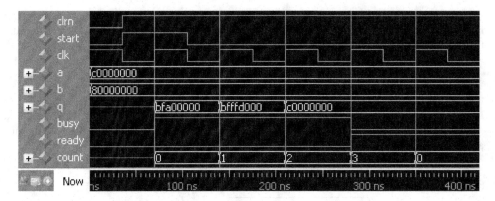

Figure 3.27 Waveform of the Newton–Raphson divider

```
    end
    function [7:0] rom;                              // a rom table
        input [3:0] b;
        case (b)
            4'h0: rom = 8'hff;          4'h1: rom = 8'hdf;
            4'h2: rom = 8'hc3;          4'h3: rom = 8'haa;
            4'h4: rom = 8'h93;          4'h5: rom = 8'h7f;
            4'h6: rom = 8'h6d;          4'h7: rom = 8'h5c;
            4'h8: rom = 8'h4d;          4'h9: rom = 8'h3f;
            4'ha: rom = 8'h33;          4'hb: rom = 8'h27;
            4'hc: rom = 8'h1c;          4'hd: rom = 8'h12;
            4'he: rom = 8'h08;          4'hf: rom = 8'h00;
        endcase
    endfunction
endmodule
```

Figure 3.27 shows the simulation waveform of the Newton–Raphson divider where $a = 0.75$, $b = 0.5$, and $q = 1.5$. If we use a big ROM table to get x_0 of 8 accurate bits, two iterations are enough: x_1 will have 16 accurate bits and x_2 will have 32 accurate bits.

The multiplication can be done with the Wallace tree and it takes two clock cycles. An addition takes one cycle. Including searching the ROM table, the final multiplication, and the rounding step, the 32-bit Newton–Raphson divider takes $1 + (2 + 1 + 2) \times 3 + 2 + 1 = 19$ clock cycles.

3.5 Binary Square Root Algorithms

Beside the essential addition, subtraction, multiplication, and division, the square root is also an important calculation as computer graphics becomes popular because the square root is required for calculating the unit vectors. Therefore, almost all the modern CPUs implement square root instructions.

The square root calculation is little more complex than the division calculation. Given a radicand d, the square root calculates q so that $q^2 + r = d$, where r is a remainder with $r \leq 2q$. The reason why $r \leq 2q$ is that $q^2 + r = d < (q + 1)^2 = q^2 + 2q + 1$, which says $r < 2q + 1$. That is, $r \leq 2q$, because both q and r are integers. For example, $15 = 3^2 + 6$.

This section introduces binary square root algorithms, including the restoring algorithm, the nonrestoring algorithm, the Goldschmidt algorithm, and the Newton–Raphson algorithm. The Verilog HDL codes that implement these algorithms and their simulation waveforms are also given.

3.5.1 Restoring Square Root Algorithm

Let's take an example to show how to calculate the square root of a decimal number by hand and paper. Assume $d = 3\ 00\ 00\ 00\ 00\ 00$, calculate $q = \sqrt{d}$. We marked off pairs of digits of d. There are six pairs and q will have six digits: $q = q_1 q_2 q_3 q_4 q_5 q_6$. The calculation steps are shown below.

		q_1	q_2	q_3	q_4	q_5	q_6			
		1	7	3	2	0	5	; Square root		
	1	$\sqrt{\ }$ 03	00	00	00	00	00	; Radicand		
$(q_1 = 1)$	1		1					; $1 \times 1 = 1$	(guess 1)	
	27		2	00				; $1 + 1 = 2$		
$(q_2 = 7)$	7		1	89				; $27 \times 7 = 189$	(guess 7)	
	343			11	00			; $27 + 7 = 34$		
$(q_3 = 3)$	3			10	29			; $343 \times 3 = 1029$	(guess 3)	
	3462				71	00		; $343 + 3 = 346$		
$(q_4 = 2)$	2				69	24		; $3462 \times 2 = 6924$	(guess 2)	
	34640				1	76	00	; $3462 + 2 = 3464$		
$(q_5 = 0)$	0					0		; $34640 \times 0 = 0$	(guess 0)	
	346405				1	76	00	00	; $34640 + 0 = 34640$	
$(q_6 = 5)$	5				1	73	20	25	; $346405 \times 5 = 1732025$	(guess 5)
						2	79	75	; Remainder	

Thus we got answers: $q = 173,205$ and $r = 27,975$ for $d = 30,000,000,000$. Actually, we can continue the calculation to get any number of digits by considering that there is a decimal point right to d and dropping 00 at each step.

The square root calculation for binary numbers is much simpler than that for decimal numbers because we can guess a root bit easily: it is either a 0 or a 1; see the following example.

		q_1	q_2	q_3	q_4			
		1	0	1	1	; Square root (11_{10})		
	1	$\sqrt{\ }$ 01	11	11	11	; Radicand (127_{10})		
$(q_1 = 1)$	1		1			; $1 \times 1 = 1$	(guess 1)	
	100		0	11		; $1 + 1 = 10$		
$(q_2 = 0)$	0		0			; $100 \times 0 = 0$	(guess 0)	
	1001			11	11	; $100 + 0 = 100$		
$(q_3 = 1)$	1			10	01	; $1001 \times 1 = 1001$	(guess 1)	
	10101			1	10	11	; $1001 + 1 = 1010$	
$(q_4 = 1)$	1			1	01	01	; $10101 \times 1 = 10101$	(guess 1)
					01	10	; Remainder (6_{10})	

For a 32-bit radicand d, assume that q is a partial root and r is a partial remainder. Initially, let $q = 0$ and let r be the two MSBs of d. The restoring square root algorithm subtracts $q01$ from r, where $q01$ means $q \ll 2 + 1$, or $4q + 1$. If the result of the subtraction is not negative, we set the root bit to 1. Otherwise, the root bit is reset to 0 and $q01$ is added back to the result to restore the original r. Then r is obtained by concatenating the next two MSBs of d. This procedure is repeated until all the bits of d are dealt with.

Figure 3.28 shows the schematic diagram of a restoring square root circuit. There are three registers: reg_d, reg_r, and reg_q, for storing the radicand d, remainder r, and root q, respectively. A subtracter is used to subtract $q01$ from the partial remainder. The MSB of the output of the subtracter is used to determine whether the result of the subtraction is negative or not. The multiplexer over reg_d is used to load d initially and to shift the content of reg_d to the left later. The multiplexer over the reg_r

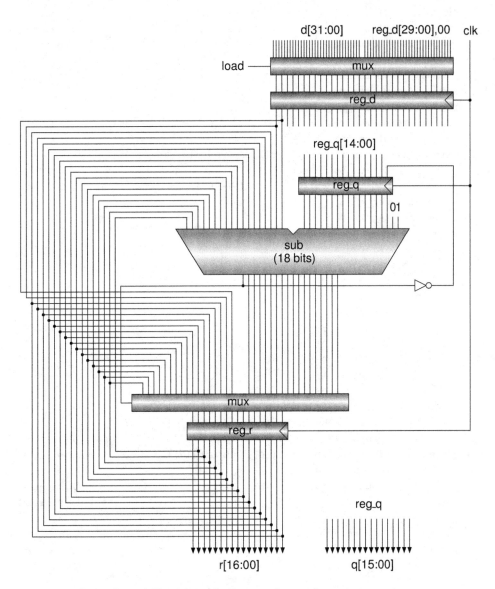

Figure 3.28   Schematic diagram of a restoring square rooter

implements the restoring. If the result of the subtraction is negative, the multiplexer selects the original
partial remainder (restoring). Otherwise, it selects the result of the subtraction.

The following Verilog HDL code implements the circuit shown in Figure 3.28. The signal load starts
the square root; busy indicates that the circuit is busy; ready indicates that the root and remainder are
available; and count is the output of a counter that is used to control the iterations of the square root.

```verilog
module root_restoring (d,load,clk,clrn,q,r,busy,ready,count);
    input   [31:0] d;                                            // radicand
    input          load;                                         // start
    input          clk, clrn;                                    // clk,reset
    output [15:0] q;                                             // root
    output [16:0] r;                                             // remainder
    output reg    busy;                                          // busy
    output reg    ready;                                         // ready
    output  [3:0] count;                                         // counter
    reg    [31:0] reg_d;
    reg    [15:0] reg_q;
    reg    [16:0] reg_r;
    reg     [3:0] count;
    wire   [17:0] sub_out = {reg_r[15:0],reg_d[31:30]} - {reg_q,2'b1}; // -
    wire   [16:0] mux_out = sub_out[17]?                        // restoring
                  {reg_r[14:0],reg_d[31:30]} : sub_out[16:0]; // or not
    assign q = reg_q;
    assign r = reg_r;
    always @ (posedge clk or negedge clrn) begin
        if (!clrn) begin
            busy  <= 0;
            ready <= 0;
        end else begin
            if (load) begin
                reg_d <= d;                                      // load d
                reg_q <= 0;
                reg_r <= 0;
                busy  <= 1;
                ready <= 0;
                count <= 0;
            end else if (busy) begin
                reg_d <= {reg_d[29:0],2'b0};                     // << 2
                reg_q <= {reg_q[14:0],~sub_out[17]};            // << 1
                reg_r <= mux_out;
                count <= count + 4'b1;                           // count++
                if (count == 4'hf) begin                         // finish
                    busy  <= 0;
                    ready <= 1;                                   // q,r ready
                end
            end
        end
    end
endmodule
```

Figure 3.29 shows the simulation waveform of the code listed above with $d = $ 0xc0000000 (32 bits). We get $q = $ 0xddb3 (16 bits) and $r = $ 0x174d7 (17 bits).

3.5.2 Nonrestoring Square Root Algorithm

In the restoring square root algorithm described in the previous section, if the result of the subtraction r is negative, $q01$ is added back to r. That is, the remainder is restored by $r + q01$, where r is the partial remainder in the current iteration. The restored remainder $r + q01$ is then shifted to the left by two bits, that is, $4(r + q01)$. Then $q01$ is subtracted from the shifted remainder, that is, $4(r + q01) - q01$. Note that $q01$ means $4q + 1$. Because the root bit is 0, we have

$$4(r_i + 4q_i + 1) - (4q_{i+1} + 1)$$

$$= 4(r_i + 4q_i + 1) - (8q_i + 1)$$

$$= 4r_i + (8q_i + 3)$$

$$= 4r_i + (4q_{i+1} + 3)$$

Therefore, we can use the negative remainder r directly for the calculation of the next iteration. This is the idea of the nonrestoring square root algorithm: if the partial remainder is negative, shift it to the left by two bits directly and add $q11$ to the shifted partial remainder. If the partial remainder in not negative, shift it to the left by two bits and subtract $q01$ from the shifted partial remainder (same as the restoring algorithm).

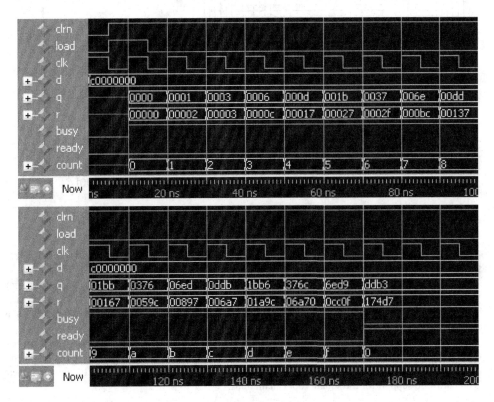

Figure 3.29 Waveform of the restoring square rooter

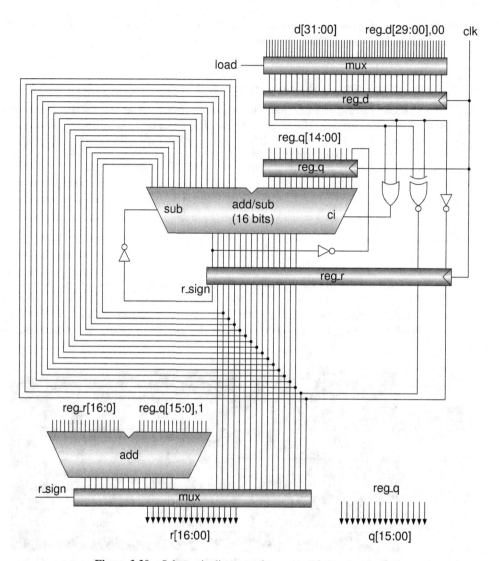

Figure 3.30   Schematic diagram of a nonrestoring square rooter

Figure 3.30 shows the schematic diagram of the nonrestoring square root circuit from which we can see that there is no multiplexer over the register reg_r.

Because the final remainder has 17 bits, an 18-bit adder/subtracter must be used for checking its sign. We can simplify the two least significant bits' calculations of −01 and +11 just by using three gates based on Table 3.5, where x and y are the two bits of d that will be subtracted by 01 or added by 11, c is the carry in of the adder/subtracter, and a and b are the two LSBs of r.

From the table, we can get the expressions $c = x + y$; $a = x \oplus y$; and $b = \overline{y}$. With these simplifications, the calculations can be done by a 16-bit adder/subtracter. The following is a program in C that implements the nonrestoring square root algorithm. It can be used in a very low cost embedded CPU in which there is no hardware support for the square root calculations.

Table 3.5 Simplifying the two LSB calculation

x	y	Calculation	c	a	b	c is active
0	0	+11	0	1	1	High
0	1	+11	1	0	0	High
1	0	+11	1	0	1	High
1	1	+11	1	1	0	High
0	0	−01	0	1	1	Low
0	1	−01	1	0	0	Low
1	0	−01	1	0	1	Low
1	1	−01	1	1	0	Low

```
// Non-Restoring Square Root, Copyright by Yamin Li, yamin@ieee.org
#include <stdio.h>                              // root_nonrestoring.c
unsigned squart(unsigned d, int *remainder) {
    unsigned q = 0;                             // q: 16-bit root
    int r = 0;                                  // r: 17-bit remainder
    int i;
    for (i = 15; i >= 0; i-- ) {
        printf("%02d: q=%04x ",i,q);
        if (r >= 0) {
            r = ((r << 2) | ((d >> (i+i)) & 3)) - ((q << 2) | 1);    // -q01
            printf("r=((r<<2)|((d>>(i+i))&3))-((q<<2)|1)=%08x ",r);
        } else {
            r = ((r << 2) | ((d >> (i+i)) & 3)) + ((q << 2) | 3);    // +q11
            printf("r=((r<<2)|((d>>(i+i))&3))+((q<<2)|3)=%08x ",r);
        }
        if (r <= 0) {q = (q << 1) | 1; printf("q=q*2+1=%04x\n",q);}
        else        {q = (q << 1) | 0; printf("q=q*2+0=%04x\n",q);}
    }
    if (r < 0) {r = r + ((q << 1) | 1);}        // remainder adjusting
    *remainder = r;                             // return remainder
    return(q);                                  // return root
}
int main(void){
    unsigned radicand,root;
    int remainder;
    printf("Input an unsigned integer in hex (i.e. c0000000): d = ");
    scanf("%x", &radicand);                     // read a radicand
    root = squart(radicand, &remainder);
    printf("hex: q = %08x, r = %08x, d = q*q + r = %08x\n",
           root,remainder,root*root + remainder);
    printf("dec: q = %08d, r = %08d, d = q*q + r = %08u\n",
           root,remainder,root*root + remainder);
}
```

The compilation command and execution example are listed below.

```
[cpu_verilog]$ gcc root_nonrestoring.c -o root_nonrestoring
[cpu_verilog]$ ./root_nonrestoring
Input an unsigned integer in hex (i.e. c0000000): d = c0000000
15: q=0000 r=((r<<2)|((d>>(i+i))&3))-((q<<2)|1)=00000002 q=q*2+1=0001
14: q=0001 r=((r<<2)|((d>>(i+i))&3))-((q<<2)|1)=00000003 q=q*2+1=0003
13: q=0003 r=((r<<2)|((d>>(i+i))&3))-((q<<2)|1)=ffffffff q=q*2+0=0006
12: q=0006 r=((r<<2)|((d>>(i+i))&3))+((q<<2)|3)=00000017 q=q*2+1=000d
11: q=000d r=((r<<2)|((d>>(i+i))&3))-((q<<2)|1)=00000027 q=q*2+1=001b
10: q=001b r=((r<<2)|((d>>(i+i))&3))-((q<<2)|1)=0000002f q=q*2+1=0037
09: q=0037 r=((r<<2)|((d>>(i+i))&3))-((q<<2)|1)=ffffffdf q=q*2+0=006e
08: q=006e r=((r<<2)|((d>>(i+i))&3))+((q<<2)|3)=00000137 q=q*2+1=00dd
07: q=00dd r=((r<<2)|((d>>(i+i))&3))-((q<<2)|1)=00000167 q=q*2+1=01bb
06: q=01bb r=((r<<2)|((d>>(i+i))&3))-((q<<2)|1)=ffffeaf q=q*2+0=0376
05: q=0376 r=((r<<2)|((d>>(i+i))&3))+((q<<2)|3)=00000897 q=q*2+1=06ed
04: q=06ed r=((r<<2)|((d>>(i+i))&3))-((q<<2)|1)=000006a7 q=q*2+1=0ddb
03: q=0ddb r=((r<<2)|((d>>(i+i))&3))-((q<<2)|1)=ffffe32f q=q*2+0=1bb6
02: q=1bb6 r=((r<<2)|((d>>(i+i))&3))+((q<<2)|3)=fffffb97 q=q*2+0=376c
01: q=376c r=((r<<2)|((d>>(i+i))&3))+((q<<2)|3)=0000cc0f q=q*2+1=6ed9
00: q=6ed9 r=((r<<2)|((d>>(i+i))&3))-((q<<2)|1)=000174d7 q=q*2+1=ddb3
hex: q = 0000ddb3, r = 000174d7, d = q*q + r = c0000000
dec: q = 00056755, r = 00095447, d = q*q + r = 3221225472
```

The following Verilog HDL code implements the 32-bit nonrestoring square root algorithm. It invokes clas16, a 16-bit carry-lookahead adder/subtracter.

```
module root_nonrestoring (d,load,clk,clrn,q,r,busy,ready,count);
    input   [31:0] d;                                        // radicand
    input          load;                                     // start
    input          clk, clrn;                                // clk,reset
    output  [15:0] q;                                        // root
    output  [16:0] r;                                        // remainder
    output reg     busy;                                     // busy
    output reg     ready;                                    // ready
    output  [3:0] count;                                     // counter
    reg     [31:0] reg_d;
    reg     [15:0] reg_q;
    reg     [17:0] reg_r;
    reg      [3:0] count;
    wire    [15:0] add_sub;
    wire           g_or  = reg_d[31]  | reg_d[30];           // or   gate
    wire           g_xnor = reg_d[31] ~^ reg_d[30];          // xnor gate
    wire           g_not  = ~reg_d[30];                      // not  gate
    wire    [17:0] r18 = {add_sub,g_xnor,g_not};
    // clas16 (sub,        a,          b,      ci,    s)
    clas16 as (~reg_r[17],reg_r[15:0],reg_q,g_or,add_sub);
```

```
assign q = reg_q;
assign r = reg_r[17]?                                    // adjust
        reg_r[16:0] + {reg_q[15:0],1'b1} : reg_r[16:0];  // remainder
always @ (posedge clk or negedge clrn) begin
    if (!clrn) begin
        busy  <= 0;
        ready <= 0;
    end else begin
        if (load) begin
            reg_d <= d;                                  // load d
            reg_q <= 0;
            reg_r <= 0;
            busy  <= 1;
            ready <= 0;
            count <= 0;
        end else if (busy) begin
            reg_d <= {reg_d[29:0],2'b0};                 // << 2
            reg_q <= {reg_q[14:0],~add_sub[15]};         // << 1
            reg_r <= r18;
            count <= count + 4'b1;                       // count++
            if (count == 4'hf) begin                     // finish
                busy  <= 0;
                ready <= 1;                              // q,r ready
            end
        end
    end
end
endmodule
```

The module of the 16-bit carry-lookahead adder/subtracter is listed below. It invokes the module of cla_16, a 16-bit CLA which was already given before. Note that the input ci must be active-low for subtraction.

```
module clas16 (sub,a,b,ci,s);        // 16-bit carry lookahead adder/subtracter
    input         sub;               // 1: sub; 0: add
    input  [15:0] a, b;              // inputs: a, b
    input         ci;                // active low for sub
    output [15:0] s;                 // output: sum
    wire          g_out, p_out;      // internal wires
    // cla_16   (a,b,        c_in,g_out,p_out,s);
    cla_16 cla (a,b^{16{sub}},ci,  g_out,p_out,s);   // use cla_16 module
endmodule
```

Figure 3.31 shows the simulation waveform of the nonrestoring square root code with also $d = 0xc0000000$. We get $q = 0xddb3$ and $r = 0x174d7$.

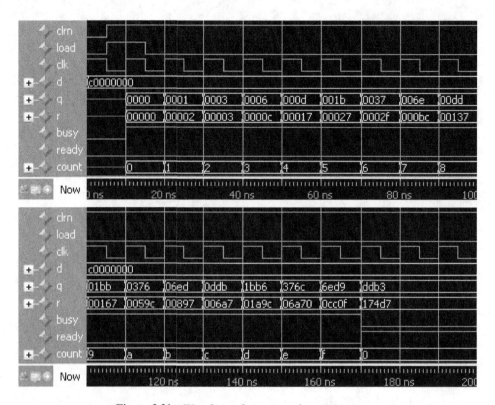

Figure 3.31   Waveform of a nonrestoring square rooter

3.5.3   Goldschmidt Square Root Algorithm

Given a radicand d in the format $0.1xx \cdots x$ or $0.01x \cdots x$, that is, $1/4 \leq d < 1$, the Goldschmidt square root algorithm uses multiplications to get the root $q = \sqrt{d}$, as given below. Consider the fraction

$$\frac{d \times r_0 \times r_1 \times r_2 \times \cdots \times r_{n-1}}{d \times r_0^2 \times r_1^2 \times r_2^2 \times \cdots \times r_{n-1}^2}$$

If the denominator converges to 1, that is, $r_0^2 \times r_1^2 \times r_2^2 \times \cdots \times r_n^2$ converges to $1/d$, or $r_0 \times r_1 \times r_2 \times \cdots \times r_n$ converges to $\sqrt{1/d}$, then the numerator converges to $d \times \sqrt{1/d} = \sqrt{d}$. Define $x_0 = d_0 = d$. Calculate

$$r_i = 1 + (1 - x_i)/2$$

$$d_{i+1} = d_i \times r_i$$

$$x_{i+1} = x_i \times r_i^2$$

for $i = 0, 1, \ldots n - 1$ so that x_n converges to 1. Then d_n converges to $\sqrt{d}$. The division by 2 in the expression of r_i can be implemented by a wire shift. The reason why $x_n \to 1$ is that, because $1/4 \leq d < 1$, we have $d = 1 - \delta$ with $0 < \delta \leq 3/4$. Let $r_0 = 1 + (1 - d)/2 = 1 + \delta/2$; then $x_1 = x_0 \times r_0^2 = (1 - \delta) \times (1 + \delta/2)^2 = (1 - \delta) \times (1 + \delta + \delta^2/4) \approx (1 - \delta) \times (1 + \delta) = 1 - \delta^2$.

Generally, $r_i = 1 + (1 - x_i)/2 \approx 1 + \delta^{2^i}/2$, $x_{i+1} = x_i \times r_i^2 = (1 - \delta^{2^i}) \times (1 + \delta^{2^i}/2)^2 = (1 - \delta^{2^i}) \times (1 + \delta^{2^i} + \delta^{2^{i+1}}/4) \approx (1 - \delta^{2^i}) \times (1 + \delta^{2^i}) = 1 - \delta^{2^{i+1}}$. Because $0 < \delta \leq 3/4$, we have $x_{i+1} = 1 - \delta^{2^{i+1}} \to 1$.

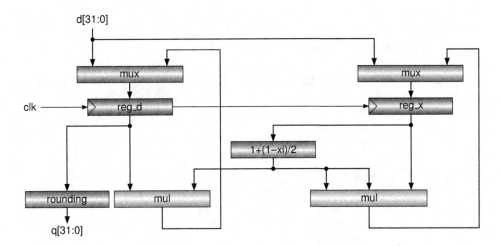

Figure 3.32   Schematic diagram of the Goldschmidt square rooter

The schematic diagram of a 32-bit Goldschmidt square root circuit is shown in Figure 3.32. Two registers `reg_d` and `reg_x` store d_i and x_i, respectively. The `mul` module in the left side of the figure calculates d_{i+1}, and the `mul` module in the right calculates x_{i+1} that has three inputs.

The following Verilog HDL code implements the 32-bit Goldschmidt square rooter. Because in the expression of x_i there is a term $\delta^{2^i}/4$, which may result in x_i exceeding 1.0, in the code we adjusted the x_i.

```
module root_goldschmidt (d,start,clk,clrn,q,busy,ready,count,xn);
    input   [31:0] d;                                    // radicand:
    input          start;                                //         .1xx...x
    input          clk, clrn;                            //    or: .01x...x
    output  [31:0] q;                                    // root: .1xx...x
    output reg     busy;                                 // busy
    output reg     ready;                                // ready
    output  [2:0]  count;                                // counter
    output  [31:0] xn;                                   // 0.1111...1
    reg     [63:0] reg_d;                                // 0.1xx or 0.01x
    reg     [63:0] reg_x;                                // 0.1xx or 0.01x
    reg     [2:0]  count;
    wire    [63:0] ri  = 64'hc000000000000000 -          // 1+(1-xi)/2 =
                         {1'b0,reg_x[63:1]};             // (3-xi)/2  1.xx
    wire    [127:0] ci  = ri    * ri;                    // ri*ri     0x.xx
    wire    [127:0] dr1 = reg_d * ri;                    // d*ri      0x.xx
    wire    [127:0] xr2 = reg_x * ci[126:63];            // x*ci      0x.xx
    wire    [63:0] xi  = {1'b0,{63{xr2[126]}}|xr2[125:63]}; // let xi<1  0.xx
    assign         q   = reg_d[62:31] + | reg_d[30:0];   // rounding up
    assign         xn  = reg_x[62:31];
    always @ (posedge clk or negedge clrn) begin
        if (!clrn) begin
            busy  <= 0;
```

```
                ready <= 0;
        end else begin
            if (start) begin
                reg_d <= {1'b0,d,31'b0};              // 0.1xx...x0. or
                reg_x <= {1'b0,d,31'b0};              // 0.01x...x0...0
                busy  <= 1;
                ready <= 0;
                count <= 0;
            end else begin
                reg_d <= dr1[126:63];                 // x.xxx...x
                reg_x <= xi;                          // 0.xxx...x
                count <= count + 3'b1;                // count++
                if (count == 3'h5) begin              // finish
                    busy  <= 0;
                    ready <= 1;                       // q is ready
                end
            end
        end
    end
end
endmodule
```

The speed of $x_n \to 1$ depends on δ. The slowest case is $\delta = 3/4$ or $d = 1/4$. Figure 3.33 shows the simulation waveform of the Goldschmidt square root with $d = 0.25$ (the slowest case). We get $q = 0.5$. From the figure, we know that the number of iterations must be at least 6. The second example shows the case of $d = 0.75$; q is ready after four iterations.

The multiplication can be done with Wallace tree and takes two clock cycles. An addition takes one cycle. Including the rounding step, the 32-bit Goldschmidt square rooter takes $(4 + 1) \times 6 + 1 = 31$ cycles. If the three-input multiplier can be done in three clock cycles, then the square rooter takes $(3 + 1) \times 6 + 1 = 25$ cycles.

3.5.4   Newton–Raphson Square Root Algorithm

Given a radicand d with $1/4 \leq d < 1$, that is $d = 0.1xx \cdots x$ or $d = 0.01x \cdots x$, the Newton–Raphson square root algorithm calculates $q = \sqrt{d}$ as follows: let $f(x) = 1/x^2 - d$; then $f(x) = 0$ at $x = 1/\sqrt{d}$. Applying $x_{i+1} = x_i - f(x_i)/f'(x_i)$ as described in the Newton–Raphson division algorithm, we have

$$x_{i+1} = x_i(3 - x_i^2 d)/2$$

where x_i is an approximation of $1/\sqrt{d}$. By repeating the iteration to get x_n until it has enough accuracy, we can calculate $\sqrt{d}$ by $d \times x_n$.

Figure 3.34 shows the schematic diagram of a 32-bit Newton–Raphson square rooter. The two registers reg_d and reg_x store the radicand d and the iteration variable x_i, respectively. x_0 can be obtained from a ROM table. The following Verilog HDL code implements the circuit shown in Figure 3.34.

```
module root_newton (d,start,clk,clrn,q,busy,ready,count);
    input   [31:0] d;                            // radicand:
    input          start;                        //           .1xxx...x
    input          clk, clrn;                    //      or: .01xx...x
    output  [31:0] q;                            // root: .1xxx...x
```

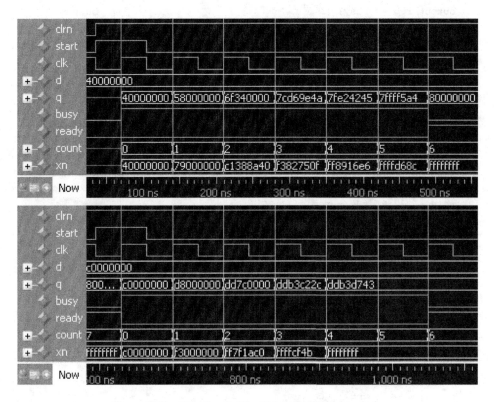

Figure 3.33 Waveform of the Goldschmidt square rooter

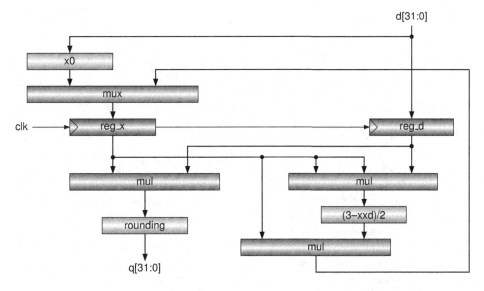

Figure 3.34 Schematic diagram of the Newton–Raphson square rooter

```
output reg     busy;                                    // busy
output reg     ready;                                   // ready
output  [2:0] count;                                    // counter
reg     [31:0] reg_d;
reg     [33:0] reg_x;
reg     [1:0] count;
// x_{i+1} = x_i * (3 - x_i * x_i * d) / 2
wire    [67:0] x_2 =  reg_x * reg_x;                     // xxxx.xxxxx...x
wire    [67:0] x2d =  reg_d * x_2[67:32];                // xxxx.xxxxx...x
wire    [33:0] b34 = 34'h300000000 - x2d[65:32];        //   xx.xxxxx...x
wire    [67:0] x68 =  reg_x * b34;                       // xxxx.xxxxx...x
wire    [65:0] d_x = reg_d * reg_x;                      //   xx.xxxxx...x
wire    [7:0] x0  = rom(d[31:27]);
assign       q   = d_x[63:32] + |d_x[31:0];             // rounding up
always @ (posedge clk or negedge clrn) begin
    if (!clrn) begin
        busy  <= 0;
        ready <= 0;
    end else begin
        if (start) begin
            reg_d <= d;                                 // load d
            reg_x <= {2'b1,x0,24'b0};                   // 01.xxxx0...0
            busy  <= 1;
            ready <= 0;
            count <= 0;
        end else begin
            reg_x <= x68[66:33];                        // /2
            count <= count + 2'b1;                      // count++
            if (count == 2'h2) begin                    // 3 iterations
                busy  <= 0;
                ready <= 1;                             // q is ready
            end
        end
    end
end
function [7:0] rom;                                     // about 1/d^{1/2}
    input [4:0] d;
    case (d)
        5'h08: rom = 8'hff;          5'h09: rom = 8'he1;
        5'h0a: rom = 8'hc7;          5'h0b: rom = 8'hb1;
        5'h0c: rom = 8'h9e;          5'h0d: rom = 8'h9e;
        5'h0e: rom = 8'h7f;          5'h0f: rom = 8'h72;
        5'h10: rom = 8'h66;          5'h11: rom = 8'h5b;
        5'h12: rom = 8'h51;          5'h13: rom = 8'h48;
        5'h14: rom = 8'h3f;          5'h15: rom = 8'h37;
        5'h16: rom = 8'h30;          5'h17: rom = 8'h29;
        5'h18: rom = 8'h23;          5'h19: rom = 8'h1d;
```

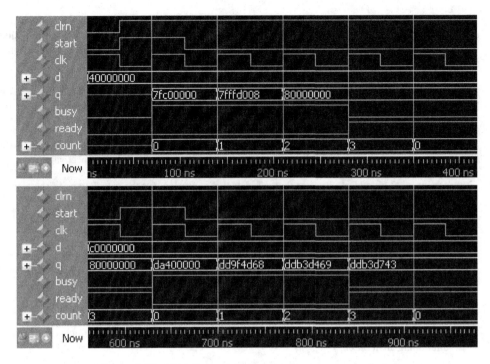

Figure 3.35 Waveform of the Newton–Raphson square rooter

```
5'h1a: rom = 8'h17;              5'h1b: rom = 8'h12;
5'h1c: rom = 8'h0d;              5'h1d: rom = 8'h08;
5'h1e: rom = 8'h04;              5'h1f: rom = 8'h00;
default: rom = 8'hff;                         // 0 - 7: not used
      endcase
   endfunction
endmodule
```

Figure 3.35 shows the simulation waveforms of the Newton–Raphson square rooter with $d = 0x0.40000000$ and $d = 0x0.c0000000$, and whose results are $q = 0x0.80000000$ and $q = 0x0.ddb3d743$, respectively.

The multiplication can be done with Wallace tree and takes two clock cycles. An addition takes one cycle. Including searching ROM table, the final multiplication, and the rounding step, the 32-bit Newton–Raphson square rooter takes $1 + (6 + 1) \times 3 + 2 + 1 = 25$ cycles.

Table 3.6 Cycles required by the Goldschmidt and Newton–Raphson algorithms

Algorithm	Divide	Square root
Goldschmidt	$3 \times 5 + 1 = 16$	$5 \times 6 + 1 = 31$
Newton–Raphson	$1 + 5 \times 3 + 2 + 1 = 19$	$1 + 7 \times 3 + 2 + 1 = 25$

Table 3.6 compares the clock cycles taken by the Goldschmidt and Newton–Raphson algorithms for the division and square root calculations. In the design of floating point units (FPUs), the last rounding step of the algorithm can be removed because there will be a normalization step in the FPU design that converts the result to the formats of IEEE 754 floating-point numbers.

Exercises

3.1 Let a, b, and s be n-bit numbers in 2's complement representation, and ci be a 1-bit carry in. Prove $s = a + \overline{b} + \overline{ci} = a - b - ci$, where $\overline{b}$ and $\overline{ci}$ are the bit inverses of b and ci, respectively.

3.2 Design a 32-bit adder/subtracter that can detect the overflow for the calculations of the signed numbers.

3.3 Design a 64-bit CLA in which the carry bits are generated with a three-level 4-tree ($64 = 4^3$).

3.4 Implement the iterative multiplication in Verilog HDL.

3.5 Write the Verilog HDL code to implement an 8×8 signed Wallace tree.

3.6 Investigate the Booth's multiplication algorithm and implement it in Verilog HDL.

3.7 Prove that the Goldschmidt and Newton–Raphson algorithms double the accuracy of the result after doing an iteration.

3.8 Write programs to calculate the contents of the ROM tables for the Newton–Raphson division and square root algorithms.

3.9 Prove the correctness of the decimal square root algorithm.

3.10 Try to use a bigger ROM table so that the division or square root can be calculated with two-time Newton–Raphson iterations.

3.11 Explain the reason why the Goldschmidt square root calculation uses one more iteration than the division calculation in the worst case.

3.12 Try to improve Goldschmidt algorithms by using ROM tables.

3.13 Investigate the SRT (Sweeney, Robertson, and Tocher) division and square root algorithms and implement them in Verilog HDL.

4

Instruction Set Architecture and ALU Design

Computers can execute binary programs. These programs consist of data and instructions of a particular CPU (central processing unit). A binary program is usually generated by a compiler that compiles a program written in high-level programming languages such as C, or by an assembler that translates a program written in the assembly programming language of a particular CPU, such as x86 or MIPS (microprocessor without interlocked pipeline stages).

This chapter introduces the instruction set architecture (ISA), some MIPS instructions, an assembler and simulator of MIPS integer instructions, and the design of an arithmetic logic unit (ALU) which calculates the operation results of some MIPS integer instructions.

4.1 Instruction Set Architecture

ISA is an important issue in hardware/software codesign. An ISA tells compiler developers "what a CPU can do," and tells CPU designers "what a CPU should do." Compiler developers use the ISA to develop compilers, and CPU designers design a CPU to implement the ISA. That is, an ISA is an interface between software and hardware, as shown in Figure 4.1.

An ISA defines the formats of instructions, the operations of instructions, the types of operands, the memory and registers the instructions can access, the byte ordering, and the addressing modes. Some popular ISAs include the Intel's x86, SGI/MIPS's MIPS32/MIPS64, IBM's PowerPC, SUN Microsystems' SPARC, HP's HP-PA, and ARM's ARM.

4.1.1 Operand Types

The main task of instructions is calculations on operands of different types. Table 4.1 lists some common types of operands. The corresponding keywords in C are also given in the table.

A byte is always 8 bits but the length of a word depends on the bit-processing ability of the CPU: for example, it has 16 bits in the 8086 CPU. Here we use 32 bits as the length of a word.

If an operand on which an instruction of a 32-bit CPU calculates is not 32 bits in length, a byte or a half word for example, the operand will be extended to 32 bits. If the operand is an unsigned type, zeroes will be appended on the left side of the operand. We call this a zero extension. If the operand is a signed type,

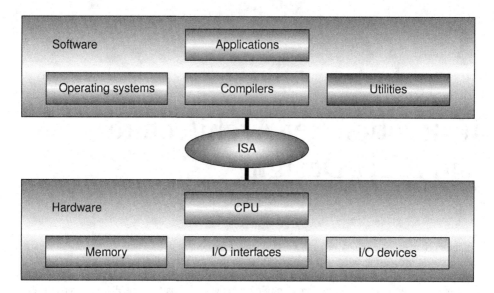

Figure 4.1   ISA as an interface between software and hardware

Table 4.1   Operand types

Operand type	Bits	Value range	Corresponding to C
Byte	8	−128 to +127	signed char
Unsigned byte	8	0 to 255	unsigned char
Half word	16	−32,768 to +32,767	short int
Unsigned half word	16	0 to 65,535	unsigned short int
Word	32	−2,147,483,648 to +2,147,483,647	int
Unsigned word	32	0 to 4,294,967,295	unsigned int
Single-precision FP	32		float
Double-precision FP	64		double

the sign extension will be performed by filling each extra bit in the left side with the most significant bit (the sign bit) of the operand.

4.1.2   Little Endian and Big Endian

Little endian and big endian define the order in which a sequence of bytes of a word is stored in memory. Little endian is an order in which the least significant byte (little end) is stored first (the lowest address). Big endian is an order in which the most significant byte (big end) is stored first.

Figure 4.2 shows an example. For a 4-byte word 0x76543210, the least significant byte 0x10 is stored in address xxxxxxx0 for a little endian machine, while in a big endian machine the most significant byte 0x76 is stored in address xxxxxxx0.

We can write a program to detect which endian is being used in the machine. The following C code uses a pointer to detect the endian. A pointer is a variable that stores the address of another variable.

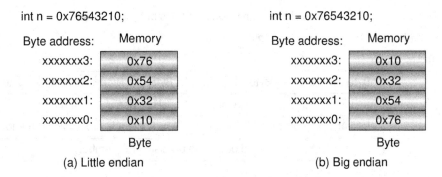

Figure 4.2   Little endian and big endian

```
main () {                                  // endian_pointer.c
    int n = 0x76543210;                    // a word
    if (*(unsigned char *)&n == 0x10)      // char starting address
        printf("little endian\n");         // stored 0x10
    else printf("big endian\n");           // stored 0x76
}
```

Also we can use a union structure to do it: see the following C code. Just as a person has different names, all members of a union define different manifestations of the same variable.

```
main() {                                          // endian_union.c
    union {                                        // union:
        int intword;                               // to assess a word
        unsigned char characters[sizeof (int)];    // with different names
    } u;                                           // union u
    u.intword = 0x76543210;                        // access as an integer
    if (u.characters[sizeof (int) - 1] == 0x76)    // access as array of byte
        printf("little endian\n");                 // high byte in high address
    else printf("big endian\n");                   // high byte in low address
}
```

A byte is the smallest unit of memory access. It can be stored at any location of the memory. For other data types, there is a problem of data alignment. The data alignment refers to where the data is located in the memory. Referring to Figure 4.3, a memory address a is said to be n-byte aligned when n is a power of 2 and a is a multiple of n bytes.

Aligned data can be accessed faster. For example, the 4 bytes of an aligned word can be written to or read from memory with one access by using the word address (all the 4 bytes have a same word address). On the other hand, we must access to memory two times if the word is unaligned. Some ISAs allow the unaligned data placement. If it is not allowed and there is an unaligned access, an alignment fault exception will occur.

4.1.3   Instruction Types

An instruction type refers to what kind of operation is done by an instruction. The instruction types differ between ISAs. Some instruction types that commonly appear in the modern ISAs are described below.

Three least significant bits of byte address

| Big endian: | 0 | 1 | 2 | 3 | 4 | 5 | 6 | 7 |
| Little endian: | 7 | 6 | 5 | 4 | 3 | 2 | 1 | 0 |

Byte	Byte	Byte	Byte	Byte	Byte	Byte	Byte
Half word		Half word		Half word		Half word	
Word or single precision FP number				Word or single precision FP number			
Long word or double precision FP number							

Figure 4.3   Data alignment in memory locations

4.1.3.1   Arithmetic Operation Type

Instructions of this type perform arithmetic operations on integers, such as addition, subtraction, multiplication, division, and square root.

4.1.3.2   Logic Operation Type

Logic operations include bitwise logical AND, OR, and NOT. Most ISAs provide NOR (NOT OR) or XOR (exclusive OR) instead of NOT.

4.1.3.3   Shift Operation Type

There are mainly three types of shift instructions: shift left, logical shift right, and arithmetic shift right. Other shift instructions include arithmetic shift left, rotate shift, and rotate shift with carry.

4.1.3.4   Memory Access Type

Memory access instructions transfer data between the memory and registers inside the CPU. A load instruction loads memory data to the register. A store instruction writes register data to the memory. All RISC (reduced instruction set computer) type ISAs have these two instructions but CISC (complex instruction set computer) type ISAs may combine them into computational instructions.

4.1.3.5   Input/Output Access Type

Input/output instructions transfer data between I/O and CPU registers. In most RISC type ISAs, some addresses of the virtual memory are assigned to the I/O space so that the load and store instructions can be used to access I/O. This is called a memory-mapped I/O. CISC type ISAs may have a dedicated I/O space, therefore the input and output instructions must be prepared for accessing the I/O.

4.1.3.6   Control Transfer Type

Control transfer instructions alter the order of the instruction execution. Conditional branch instructions confirm a condition to determine whether to branch to a target address. A jump instruction jumps to a target address unconditionally. A subroutine call instruction jumps to the entry of the subroutine and saves the return address to somewhere. And a return instruction returns from the subroutine.

4.1.3.7 Floating-Point Calculation Type

All the computational instructions described above operate on integers. Floating-point instructions perform arithmetic operations on floating-point numbers. There are also some instructions that convert data formats between integers and floating-point numbers.

4.1.3.8 System Control Type

System control instructions include system calls, return from exceptions, and instructions that read CPU state registers or write CPU control registers.

4.1.4 Instruction Architecture

The instruction architectures decide how the CPU will store data. Figure 4.4 shows four instruction architectures that are used in modern CPUs.

In a stack architecture, the operands are implicitly on top of the stack. Two source operands are popped from the top of the stack, and the result is pushed onto the stack. The stack has a feature of first-in last-out or last-in first-out; it cannot be accessed randomly. JVM (Java virtual machine) Bytecode uses this architecture.

In an accumulator architecture, one operand is implicitly in the accumulator and the other operand is in the memory or register. The operation result is stored in accumulator implicitly. Z80 and 6502 ISAs use this architecture.

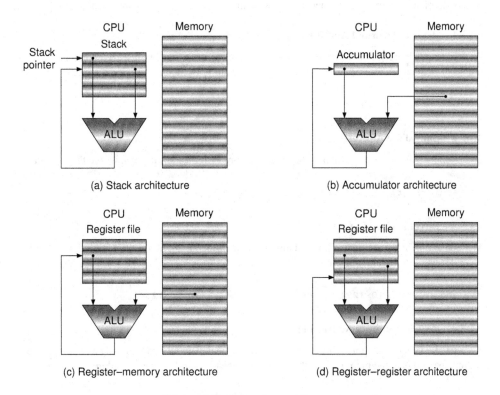

(a) Stack architecture

(b) Accumulator architecture

(c) Register–memory architecture

(d) Register–register architecture

Figure 4.4 Instruction architecture

Table 4.2   Code examples of instruction architectures

Stack		Accumulator		General-purpose register					
				Register–memory			Register–register		
push	x	load	x	load	r1,	x	load	r1,	x
push	y	add	y	add	r1,	y	load	r2,	y
add		store	z	store	z,	r1	add	r3,	r1,	r2
pop	z						store	z,	r3

In a register–memory architecture, one operand is explicitly in a general-purpose register file and the other operand is in the memory. The operation result is stored in the register. The Intel x86 ISA uses this architecture.

In a register–register architecture, all operands are explicitly in a general-purpose register file. The operation result is also stored in the register. Almost all RISC type ISAs use this architecture. The data transfer between the register file and memory is performed by load and store instructions.

Table 4.2 gives the instruction code examples that implement $z = x + y$ in the architectures described above, where x, y, and z are in memory, and r1, r2, and r3 are registers. The **add** instruction performs the addition.

4.1.5   Addressing Modes

The addressing modes define how the instructions get operands. An operand can be in a register, in the memory, or in the instruction (immediate). Some simple addressing modes are given below.

4.1.5.1   Register Operand Addressing

The operand is in a register of the register file. In the following example, the two source operands are in register 1 (r1) and register 2 (r2), respectively. The result of the addition is stored in register 3 (r3).

```
    add  r3, r1, r2          ; r3 <-- r1 + r2
```

4.1.5.2   Immediate Addressing

The operand is in instruction. In the following example, -1 (constant) is given within the instruction.

```
    addi r1, r1, -1          ; r1 <-- r1 - 1
```

4.1.5.3   Direct Addressing

The operand is in the memory whose address is given in the instruction. In the following example, the second source operand is in the memory location of 0x1234.

```
    add r3, r1, [0x1234]     ; r3 <-- r1 + Memory[0x1234]
```

4.1.5.4   Register Indirect Addressing

The operand is in the memory whose address is given by a register. In the following example, the second source operand is in the memory, and the content in register r2 is the memory address.

```
    add r3, r1, (r2)         ; r3 <-- r1 + Memory[r2]
```

4.1.5.5 Offset Addressing

The operand is in the memory whose address is the sum of an offset and the content of a register. In the following example, the second source operand is in the memory and the memory address is the sum of 0x1234 and the content in register r2.

```
add r3, r1, 0x1234(r2)   ; r3 <-- r1 + Memory[0x1234+r2]
```

4.2 MIPS Instruction Format and Registers

This section gives an example of the MIPS ISAs. MIPS has two ISA versions—MIPS32 and MIPS64. This book focuses on MIPS32.

4.2.1 MIPS Instruction Format

All the MIPS instructions are 32 bits in length. Figure 4.5 gives three formats of the MIPS instructions. op is the opcode (operation code) of the instruction. func stands for function.

The op of the R (register) format instructions is 0, and the operation of an instruction is specified by func. Each of rs, rt, and rd is a 5-bit register number. Shift instructions use sa to specify the shift amount (a constant). The content in register rt will be shifted, and the result of the shift will be written to register rd. Other R format instructions read two source operands from registers rs and rt, respectively, operate on the two operands, and write the result to the register rd.

The computational instructions of the I (immediate) format use immediate as the second source operand. The 16-bit immediate must be sign- or zero-extended into 32 bits. The first source operand is in register rs, and the result is written to register rt.

The conditional branch instructions of the I format compare the two operands located in registers rs and rt, respectively, and determine whether to branch to the target address. immediate is a signed word-offset and is used to calculate the branch target address.

The load instructions of the I format load memory data to the register rt. The store instructions store the register rt data to the memory. For both load and store instructions, the memory address is the sum of the sign-extended immediate and the operand in the register rs.

The address in the J (jump) format instructions will be shifted to the left by 2 bits to form the low 28 bits of the jump target address.

4.2.2 MIPS General-Purpose Registers

A register number of MIPS instructions has 5 bits. Therefore, there are 32 registers in the integer register file. Table 4.3 lists the names and usages of these registers.

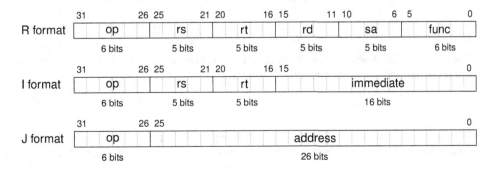

Figure 4.5 MIPS instruction formats

Table 4.3   MIPS general-purpose registers

Register name	Register number	Use
$zero	0	Constant 0
$at	1	Assembler temporary
$v0 to $v1	2 to 3	Function return value
$a0 to $a3	4 to 7	Function parameters
$t0 to $t7	8 to 15	Temporaries
$s0 to $s7	16 to 23	Saved temporaries
$t8 to $t9	24 to 25	Temporaries
$k0 to $k1	26 to 27	Reserved for OS kernel
$gp	28	Global pointer
$sp	29	Stack pointer
$fp	30	Frame pointer
$ra	31	Return address

Note that the register 0 is not a register; its content is always a constant 0. Although Table 4.3 gives the convention on the usage of the registers, there are no differences among the remaining 31 registers from the hardware point of view. This usage gives the rules for preparing MIPS assembly language programs. gcc, a GNU C compiler, for MIPS machines follows these rules very well.

4.3   MIPS Instructions and AsmSim Tool

This section describes some MIPS instructions in detail and introduces a MIPS assembler and simulator, called AsmSim.

4.3.1   Some Typical MIPS Instructions

Table 4.4 lists some MIPS instructions that will be implemented in our CPU. We selected 20 typical and essential MIPS integer instructions that cover the three instruction formats. Each of these instructions is explained below.

```
add/sub/and/or/xor rd, rs, rt   # rd <-- rs op rt
```

These five instructions have the same format and perform addition, subtraction, AND, OR, and XOR, respectively. rs and rt are the two source register numbers, and rd is the destination register number.

```
sll/srl/sra rd, rt, sa          # rd <-- rt shift sa
```

These are three shift instructions: shift left logical (sll), shift right logical (srl), and shift right arithmetic (sra). The shift amount is given by the 5-bit sa.

```
addi rt, rs, immediate          # rt <-- rs + (sign)immediate
```

The addi (add immediate) instruction adds a 16-bit signed immediate to the 32-bit value in register rs. The result of the addition is saved to register rt. The 16-bit immediate is sign-extended into 32 bits.

```
andi/ori/xori rt, rs, immediate # rt <-- rs op (zero)immediate
```

Table 4.4 Twenty MIPS integer instructions

Inst.	[31:26]	[25:21]	[20:16]	[15:11]	[10:6]	[5:0]	Meaning
add	000000	rs	rt	rd	00000	100000	Register add
sub	000000	rs	rt	rd	00000	100010	Register subtract
and	000000	rs	rt	rd	00000	100100	Register AND
or	000000	rs	rt	rd	00000	100101	Register OR
xor	000000	rs	rt	rd	00000	100110	Register XOR
sll	000000	00000	rt	rd	sa	000000	Shift left
srl	000000	00000	rt	rd	sa	000010	Logical shift right
sra	000000	00000	rt	rd	sa	000011	Arithmetic shift right
jr	000000	rs	00000	00000	00000	001000	Register jump
addi	001000	rs	rt	Immediate			Immediate add
andi	001100	rs	rt	Immediate			Immediate AND
ori	001101	rs	rt	Immediate			Immediate OR
xori	001110	rs	rt	Immediate			Immediate XOR
lw	100011	rs	rt	offset			Load memory word
sw	101011	rs	rt	offset			Store memory word
beq	000100	rs	rt	offset			Branch on equal
bne	000101	rs	rt	offset			Branch on not equal
lui	001111	00000	rt	immediate			Load upper immediate
j	000010		address				Jump
jal	000011		address				Call

The andi (and immediate), ori (or immediate), and xori (exclusive or immediate) instructions perform bitwise logical AND, OR, and XOR, respectively, on the value in register rs and a 16-bit unsigned immediate. The result is written to register rt. The 16-bit immediate is zero-extended into 32 bits.

```
lui rt, immediate                 # rt <-- immediate << 16
```

The lui (load upper immediate) instruction shifts immediate to the left by 16 bit. By using lui and ori instructions, we can set a register to an any 32-bit constant (lui sets high 16 bits and ori sets low 16 bits).

```
lw rt, offset(rs)                 # rt <-- memory[rs + offset]
```

The lw (load word) instruction loads a 32-bit word from memory. The memory address is calculated by adding a 16-bit signed immediate (offset) to the 32-bit value in register rs. The loaded word is saved to register rt.

```
sw rt, offset(rs)                 # memory[rs + offset] <-- rt
```

The sw (store word) instruction stores a 32-bit word to memory. The memory address is calculated by adding a 16-bit signed offset to the 32-bit value in register rs. The word to be stored is in register rt.

```
beq rs, rt, label                 # if (rs == rt) PC <-- label
```

The beq (branch on equal) instruction transfers control to a PC-relative target address (label) if the values in registers rs and rt are equal. The branch target address is calculated by adding an 18-bit signed constant (the 16-bit offset shifted to the left by 2 bits) to the address of the instruction following beq.

```
    bne rs, rt, label                    # if (rs != rt) PC <-- label
```

The bne (branch on not equal) instruction transfers control to a PC-relative target address (label) if the values in registers rs and rt are not equal. The branch target address is calculated by adding an 18-bit signed constant (the 16-bit offset shifted to the left by 2 bits) to the address of the instruction following bne.

```
    j target                             # PC <-- target
```

The j (jump) instruction transfers control to a target address whose low 28 bits are the 26-bit target shifted to the left by 2 bits; the remaining upper bits are the corresponding bits of the address of the instruction following the j instruction.

```
    jal target                           # $31 <-- PC + 8; PC <-- target
```

The jal (jump and link) is a subroutine call instruction that does the same job as the j instruction and meanwhile saves the return address to register $31. The return address is the location at which execution continues after returning from the subroutine. MIPS uses the delayed branch technique; the return address is PC + 8.

```
    jr rs                                # PC <-- rs
```

The jr (jump register) instruction transfers control to a target address given in register rs. The return from a subroutine can be done by letting rs to be 31.[1]

In MIPS ISA, there is no subi (subtract immediate) instruction. Such instruction can be implemented by addi because the immediate is represented with a signed number: $x - i = x + (-i)$.

MIPS assembler supports several pseudo-instructions. A pseudo-instruction will be translated into an actual instruction or a sequence of instructions by the assembler. For example, a nop (no operation) pseudo-instruction is translated into sll $0, $0, 0, which shifts the content of register $0 to the left by 0 bit and writes the result to register $0. The 32-bit encoding of this instruction is 0.

Another pseudo-instruction is li (load immediate), which can load a 32-bit immediate to a register. For example, li $4, 0x1234abcd loads the 32-bit immediate 0x1234abcd to register $4. Because an MIPS instruction is 32 bits in length and its opcode field takes 6 bits, loading a 32-bit immediate into a register cannot be done by only one instruction. The assembler translates li $4, 0x1234abcd into a sequence of two instructions: lui $4, 0x1234 and ori $4, $4, 0xabcd. The first instruction writes 0x12340000 to register $4, and the second instruction writes the result of 0x12340000 OR 0x0000abcd, or 0x1234abcd, to register $4.

Table 4.5 lists some MIPS pseudo-instructions and the corresponding actual instructions. A pseudo-instruction can be translated in several ways. For example, the move $4, $2 (the content of

---

[1] The behavior of the subroutine call and return is likely similar to that a person on a business trip. When the author teaches students in Japan about the MIPS jal and jr instructions, the following example is used: when the person goes to the destination city on a business trip, he or she takes a fight of JAL (Japan Airline) because he or she must go there hurriedly. When returning to his or her home city, he or she can take a train of JR (Japan Railway) without haste. One more thing is that he or she must know where the home is. This information is in the person's brain memory. MIPS CPU remembers it by recording the return address into the $31 register.

Table 4.5   Some MIPS pseudo-instruction examples

Pseudo-instruction	Actual instruction(s)	Meaning
nop	sll $0, $0, 0	No operation
clear $4	sll $4, $0, 0	Clear
move $4, $5	sll $4, $5, 0	Copy content of $5 to $4
not $4, $5	nor $4, $5, $0	Write inverse of $5 to $4
subi $4, $5, 1	addi $4, $5, -1	Subtract a 16-bit immediate
li $4, 0x1234abcd	lui $4, 0x1234	Load a 32-bit immediate
	ori $4, $4, 0xabcd	
la $4, label	lui $4, %hi(label)	Load a label address
	ori $4, $4, %lo(label)	
b label	beq $0, $0, label	Unconditional branch
bz $4, label	beq $4, $0, label	Branch on zero
bnz $4, label	bne $4, $0, label	Branch on not zero

Table 4.6   Code sequences for unimplemented instructions

Unimplemented instruction	Code sequence	Meaning
bltz $4, label	srl $at, $4, 31	Branch on negative
	bne $at, $0, label	
bgez $4, label	srl $at, $4, 31	Branch on not negative
	beq $at, $0, label	

$2 is copied to $4) pseudo-instruction can be translated into sll/srl/sra $4, $2, 0; it can be also translated into add/sub/or/xor $4, $2, $0, or addi/ori/xori $4, $2, 0.

The not pseudo-instruction can be translated into nor instruction, but our CPU will not implement the nor. To perform not $4, $5, we can use a sequence of two instructions: addi $4, $0, -1 and xor $4, $5, $4. The first instruction loads 0xffffffff to register $4; the second instruction performs an XOR operation on the content of $5 and 0xffffffff and writes the result to register $4. By using only the 20 instructions listed in Table 4.4, we can implement other instructions. Table 4.6 lists two examples. The register $at ($1) is used in the sequences.

4.3.2   Supporting Subroutine Call and Pointer

Subroutine call and pointer are two important concepts in high-level programming languages. We use examples to show how MIPS implements these concepts.

4.3.2.1   Subroutine Call

The MIPS CPU uses jal and jr instructions to implement a subroutine call and return, respectively. The jal instruction saves the return address (PC + 8 in pipelined CPU; here we assume PC + 4) into register $ra ($31) and jumps to the entry of the subroutine. jr $ra writes the content of $ra (return address) into the PC. Let's see the following C code.

```
int sum(int *array, int n) {                      // subroutine_call.c
    int i;
    int total = 0;
    for (i = 0; i < n; i++) {
        total += array[i];                        // sum of n array elements
    }
    return(total);                                // return sum
}
int main() {
    int a[] = {1, 2, 3};                          // initialize the array
    printf ("The sum is %d\n", sum(a, 3));        // call subroutine
    return(0);
}
```

In the main program, the integer array a holds three elements: a[0] = 1, a[1] = 2, and a[2] = 3. The main program calls the function sum (a subroutine is named as a function in C and we use the both names in the text). It passes two parameters to sum: the start address of a and the number of elements. The function sum calculates the sum of the elements with a for loop and returns the result total to the main program. Actually, the main program can be also considered as a subroutine. The operating system (OS) calls it, and after the execution is finished, the control will be returned to the OS.

The C code can be compiled to the following assembly program. Because register $ra contains the return address to the OS initially, before calling function sum the content of $ra must be saved somewhere; otherwise, it will be overwritten by the jal sum instruction. MIPS saves the content of $ra to a memory stack. Stack uses the main memory, and it has a stack pointer (register $sp) which points to the top of the stack. The local variables, the array a for instance, are also saved in the stack. If some registers are required to have the same values before and after the function call, their contents must be also saved in the stack.

```
 1:  .data                       # data segment
 2:  $LC0:                       # address of array a
 3:        .word  1, 2, 3        # elements of array a
 4:  $LC1:                       # address of string
 5:        .ascii "The sum is %d\n" # string
 6:  .text                       # code segment
 7:  sum:                        # entry of subroutine sum
 8:       subi  $sp, $sp, 8      # reserve stack space
 9:       move  $3, $0           # i = 0
10:       move  $6, $0           # total = 0
11:       blez  $5, $L3          # goto $L3 if n <= 0
12:  $L5:                        # for loop
13:       sll   $2, $3, 2        # i * 4 (4 bytes per word)
14:       add   $2, $2, $4       # base address + i * 4
15:       lw    $2, 0($2)        # load a[i]
16:       addi  $3, $3, 1        # i++
17:       add   $6, $6, $2       # total += a[i]
18:       bne   $3, $5, $L5      # goto $L5 if i != n
19:  $L3:                        # end of loop
20:       move  $2, $6           # move total to $2
```

```
21:      addi    $sp, $sp, 8      # release stack space
22:      jr      $ra              # return from subroutine
23: main:                         # program entry
24:      subi    $sp, $sp, 40     # reserve stack space
25:      sw      $ra, 32($sp)     # save return address
26:      la      $5, $LC0         # address of array a
27:      lw      $2, 0($5)        # load a[0]
28:      lw      $3, 4($5)        # load a[1]
29:      lw      $4, 8($5)        # load a[2]
30:      sw      $2, 16($sp)      # store a[0] to stack
31:      sw      $3, 20($sp)      # store a[1] to stack
32:      sw      $4, 24($sp)      # store a[2] to stack
33:      addi    $4, $sp, 16      # $4: address of a[0]
34:      li      $5, 3            # $5: n = 3
35:      jal     sum              # call subroutine sum
36:      la      $4, $LC1         # $4: address of string
37:      move    $5, $2           # $5: total
38:      jal     printf           # call printf
39:      move    $2, $0           # return value 0
40:      lw      $ra, 32($sp)     # restore return address
41:      addi    $sp, $sp, 40     # release stack space
42:      jr      $ra              # return to operating system
43: .end                         # end of program
```

Referring to Figure 4.6(a), in MIPS the stack grows downward and the register $sp ($29) points to the top location of the stack. In line 24 of the assembly program (the entry of the main program), a 40-byte (or 10-word) stack space is reserved. The instruction in line 25 saves the return address in register $ra to the stack. The instructions in lines 26–32 push the local variables (three array elements) onto the stack. Before calling the subroutine, the parameters for the subroutine are putted to registers $4 and $5. Register $4 holds the starting address of the array a (line 33) and register $5 holds a 3, which is the number of array elements (line 34). Then the subroutine sum is called (line 35). The result calculated by the subroutine is in register $2; it is copied to $5 (line 37). The string address of a string that will be printed out is written to register $4 (line 36). Then the program calls a system function printf to display the result (line 38). The user's work is done. Before returning to the OS (line 42), the return value in $2 is cleared (line 39) for return(0) in the C code, the original return address is restored to register $ra from the stack (line 40), and the stack space is released (line 41).

Referring to Figure 4.6(b), in the subroutine of the assembly program an 8-byte stack space is reserved (line 8). Because there is no further subroutine call, the return address need not be saved. The local variables within the subroutine should use the stack, but there are only two local variables: i and total, that directly use registers $3 (line 9) and register $6 (line 10), respectively. Therefore, the reserved stack was not used at all. Both variables are cleared to 0 initially. The instructions in lines 11–18 do the for loop. Then the result is moved to the register $2 (line 20). After releasing the stack (line 21), the subroutine is completed by returning to the main program (line 22).

4.3.2.2 Pointer

Now, we show how to use MIPS instructions to implement the pointer operations in C programs. Simply we can define a variable of a certain type (int, float, or double) with a name. The value of a variable

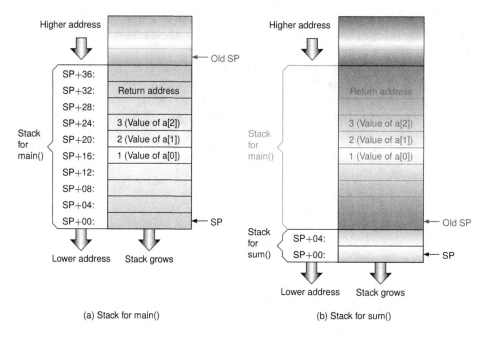

Figure 4.6 Stack for function call

can vary. The value is stored in the memory. A pointer can be considered as the memory address; in that location, the value of a variable is stored.

In the following C code example, we first define a variable num whose type is int (integer) and a variable ptr whose type is a pointer. ptr points to a location where an integer (*ptr) is stored. Then, we let the value of the variable num to be a 1, and let the value of the variable ptr to be the address of num (&num). Finally, we display the value of the ptr with a pointer format (%p).

```
main() {                      // pointer.c
    int  num;                 // an integer variable num
    int *ptr;                 // a pointer variable ptr pointing to an int
    num =  1;                 // let the value of num be a 1
    ptr = &num;               // let the value of ptr be the address of num
    printf ("The address of num is %p\n", ptr);          // print ptr out
}
```

Before running this program, we knew that the value of the variable num is a 1, but we didn't know the value of the variable ptr. That is, we didn't know where the value of num was stored. The following assembly codes implement the C codes above.

```
1: .data                                  # data segment
2: $LC0:                                   # address of string
3:     .ascii "The address of num is %p\n" # string
4: .text                                   # code segment
5: main:                                   # program entry
```

```
 6:      subi   $sp, $sp, 32            # reserve stack space
 7:      sw     $ra, 24($sp)           # save return address
 8:      li     $2,  0x00000001        # num = 1
 9:      sw     $2,  16($sp)           # store num into stack
10:      la     $4,  $LC0              # $4: address of string
11:      addi   $5,  $sp, 16           # $5: ptr = &num
12:      jal    printf                 # call printf
13:      lw     $ra, 24($sp)           # restore return address
14:      addi   $sp, $sp, 32           # release stack space
15:      jr     $ra                    # return to operating system
16: .end                              # end of program
```

The ASCII string in line 3 will be printed out. %p is the pointer format directive used for displaying the value of ptr. The label address of the string is $LC0 (line 2). In line 6, a 32-byte (or 8-word) stack space is reserved. The instruction in line 7 saves the return address to the stack. The value 1 is loaded into register $2 (line 8). It is then stored in a location of the stack memory; the memory location is the content of register $sp plus 16 (line 9), which is the value of ptr, as shown in Figure 4.7.

For printing the value of ptr out (line 12), the address of the string is putted into register $4 (line 10), and the value of the pointer ptr is putted into register $5 (line 11), which is the location where the value of num is stored. Before finishing the program, the original return address is restored to register $ra from the stack (line 13), and the stack space is released (line 14).

4.3.3 AsmSim—A MIPS Assembler and Simulator

A better way to understand the operation of each instruction is to see its execution results. This section introduces a Web-based graphical tool, named AsmSim, that contains a MIPS assembler and an execution simulator for the MIPS integer instructions.

The main features of the tool include the supports of the common function calls that are usually used in C programs, the capability of showing images on a virtual video graphics array (VGA) display, and the

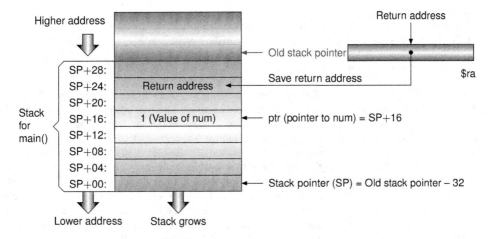

Figure 4.7 Pointer of a variable

auto-generation of Xilinx COE files and Altera MIF files. In detail, the AsmSim provides the following functions:

1. The basic MIPS assembly programs' assembling and simulation;
2. The basic input and output system function calls, such as scanf() and printf();
3. Interrupt mechanism, like addKeyListener() and removeKeyListener() in Java, for handling the keyboard interrupt request;
4. Manipulating a standard VGA (640 × 480 pixels) window. The direct video random access memory (VRAM) access function allows the user program to read/write a pixel (24-bit RGB pattern) from/to the VRAM. It is also possible for the user program to define the starting address of the VRAM;
5. Graphics object draw and fill. The currently supported objects include line, oval, rectangle, and string. The generated image is stored in an off-screen buffer, and the paint function call will display the image on the graphics display by writing the image to the VRAM;
6. Automatically generating Xilinx COE and Altera MIF memory initialization files for the purpose of using Xilinx and Altera FPGA boards;
7. Some other useful function calls, like "get a random number," "get the timer," "sleep," and "get calendar." AsmSim also supports the data structure of the linked list for implementing graph algorithms.

The current version of the AsmSim implements almost all the MIPS integer instructions. It accepts the assembly program. The three windows, namely the main window, the program edit window, and the graphics console window, are launched.

The main window (Figure 4.8) shows the instructions, the values of the registers, the data of the data memory, and the control buttons. The instructions are displayed with memory addresses, instruction encodings, and the user source codes. The values of the registers are displayed in hexadecimal format. In addition to the register names, the register numbers are also shown in the window. There are 11 buttons in the main window, whose functions are described below (the last three buttons are not shown in the figure because of space limitation).

1. [edit]: opens the program edit window (in case it was closed);
2. [step]: executes one instruction that is highlighted currently;
3. [goto]: executes instructions until a break-point is reached.
4. [ascii]: displays the ASCII of the selected contents in the data memory;
5. [restart]: reloads the user program;
6. [run]: executes user program;
7. [stop]: stops the execution;
8. [quit]: quits from debug mode and enters command line mode;
9. [inst]: displays the encodings and formats of the MIPS instructions;
10. [xilinx]: generates Xilinx COE file;
11. [altera]: generates Altera MIF file.

Referring to Figure 4.8, the highlighted instruction is the one that will be executed next. The figure shows the image after the first two instructions were executed. The subi $sp, $sp, 24 instruction reserves stack space of 24 bytes for the main function. The content of the $sp register is highlighted. The sw $ra, 20($sp) instruction saves the return address (the content of $31 register) into the stack. After the execution of this instruction, the memory location of 0x0000ff3c (0x0000ff28 + 0x14) contains 0x03ffff00. This content is also highlighted but is not shown in the figure.

The program edit window (Figure 4.9) is used to edit the user program. We followed the rule of the gcc for the MIPS register usages. The assembler supports two kinds formats of the comments: /**/ and #. We can use control_s for searching text, just like emacs. The figure shows the case of searching "print." The MIPS assembly program shown in the figure prints out "Hello, World!" on the console window. The address of the hello message is loaded into the $4 register. Then, the jal printf instruction invokes a system call to display the message.

The keyword .text indicates the start of the code segment. We can also use .code. .data indicates the start of data segment. Within the data segment, we can use .ascii or .asciiz to declare the strings.

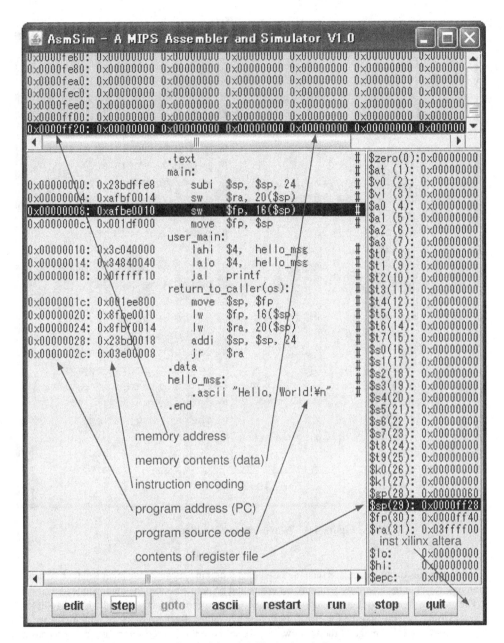

Figure 4.8   AsmSim main window

Other directives include .word and .comm. .word supports the declaration of multiple words by separating each word with a comma. .comm reserves memory space. It has the format .comm array_name, bytes, where array_name is a variable name of an array and bytes defines the memory space length in bytes. Asmsim also supports .comm array_name, bytes, align, where align is the alignment size in bytes.

The assembler supports the use of some pseudo-instructions, for example, the la (load address) instruction in the figure. The text in the editor window is highlighted with different colors. The Assemble

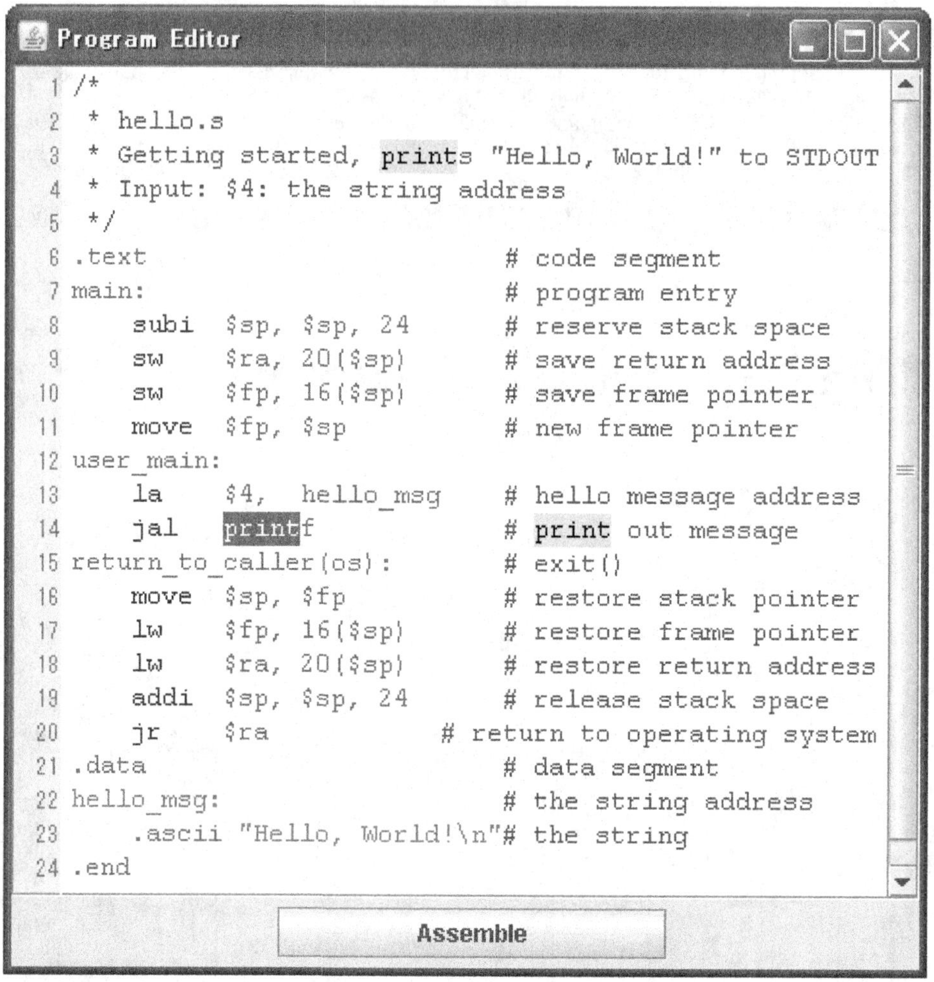

```
 Program Editor                                    [_][□][X]
 1 /*
 2  * hello.s
 3  * Getting started, prints "Hello, World!" to STDOUT
 4  * Input: $4: the string address
 5  */
 6 .text                          # code segment
 7 main:                          # program entry
 8      subi  $sp, $sp, 24        # reserve stack space
 9      sw    $ra, 20($sp)        # save return address
10      sw    $fp, 16($sp)        # save frame pointer
11      move  $fp, $sp            # new frame pointer
12 user_main:
13      la    $4,  hello_msg      # hello message address
14      jal   printf              # print out message
15 return_to_caller(os):         # exit()
16      move  $sp, $fp            # restore stack pointer
17      lw    $fp, 16($sp)        # restore frame pointer
18      lw    $ra, 20($sp)        # restore return address
19      addi  $sp, $sp, 24        # release stack space
20      jr    $ra              # return to operating system
21 .data                          # data segment
22 hello_msg:                     # the string address
23      .ascii "Hello, World!\n"# the string
24 .end
```

Assemble

Figure 4.9   AsmSim program editor window

button assembles the program, and then the assembled program appears in the main window. If there are grammar errors in the assembly program, the errors will be pointed out in the main window.

The graphics console window (Figure 4.10) has two parts: the upper part is a standard VGA (640 × 480 pixels) virtual window and the lower part is a console window. Because of space limitations, we have shown only the partial image of the window. The console window is used for inputting commands/data and outputting messages. The VGA window is used for showing image in a VRAM (video random access memory). The user program can write true-color pixels into the VRAM directly with sw instruction. We also prepared some function calls to draw graphics objects and character string, which will be described later in detail.

Once a program is loaded or the program in the edit window is assembled, AsmSim enters the debug mode. In the debug mode, we can input the following debug commands in the console: step, run, restart, ascii, and quit. The quit command lets AsmSim to quit from the debug mode and enter the command line mode in which the asmsim command can be executed. AsmSim is provided with some MIPS assembly program examples, as listed below.

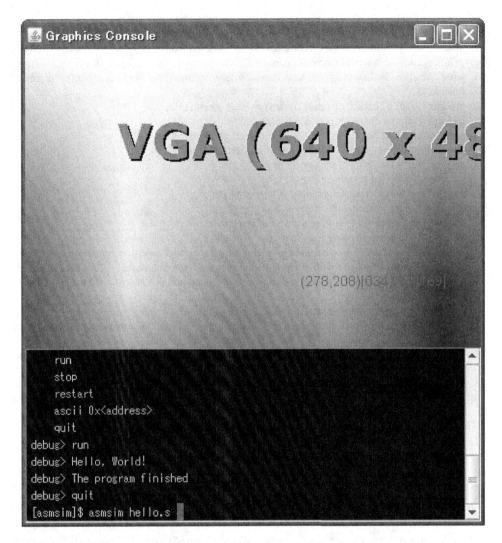

Figure 4.10 AsmSim graphics console window

```
[asmsim]$ asmsim
Usage: asmsim file
   ex., asmsim hello.s
      file:
         hello.s printf.s scanf.s getchar.s putchar.s sprintf.s getrandom.s
         gettimer.s getarrow.s getcal.s getcals.s linked_list.s sleep.s
         recursive.s malloc.s dfs.s bfs.s graphics.s key_event.s vram.s fonts.s
         kanji.s picture.s subroutine_call.s pointer.s root_nonrestoring.s
You can edit your own program in the Program Editor window, and click
Assembly button.
[asmsim]$ asmsim hello.s
```

The following describes briefly the basic operations of these examples.

1. **hello.s** displays hello message on console window as described before.
2. **printf.s** prints out a word in three formats: "%d" (decimal), "%x" (hexadecimal), and "%s" (string).
3. **scanf.s** reads a decimal integer, a hexadecimal integer, and a string from keyboard (need pressing Enter key).
4. **getchar.s** reads a character from keyboard (no need pressing Enter key).
5. **putchar.s** prints a character out on console window.
6. **sprintf.s** is similar to printf.s but it writes contents to a buffer.
7. **getrandom.s** gets a random number between 0 and 99.
8. **gettimer.s** gets the value of the system timer.
9. **getarrow.s** waits for pressing an arrow key.
10. **getcal.s** gets the calendar information in integer format.
11. **getcals.s** gets the calendar information in string format.
12. **linked_list.s** shows the data structure of the linked list.
13. **sleep.s** demonstrates sleeping for a moment.
14. **recursive.s** calculates Fibonacci numbers recursively. Generally, recursive programs are difficult to understand even if they are written in high-level programming languages, such as C and Java. By a single-step instruction execution, we can track the stack and see how the recursive program is implemented.
15. **malloc.s** allocates memory space.
16. **dfs.s** implements the depth-first search (DFS) algorithm. The graph is represented with the linked list.
17. **bfs.s** implements the breadth-first search (BFS) algorithm. The graph is represented with the linked list.
18. **graphics.s** draws graphics objects and string on graphics window.
19. **key_event.s** is a simple game program that demonstrates the keyboard interrupt. The game program uses four arrow keys to keep a randomly moved object to the center of graphics window.
20. **vram.s** writes true-color pixels into VRAM directly with sw instruction.
21. **fonts.s** shows ASCII fonts using a user-defined font table.
22. **kanji.s** displays some Japanese characters using a user-defined font table.
23. **picture.s** shows a bit-mapped picture image on graphics window.
24. **subroutine_call.s** shows how the function call in C is implemented.
25. **pointer.s** shows how the pointer in C is implemented.
26. **root_nonrestoring.s** implements a nonrestoring square root algorithm.

AsmSim supports a set of system function calls that can be invoked by the jal instruction. For example, the jal printf in the following program

```
.data                                    # data segment
print_dec:                               # string address
    .ascii "0x%x in decimal: %d\n"       # string
a_word:                                  # data address
    .word 0x476f6f64                     # data
.text                                    # code segment
    la    $4,   print_dec                # prepare to print
    la    $6,   a_word                   # address of data
    lw    $5,   0($6)                     # load the word
    move  $6,   $5                        # the same
    jal   printf                         # print out sum
.end
```

will print out the following message on the console:

```
debug> 0x476f6f64 in decimal: 1198485348
```

la (load address) is a pseudo-instruction that loads a 32-bit address to a register. It is translated into lui (loading upper 16 bits of the address) and ori (loading lower 16 bits of the address) instructions. We follow the gcc's agreements to arrange the registers for the arguments. In the example above, register $4 holds the address of the string. In the string, the two format directives %x and %d indicate to print data in hexadecimal format and in decimal format, respectively. The corresponding data are putted into registers $5 and $6. If we use %s (printing in string) in the string, it will print out "good," that is, the word 0x476f6f64 is explained as ASCIIs in the big endian format.

AsmSim supports several system calls that are commonly used in C or Java programs. We give some system calls and the register arrangements for the function parameters below.

1. **getchar()** gets a character from STDIN. The ASCII of the character will be in register $2 after the invocation.
2. **putchar()** puts a character to STDOUT. Register $4 holds the ASCII of the character which will be outputted.
3. **scanf()** reads an input from STDIN. Register $4 holds the address of a string which can be "%i," "%d," "%u," "%x," or "%s," indicating the format of the input data. Register $5 holds the memory address where the inputted value will be stored.
4. **printf()** writes output(s) to STDOUT. Register $4 holds the address of a string which may contain multiple directives of "%i," "%d," "%u," "%x," "%X," "%p," and/or "%s." The multiple output data corresponding to the directives are putted into registers $5, $6, $7, memory[reg[$sp]+16], memory[reg[$sp]+20], … , in sequence.
5. **sprintf()** writes output(s) to a buffer. Register $4 holds the address of the buffer. Register $5 holds the address of a string which may contain multiple directives of "%i," "%d," "%u," "%x," "%X," "%p," and/or "%s." The multiple output data corresponding to the directives are putted into registers $6, $7, memory[reg[$sp]+16], memory[reg[$sp]+20], … , in sequence.
6. **malloc()** allocates a block of memory. Register $4 holds the block size in bytes. The starting address of the allocated memory block will be in register $2 after the invocation.
7. **gettimer()** gets the system timer in ms (milliseconds). The timer value will be in register $2 after the invocation.
8. **getrandom()** gets a random integer number r with $0 \leq r < n$, where n is given in register $4. The random number will be in register $2 after the invocation.
9. **getarrow()** gets an arrow key's information. The information of the pressed arrow key will be in register $2 after the invocation: 0x8000 for the Up Arrow key; 0x8100 for the Left Arrow key; 0x8200 for the Down Arrow key; and 0x8300 for the Right Arrow key. We also defined 0xff00 for the Escape key.
10. **getcal()** gets the calendar in integers. Register $4 holds the starting address of an integer array where the month, date, year, day, hour, minute, and second will be stored in sequence.
11. **getcals()** gets the calendar in string. Register $4 holds the address of the string. The format of the string is "Month Date Year Day Hour:Minute:Second."
12. **sleep()** suspends execution for an interval of time. Register $4 holds the interval in milliseconds.
13. **key_event_ena()** enables keyboard interrupt. Pressing a key will generate an interrupt and cause a transfer of control to a predefined exception handler.
14. **key_event_dis()** disables keyboard interrupt.
15. **drawline()** draws a color line on an off-screen buffer. Register $4 holds the starting address of an integer buffer where the line color, $x1$, $y1$, $x2$, and $y2$ are stored in sequence. The line will be drawn from $(x1, y1)$ to $(x2, y2)$. The line color is defined by a 24-bit integer: 8 bits for each of red, green, and blue, from MSB to LSB.

16. **drawoval()** draws a color oval which fits into a specified rectangle on an off-screen buffer. Register $4 holds the starting address of an integer buffer where the line color, x, y, w, and h are stored in sequence. (x, y) is the up-left coordinate of the rectangle; w and h are the width and height of the rectangle, respectively. The color is specified by a 24-bit integer.

17. **drawrect()** draws a color rectangle on an off-screen buffer. Register $4 holds the starting address of an integer buffer where the line color, x, y, w, and h are stored in sequence. (x, y) is the up-left coordinate of the rectangle; w and h are the width and height of the rectangle, respectively. The color is specified by a 24-bit integer.

18. **drawrect3d()** is similar to drawrect() but the rectangle appears to be raised above the surface (3D). The parameters are the same as for drawrect().

19. **drawstring()** draws specified text at specified location on an off-screen buffer. Register $4 holds the starting address of a buffer where the text color, x, y, font information, and the text are stored in sequence. (x, y) is the coordinate where the text will be drawn; the font information is given by a 24-bit integer that specifies the font name (8 bits), font type (8 bits), and font size (8 bits). The supported font names in this simulator include "Times New Roman" (0), "Arial" (1), "Courier New" (2), and "Colonna MT" (3). The supported font types are "Plain" (0), "Bold" (1), and "Italic" (2).

20. **filloval()** fills a color oval on an off-screen buffer. The parameters are the same as for drawoval().

21. **fillrect()** fills a color rectangle on an off-screen buffer. The parameters are the same as for drawrect().

22. **fillrect3d()** fills a color 3D rectangle on an off-screen buffer. The parameters are the same as for drawrect3d().

23. **refresh_vga_auto()** enables the automatic VRAM display. In this mode, the VGA image is updated whenever there is any change in the off-screen buffer.

24. **refresh_vga_manu()** means refreshing the VGA manually. It inhibits the automatic VRAM display. The VGA image is updated only when the user program executes the system call "paint()."

25. **paint()** shows the image in the off-screen buffer on the VGA. The drawing of objects and filling of objects described above are done in the off-screen buffer.

Using key_event_ena(), gettimer(), getrandom(), sleep(), and graphics draw and fill function calls, we can develop some interesting games, Tetris and Othello for example. A code fragment listed below illustrates how to use the fillrect3d() and drawstring() system calls to fill a 3D rectangle and draw a string, respectively.

```
.text                               # code segment
main:                               # program entry
    subi  $sp, $sp, 64              # reserve stack space
    sw    $ra, 60($sp)              # save return address
    sw    $fp, 56($sp)              # save frame pointer
    move  $fp, $sp                  # new frame pointer
fill_back_ground:
    la    $4,  g_bgcolor            # back ground color
    jal   fillrect                  # fill rectangle
fill_3d_rect:
    la    $4,  g_fillrect3d         # fill a rect parameters
    jal   fillrect3d                # fill a rect
draw_my_canvas_string:
    la    $4,  g_drawstring         # draw the string parameters
    jal   drawstring                # draw the string
paint_on_canvas:
    jal   paint                     # paint off-screen buffer
return_to_caller(os):               # exit();
```

```
        move    $sp, $fp                          # restore stack pointer
        lw      $fp, 56($sp)                      # restore frame pointer
        lw      $ra, 60($sp)                      # restore return address
        addi    $sp, $sp,  64                     # release stack space
        jr      $ra                               # return to operating system
.data                                             # data segment
g_bgcolor:
        .word   0xffffff                          # color
        .word      0,   0                         # x, y
        .word   640, 480                          # width, height
g_drawstring:
        .word   0x000000                          # color
        .word   230, 180                          # x, y
        .word   0x020128                          # font name_type_size
        .ascii  "myCanvas"                        # string
g_fillrect3d:
        .word   0x007f00                          # color
        .word   350, 220                          # x, y
        .word    80, 50                           # width, height
.end
```

We implemented the interrupt mechanism in the AsmSim that performs functions such as addKeyListener() and removeKeyListener() in Java, for handling the keyboard interrupt. The key_event_ena() system call enables keyboard interrupt. Once the interrupt is enabled, pressing a key will cause a transfer of control to an exception handler. The system call of key_event_dis() disables keyboard interrupt.

Many applications, especially game programs, require the use of arrow keys. Because there are no ASCII codes for arrow keys, we defined 16-bit codes for four arrow keys and Escape key as described above. Once the interrupt is enabled, the 16-bit code of the pressed key is putted in register $2 (the ASCII of an ordinary key is putted in the lower byte and the upper byte is reset to zero). The following code fragment illustrates the structure of the interrupt handler.

```
.text                                             # code segment
__Key_Event:                                      # key event entry
        subi    $sp, $sp, 40                      # reserve stack space
        sw      $ra, 36($sp)                      # save return address
        sw      $fp, 32($sp)                      # save frame pointer
        sw      $3,  28($sp)                      # save $3
        sw      $4,  24($sp)                      # save $4
        sw      $5,  20($sp)                      # save $5
        move    $fp, $sp                          # new frame pointer
key_code:
        move    $5,  $2                           # key code in $2
        la      $4,  x                            # address of x_axis
test_escape:
        li      $3,  0xff00                       # escape?
        bne     $3,  $5, not_esc                  # no, check next
        sw      $3,  0($4)                        # 0xff00: outside x
        j       return                            # return
```

```
not_esc:
    li      $3,    0x8100                          # left arrow?
    bne     $3,    $5,  not_left
    ...
not_left:
    li      $3,    0x8300                          # right arrow?
    bne     $3,    $5,  not_right
    ...
    move    $sp, $fp                               # restore stack pointer
    lw      $5,    20($sp)                         # restore $5
    lw      $4,    24($sp)                         # restore $4
    lw      $3,    28($sp)                         # restore $3
    lw      $fp,   32($sp)                         # restore frame pointer
    lw      $ra,   36($sp)                         # restore return address
    addu    $sp, $sp, 40                           # release stack space
    eret                                           # return from exception
main:
    ...
    jal     key_event_ena                          # enable key event
    ...
    jr      $ra                                    # return to OS
.data                                              # data segment
    ...
.end
```

In the main program, the keyboard interrupt is enabled with the instruction `jal key_event_ena`. The interrupt handler entry is `__Key_Event`, which is a predefined label. Once entered into the handler, further interrupt is disabled automatically. In the beginning of the handler, some important registers are saved into the stack. Then we can determine which key is pressed by checking the value in register $2, and do something according to the pressed key. Before returning from the interrupt handler, the contents of the saved registers must be restored. The instruction for returning from interrupt is `eret` (exception return).

AsmSim provides a standard VGA window which has 640 pixels per horizontal line and 480 lines. In addition to the graphics draw and fill function calls, AsmSim allows the user program to write pixels to VRAM directly. The VRAM is organized as a one-dimensional array. Suppose we want to write a pixel at the (x, y) position of the VGA, where x and y are the horizontal and vertical coordinates, respectively; then we can calculate the address of the VRAM as $a = 640 \times y + x = (512 + 128) \times y + x = (y \ll 9) + (y \ll 7) + x$.

AsmSim allows the user program to define the starting address of the VRAM. The following code illustrates how to write a pixel to VRAM. The stating address of the VRAM is defined as 0xd0000000 (to 0xd007ffff). x is in $7, y is in $6, and the pixel is in $5. The pixel is written to VRAM with the `sw` instruction.

```
`define _Video_RAM 0xd0000000                      # - 0xd007ffff
.text                                               # code segment
    ...
    la      $4,    _Video_RAM                       # vram start address
    sll     $8,    $6,   9                          # y << 9
    add     $9,    $8,   $4                         # + base of vram
```

```
sll   $8,  $6,  7                          # y << 7
add   $8,  $8,  $7                         # + x
add   $9,  $8,  $9                         # vram address
sw    $5,  0($9)                           # write pixel
```

We can use lw and sw instructions to read from and write to the VRAM, respectively. The default mode is that, once a pixel is written into VRAM, it appears on the VGA immediately. But this mode will take a long time when a large number of pixels need to be written into the VRAM continuously. Therefore, we provide a system call, namely refresh_vga_manu(), to inhibit automatic display. Under this mode, the VGA image is updated only when the user program executes the system call paint(). Another system call, refresh_vga_auto(), resumes the default mode.

AsmSim also supports the user-defined font tables. For example, the following data define the 8×8 fonts of character "A."

```
ascii_font_table:
      .word 0x3c66667e, 0x66666600              # 'A', ASCII: 0x41
```

The VGA can display $640/8 \times 480/8 = 80 \times 60$ characters. The font table defines the shape of characters, and we can use different colors to show the background and foreground of the characters. The user program may maintain a character position (x, y) in VGA for $0 \le x < 80$ and $0 \le y < 60$, and calculate the VRAM addresses according to (x, y) to put the font pixels. Figure 4.11 shows the execution result of the fonts.s program, which displays the ASCII characters on graphics window.

Graph searching algorithms are subjects in the course of discrete systems, graph theory, or data structure and algorithms. There are many methods to represent a graph in computers, such as the incidence matrix, adjacency matrix, and linked list. The incidence matrix representation will lose the cycle edge information, and the adjacency matrix representation may lose the parallel edges information. AsmSim supports the data structure of the linked list. Figure 4.12 gives a graph example in which there is a cycle edge at node v_1.

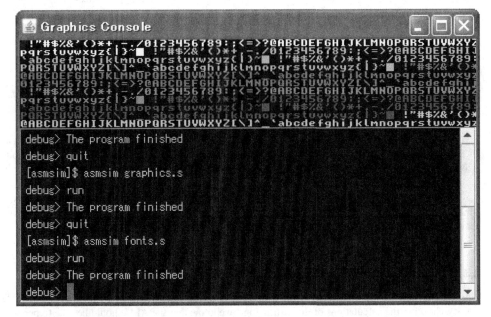

Figure 4.11   AsmSim graphics window output when executing fonts.s

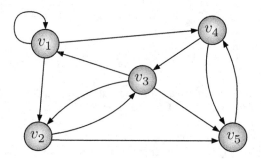

Figure 4.12   A graph that will be represented with linked lists

The graph can be represented with the following linked-list data structure. AsmSim supports the labels (pointers) in the fields of the word value.

```
'define NULL 0x0
.data
v1:   .word    1, p1                              # value, pointer
p1:   .word   v1, p2                              # value, pointer
p2:   .word   v2, p3                              # value, pointer
p3:   .word   v4, NULL                            # value, pointer
v2:   .word    2, p4                              # value, pointer
p4:   .word   v5, p5                              # value, pointer
p5:   .word   v3, NULL                            # value, pointer
v3:   .word    3, p6                              # value, pointer
p6:   .word   v1, p7                              # value, pointer
p7:   .word   v2, p8                              # value, pointer
p8:   .word   v5, NULL                            # value, pointer
v4:   .word    4, p9                              # value, pointer
p9:   .word   v5, p10                             # value, pointer
p10:  .word   v3, NULL                            # value, pointer
v5:   .word    5, p11                             # value, pointer
p11:  .word   v4, NULL                            # value, pointer
```

Then, we can use the assembly programming language to implement some search algorithms, such as DFS and BFS, which are generally implemented in C or Java. Suppose that the search starts from node v_1 of the graph shown in Figure 4.12. DFS algorithm visits the nodes of the graph in the order v_1, v_4, v_3, v_5, and v_2; and the BFS algorithm visits the nodes of the graph in the order v_1, v_4, v_2, v_3, and v_5.

With the editor window, you can write your own program, then assemble and execute it in the main window as well as the graphics console window. AsmSim can be executed entirely online with a web browser. There is also a jar (Java archive) version, which can be downloaded from Wiley's page.

4.4   ALU Design

This section describes how to design an ALU that performs the calculations of the instructions in Table 4.4. By checking the calculations of the instructions, we know that the ALU should be able to perform the following operations:

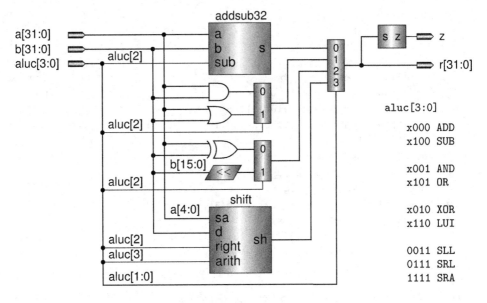

Figure 4.13   Schematic diagram of ALU

1. ADD (addition) for instructions of `add`, `addi`, `lw`, and `sw`;
2. SUB (subtraction) for instructions of `sub`, `beq`, and `bne`;
3. AND (bitwise logical and) for instructions of `and` and `andi`;
4. OR (bitwise logical or) for instructions of `or` and `ori`;
5. XOR (bitwise logical exclusive or) for instructions of `xor` and `xori`;
6. LUI (load upper immediate) for `lui` instruction;
7. SLL (shift left logical) for `sll` instruction;
8. SRL (shift right logical) for `srl` instruction;
9. SRA (shift right arithmetic) for `sra` instruction;

`jr`, `j`, and `jal` instructions do not require ALU to perform any operations. We have given almost all the components that can perform the above operations in the previous two chapters. The idea of designing ALU is to prepare these components and use a multiplexer to select an output from all the outputs of the components based on an arithmetic logic unit control (ALUC) signal which specifies what operation should be done. Figure 4.13 shows a possible circuit of ALU.

The inputs of ALU include two 32-bit data a and b, and a 4-bit aluc (ALUC). Three multiplexers are used: two 2-to-1 multiplexers and a 4-to-1 multiplexer. Their selection signals use some bit(s) of `aluc`. The component of `addsub32` performs ADD or SUB based on `aluc[2]`. The component of `shift` performs SLL, SRL, or SRA, based on `aluc[3:2]`. Three kinds of logic gates perform AND, OR, and XOR, respectively. LUI can be done by wiring low 16 bits to high 16 bits. The outputs of the ALU are a 32-bit r (result) and a 1-bit z (zero) flag. If the ALU result is 0, the zero flag will be a 1; otherwise it outputs a 0.

The Verilog HDL code implementing the ALU circuit shown in Figure 4.13 is given below. The corresponding operations of the `aluc` are listed in the right side of the code where an x denotes a don't care input.

```
module alu (a,b,aluc,r,z);      // 32-bit alu with a zero flag
    input   [31:0] a, b;        // inputs: a, b
    input   [3:0] aluc;         // input:  alu control: // aluc[3:0]:
```

```
output  [31:0] r;                       // output: alu result   // x 0 0 0  ADD
output         z;                       // output: zero flag     // x 1 0 0  SUB
wire    [31:0] d_and = a & b;                                    // x 0 0 1  AND
wire    [31:0] d_or  = a | b;                                    // x 1 0 1  OR
wire    [31:0] d_xor = a ^ b;                                    // x 0 1 0  XOR
wire    [31:0] d_lui = {b[15:0],16'h0};                          // x 1 1 0  LUI
wire    [31:0] d_and_or  = aluc[2]? d_or  : d_and;               // 0 0 1 1  SLL
wire    [31:0] d_xor_lui = aluc[2]? d_lui : d_xor;               // 0 1 1 1  SRL
wire    [31:0] d_as, d_sh;                                       // 1 1 1 1  SRA
// addsub32    (a,b,sub,    s);
addsub32 as32 (a,b,aluc[2],d_as);                               // add/sub
// shift       (d,sa,    right,   arith,   sh);
shift shifter (b,a[4:0],aluc[2],aluc[3],d_sh);                  // shift
// mux4x32     (a0,   a1,      a2,      a3,   s,         y);
mux4x32 res (d_as,d_and_or,d_xor_lui,d_sh,aluc[1:0],r);  // alu result
assign z = ~|r;                                                 // z = (r == 0)
endmodule
```

The modules of mux4x32 (a 32-bit 4-to-1 multiplexer) and shift were already given in Chapter 2. The following Verilog HDL code implements the module of addsub32 that performs an addition or a subtraction by invoking the module of cla32, a 32-bit carry-lookahead adder, which can be found in the Chapter 3.

```
module addsub32 (a,b,sub,s);              // 32-bit adder/subtracter
    input   [31:0] a, b;                  // inputs: a, b
    input          sub;                   // sub == 1: s = a - b
                                          // sub == 0: s = a + b
    output  [31:0] s;                     // output sum s
    // sub == 1: a - b = a + (-b) = a + not(b) + 1 = a + (b xor sub) + sub
    // sub == 0: a + b = a +    b  = a +    b      + 0 = a + (b xor sub) + sub
    wire    [31:0] b_xor_sub = b ^ {32{sub}};    // (b xor sub)
    // cla32    (a, b,        ci,   s);
    cla32 as32 (a, b_xor_sub, sub, s);           // b: (b xor sub); ci: sub
endmodule
```

The simulation waveform of the ALU is shown in Figure 4.14. In sequence, we simulated ADD (aluc = 0), SUB (aluc = 4), AND (aluc = 1), OR (aluc = 5), XOR (aluc = 2), LUI (aluc = 6), SLL (aluc = 3), SRL (aluc = 7), SRA (aluc = f), SUB, and ADD. The zero flag z outputs 1 when r is 0.

In addition to the 20 instructions listed in Table 4.4, the instructions of exception handling, TLB management, and calculations on floating-point numbers, as listed in Table 4.7, will be also implemented in our CPU designs. These instructions will be described in the following chapters.

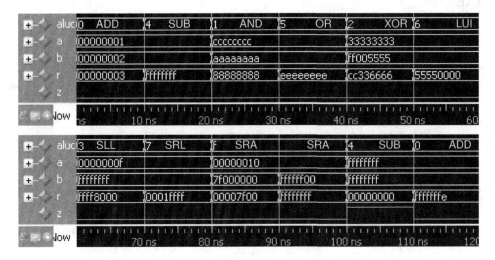

Figure 4.14   Waveform of ALU

Table 4.7   MIPS instructions related to interrupt/exception, TLB, and FP

Inst.	[31:26]	[25:21]	[20:16]	[15:11]	[10:6]	[5:0]	Meaning
syscall	000000	00000	00000	00000	00000	001100	System call
eret	010000	10000	00000	00000	00000	011000	Return from exception
tlbwi	010000	10000	00000	00000	00000	000010	Write indexed TLB entry
tlbwr	010000	10000	00000	00000	00000	000110	Write random TLB entry
mfc0	010000	00000	rt	rd	00000	000000	Load control word
mtc0	010000	00100	rt	rd	00000	000000	Store control word
lwc1	110001	rs	ft	offset			Load FP word
swc1	111001	rs	ft	offset			Store FP word
add.s	010001	10000	ft	fs	fd	000000	FP add
sub.s	010001	10000	ft	fs	fd	000001	FP subtract
mul.s	010001	10000	ft	fs	fd	000010	FP multiply
div.s	010001	10000	ft	fs	fd	000011	FP division
sqrt.s	010001	10000	00000	fs	fd	000100	FP square root

Exercises

4.1 Using only the instructions listed in Table 4.4, write code sequences to implement the following instructions.

```
bgez   rs,   label          # branch on greater than or equal to zero
bgtz   rs,   label          # branch on greater than zero
blez   rs,   label          # branch on less than or equal to zero
bltz   rs,   label          # branch on less than zero
```

4.2 Redesign the ALU in the behavioral-style Verilog HDL (you can use `casex` statement).

4.3 Add an output v (a flag of overflow) to ALU. When the result of addition or subtraction on the 2's complement numbers overflows, v outputs a 1.

4.4 Design an ALU that can perform a `sla` (shift left arithmetic) instruction (keeping the sign bit unchanged).

4.5 Try to understand the following MIPS assembly program, add comments after #, and execute it with AsmSim.

```
.data
$LC0:    .ascii  "%02d: q=%04x "
$LC1:    .ascii  "r=((r<<2)|((d>>(i+i))&3))-((q<<2)|1)=%08x "
$LC2:    .ascii  "r=((r<<2)|((d>>(i+i))&3))+((q<<2)|3)=%08x "
$LC3:    .ascii  "q=q*2+1=%04x\n"
$LC4:    .ascii  "q=q*2+0=%04x\n"
$LC5:    .ascii  "Input an unsigned integer in hex (i.e. c0000000): d = "
$LC6:    .ascii  "%x"
$LC7:    .ascii  "hex: q = %08x, r = %08x, d = q*q + r = %08x\n"
$LC8:    .ascii  "dec: q = %08d, r = %08d, d = q*q + r = %u\n"
.text
squart: subi  $sp,  $sp,   40          #
        sw    $19,  28($sp)            #
        move  $19,  $4                 #
        sw    $20,  32($sp)            #
        move  $20,  $5                 #
        sw    $17,  20($sp)            #
        move  $17,  $0                 #
        sw    $16,  16($sp)            #
        move  $16,  $0                 #
        sw    $18,  24($sp)            #
        li    $18,  0x0000000f         #
        sw    $31,  36($sp)            #
$L5:    la    $4,   $LC0               #
        move  $5,   $18                #
        move  $6,   $17                #
        jal   printf                   #
```

```
        bltz  $16,  $L6              #
        sll   $2,   $16,  2          #
        sll   $3,   $18,  1          #
        srl   $3,   $19,  $3         #
        andi  $3,   $3,   0x0003     #
        or    $2,   $2,   $3         #
        sll   $3,   $17,  2          #
        ori   $3,   $3,   0x0001     #
        subu  $16,  $2,   $3         #
        la    $4,   $LC1             #
        j     $L12                   #
$L6:    sll   $2,   $16,  2          #
        sll   $3,   $18,  1          #
        srl   $3,   $19,  $3         #
        andi  $3,   $3,   0x0003     #
        or    $2,   $2,   $3         #
        sll   $3,   $17,  2          #
        ori   $3,   $3,   0x0003     #
        addu  $16,  $2,   $3         #
        la    $4,   $LC2             #
$L12:   move  $5,   $16              #
        jal   printf                #
        bltz  $16,  $L8              #
        sll   $2,   $17,  1          #
        ori   $17,  $2,   0x0001     #
        la    $4,   $LC3             #
        j     $L13                   #
$L8:    sll   $17,  $17,  1          #
        la    $4,   $LC4             #
$L13:   move  $5,   $17              #
        jal   printf                #
        subi  $18,  $18,  1          #
        bgez  $18,  $L5              #
        bgez  $16,  $L11             #
        sll   $2,   $17,  1          #
        ori   $2,   $2,   0x0001     #
        addu  $16,  $16,  $2         #
$L11:   sw    $16,  0($20)           #
        move  $2,   $17              #
        lw    $31,  36($sp)          #
        lw    $20,  32($sp)          #
        lw    $19,  28($sp)          #
        lw    $18,  24($sp)          #
        lw    $17,  20($sp)          #
        lw    $16,  16($sp)          #
        addi  $sp,  $sp,  40         #
        jr    $31                    #
```

```
main:   subi  $sp,   $sp,   40        #
        sw    $31,   32($sp)          #
        sw    $17,   28($sp)          #
        sw    $16,   24($sp)          #
        la    $4,    $LC5             #
        jal   printf                 #
        la    $4,    $LC6             #
        addi  $5,    $sp,   16        #
        jal   scanf                  #
        lw    $4,    16($sp)          #
        addi  $5,    $sp,   20        #
        jal   squart                 #
        move  $16,   $2               #
        mult  $16,   $16              #
        mflo  $17                     #
        lw    $7,    20($sp)          #
        la    $4,    $LC7             #
        move  $5,    $16              #
        move  $6,    $7               #
        addu  $7,    $17,   $7        #
        jal   printf                 #
        lw    $7,    20($sp)          #
        la    $4,    $LC8             #
        move  $5,    $16              #
        move  $6,    $7               #
        addu  $7,    $17,   $7        #
        jal   printf                 #
        lw    $31,   32($sp)          #
        lw    $17,   28($sp)          #
        lw    $16,   24($sp)          #
        addi  $sp,   $sp,   40        #
        jr    $31                     #
    .end
```

4.6 Write an MIPS assembly program to show your photograph on VGA and execute it with AsmSim. You may write a program in C or Java to extract data words from the photograph file.

4.7 Design an ALU that can perform all the calculations of the MIPS integer instructions.

5

Single-Cycle CPU Design in Verilog HDL

Most computers in use today are synchronous computers. A synchronous computer is one in which all operations are controlled by a clock.

Referring to Figure 5.1, a clock is a periodic signal that oscillates instantaneously between 0 and 1. The transition from 0 to 1 is called a rising edge (or positive edge) and the transition from 1 to 0 is called a falling edge (or negative edge). A clock cycle is defined as the time from one rising edge to the next rising edge.

A single-cycle CPU (central processing unit) executes each instruction in one clock cycle. CPU states (program counter (PC) and registers) are updated at the rising edges of the clock. The cycle time is determined so that the most complex instruction can complete the execution correctly in one clock cycle. Compared to other CPUs, the single-cycle CPU is the most cost-effective but time-consuming CPU.

This chapter describes the design method of the single-cycle CPU and gives the Verilog HDL (hardware description language) code as well as the simulation waveforms.

5.1 The Circuits Required for Executing an Instruction

Instructions are executed by hardware circuits. This section describes what kinds of circuits are required for executing the instructions listed in Table 4.4.

5.1.1 The Circuits Required by Instruction Fetch

Instructions are stored in the memory. CPU must fetch instruction from the memory first. There is a PC in CPU that points to a memory location. The instruction in that location will be executed by the CPU. That is, CPU uses PC as the address to read an instruction from the memory, as shown in Figure 5.2.

If the fetched instruction is neither a branch instruction nor a jump instruction, the PC will be incremented by 4, pointing to the next instruction, because an instruction has 32 bits (4 bytes) while the PC is a byte address. The multiplexer (mux) in the figure is used to select the next PC, which will be written to the PC at the rising edge of the clock. The other inputs of the multiplexer will be described later.

5.1.2 The Circuits Required by Instruction Execution

After fetching an instruction, the CPU will execute it. The main components for executing an instruction include an arithmetic logic unit (ALU), a register file, and a control unit.

Computer Principles and Design in Verilog HDL, First Edition. Yamin Li.
© 2015 Tsinghua University Press. All rights reserved. Published 2015 by John Wiley & Sons Singapore Pte Ltd.
Companion Website: www.wiley.com/go/li/verilog

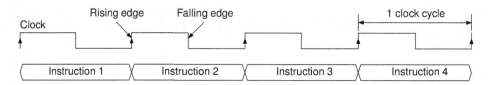

Figure 5.1 Clock and clock cycle

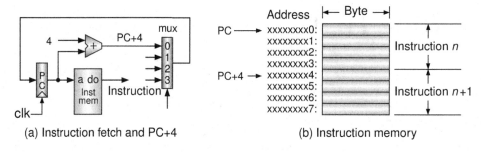

(a) Instruction fetch and PC+4   (b) Instruction memory

Figure 5.2 Block diagram of instruction fetch

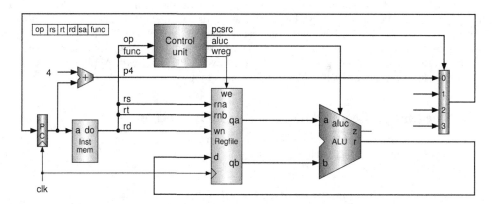

Figure 5.3 Block diagram for R-format arithmetic and logic instructions

5.1.2.1   The Circuits Required by R-Format Instructions

Figure 5.3 shows the circuits required for executing the following instructions:

```
add rd, rs, rt          # rd <-- rs ADD rt
sub rd, rs, rt          # rd <-- rs SUB rt
and rd, rs, rt          # rd <-- rs AND rt
or  rd, rs, rt          # rd <-- rs OR  rt
xor rd, rs, rt          # rd <-- rs XOR rt
```

Two source operands are read from qa and qb of the register file based on rs and rt, respectively. These two operands are sent to the inputs of the ALU for calculation. The difference between these five instructions is only on the ALU's operations. A control unit generates the control signals. pcsrc

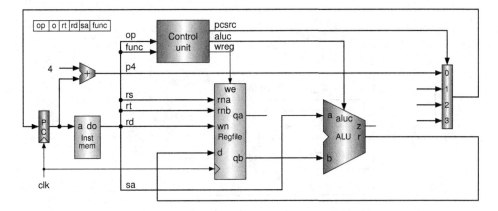

Figure 5.4   Block diagram for R-format shift instructions

(PC source) is the selection signal of the multiplexer. `aluc` (ALU control) controls the operations of ALU. `wreg` (write register) is a write enable signal. If `wreg` is a 1, the result calculated by the ALU will be written to the register `rd` of the register file.

Figure 5.4 shows the circuits required for executing the following instructions:

```
sll rd, rt, sa          # rd <-- rt SLL sa
srl rd, rt, sa          # rd <-- rt SRL sa
sra rd, rt, sa          # rd <-- rt SRA sa
```

Different from Figure 5.3, the input a of the ALU in Figure 5.4 is the 5-bit `sa` (shift amount). The high 27 bits of ALU's input a can be any value which will be ignored by the shift operations. The `rs` field is not used. We will use a multiplexer to select `sa` or `qa` of the register file based on the instruction.

5.1.2.2   The Circuits Required by I-Format Instructions

Figure 5.5 shows the circuits required for executing the following instructions:

```
addi rt, rs, immediate  # rt <-- rs ADD (sign)immediate
andi rt, rs, immediate  # rt <-- rs AND (zero)immediate
ori  rt, rs, immediate  # rt <-- rs OR  (zero)immediate
xori rt, rs, immediate  # rt <-- rs XOR (zero)immediate
lui  rt,     immediate  # rt <--     LUI        immediate
```

These five instructions use immediate as the input b of ALU. The 16-bit `immediate` must be extended to 32 bits. Similarly, a multiplexer will be inserted in front of the ALU's input b. `rt` is the destination register number.

5.1.2.3   The Circuits Required by Load/Store Instructions

Figure 5.6 shows the circuits required for executing the following instructions:

```
lw rt, offset(rs)       # rt <-- memory[rs + offset]
sw rt, offset(rs)       # memory[rs + offset] <-- rt
```

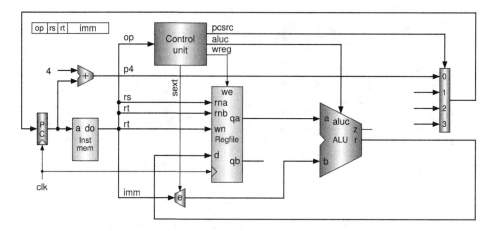

Figure 5.5   Block diagram for I-format arithmetic and logic instructions

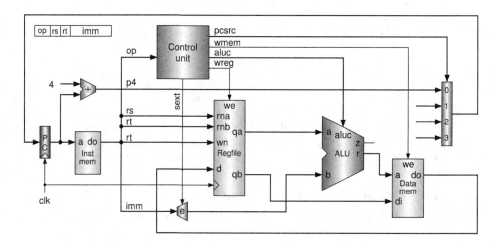

Figure 5.6   Block diagram for I-format load and store instructions

These two instructions access the data memory (Data mem component in the figure). The memory address is calculated by the ALU in the same way as the addi instruction does. For the lw instruction, the data word read from the data memory is written to the rt register of the register file. The sw instruction writes the data held in the rt register to the data memory. wmem (write memory), one of the control signals generated by the control unit, is the memory write enable signal.

5.1.2.4   The Circuits Required by Conditional Branch Instructions

Figure 5.7 shows the circuits required for executing the following instructions:

```
beq rs, rt, label        # if (rs == rt) PC <-- label
bne rs, rt, label        # if (rs != rt) PC <-- label
```

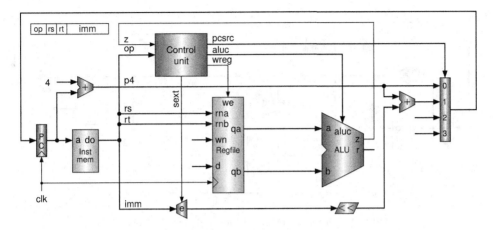

Figure 5.7 Block diagram for I-format branch instructions

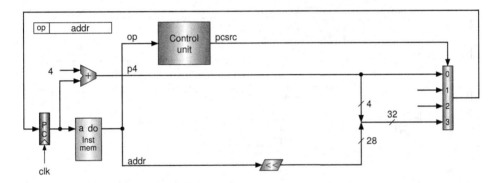

Figure 5.8 Block diagram for J-format jump instruction

The two operands in registers rs and rt are compared by the ALU. The result of the comparison z is sent to the control unit to determine whether to branch or not. If the branch is taken, the 2-bit pcsrc signal will be 01 to select the branch target address. An additional 32-bit adder is used to calculate the branch target address, which is the sum of the PC+4 (p4 in the figure) and the 2-bit left-shifted sign-extended immediate. These two instructions do not write the register file (wreg = 0).

5.1.2.5 The Circuits Required by Jump Instructions

Figure 5.8 shows the circuits required for executing the jump instruction:

```
    j target                    # PC <-- target
```

This instruction uses neither the ALU nor the register file. The jump target address is obtained by the concatenation of the high 4 bits of PC+4 and 2-bit left-shifted addr. pcsrc will be 11 to select the jump target address.

Figure 5.9 shows the circuits required for executing the jal instruction:

```
    jal target                  # $31 <-- PC + 4; PC <-- target
```

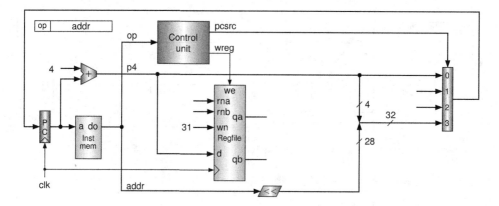

Figure 5.9   Block diagram for J-format jump-and-link instruction

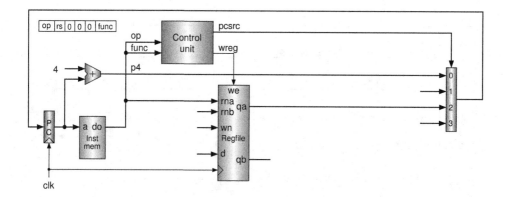

Figure 5.10   Block diagram for R-format jump-register instruction

In addition to what the `j` instruction does, the `jal` instruction saves the return address to register `$31` of the register file. The number 31 does not appear in the instruction; we must create it by hardwire. We used PC+4 as the return address in our single-cycle CPU design. MIPS ISA defines it as PC+8. We will use PC+8 as the return address in the design of a pipelined CPU in Chapter 8.

Figure 5.10 shows the circuits required for executing the `jr` instruction:

```
jr rs  # PC <-- rs
```

The jump target address is obtained from register `rs` of the register file. `pcsrc` will be 10 to select the jump target address.

The circuits required by executing the 20 instructions have been given separately. We will describe how to combine these circuits together to form a CPU. Before that, we describe how to design the register file.

5.2   Register File Design

The register file contains 32 general-purpose registers. Figure 5.11 shows the symbol of the register file that will be used in our CPU design. It has two read ports (port A and port B) and a write port. Each port

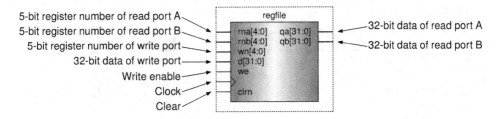

Figure 5.11 Symbol of the register file

has a 5-bit address input (register number). A 32-bit data input and a write enable are provided for the write port.

This section first gives the schematic circuit of the register file and its Verilog HDL code. Then a behavioral-style Verilog HDL code is given which also implements the register file but is short.

5.2.1 Register File Schematic Circuit

A register in the register file has 32 bits. It can be designed with 32 D flip-flops, as shown in Figure 5.12. We name it dffe32. The input e is the write enable.

The register file has 32 registers but the register $0 always contains a constant 0. Thus we can use 31 dffe32s to implement registers $1–$31, and use a gnd (ground) for register $0, as shown in Figure 5.13. We name it reg32. It has thirty-two 32-bit outputs: there are 32 registers, and each register has a 32-bit output.

The final schematic circuit of the register file is shown in Figure 5.14. reg32 is the storage body. It has thirty-two 32-bit outputs. Each read port uses a 32-bit 32-to-1 multiplexer (mux32x32) to select one register output according to its 5-bit read register number. The two read ports are independent of each other.

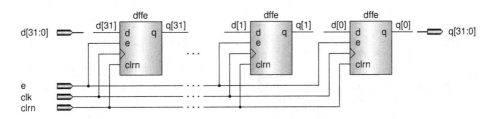

Figure 5.12 Block diagram of 32-bit D flip-flops with enable control

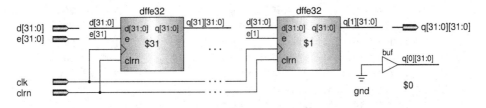

Figure 5.13 Block diagram of thirty-two 32-bit registers

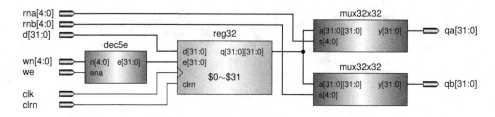

Figure 5.14   Block diagram of the register file

A 5-32 decoder (dec5e) generates at most one output that is a 1 among 32 outputs according to the 5-bit write register number. Each output of the decoder is connected to the enable input of the corresponding register so that at most one register will be written with the 32-bit input data at the rising edge of the clock. If the we signal is 0, no register is updated.

5.2.2 Register File Design in Dataflow-Style Verilog HDL

The following Verilog HDL code implements the schematic circuit shown in Figure 5.14. It invokes the dec5e (5-32 decoder with enable control) module and the dffe32 (32-bit register with enable control) module, which will be given next.

```
module regfile_dataflow (rna,rnb,d,wn,we,clk,clrn,qa,qb);
    input    [4:0] rna,rnb,wn;                         // reg numbers
    input    [31:0] d;                                 // write data
    input          we;                                 // write enable
    input          clk, clrn;                          // clock, reset
    output   [31:0] qa,qb;                             // read ports
    wire     [31:0] e;                                 // enables
    wire     [31:0] r00,r01,r02,r03,r04,r05,r06,r07;
    wire     [31:0] r08,r09,r10,r11,r12,r13,r14,r15;
    wire     [31:0] r16,r17,r18,r19,r20,r21,r22,r23;
    wire     [31:0] r24,r25,r26,r27,r28,r29,r30,r31;
    dec5e decoder (wn,we,e);                           // wn decoder
    assign        r00 = 0;                             // $0
    dffe32 reg01  (d,clk,clrn,e[01],r01);              // $1
    dffe32 reg02  (d,clk,clrn,e[02],r02);              // $2
    dffe32 reg03  (d,clk,clrn,e[03],r03);              // $3
    dffe32 reg04  (d,clk,clrn,e[04],r04);              // $4
    dffe32 reg05  (d,clk,clrn,e[05],r05);              // $5
    dffe32 reg06  (d,clk,clrn,e[06],r06);              // $6
    dffe32 reg07  (d,clk,clrn,e[07],r07);              // $7
    dffe32 reg08  (d,clk,clrn,e[08],r08);              // $8
    dffe32 reg09  (d,clk,clrn,e[09],r09);              // $9
    dffe32 reg10  (d,clk,clrn,e[10],r10);              // $10
    dffe32 reg11  (d,clk,clrn,e[11],r11);              // $11
    dffe32 reg12  (d,clk,clrn,e[12],r12);              // $12
    dffe32 reg13  (d,clk,clrn,e[13],r13);              // $13
```

```
dffe32 reg14   (d,clk,clrn,e[14],r14);                    // $14
dffe32 reg15   (d,clk,clrn,e[15],r15);                    // $15
dffe32 reg16   (d,clk,clrn,e[16],r16);                    // $16
dffe32 reg17   (d,clk,clrn,e[17],r17);                    // $17
dffe32 reg18   (d,clk,clrn,e[18],r18);                    // $18
dffe32 reg19   (d,clk,clrn,e[19],r19);                    // $19
dffe32 reg20   (d,clk,clrn,e[20],r20);                    // $20
dffe32 reg21   (d,clk,clrn,e[21],r21);                    // $21
dffe32 reg22   (d,clk,clrn,e[22],r22);                    // $22
dffe32 reg23   (d,clk,clrn,e[23],r23);                    // $23
dffe32 reg24   (d,clk,clrn,e[24],r24);                    // $24
dffe32 reg25   (d,clk,clrn,e[25],r25);                    // $25
dffe32 reg26   (d,clk,clrn,e[26],r26);                    // $26
dffe32 reg27   (d,clk,clrn,e[27],r27);                    // $27
dffe32 reg28   (d,clk,clrn,e[28],r28);                    // $28
dffe32 reg29   (d,clk,clrn,e[29],r29);                    // $29
dffe32 reg30   (d,clk,clrn,e[30],r30);                    // $30
dffe32 reg31   (d,clk,clrn,e[31],r31);                    // $31
assign qa = select(r00,r01,r02,r03,r04,r05,r06,r07,      // read port a
                   r08,r09,r10,r11,r12,r13,r14,r15,
                   r16,r17,r18,r19,r20,r21,r22,r23,
                   r24,r25,r26,r27,r28,r29,r30,r31,rna);
assign qb = select(r00,r01,r02,r03,r04,r05,r06,r07,      // read port b
                   r08,r09,r10,r11,r12,r13,r14,r15,
                   r16,r17,r18,r19,r20,r21,r22,r23,
                   r24,r25,r26,r27,r28,r29,r30,r31,rnb);
function [31:0] select;
    input [31:0] r00,r01,r02,r03,r04,r05,r06,r07,
                 r08,r09,r10,r11,r12,r13,r14,r15,
                 r16,r17,r18,r19,r20,r21,r22,r23,
                 r24,r25,r26,r27,r28,r29,r30,r31;
    input [4:0]  s;                                       // reg number
    case (s)
        5'd00: select = r00;        5'd01: select = r01;
        5'd02: select = r02;        5'd03: select = r03;
        5'd04: select = r04;        5'd05: select = r05;
        5'd06: select = r06;        5'd07: select = r07;
        5'd08: select = r08;        5'd09: select = r09;
        5'd10: select = r10;        5'd11: select = r11;
        5'd12: select = r12;        5'd13: select = r13;
        5'd14: select = r14;        5'd15: select = r15;
        5'd16: select = r16;        5'd17: select = r17;
        5'd18: select = r18;        5'd19: select = r19;
        5'd20: select = r20;        5'd21: select = r21;
        5'd22: select = r22;        5'd23: select = r23;
        5'd24: select = r24;        5'd25: select = r25;
        5'd26: select = r26;        5'd27: select = r27;
```

```
                5'd28: select = r28;              5'd29: select = r29;
                5'd30: select = r30;              5'd31: select = r31;
            endcase
        endfunction
    endmodule
```

The following Verilog HDL code implements the 5-32 decoder with enable control.

```
module dec5e (n,ena,e);                 // 5-32 decoder with an enable
    input    [4:0] n;                   // 5-bit number
    input          ena;                 // master enable
    output [31:0] e;                    // 32-bit enables
    assign         e = ena? decoder(n)  : 32'h00000000;
    function [31:0] decoder;
        input [4:0] n;
        case (n)
            5'd00: decoder=32'h00000001; // 00000000000000000000000000000001
            5'd01: decoder=32'h00000002; // 00000000000000000000000000000010
            5'd02: decoder=32'h00000004; // 00000000000000000000000000000100
            5'd03: decoder=32'h00000008; // 00000000000000000000000000001000
            5'd04: decoder=32'h00000010; // 00000000000000000000000000010000
            5'd05: decoder=32'h00000020; // 00000000000000000000000000100000
            5'd06: decoder=32'h00000040; // 00000000000000000000000001000000
            5'd07: decoder=32'h00000080; // 00000000000000000000000010000000
            5'd08: decoder=32'h00000100; // 00000000000000000000000100000000
            5'd09: decoder=32'h00000200; // 00000000000000000000001000000000
            5'd10: decoder=32'h00000400; // 00000000000000000000010000000000
            5'd11: decoder=32'h00000800; // 00000000000000000000100000000000
            5'd12: decoder=32'h00001000; // 00000000000000000001000000000000
            5'd13: decoder=32'h00002000; // 00000000000000000010000000000000
            5'd14: decoder=32'h00004000; // 00000000000000000100000000000000
            5'd15: decoder=32'h00008000; // 00000000000000001000000000000000
            5'd16: decoder=32'h00010000; // 00000000000000010000000000000000
            5'd17: decoder=32'h00020000; // 00000000000000100000000000000000
            5'd18: decoder=32'h00040000; // 00000000000001000000000000000000
            5'd19: decoder=32'h00080000; // 00000000000010000000000000000000
            5'd20: decoder=32'h00100000; // 00000000000100000000000000000000
            5'd21: decoder=32'h00200000; // 00000000001000000000000000000000
            5'd22: decoder=32'h00400000; // 00000000010000000000000000000000
            5'd23: decoder=32'h00800000; // 00000000100000000000000000000000
            5'd24: decoder=32'h01000000; // 00000001000000000000000000000000
            5'd25: decoder=32'h02000000; // 00000010000000000000000000000000
            5'd26: decoder=32'h04000000; // 00000100000000000000000000000000
            5'd27: decoder=32'h08000000; // 00001000000000000000000000000000
            5'd28: decoder=32'h10000000; // 00010000000000000000000000000000
            5'd29: decoder=32'h20000000; // 00100000000000000000000000000000
            5'd30: decoder=32'h40000000; // 01000000000000000000000000000000
```

```
            5'd31: decoder=32'h80000000; // 10000000000000000000000000000000
        endcase
    endfunction
endmodule
```

The following Verilog HDL code implements the 32-bit register with enable control.

```
module dffe32 (d,clk,clrn,e,q);              // a 32-bit register
    input       [31:0] d;                    // input d
    input            e;                      // e: enable
    input              clk, clrn;            // clock and reset
    output reg [31:0] q;                     // output q
    always @(negedge clrn or posedge clk)
        if (!clrn)   q <= 0;                 // q = 0 if reset
        else if (e) q <= d;                  // save d if enabled
endmodule
```

5.2.3   Register File Design in Behavioral-Style Verilog HDL

The code given in the previous section is too "hard"; it is almost the same as the hardware circuit shown in Figure 5.14. In this section, we show that the register file can be designed with Verilog HDL in a "soft" way.

The Verilog HDL supports a two-dimensional array (matrix) statement. For example, the statement reg [31:0] register [0:31] defines a 32 × 32-bit storage: [31:0] defines a register that has 32 bits and [0:31] declares that there are 32 such registers. Then we can use an index to read or write a 32-bit register. A behavioral-style Verilog HDL code that implements the register file is listed below.

```
module regfile (rna,rnb,d,wn,we,clk,clrn,qa,qb);   // 32x32 regfile
    input   [31:0] d;                              // data of write port
    input   [4:0] rna;                             // reg # of read port A
    input   [4:0] rnb;                             // reg # of read port B
    input   [4:0] wn;                              // reg # of write port
    input        we;                               // write enable
    input          clk, clrn;                      // clock and reset
    output [31:0] qa, qb;                          // read ports A and B
    reg    [31:0] register [1:31];                 // 31 32-bit registers
    assign qa = (rna == 0)? 0 : register[rna];     // read port A
    assign qb = (rnb == 0)? 0 : register[rnb];     // read port B
    integer i;
    always @(posedge clk or negedge clrn)          // write port
        if (!clrn)
            for (i = 1; i < 32; i = i + 1)
                register[i]   <= 0;                // reset
        else
            if ((wn != 0) && we)                   // not reg[0] & enabled
                register[wn] <= d;                 // write d to reg[wn]
endmodule
```

This register file has one write port. In Chapter 10, we will give the Verilog HDL code for implementing a register file with two write ports.

5.3   Single-Cycle CPU Datapath Design

A CPU consists of a datapath and a control unit. A datapath is a collection of functional units, such as the ALU, register file, and multiplexers, that perform data processing operations. A control unit is a component that manages the operations of the datapath. This section introduces the design method of the datapath of the single-cycle CPU.

5.3.1   Using Multiplexers

If there are two or more signals that will be connected to the same input of a component, a multiplexer can be used for selecting a signal based on the instruction.

5.3.1.1   Selection for Next PC (Selection Signal: pcsrc)

The PC will be updated on the rising edge of the clock in the single-cycle CPU. If the current instruction is neither a branch nor a jump, PC+4 will be written to the PC. But the following five instructions may transfer control to a branch or jump target address that is not the PC+4.

```
beq/bne  rs, rt, label  # if (eq/ne) pc <-- label     op  rs  rt      offset

jr       rs             # pc <-- rs                    op  rs  0   0   0  func

j/jal    address        # pc <-- address << 2          op         address
```

A 4-to-1 multiplexer can be used to select an address for the next PC as shown in Figure 5.15. The input 0 of the multiplexer is PC+4 (neither branch nor jump); input 1 is the branch target address; input 2 is the jump target address coming from the register rs of the register file; and input 3 is the jump target address coming from addr and PC+4. The 2-bit pcsrc (PC source) is the selection signal of the multiplexer.

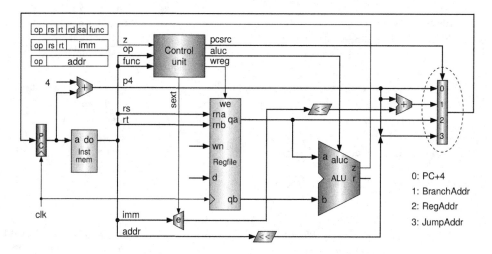

Figure 5.15   Using multiplexer for the next PC selection

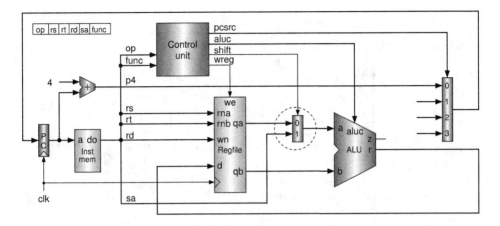

Figure 5.16   Using a multiplexer for ALU A-input selection

5.3.1.2   Selection for ALU Input A (Selection Signal: shift)

Shift instructions use sa as the shift amount that will be sent to the input a of the ALU for the operation. Other instructions may use the value in the register rs of the register file; see the following two examples.

add rd, rs, rt	# rd <-- rs + rt

op	rs	rt	rd	0	func

sll rd, rt, sa	# rd <-- rt << sa

op	0	rt	rd	sa	func

sa or the value of the register rs can be selected by a 2-to-1 multiplexer, as shown in Figure 5.16. The 1-bit shift is the selection signal of the multiplexer. If shift is a 1, sa is selected. Otherwise, qa of the register file is selected.

5.3.1.3   Selection for ALU Input B (Selection Signals: aluimm and regrt)

The immediate (imm) of the I-format computational instructions will be sent to the input b of the ALU for the calculation. Other instructions may use the value in the register rt of the register file; see the following two examples.

add rd, rs, rt	# rd <-- rs + rt

op	rs	rt	rd	0	func

addi rt, rs, imm	# rt <-- rs + imm

op	rs	rt	imm

In addition to the input b of the ALU, the destination register number also needs to be selected from rd and rt. Figure 5.17 shows these selections by two 2-to-1 multiplexers. The multiplexer in the front of the ALU input b has 32 bits and its selection signal is aluimm. If aluimm is a 1, the extended immediate is selected. Otherwise, qb of the register file is selected. The multiplexer in front of the register file's input wn has 5 bits, and its selection signal is regrt. If regrt is a 1, rt is selected. Otherwise, rd is selected.

5.3.1.4   Selection for Register File Inputs (Selection Signals: m2reg and jal)

The data that will be written to the register file may be the output of the ALU, the data in the data memory, or the return address (for jal instruction). The destination register number may be rd, rt, or a constant 31 (for jal instruction).

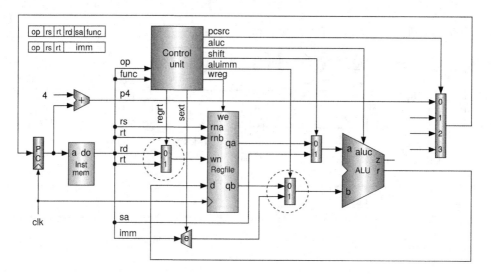

Figure 5.17   Using a multiplexer for ALU B-input selection

The load (`lw`) instruction writes the data read from the memory to the register `rt` of the register file. Other instructions may write the output of the ALU to the register `rd` or `rt` of the register file. The subroutine call `jal` instruction saves the return address to register `$31`. See the following three examples.

```
add  rd, rs, rt      # rd <-- rs + rt
```
op	rs	rt	rd	0	func

```
lw   rt, imm(rs)      # rt <-- mem[rs+imm]
```
op	rs	rt	imm

```
jal address # $31 <-- pc + 4; pc <-- address << 2
```
op	address

Two 32-bit 2-to-1 multiplexers are used for selecting the data that will be written to the register file, as shown in Figure 5.18. The selection signal of the multiplexer on the right side of the figure is `m2reg`. If `m2reg` is a 1, the data read from the memory is selected. The selection signal of the multiplexer on the left side is `jal`. If `jal` is a 1, the return address `p4` is selected.

For the `jal` instruction, because the destination register number 31 does not appear in the instruction, we must generate it by hardware (the component `f` in the figure). The following Verilog HDL code is a hardware implementation for generating the 5-bit destination register number:

```
assign wn = reg_dest | {5{jal}};
```

It performs a bitwise logical OR operation. If `jal` is a 1, a 5-bit pattern 11111 (decimal 31) will be assigned to the destination register number `wn`. Otherwise, `reg_dest`, which is `rd` or `rt`, will be assigned to `wn`.

5.3.2   Single-Cycle CPU Schematic Circuit

By summarizing the discussions in the previous section, we got the schematic circuit of the single-cycle CPU plus the instruction memory and the data memory, as shown in Figure 5.19, which can execute all the instructions listed in Table 4.4.

Putting the memory modules outside, Figure 5.20 shows the single-cycle computer that consists of a single-cycle CPU and two memory modules.

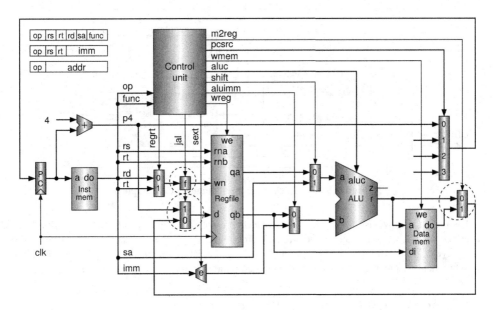

Figure 5.18   Using a multiplexer for register file written data selection

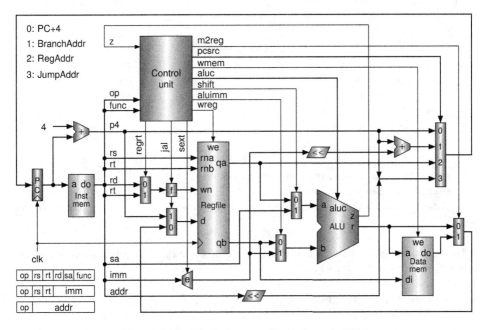

Figure 5.19   Block diagram of a single-cycle CPU

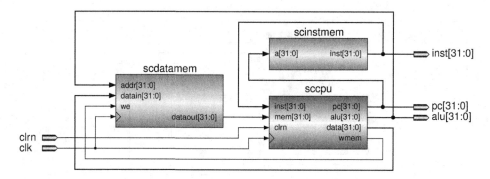

Figure 5.20   Block diagram of single-cycle computer

   The reason why we use separate instruction memory and data memory is that the single-cycle CPU completes the execution of an instruction, including instruction fetch and data memory access, in one clock cycle. We can implement an instruction cache and a data cache inside the CPU so that only one off-chip memory module can be used for storing both the program (instructions) and data.

5.3.3   Single-Cycle CPU Verilog HDL Codes

The following Verilog HDL code implements the single-cycle computer shown in Figure 5.20. It invokes sccpu (single-cycle CPU), scinstmem (instruction memory), and scdatamem (data memory). The first module is given next, and the last two modules will be given later.

```
module sccomp (clk,clrn,inst,pc,aluout,memout);    // single cycle computer
    input           clk, clrn;                     // clock and reset
    output [31:0] pc;                              // program counter
    output [31:0] inst;                            // instruction
    output [31:0] aluout;                          // alu output
    output [31:0] memout;                          // data memory output
    wire   [31:0] data;                           // data to data memory
    wire          wmem;                           // write data memory
    sccpu cpu (clk,clrn,inst,memout,pc,wmem,aluout,data);   // cpu
    scinstmem imem (pc,inst);                      // inst memory
    scdatamem dmem (clk,memout,data,aluout,wmem);  // data memory
endmodule
```

   The following Verilog HDL code that implements a single-cycle CPU is almost identical to the schematic circuit shown in Figure 5.19.

```
module sccpu (clk,clrn,inst,mem,pc,wmem,alu,data);
    input  [31:0] inst;                           // inst from inst memory
    input  [31:0] mem;                            // data from data memory
    input         clk, clrn;                      // clock and reset
    output [31:0] pc;                             // program counter
    output [31:0] alu;                            // alu output
    output [31:0] data;                           // data to data memory
    output        wmem;                           // write data memory
```

```
// instruction fields
wire    [5:0] op   = inst[31:26];          // op
wire    [4:0] rs   = inst[25:21];          // rs
wire    [4:0] rt   = inst[20:16];          // rt
wire    [4:0] rd   = inst[15:11];          // rd
wire    [5:0] func = inst[05:00];          // func
wire   [15:0] imm  = inst[15:00];          // immediate
wire   [25:0] addr = inst[25:00];          // address
// control signals
wire    [3:0] aluc;                        // alu operation control
wire    [1:0] pcsrc;                       // select pc source
wire          wreg;                        // write regfile
wire          regrt;                       // dest reg number is rt
wire          m2reg;                       // instruction is an lw
wire          shift;                       // instruction is a shift
wire          aluimm;                      // alu input b is an i32
wire          jal;                         // instruction is a jal
wire          sext;                        // is sign extension
// datapath wires
wire   [31:0] p4;                          // pc+4
wire   [31:0] bpc;                         // branch target address
wire   [31:0] npc;                         // next pc
wire   [31:0] qa;                          // regfile output port a
wire   [31:0] qb;                          // regfile output port b
wire   [31:0] alua;                        // alu input a
wire   [31:0] alub;                        // alu input b
wire   [31:0] wd;                          // regfile write port data
wire   [31:0] r;                           // alu out or mem
wire   [31:0] sa  = {27'b0,inst[10:6]};    // shift amount
wire   [15:0] s16 = {16{sext & inst[15]}}; // 16-bit signs
wire   [31:0] i32 = {s16,imm};             // 32-bit immediate
wire   [31:0] dis = {s16[13:0],imm,2'b00}; // word distance
wire   [31:0] jpc = {p4[31:28],addr,2'b00};// jump target address
wire    [4:0] reg_dest;                    // rs or rt
wire    [4:0] wn  = reg_dest | {5{jal}};   // regfile write reg #
wire          z;                           // alu, zero tag
// control unit
sccu_dataflow cu (op,func,z,wmem,wreg,     // control unit
                  regrt,m2reg,aluc,shift,
                  aluimm,pcsrc,jal,sext);
// datapath
dff32 i_point (npc,clk,clrn,pc);           // pc register
cla32 pcplus4 (pc,32'h4,1'b0,p4);          // pc + 4
cla32 br_addr (p4,dis,1'b0,bpc);           // branch target address
mux2x32 alu_a (qa,sa,shift,alua);          // alu input a
mux2x32 alu_b (qb,i32,aluimm,alub);        // alu input b
mux2x32 alu_m (alu,mem,m2reg,r);           // alu out or mem
```

```
    mux2x32 link   (r,p4,jal,wd);                    // r or p4
    mux2x5 reg_wn (rd,rt,regrt,reg_dest);            // rs or rt
    mux4x32 nextpc(p4,bpc,qa,jpc,pcsrc,npc);         // next pc
    regfile rf (rs,rt,wd,wn,wreg,clk,clrn,qa,qb);    // register file
    alu alunit (alua,alub,aluc,alu,z);               // alu
    assign data = qb;                                // regfile output port b
endmodule
```

The Verilog HDL code of the module `sccu_dataflow` (control unit) invoked by `sccpu_dataflow` (CPU) will be given in the following section.

5.4   Single-Cycle CPU Control Unit Design

As mentioned in the previous section, a CPU consists of a datapath and a control unit. A control unit is a component that directs the operations of the datapath. This section introduces the control unit design of the single-cycle CPU.

5.4.1   Logic Design of the Control Unit

A control unit generates control signals based on the instruction that is currently being executed. The first step of designing the control unit is to decode the instruction based on the instruction's opcode (op), as listed in Table 5.1. The function code (func) needs to be checked also if the instruction has an R-format. In order to avoid conflict with the keywords of Verilog HDL, an `i_` is prefixed to the name of each instruction.

From the table, we can get each instruction decode expression as shown below. A temporary wire, `rtype`, is used for decoding all the R-format instructions.

$$\text{rtype} = \overline{\text{op}[5]}\ \overline{\text{op}[4]}\ \overline{\text{op}[3]}\ \overline{\text{op}[2]}\ \overline{\text{op}[1]}\ \overline{\text{op}[0]};$$

$$\text{i\_add} = \text{rtype}\ \text{func}[5]\ \overline{\text{func}[4]}\ \overline{\text{func}[3]}\ \overline{\text{func}[2]}\ \overline{\text{func}[1]}\ \overline{\text{func}[0]};$$

$$\text{i\_sub} = \text{rtype}\ \text{func}[5]\ \overline{\text{func}[4]}\ \overline{\text{func}[3]}\ \text{func}[2]\ \overline{\text{func}[1]}\ \overline{\text{func}[0]};$$

Table 5.1   Instruction decode

R-format			I- and J-format	
Inst.	op[5:0]	func[5:0]	Inst.	op[5:0]
i_add	000000	100000	i_addi	001000
i_sub	000000	100010	i_andi	001100
i_and	000000	100100	i_ori	001101
i_or	000000	100101	i_xori	001110
i_xor	000000	100110	i_lw	100011
i_sll	000000	000000	i_sw	101011
i_srl	000000	000010	i_beq	000100
i_sra	000000	000011	i_bne	000101
i_jr	000000	001000	i_lui	001111
			i_j	000010
			i_jal	000011

$i\_and = \text{rtype } func[5]\ \overline{func[4]}\ \overline{func[3]}\ func[2]\ \overline{func[1]}\ \overline{func[0]};$

$i\_or = \text{rtype } func[5]\ \overline{func[4]}\ \overline{func[3]}\ func[2]\ \overline{func[1]}\ func[0];$

$i\_xor = \text{rtype } func[5]\ \overline{func[4]}\ \overline{func[3]}\ func[2]\ func[1]\ \overline{func[0]};$

$i\_sll = \text{rtype } \overline{func[5]}\ \overline{func[4]}\ \overline{func[3]}\ \overline{func[2]}\ \overline{func[1]}\ \overline{func[0]};$

$i\_srl = \text{rtype } \overline{func[5]}\ \overline{func[4]}\ \overline{func[3]}\ \overline{func[2]}\ func[1]\ \overline{func[0]};$

$i\_sra = \text{rtype } \overline{func[5]}\ \overline{func[4]}\ \overline{func[3]}\ \overline{func[2]}\ func[1]\ func[0];$

$i\_jr = \text{rtype } \overline{func[5]}\ \overline{func[4]}\ func[3]\ \overline{func[2]}\ \overline{func[1]}\ \overline{func[0]};$

$i\_addi = \overline{op[5]}\ \overline{op[4]}\ op[3]\ \overline{op[2]}\ \overline{op[1]}\ \overline{op[0]};$

$i\_andi = \overline{op[5]}\ \overline{op[4]}\ op[3]\ op[2]\ \overline{op[1]}\ \overline{op[0]};$

$i\_ori = \overline{op[5]}\ \overline{op[4]}\ op[3]\ op[2]\ \overline{op[1]}\ op[0];$

$i\_xori = \overline{op[5]}\ \overline{op[4]}\ op[3]\ op[2]\ op[1]\ \overline{op[0]};$

$i\_lw = op[5]\ \overline{op[4]}\ \overline{op[3]}\ \overline{op[2]}\ op[1]\ op[0];$

$i\_sw = op[5]\ \overline{op[4]}\ op[3]\ \overline{op[2]}\ op[1]\ op[0];$

$i\_beq = \overline{op[5]}\ \overline{op[4]}\ \overline{op[3]}\ op[2]\ \overline{op[1]}\ \overline{op[0]};$

$i\_bne = \overline{op[5]}\ \overline{op[4]}\ \overline{op[3]}\ op[2]\ \overline{op[1]}\ op[0];$

$i\_lui = \overline{op[5]}\ \overline{op[4]}\ op[3]\ op[2]\ op[1]\ op[0];$

$i\_j = \overline{op[5]}\ \overline{op[4]}\ \overline{op[3]}\ \overline{op[2]}\ op[1]\ \overline{op[0]};$

$i\_jal = \overline{op[5]}\ \overline{op[4]}\ \overline{op[3]}\ \overline{op[2]}\ op[1]\ op[0];$

The control signals can be classified into the following four categories: (i) the selection signals of the multiplexers; (ii) the ALU operation control signal; (iii) the register file and memory write enables; and (iv) others such as sign or zero extension. Table 5.2 lists all the control signals of the single-cycle CPU.

Table 5.2 Control signals of single-cycle CPU

Signal	Meaning	Action
wreg	Write register	1: write; 0: do not write
regrt	Destination register is rt	1: select rt; 0: select rd
jal	Subroutine call	1: is jal; 0: is not jal
m2reg	Save memory data	1: select memory data; 0: select ALU result
shift	ALU A uses sa	1: select sa; 0: select register data
aluimm	ALU B uses immediate	1: select immediate; 0: select register data
sext	Immediate sign extend	1: sign-extend; 0: zero extend
aluc[3:0]	ALU operation control	x000: ADD; x100: SUB; x001: AND x101: OR; x010: XOR; x110: LUI 0011: SLL; 0111: SRL; 1111: SRA
wmem	Write memory	1: write memory; 0: do not write
pcsrc[1:0]	Next instruction address	00: select PC+4; 01: branch address 10: register data; 11: jump address

Table 5.3 Truth table of the control signals

Inst.	z	wreg	regrt	jal	m2reg	shift	aluimm	sext	aluc[3:0]	wmem	pcsrc[1:0]
i_add	x	1	0	0	0	0	0	x	x 0 0 0	0	0 0
i_sub	x	1	0	0	0	0	0	x	x 1 0 0	0	0 0
i_and	x	1	0	0	0	0	0	x	x 0 0 1	0	0 0
i_or	x	1	0	0	0	0	0	x	x 1 0 1	0	0 0
i_xor	x	1	0	0	0	0	0	x	x 0 1 0	0	0 0
i_sll	x	1	0	0	0	1	0	x	0 0 1 1	0	0 0
i_srl	x	1	0	0	0	1	0	x	0 1 1 1	0	0 0
i_sra	x	1	0	0	0	1	0	x	1 1 1 1	0	0 0
i_jr	x	0	x	x	x	x	x	x	x x x x	0	1 0
i_addi	x	1	1	0	0	0	1	1	x 0 0 0	0	0 0
i_andi	x	1	1	0	0	0	1	0	x 0 0 1	0	0 0
i_ori	x	1	1	0	0	0	1	0	x 1 0 1	0	0 0
i_xori	x	1	1	0	0	0	1	0	x 0 1 0	0	0 0
i_lw	x	1	1	0	1	0	1	1	x 0 0 0	0	0 0
i_sw	x	0	x	x	x	0	1	1	x 0 0 0	1	0 0
i_beq	0	0	x	x	x	0	0	1	x 0 1 0	0	0 0
i_beq	1	0	x	x	x	0	0	1	x 0 1 0	0	0 1
i_bne	0	0	x	x	x	0	0	1	x 0 1 0	0	0 1
i_bne	1	0	x	x	x	0	0	1	x 0 1 0	0	0 0
i_lui	x	1	1	0	0	x	1	x	x 1 1 0	0	0 0
i_j	x	0	x	x	x	x	x	x	x x x x	0	1 1
i_jal	x	1	x	1	x	x	x	x	x x x x	0	1 1

Table 5.3 is the truth table for the control unit. We take the `lw` instruction as an example to explain how to fill the table. The ALU performs addition to calculate the memory address (`aluc[3:0]` = x000); one operand of the addition comes from register `rs` of the register file (`shift` = 0); the other operand is the immediate (`aluimm` = 1); the immediate is sign-extended (`sext` = 1); the result will be written to register file (`wreg` = 1); it is the memory data (`m2reg` = 1); the destination register number is `rt` (`regrt` = 1); it is not a `jal` instruction (`jal` = 0); it does not write memory (`wmem` = 0); and the address of the next instruction is PC+4 (`pcsrc[1:0]` = 00).

From the truth table, we can get the logic expression of each control signal as shown below (not simplified). If there are multiple bits in a control signal, `aluc` for instance, we must write a logic expression for each bit individually.

```
wreg    = i_add + i_sub + i_and + i_or + i_xor + i_sll + i_srl + i_sra +
          i_addi + i_andi + i_ori + i_xori + i_lw + i_lui + i_jal;
regrt   = i_addi + i_andi + i_ori + i_xori + i_lw + i_lui;
m2reg   = i_lw;
shift   = i_sll + i_srl + i_sra;
aluimm  = i_addi + i_andi + i_ori + i_xori + i_lw + i_lui + i_sw;
sext    = i_addi + i_lw + i_sw + i_beq + i_bne;
aluc[3] = i_sra;
aluc[2] = i_sub + i_or + i_srl + i_sra + i_ori + i_lui;
aluc[1] = i_xor + i_sll + i_srl + i_sra + i_xori + i_beq + i_bne + i_lui;
aluc[0] = i_and + i_or + i_sll + i_srl + i_sra + i_andi + i_ori;
```

```
wmem      = i_sw;
pcsrc[1]  = i_jr + i_j + i_jal;
pcsrc[0]  = i_beq z + i_bne ‾z‾ + i_j + i_jal;
```

5.4.2 Verilog HDL Codes of the Control Unit

Based on the logic expressions given above, we can write the Verilog HDL code for implementing the
control unit of the single-cycle CPU, as listed below. The first half is the code for decoding instructions,
and the second half is for generating control signals.

```
module sccu_dataflow (op,func,z,wmem,wreg,regrt,m2reg,aluc,shift,aluimm,
                       pcsrc,jal,sext);                // control unit
  input  [5:0] op, func;                               // op, func
  input        z;                                      // alu zero tag
  output [3:0] aluc;                                   // alu operation control
  output [1:0] pcsrc;                                  // select pc source
  output       wreg;                                   // write regfile
  output       regrt;                                  // dest reg number is rt
  output       m2reg;                                  // instruction is an lw
  output       shift;                                  // instruction is a shift
  output       aluimm;                                 // alu input b is an i32
  output       jal;                                    // instruction is a jal
  output       sext;                                   // is sign extension
  output       wmem;                                   // write data memory
  // decode instructions
  wire rtype  = ~|op;                                                      // r format
  wire i_add  = rtype& func[5]&~func[4]&~func[3]&~func[2]&~func[1]&~func[0];
  wire i_sub  = rtype& func[5]&~func[4]&~func[3]&~func[2]& func[1]&~func[0];
  wire i_and  = rtype& func[5]&~func[4]&~func[3]& func[2]&~func[1]&~func[0];
  wire i_or   = rtype& func[5]&~func[4]&~func[3]& func[2]&~func[1]& func[0];
  wire i_xor  = rtype& func[5]&~func[4]&~func[3]& func[2]& func[1]&~func[0];
  wire i_sll  = rtype&~func[5]&~func[4]&~func[3]&~func[2]&~func[1]&~func[0];
  wire i_srl  = rtype&~func[5]&~func[4]&~func[3]&~func[2]& func[1]&~func[0];
  wire i_sra  = rtype&~func[5]&~func[4]&~func[3]&~func[2]& func[1]& func[0];
  wire i_jr   = rtype&~func[5]&~func[4]& func[3]&~func[2]&~func[1]&~func[0];
  wire i_addi = ~op[5]&~op[4]& op[3]&~op[2]&~op[1]&~op[0];        // i format
  wire i_andi = ~op[5]&~op[4]& op[3]& op[2]&~op[1]&~op[0];
  wire i_ori  = ~op[5]&~op[4]& op[3]& op[2]&~op[1]& op[0];
  wire i_xori = ~op[5]&~op[4]& op[3]& op[2]& op[1]&~op[0];
  wire i_lw   =  op[5]&~op[4]&~op[3]&~op[2]& op[1]& op[0];
  wire i_sw   =  op[5]&~op[4]& op[3]&~op[2]& op[1]& op[0];
  wire i_beq  = ~op[5]&~op[4]&~op[3]& op[2]&~op[1]&~op[0];
  wire i_bne  = ~op[5]&~op[4]&~op[3]& op[2]&~op[1]& op[0];
  wire i_lui  = ~op[5]&~op[4]& op[3]& op[2]& op[1]& op[0];
  wire i_j    = ~op[5]&~op[4]&~op[3]&~op[2]& op[1]&~op[0];        // j format
  wire i_jal  = ~op[5]&~op[4]&~op[3]&~op[2]& op[1]& op[0];
  // generate control signals
```

```
assign regrt    = i_addi | i_andi | i_ori  | i_xori | i_lw  | i_lui;
assign jal      = i_jal;
assign m2reg    = i_lw;
assign wmem     = i_sw;
assign aluc[3]  = i_sra;                              // refer to alu.v for aluc
assign aluc[2]  = i_sub  | i_or   | i_srl  | i_sra  | i_ori  | i_lui;
assign aluc[1]  = i_xor  | i_sll  | i_srl  | i_sra  | i_xori | i_beq |
                  i_bne  | i_lui;
assign aluc[0]  = i_and  | i_or | i_sll | i_srl | i_sra | i_andi | i_ori;
assign shift    = i_sll  | i_srl  | i_sra;
assign aluimm   = i_addi | i_andi | i_ori  | i_xori | i_lw  | i_lui | i_sw;
assign sext     = i_addi | i_lw   | i_sw   | i_beq  | i_bne;
assign pcsrc[1]= i_jr    | i_j    | i_jal;
assign pcsrc[0]= i_beq & z | i_bne &~z | i_j | i_jal;
assign wreg     = i_add  | i_sub  | i_and  | i_or   | i_xor | i_sll  |
                  i_srl  | i_sra  | i_addi | i_andi | i_ori | i_xori |
                  i_lw   | i_lui  | i_jal;
endmodule
```

5.5   Test Program and Simulation Waveform

The design of the single-cycle CPU has been completed. This section gives the test program that is used to verify the correctness of the CPU. The following Verilog HDL code implements an instruction memory with a read-only memory (ROM). Although an ROM is called a memory, it is actually a combinational circuit.

```
module scinstmem (a,inst);            // instruction memory, rom
    input   [31:0] a;                 // address
    output  [31:0] inst;              // instruction
    wire    [31:0] rom [0:31];        // rom cells: 32 words * 32 bits
    // rom[word_addr] = instruction   // (pc) label    instruction
    assign rom[5'h00] = 32'h3c010000; // (00) main:   lui  $1, 0
    assign rom[5'h01] = 32'h34240050; // (04)         ori  $4, $1, 80
    assign rom[5'h02] = 32'h20050004; // (08)         addi $5, $0,  4
    assign rom[5'h03] = 32'h0c000018; // (0c) call:   jal  sum
    assign rom[5'h04] = 32'hac820000; // (10)         sw   $2, 0($4)
    assign rom[5'h05] = 32'h8c890000; // (14)         lw   $9, 0($4)
    assign rom[5'h06] = 32'h01244022; // (18)         sub  $8, $9, $4
    assign rom[5'h07] = 32'h20050003; // (1c)         addi $5, $0,  3
    assign rom[5'h08] = 32'h20a5ffff; // (20) loop2:  addi $5, $5, -1
    assign rom[5'h09] = 32'h34a8ffff; // (24)         ori  $8, $5, 0xffff
    assign rom[5'h0A] = 32'h39085555; // (28)         xori $8, $8, 0x5555
    assign rom[5'h0B] = 32'h2009ffff; // (2c)         addi $9, $0, -1
    assign rom[5'h0C] = 32'h312affff; // (30)         andi $10,$9, 0xffff
    assign rom[5'h0D] = 32'h01493025; // (34)         or   $6, $10, $9
    assign rom[5'h0E] = 32'h01494026; // (38)         xor  $8, $10, $9
    assign rom[5'h0F] = 32'h01463824; // (3c)         and  $7, $10, $6
```

```
    assign rom[5'h10] = 32'h10a00001;    // (40)         beq  $5, $0, shift
    assign rom[5'h11] = 32'h08000008;    // (44)         j    loop2
    assign rom[5'h12] = 32'h2005ffff;    // (48) shift:  addi $5, $0, -1
    assign rom[5'h13] = 32'h000543c0;    // (4c)         sll  $8, $5, 15
    assign rom[5'h14] = 32'h00084400;    // (50)         sll  $8, $8, 16
    assign rom[5'h15] = 32'h00084403;    // (54)         sra  $8, $8, 16
    assign rom[5'h16] = 32'h000843c2;    // (58)         srl  $8, $8, 15
    assign rom[5'h17] = 32'h08000017;    // (5c) finish: j    finish
    assign rom[5'h18] = 32'h00004020;    // (60) sum:    add  $8, $0, $0
    assign rom[5'h19] = 32'h8c890000;    // (64) loop:   lw   $9, 0($4)
    assign rom[5'h1A] = 32'h20840004;    // (68)         addi $4, $4,  4
    assign rom[5'h1B] = 32'h01094020;    // (6c)         add  $8, $8, $9
    assign rom[5'h1C] = 32'h20a5ffff;    // (70)         addi $5, $5, -1
    assign rom[5'h1D] = 32'h14a0fffb;    // (74)         bne  $5, $0, loop
    assign rom[5'h1E] = 32'h00081000;    // (78)         sll  $2, $8, 0
    assign rom[5'h1F] = 32'h03e00008;    // (7c)         jr   $31
    assign inst = rom[a[6:2]];           // use word address to read rom
endmodule
```

The test data are stored in the data memory; see the following Verilog HDL code.

```
module scdatamem (clk,dataout,datain,addr,we);          // data memory, ram
    input         clk;                   // clock
    input         we;                    // write enable
    input  [31:0] datain;                // data in (to memory)
    input  [31:0] addr;                  // ram address
    output [31:0] dataout;               // data out (from memory)
    reg    [31:0] ram [0:31];            // ram cells: 32 words * 32 bits
    assign dataout = ram[addr[6:2]];     // use word address to read ram
    always @ (posedge clk)
        if (we) ram[addr[6:2]] = datain; // use word address to write ram
    integer i;
    initial begin                        // initialize memory
        for (i = 0; i < 32; i = i + 1)
            ram[i] = 0;
        // ram[word_addr] = data          // (byte_addr) item in data array
        ram[5'h14] = 32'h000000a3;        // (50)  data[0]    0 +  A3 =  A3
        ram[5'h15] = 32'h00000027;        // (54)  data[1]   a3 +  27 =  ca
        ram[5'h16] = 32'h00000079;        // (58)  data[2]   ca +  79 = 143
        ram[5'h17] = 32'h00000115;        // (5c)  data[3]  143 + 115 = 258
        // ram[5'h18] should be 0x00000258, the sum stored by sw instruction
    end
endmodule
```

Figures 5.21 and 5.22 show the simulation waveforms of the single-cycle CPU when executing the test program.

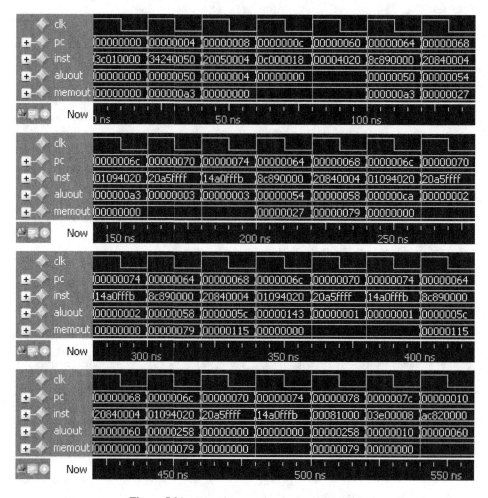

Figure 5.21   Waveform 1 of a single-cycle CPU

From the simulation waveforms, we know that our single-cycle CPU works correctly logically. The single-cycle CPU executes each instruction in one clock cycle. In the following chapters, we will design CPUs with different organizations.

Exercises

5.1   What is a single-cycle CPU? Explain its advantages and disadvantages.

5.2   Use behavioral-style Verilog HDL to design a single-cycle CPU. Hint: there is no need to have an ALU control signal; use `case` statement to deal with each instruction. The following code shows a design example of a single-cycle CPU which can execute three instructions: `add`, `lw`, and `beq`. Although the control signals are declared as `reg`, combinational circuits will be generated.

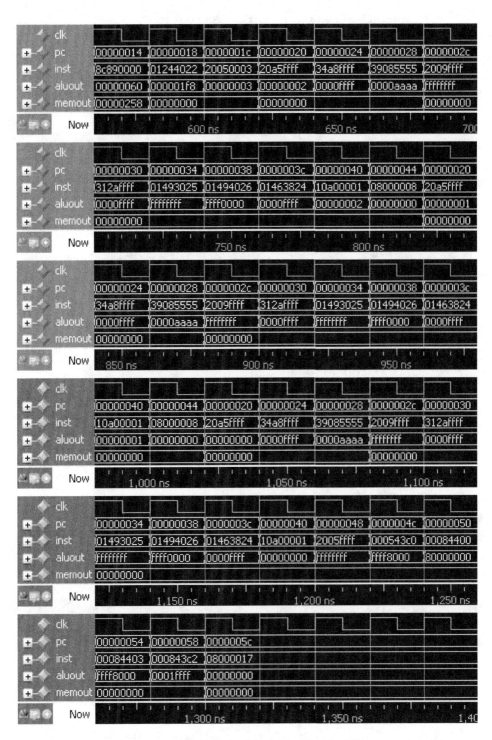

Figure 5.22 Waveform 2 of a single-cycle CPU

```
// internal signals (instruction format)
wire [05:00] opcode   = inst[31:26];
wire [04:00] rs       = inst[25:21];
wire [04:00] rt       = inst[20:16];
wire [04:00] rd       = inst[15:11];
wire [04:00] sa       = inst[10:06];
wire [05:00] func     = inst[05:00];
wire [15:00] imm      = inst[15:00];
wire [25:00] addr     = inst[25:00];
wire         sign     = inst[15];
wire [31:00] br_offset = {{14{sign}},imm,2'b00};
// instruction decode
wire i_add = (opcode == 6'h00) & (func == 6'h20);
wire i_lw  = (opcode == 6'h23);
wire i_beq = (opcode == 6'h04);
reg        wreg;     // write enable
reg [31:0] ALU_out;  // ALU output
reg  [4:0] dest_rn;  // destination register number
reg [31:0] next_pc;  // next PC
// program counter
reg [31:0] pc;
always @ (posedge clk or negedge clrn) begin
    if (!clrn) pc <= 0;
    else pc <= next_pc;
end
// data written to register file
wire [31:0] data_to_regfile = i_lw ? d_f_mem : ALU_out;
// register file
reg  [31:0] regfile [1:31];
wire [31:0] a = (rs==0) ? 0 : regfile[rs];
wire [31:0] b = (rt==0) ? 0 : regfile[rt];
always @ (posedge clk) begin
    if (wreg && (dest_rn != 0)) begin
        regfile[dest_rn] <= data_to_regfile;
    end
end
// pc + 4
wire [31:0] pc_plus_4 = pc + 4;
// output signals
assign m_addr = ALU_out;
// control signals and ALU output
// will be combinational circuit
always @(*) begin
    ALU_out = 0;
    dest_rn = rd;
    wreg    = 0;
    next_pc = pc_plus_4;
```

```
        case (1'b1)
            i_add: begin // add
                ALU_out = a + b;
                wreg    = 1; end
            i_lw: begin // lw
                ALU_out = a + {{16{sign}},imm};
                dest_rn = rt;
                wreg    = 1; end
            i_beq: begin // beq
                if (a == b)
                    next_pc = pc_plus_4 + br_offset; end
            default: ;
        endcase
    end
```

5.3 Use Xilinx's BMG (block memory generator) or Altera's LPM (library of parameterized modules) to design the instruction memory and data memory. And simulate your CPU designed in Verilog HDL of behavioral style.

5.4 Design a single-cycle CPU that can execute all the MIPS integer instructions.

6

Exceptions and Interrupts Handling and Design in Verilog HDL

An unexpected event may occur during the program execution. For example, a calculation result overflows or an I/O device requests the CPU (central processing unit) to provide service. The former is called a synchronous internal exception, and the latter is called an asynchronous external interrupt. This type of event alters the normal sequence of execution and forces the CPU to transfer control to a special program (a handler).

This chapter describes how a single-cycle CPU deals with exceptions and interrupts. The Verilog HDL (hardware description language) code and the simulation waveform of the CPU are also provided.

6.1 Exceptions and Interrupts

This section introduces the general mechanism for dealing with the exceptions and interrupts, including how to detect an exception and an interrupt, how to transfer control to the exception/interrupt handler, and how to return from the exception or the interrupt.

6.1.1 Types of Exceptions and Interrupts

A synchronous internal exception is caused by the actions of an instruction in the currently executing program. It is synchronous because the same exception will occur at the same place in the program whenever the program runs, assuming other circumstances are the same. TLB (translation lookaside buffer) miss, page fault, divided by zero, illegal instruction, unimplemented instruction, misaligned memory address, and arithmetic overflow are examples of the synchronous internal exceptions.

An asynchronous external interrupt is generated by external hardware devices. It is asynchronous because it is not tied to the execution of the currently executing program and can occur at any time. Keystroke, mouse movement, and timer interrupt are examples of asynchronous external interrupts.

When an exception or an interrupt occurs, the CPU suspends the current program and transfers control to a pre-prepared exception or interrupt service routine to deal with the exception or interrupt. This procedure is called an exception acknowledge or interrupt acknowledge. The service routine is also known as an exception handler or interrupt handler, which resides in the operating system kernel.

Depending on the type of the exception/interrupt, the best result of executing the handler is to resume execution of the suspended program, after doing whatever it needs to, as shown in Figure 6.1.

Computer Principles and Design in Verilog HDL, First Edition. Yamin Li.
© 2015 Tsinghua University Press. All rights reserved. Published 2015 by John Wiley & Sons Singapore Pte Ltd.
Companion Website: www.wiley.com/go/li/verilog

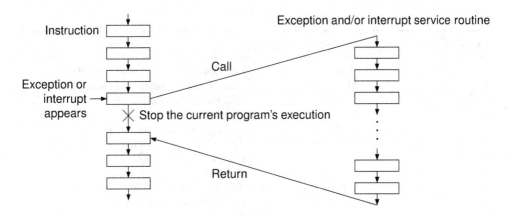

Figure 6.1 Dealing with exceptions and interrupts

Some exceptions or interrupts may cancel the continuous execution of the suspended program, display an error message on screen, and return to the operating system. The illegal instruction exception and `control_c` keystroke interrupt are examples that cancel the execution of the program.

6.1.2 Polled Interrupt and Vectored Interrupt

There are two ways to transfer control to the exception/interrupt handler: polled exception/interrupt and vectored exception/interrupt.

6.1.2.1 Polled Interrupt

When an exception or interrupt occurs, the CPU transfers control to the handler by jumping to a fixed address, which is the entry address of the handler. This address can be hardwired (not changeable) or stored in a particular register to which the CPU can write with an instruction (changeable).

The handler can read some information from somewhere to determine what exception occurred or which device made the interrupt request. The MIPS (microprocessor without interlocked pipeline stages) ISA (instruction set architecture) defines a Cause register, which contains the source of the exception and interrupt, as shown in Figure 6.2. It is a register of coprocessor 0 (CP0).

Figure 6.2 shows only the fields that are related to exceptions and interrupts. The ExcCode field indicates the source of the exception, and the IP[7:0] field indicates the source of the pending interrupt requests. The hardware has the responsibility to set up these fields with suitable values based on what exception or interrupt happened. Table 6.1 lists the meaning of each ExcCode encoding. The encoding 0 indicates that an interrupt occurs; other encodings indicate that an exception happens. It does not matter even if you do not understand them fully at present.

31	30	29	28	27	26	25	24	23	22	21	20	19	18	17	16	15 14 13 12 11 10 9 8	7	6 5 4 3 2	1 0
0	0	0	0	0	0	0	0	0	0	0	0	0	0	0	0	IP	0	ExcCode	0 0

Figure 6.2 Cause register of CP0

Table 6.1   ExcCode field of the MIPS Cause register

ExcCode	Mnemonic	Description
0	Int	Interrupt ($IP[7:0]$ indicates the interrupt source)
1	Mod	TLB hit on store but memory was not yet modified
2	TLBL	TLB miss on instruction fetch or load
3	TLBS	TLB miss on store
4	AdEL	Address error when fetch or load
5	AdES	Address error when store
6	IBE	Bus error on instruction fetch
7	DBE	Bus error on load or store
8	Sys	Executing system call instruction
9	Bp	Executing break instruction
10	RI	Attempt to execute reserved instruction
11	CpU	Coprocessor is not available
12	Ov	Arithmetic overflow
13	Tr	Executing trap instruction
14	–	Reserved
15	FPE	Floating-point exceptions
16–22	–	Reserved
23	WATCH	Virtual address matches value in Watch register
24	MCheck	TLB multiple match but not consistent
25–29	–	Reserved
30	CacheErr	Cache error
31	–	Reserved

We can use the following program fragment to check the ExcCode of the Cause register, and then let the CPU transfer control to an address for each corresponding exception, assuming that a jump table is pre-prepared.

```
lui  $k0, %hi(jump_table_base)            # jump table base address
mfc0 $k1, c0_cause                        # $27 <-- cause
andi $k1, $k1, 0x7c                       # [6:2]: exccode
add  $k0, $k0, $k1                        # plus offset
lw   $k0, %low(jump_table_base)($k0)      # get address from jump table
jr   $k0                                  # jump to the address
```

We can further write a similar program for the case of ExcCode = 0 (interrupt), to let CPU jump to an address for each corresponding interrupt, by using the value in the $IP[7:0]$ field of the Cause register and an interrupt jump table.

6.1.2.2   Vectored Interrupt

The polled interrupt method makes the control to be transferred to a common entry address and then to an individual entry address by checking the source information of the exception or interrupt. In contrast, the vectored interrupt method makes the control to be transferred directly or indirectly to the entry address of the individual exception or interrupt with a vector, as shown in Figure 6.3.

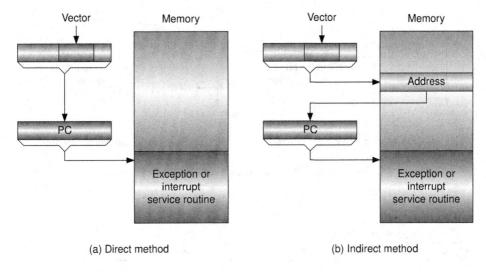

(a) Direct method   (b) Indirect method

Figure 6.3   Mechanism of vectored exception/interrupt

When an exception or an interrupt occurs, a unique vector is generated by the hardware. By attaching suitable values to the left side and right side of the vector, we get a memory address. This address can be written to the PC, as shown in Figure 6.3(a), so that the control is transferred to the individual entry address directly. SUN SPARC adopts this method.

Or, we can use the address to get a value from memory and write the value to the PC, as shown in Figure 6.3(b), so that the control is transferred to the individual entry address indirectly, just like what the Intel x86 does.

6.1.3   Return from Exception/Interrupt

In order to return from the exception or interrupt handler to the program that was suspended, the return address must be saved somewhere. Based on the ISAs, the return address can be saved to a general-purpose register, to a special register, or to the stack memory.

In MIPS ISA, an exception program counter (EPC) is prepared for storing the return address. Referring to Figure 6.4(a), the EPS contains the address of the instruction that generates the exception (this instruction may need to be executed again, in the case of a TLB miss exception, for instance). In the case of an interrupt, when an external interrupt request comes, the current instruction will be executed before transferring control to the interrupt handler. Therefore, the EPC contains the address of the next instruction, as shown in Figure 6.4(b).

The MIPS `eret` (exception return) instruction is used to return from the exception or interrupt handler to the program that was interrupted. It writes the content of EPC to PC. In the case of an exception, if the instruction that generates the exception need not be executed again, we can increment the return address by 4 before executing the `eret` instruction, as illustrated in the code below.

```
mfc0   $k0, c0_epc            # $26 <-- epc
addiu  $k0, $k0, 4            # $26 <-- $26 + 4
mtc0   $k0, c0_epc            # epc <-- $26
eret                          # return: pc  <-- epc
```

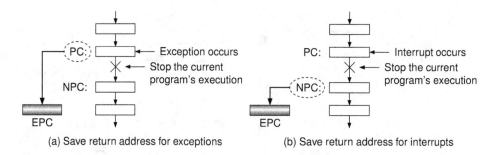

Figure 6.4   EPC contents for exceptions and interrupts

31	30	29	28	27	26	25	24	23	22	21	20	19	18	17	16	15	14	13	12	11	10	9	8	7	6	5	4	3	2	1	0
0	0	0	0	0	0	0	0	0	0	0	0	0	0	0	0				IM					0	0	0	0	0	0	0	IE

Figure 6.5   Status register of CP0

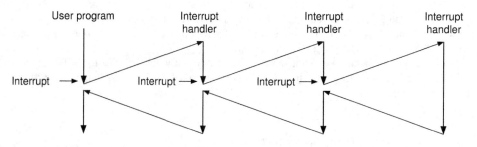

Figure 6.6   Interrupt nesting

6.1.4   Interrupt Mask and Interrupt Nesting

Because interrupts can occur anytime, further interrupt requests may come when the CPU is dealing with a current interrupt. These requests can be enabled or disabled (masked) by setting or resetting the enable bits of a special control register.

MIPS ISA provides such a control register, named the Status register. Some bits of the Status register are shown in Figure 6.5. The ith bit in the field of IM[7:0] corresponds to the ith interrupt: if it is a 0, the interrupt is disabled; otherwise the interrupt is enabled. The rightmost bit, IE, is the master control bit for all the interrupts. MIPS CPU clears all these bits when it acknowledges an interrupt. This prevents interrupt nesting automatically.

Interrupt nesting can be allowed by an interrupt handler by enabling the interrupts. Figure 6.6 shows an example of interrupt nesting.

Note that before enabling further interrupts in the interrupt handler for implementing the interrupt nesting, some states must be saved to the stack memory. For example, the MIPS EPC must be saved so that the control can be returned from an interrupt handler.

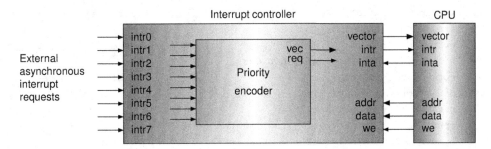

Figure 6.7   Priority interrupt controller

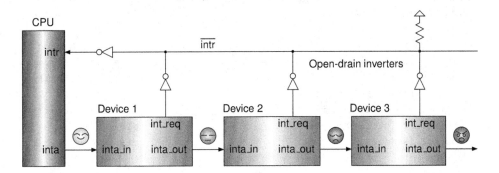

Figure 6.8   Daisy chain for priority interrupts

6.1.5   Interrupt Priority

If there are multiple interrupts, which interrupt should be serviced first? The answer is, the interrupt that has higher interrupt priority than others. For example, if the timer interrupt and a keyboard interrupt occur at the same time, the timer interrupt should be serviced first.

Because more than one device can assert the interrupt request simultaneously, some methods must be employed to ensure device priority. Figure 6.7 shows an interrupt controller in which there is an 8-3 priority encoder.

The 8-3 priority encoder was described in Chapter 2. The `req` output signal of the priority encoder is actually the `intr` which is sent to the CPU as the interrupt request signal. The 3-bit `vec` output signals of the priority encoder will be the rightmost three bits of the vector.

Daisy chain is another way to assign priorities to devices that generate level-sensitive interrupts, as shown in Figure 6.8. Any requesting device can take the interrupt line `int_req` high, and keep it asserted high until it is serviced. The inverted requests are wire-ORed to generate an active-low interrupt request signal ($\overline{intr}$). Note that the three NOT gates in the figure must be the open-drain or open-collector inverters in case their outputs are wired together.

The priority is assigned using the interrupt acknowledge signal, `inta`, generated by the CPU in response to an interrupt request. `inta` is passed through each device, from the highest priority device first, to the lowest priority device last. If device 1 generated the interrupt, it will block the `inta`. Otherwise, it will pass `inta` on to the next device in the chain. Other devices follow the same procedure.

The following code shows how to use open-drain inverters in Verilog HDL. The key component is the Altera `opndrn` primitive.

```
module high_z_oc (in1,in2,in3,out1);          // test open drain buffer
    input   in1, in2, in3;                     // three input signals
    output out1;                               // one output signal
    wire    not_out1, not_out2, not_out3;
    not     not1 (not_out1, in1);              // regular not gate
    not     not2 (not_out2, in2);
    not     not3 (not_out3, in3);
    opndrn oc1 (.in(not_out1), .out(out1));    // opndrn is an
    opndrn oc2 (.in(not_out2), .out(out1));    // open-drain
    opndrn oc3 (.in(not_out3), .out(out1));    // buffer
endmodule
```

We can also use bufif1, tri, and regular Verilog HDL statements to implement the function above; see the following three examples.

```
module high_z_buf (in1,in2,in3,out1);         // same as high_z_oc.v
    input   in1, in2, in3;                     // three input signals
    output out1;                               // one output signal
    // bufif1 (out,   in, ctl);
    bufif1 i1 (out1, 0,  in1);                 // tri-state buffer:
    bufif1 i2 (out1, 0,  in2);                 // ctl==1: out=in
    bufif1 i3 (out1, 0,  in3);                 // ctl==0: out=high_z
endmodule
```

```
module high_z_tri (in1,in2,in3,out1);         // same as high_z_oc.v
    input   in1, in2, in3;                     // three input signals
    output out1;                               // one output signal
    tri     out1;                              // tri
    assign out1 = in1 ? 0 : 1'bz;              // tri-state signal
    assign out1 = in2 ? 0 : 1'bz;              // can be assigned
    assign out1 = in3 ? 0 : 1'bz;              // multiple times
endmodule
```

```
module high_z_nor (in1,in2,in3,out1);         // same as high_z_oc.v
    input   in1, in2, in3;                     // three input signals
    output out1;                               // one output signal
    assign out1 = (in1 | in2 | in3) ? 0 : 1'bz;  // z: high-impedance
endmodule
```

6.2   Design of CPU with Exception and Interrupt Mechanism

This section describes a single-cycle CPU that can deal with interrupt and exceptions. The goal of this book is not to implement all the MIPS functions; therefore we simplified the mechanism of interrupt and exceptions, and some implementations are not compatible with the MIPS architecture. We have made the following assumptions in our CPU implementation:

1. There is only one external asynchronous interrupt request.
2. The CPU deals with only three exceptions: arithmetic signed overflow, unimplemented instruction, and system call.
3. For exceptions, the address of the current instruction that generates the exception is saved to EPC; for interrupt, the address of the next instruction is saved to EPC.
4. The polled interrupt is adopted. The entry address of the interrupt/exception handler is 0x00000008.
5. In response to an interrupt or an exception, the content of the Status register is shifted to the left by 4 bits in order to save previous settings of the Status register and disable further interrupts.
6. When returning from the interrupt/exception handler, the content of the Status register is shifted to the right by 4 bits in order to restore the previous settings of the Status register.

6.2.1 Exception/Interrupt Handling and Related Registers

Based on our assumptions, there are four things which must be done simultaneously when the CPU responds to an interrupt or an exception, as listed below.

1. Update the ExcCode field of the Cause register.
2. Shift the contents of the Status register to the left by 4 bits.
3. Save the return address to EPC.
4. Jump to the interrupt/exception handler.

For an interrupt request, the CPU also needs to issue an interrupt acknowledge signal to the device that generates the interrupt. In the interrupt/exception handler, the CPU performs what it needs to do based on the ExcCode field of the Cause register and finally executes the eret instruction to return to the interrupted program.

We redefined Cause and Status registers based on our assumptions; their formats are shown in Figure 6.9. The EPC is also shown in the figure, which contains the return address. The bits 0, 1, 2, and 3 in the IM field of the Status register are the mask control bits of interrupt, system call, unimplemented instruction, and arithmetic overflow, respectively. A 1 in IM field indicates that the corresponding interrupt or exception is enabled; a 0 indicates that it is disabled. The S field is used for saving the contents in IM field. The 2-bit ExcCode indicates one of the causes, as listed in Table 6.2.

6.2.2 Instructions Related to Exceptions and Interrupts

In our implementation, there are three instructions, add, sub, and addi, that may generate the arithmetic overflow exception. These instructions calculate on signed operands. MIPS has also instructions that calculate on unsigned numbers, such as addu (add unsigned word) and subu (subtract unsigned word), which do not generate exceptions under any circumstances.

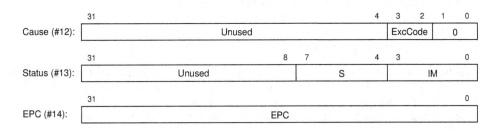

Figure 6.9   Three registers related to exceptions and interrupts

Table 6.2 Definition of ExcCode and interrupt mask

ExcCode	Mnemonic	Mask	Description
0	Int	IM[0]	External interrupt
1	Sys	IM[1]	Executing the system call instruction
2	Unimpl	IM[2]	Executing an unimplemented instruction
3	Ov	IM[3]	Arithmetic signed overflow

Table 6.3 Overflow on signed add and subtract

	aluc[3:0]	a[31]	b[31]	r[31]	v	Comment
ADD	x 0 0 0	0	0	1	1	Plus + plus, result is minus
ADD	x 0 0 0	1	1	0	1	Minus + minus, result is plus
SUB	x 1 0 0	0	1	1	1	Plus − minus, result is minus
SUB	x 1 0 0	1	0	0	1	Minus − plus, result is plus

The ALU (arithmetic logic unit) does not care about whether an instruction is add or addi. For these two instructions, the ALU just performs addition. Table 6.3 lists the cases that cause an arithmetic overflow.

aluc is the ALU's operation control signal which was given in Figure 4.13 of Chapter 4. a[31], b[31], and r[31] are the most significant bits of the 32-bit input a, input b, and output r, respectively. These bits indicate the sign of the operands in the 2's complement representation. From the table, we can write the expression of the overflow as shown below (in Verilog HDL format).

```
v = ~aluc[2] &~a[31] &~b[31] &  r[31] &~aluc[1] &~aluc[0] |
    ~aluc[2] & a[31] & b[31] &~r[31] &~aluc[1] &~aluc[0] |
     aluc[2] &~a[31] & b[31] &  r[31] &~aluc[1] &~aluc[0] |
     aluc[2] & a[31] &~b[31] &~r[31] &~aluc[1] &~aluc[0];
```

The overflow is one of the flags of the ALU. This flag can be generated with a simpler circuit than given above. Please try it.

The CPU can use the instruction mfc0 rt, rd (move from CP0) or mtc0 rt, rd (move to CP0) to read or write one of the Cause, Status, and EPC registers. rt is the register number of the general-purpose register file. rd is the CP0 register number. These two instructions must be implemented in our CPU. Other two instructions, syscall (system call) and eret (exception return), must be also implemented. The formats of these four instructions are shown in Figure 6.10.

6.2.3 Schematic of the CPU

According to the discussion above, we can add the interrupt and exception handling circuit to the single-cycle CPU that we designed in the previous chapter. The detailed schematic circuit of the single-cycle CPU with the interrupt/exception mechanism is shown in Figure 6.11.

The Cause, Status, and EPC registers are shown at the bottom of the figure. The write signals of these three registers are wcau (write cause), wsta (write status), and wepc (write EPC), respectively. data, coming from the register file, can be written to these registers if the mtc0 instruction is executed.

	31	30	29	28	27	26	25	24	23	22	21	20	19	18	17	16	15	14	13	12	11	10	9	8	7	6	5	4	3	2	1	0
mfc0	0	1	0	0	0	0	0	0	0	0	0	rt					rd					0	0	0	0	0	0	0	0	0	0	0
mtc0	0	1	0	0	0	0	0	0	1	0	0	rt					rd					0	0	0	0	0	0	0	0	0	0	0
syscall	0	0	0	0	0	0	0	0	0	0	0	0	0	0	0	0	0	0	0	0	0	0	0	0	0	0	0	0	1	1	0	0
eret	0	1	0	0	0	0	1	0	0	0	0	0	0	0	0	0	0	0	0	0	0	0	0	0	0	0	0	1	1	0	0	0

Figure 6.10 Format of exception- and interrupt-related instructions

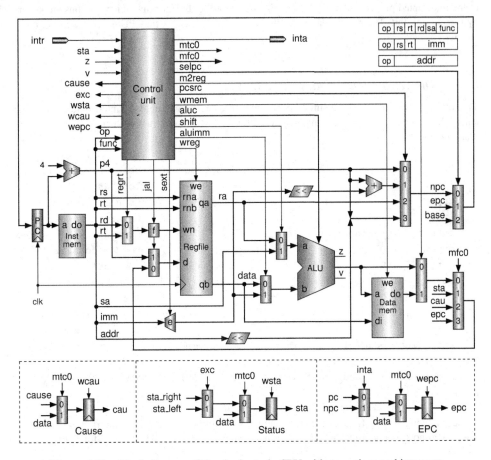

Figure 6.11 Block diagram of the single-cycle CPU with exceptions and interrupts

As described below, other data may be also written to these registers; therefore we used three multiplexers whose selection signal is named as mtc0 (move to CP0).

- Cause register: the ExcCode, generated by the control unit, is written to Cause register when an interrupt or an exception occurs.
- Status register: sta_left and sta_right are the data by shifting left and right of the Status register, respectively. When an interrupt or an exception occurs, sta_left is selected and written to

Status register for saving the previous settings and disabling further interrupt and exceptions. When the eret instruction is executed, sta_right is selected and written to the Status register for restoring the previous settings.
• EPC register: when an interrupt occurs, npc (next PC) is written to EPC; when an exception occurs, pc (current PC) is written to EPC.

The contents of the Cause, Status, and EPC registers can be read through a 4-to-1 multiplexer whose selection signal is mfc0 (move from CP0). In the right-center of the figure, a 3-to-1 multiplexer is added. When an interrupt or an exception occurs, the base is written to the PC, which is the entry address of the interrupt/exception handler. When the eret instruction is executed, epc is written to the PC for returning to the interrupted program. All the control signals related to the interrupt and exceptions are generated by the control unit. The remaining part is the same as the single-cycle CPU described in the previous chapter.

6.2.4   The Verilog HDL Codes of the CPU

The following Verilog HDL code implements the single-cycle computer with interrupt/exception mechanism. It invokes sccpu_intr (single-cycle CPU), sci_intr (instruction memory), and scd_intr (data memory).

```
module sc_interrupt (clk,clrn,inst,pc,aluout,memout,memclk,intr,inta);
    input           clk, clrn;              // clock and reset
    input           memclk;                 // synch ram clock
    input           intr;                   // interrupt request
    output          inta;                   // interrupt acknowledge
    output [31:0] pc;                       // program counter
    output [31:0] inst;                     // instruction
    output [31:0] aluout;                   // alu output
    output [31:0] memout;                   // data memory output
    wire   [31:0] data;                     // data to data memory
    wire          wmem;                     // write data memory
    sccpu_intr cpu (clk,clrn,inst,memout,pc,wmem,aluout,data,intr,inta);
    sci_intr im (pc,inst);                  // inst memory
    scd_intr dm (memout,data,aluout,wmem,memclk); // data memory
endmodule
```

The following Verilog HDL code that implements a single-cycle CPU with interrupt/exception mechanism is almost identical to the schematic circuit shown in Figure 6.11 (except for the two memory modules).

```
module sccpu_intr (clk,clrn,inst,mem,pc,wmem,alu,data,intr,inta);
    input  [31:0] inst;                     // inst from inst memory
    input  [31:0] mem;                      // data from data memory
    input         intr;                     // interrupt request
    input         clk, clrn;                // clock and reset
    output [31:0] pc;                       // program counter
    output [31:0] alu;                      // alu output
    output [31:0] data;                     // data to data memory
```

```
output          wmem;                              // write data memory
output          inta;                              // interrupt acknowledge
parameter       BASE = 32'h00000008;               // exc/int handler entry
parameter       ZERO = 32'h00000000;               // zero
// instruction fields
wire    [5:0] op   = inst[31:26];                   // op
wire    [4:0] rs   = inst[25:21];                   // rs
wire    [4:0] rt   = inst[20:16];                   // rt
wire    [4:0] rd   = inst[15:11];                   // rd
wire    [5:0] func = inst[05:00];                   // func
wire    [15:0] imm = inst[15:00];                   // immediate
wire    [25:0] addr = inst[25:00];                  // address
// control signals
wire    [3:0] aluc;                                 // alu operation control
wire    [1:0] pcsrc;                                // select pc source
wire          wreg;                                 // write regfile
wire          regrt;                                // dest reg number is rt
wire          m2reg;                                // instruction is an lw
wire          shift;                                // instruction is a shift
wire          aluimm;                               // alu input b is an i32
wire          jal;                                  // instruction is a jal
wire          sext;                                 // is sign extension
wire    [1:0] mfc0;                                 // move from c0 regs
wire    [1:0] selpc;                                // select for pc
wire          v;                                    // overflow
wire          exc;                                  // exc or int occurs
wire          wsta;                                 // write status reg
wire          wcau;                                 // write cause reg
wire          wepc;                                 // write epc reg
wire          mtc0;                                 // move to c0 regs
// datapath wires
wire    [31:0] p4;                                  // pc+4
wire    [31:0] bpc;                                 // branch target address
wire    [31:0] npc;                                 // next pc, not exc/int
wire    [31:0] qa;                                  // regfile output port a
wire    [31:0] qb;                                  // regfile output port b
wire    [31:0] alua;                                // alu input a
wire    [31:0] alub;                                // alu input b
wire    [31:0] wd;                                  // regfile write port data
wire    [31:0] r;                                   // alu out or mem
wire    [31:0] sa  = {27'b0,inst[10:6]};            // shift amount
wire    [15:0] s16 = {16{sext & inst[15]}};         // 16-bit signs
wire    [31:0] i32 = {s16,imm};                     // 32-bit immediate
wire    [31:0] dis = {s16[13:0],imm,2'b00};         // word distance
wire    [31:0] jpc = {p4[31:28],addr,2'b00};        // jump target address
wire    [4:0] reg_dest;                             // rs or rt
wire    [4:0] wn  = reg_dest | {5{jal}};            // regfile write reg #
```

```
    wire          z;                              // alu zero tag
    wire  [31:0] sta;                             // output of status reg
    wire  [31:0] cau;                             // output of cause reg
    wire  [31:0] epc;                             // output of epc reg
    wire  [31:0] sta_in;                          // data in for status reg
    wire  [31:0] cau_in;                          // data in for cause reg
    wire  [31:0] epc_in;                          // data in for epc reg
    wire  [31:0] sta_lr;                          // status left/right shift
    wire  [31:0] pc_npc;                          // pc or npc
    wire  [31:0] cause;                           // exc/int cause
    wire  [31:0] res_c0;                          // r or c0 regs
    wire  [31:0] n_pc;                            // next pc
    wire  [31:0] sta_r = {4'h0,sta[31:4]};        // status >> 4
    wire  [31:0] sta_l = {sta[27:0],4'h0};        // status << 4
    // control unit
    sccu_intr cu (op,rs,rd,func,z,wmem,wreg,      // control unit
                  regrt,m2reg,aluc,shift,
                  aluimm,pcsrc,jal,sext,
                  intr,inta,v,sta,                 // exc/int signals
                  cause,exc,wsta,wcau,
                  wepc,mtc0,mfc0,selpc);
    // datapath
    dff32 i_point (n_pc,clk,clrn,pc);             // pc register
    cla32 pcplus4 (pc,32'h4,1'b0,p4);             // pc + 4
    cla32 br_addr (p4,dis,1'b0,bpc);              // branch target address
    mux2x32 alu_a (qa,sa,shift,alua);             // alu input a
    mux2x32 alu_b (qb,i32,aluimm,alub);           // alu input b
    mux2x32 alu_m (alu,mem,m2reg,r);              // alu out or mem
    mux2x32 link  (res_c0,p4,jal,wd);             // res_c0 or p4
    mux2x5 reg_wn (rd,rt,regrt,reg_dest);         // rs or rt
    mux4x32 nextpc(p4,bpc,qa,jpc,pcsrc,npc);      // next pc, not exc/int
    regfile rf (rs,rt,wd,wn,wreg,clk,clrn,qa,qb); // register file
    alu_ov alunit (alua,alub,aluc,alu,z,v);       // alu_ov, z and v tags
    dffe32  c0sta (sta_in,clk,clrn,wsta,sta);     // c0 status register
    dffe32  c0cau (cau_in,clk,clrn,wcau,cau);     // c0 cause register
    dffe32  c0epc (epc_in,clk,clrn,wepc,epc);     // c0 epc register
    mux2x32 cau_x (cause,qb,mtc0,cau_in);         // mux  for cause reg
    mux2x32 sta_1 (sta_r,sta_l,exc,sta_lr);       // mux1 for status reg
    mux2x32 sta_2 (sta_lr,qb,mtc0,sta_in);        // mux2 for status reg
    mux2x32 epc_1 (pc,npc,inta,pc_npc);           // mux1 for epc reg
    mux2x32 epc_2 (pc_npc,qb,mtc0,epc_in);        // mux2 for epc reg
    mux4x32 nxtpc (npc,epc,BASE,ZERO,selpc,n_pc); // mux for pc
    mux4x32 fr_c0 (r,sta,cau,epc,mfc0,res_c0);    // r or c0 regs
    assign data = qb;                             // regfile output port b
endmodule
```

The following Verilog HDL code implements the control unit of the single-cycle CPU. Control signals related to interrupt and exceptions are added.

```verilog
module sccu_intr (op,op1,rd,func,z,wmem,wreg,regrt,m2reg,aluc,shift,aluimm,
                  pcsrc,jal,sext,intr,inta,v,sta,cause,exc,wsta,wcau,wepc,
                  mtc0,mfc0,selpc);                      // control unit
  input  [31:0] sta;                                     // c0 status
  input  [5:0] op, func;                                 // op, func
  input  [4:0] op1, rd;                                  // op1, rd
  input        z, v;                                     // z, v flags
  input        intr;                                     // interrupt request
  output [31:0] cause;                                   // c0 cause
  output [3:0] aluc;                                     // alu control
  output [1:0] mfc0;                                     // move from c0 regs
  output [1:0] selpc;                                    // select for pc
  output [1:0] pcsrc;                                    // select pc source
  output       wreg,regrt,jal,m2reg,shift,aluimm,sext,wmem;
  output       inta;                                     // interrupt ack
  output       exc;                                      // exc or int occurs
  output       wsta;                                     // write status reg
  output       wcau;                                     // write cause reg
  output       wepc;                                     // move to c0 regs
  output       mtc0;                                     // move to c0 regs
  wire rtype   = ~|op;                                                 // r format
  wire i_add   = rtype& func[5]&~func[4]&~func[3]&~func[2]&~func[1]&~func[0];
  wire i_sub   = rtype& func[5]&~func[4]&~func[3]&~func[2]& func[1]&~func[0];
  wire i_and   = rtype& func[5]&~func[4]&~func[3]& func[2]&~func[1]&~func[0];
  wire i_or    = rtype& func[5]&~func[4]&~func[3]& func[2]&~func[1]& func[0];
  wire i_xor   = rtype& func[5]&~func[4]&~func[3]& func[2]& func[1]&~func[0];
  wire i_sll   = rtype&~func[5]&~func[4]&~func[3]&~func[2]&~func[1]&~func[0];
  wire i_srl   = rtype&~func[5]&~func[4]&~func[3]&~func[2]& func[1]&~func[0];
  wire i_sra   = rtype&~func[5]&~func[4]&~func[3]&~func[2]& func[1]& func[0];
  wire i_jr    = rtype&~func[5]&~func[4]& func[3]&~func[2]&~func[1]&~func[0];
  wire i_addi  = ~op[5] &~op[4] & op[3] &~op[2] &~op[1] &~op[0];       // i format
  wire i_andi  = ~op[5] &~op[4] & op[3] & op[2] &~op[1] &~op[0];
  wire i_ori   = ~op[5] &~op[4] & op[3] & op[2] &~op[1] & op[0];
  wire i_xori  = ~op[5] &~op[4] & op[3] & op[2] & op[1] &~op[0];
  wire i_lw    =  op[5] &~op[4] &~op[3] &~op[2] & op[1] & op[0];
  wire i_sw    =  op[5] &~op[4] & op[3] &~op[2] & op[1] & op[0];
  wire i_beq   = ~op[5] &~op[4] &~op[3] & op[2] &~op[1] &~op[0];
  wire i_bne   = ~op[5] &~op[4] &~op[3] & op[2] &~op[1] & op[0];
  wire i_lui   = ~op[5] &~op[4] & op[3] & op[2] & op[1] & op[0];
  wire i_j     = ~op[5] &~op[4] &~op[3] &~op[2] & op[1] &~op[0];       // j format
  wire i_jal   = ~op[5] &~op[4] &~op[3] &~op[2] & op[1] & op[0];
  wire c0type  = ~op[5] & op[4] &~op[3] &~op[2] &~op[1] &~op[0];
  wire i_mfc0  = c0type &~op1[4] &~op1[3] &~op1[2] &~op1[1] &~op1[0];
  wire i_mtc0  = c0type &~op1[4] &~op1[3] & op1[2] &~op1[1] &~op1[0];
```

```verilog
wire i_eret = c0type & op1[4] &~op1[3] &~op1[2] &~op1[1] &~op1[0] &
              ~func[5] & func[4] & func[3] &~func[2] &~func[1] &~func[0];
wire i_syscall = rtype &
              ~func[5] &~func[4] & func[3] & func[2] &~func[1] &~func[0];
wire unimplemented_inst = ~(i_mfc0 | i_mtc0 | i_eret | i_syscall |
    i_add | i_sub | i_and | i_or | i_xor | i_sll | i_srl | i_sra |
    i_jr | i_addi | i_andi | i_ori | i_xori | i_lw | i_sw | i_beq |
    i_bne| i_lui| i_j| i_jal);
wire rd_is_status = (rd == 5'd12);                    // is cp0 status reg
wire rd_is_cause  = (rd == 5'd13);                    // is cp0 cause reg
wire rd_is_epc    = (rd == 5'd14);                    // is cp0 epc reg
wire overflow = v & (i_add | i_sub | i_addi);         // overflow
wire int_int  = sta[0] & intr;                        // sta[0]: enable
wire exc_sys  = sta[1] & i_syscall;                   // sta[1]: enable
wire exc_uni  = sta[2] & unimplemented_inst;          // sta[2]: enable
wire exc_ovr  = sta[3] & overflow;                    // sta[3]: enable
assign inta   = int_int;                              // interrupt ack
// exccode: 0 0 : intr                                // generate exccode
//          0 1 : i_syscall
//          1 0 : unimplemented_inst
//          1 1 : overflow
wire exccode0 = i_syscall | overflow;
wire exccode1 = unimplemented_inst | overflow;
// mfc0:    0 0 : alu_mem                             // generate mux sel
//          0 1 : sta
//          1 0 : cau
//          1 1 : epc
assign mfc0[0] = i_mfc0 & rd_is_status | i_mfc0 & rd_is_epc;
assign mfc0[1] = i_mfc0 & rd_is_cause  | i_mfc0 & rd_is_epc;
// selpc:   0 0 : npc                                 // generate mux sel
//          0 1 : epc
//          1 0 : exc_base
//          1 1 : x
assign selpc[0] = i_eret;
assign selpc[1] = exc;
assign cause = {28'h0,exccode1,exccode0,2'b00};       // cause
assign exc   = int_int | exc_sys | exc_uni | exc_ovr; // exc or int occurs
assign mtc0  = i_mtc0;                                // highest priority
assign wsta  = exc | mtc0 & rd_is_status | i_eret;    // write status reg
assign wcau  = exc | mtc0 & rd_is_cause;              // write cause reg
assign wepc  = exc | mtc0 & rd_is_epc;                // write epc reg
assign regrt   = i_addi| i_andi| i_ori| i_xori| i_lw | i_lui| i_mfc0;
assign jal     = i_jal;
assign m2reg   = i_lw;
assign wmem    = i_sw;
assign aluc[3] = i_sra;                               // refer to alu_ov.v
assign aluc[2] = i_sub| i_or| i_srl| i_sra| i_ori| i_lui;
```

```
assign aluc[1]   = i_xor| i_sll| i_srl| i_sra| i_xori| i_beq| i_bne| i_lui;
assign aluc[0]   = i_and| i_or| i_sll| i_srl| i_sra| i_andi| i_ori;
assign shift     = i_sll | i_srl | i_sra;
assign aluimm    = i_addi| i_andi| i_ori| i_xori| i_lw | i_lui| i_sw;
assign sext      = i_addi| i_lw  | i_sw | i_beq | i_bne;
assign pcsrc[1]  = i_jr  | i_j   | i_jal;
assign pcsrc[0]  = i_beq & z | i_bne &~z | i_j | i_jal;
assign wreg = i_add | i_sub | i_and| i_or  | i_xor| i_sll| i_srl| i_sra|
              i_addi| i_andi| i_ori| i_xori| i_lw  | i_lui| i_jal| i_mfc0;
endmodule
```

The following Verilog HDL code implements the ALU. Overflow flag output v is added to the ALU.

```
module alu_ov (a,b,aluc,r,z,v);      // 32-bit alu with zero and overflow flags
    input   [31:0] a, b;             // inputs:  a, b
    input   [3:0] aluc;              // input:   alu control:   // aluc[3:0]:
    output  [31:0] r;                // output:  alu result     // x 0 0 0   ADD
    output         z, v;             // outputs: zero, overflow // x 1 0 0   SUB
    wire    [31:0] d_and = a & b;                               // x 0 0 1   AND
    wire    [31:0] d_or  = a | b;                               // x 1 0 1   OR
    wire    [31:0] d_xor = a ^ b;                               // x 0 1 0   XOR
    wire    [31:0] d_lui = {b[15:0],16'h0};                     // x 1 1 0   LUI
    wire    [31:0] d_and_or  = aluc[2]? d_or  : d_and;          // 0 0 1 1   SLL
    wire    [31:0] d_xor_lui = aluc[2]? d_lui : d_xor;          // 0 1 1 1   SRL
    wire    [31:0] d_as,d_sh;                                   // 1 1 1 1   SRA
    // addsub32    (a,b,sub,    s);
    addsub32 as32 (a,b,aluc[2],d_as);                           // add/sub
    // shift       (d,sa,    right,   arith,   sh);
    shift shifter (b,a[4:0],aluc[2],aluc[3],d_sh);              // shift
    // mux4x32     (a0,   a1,      a2,      a3, s,         y);
    mux4x32 res (d_as,d_and_or,d_xor_lui,d_sh,aluc[1:0],r);     // alu result
    assign z = ~|r;                                            // z = (r == 0)
    assign v = ~aluc[2] &~a[31] &~b[31] & r[31] &~aluc[1] &~aluc[0] |
                ~aluc[2] & a[31] & b[31] &~r[31] &~aluc[1] &~aluc[0] |
                 aluc[2] &~a[31] & b[31] & r[31] &~aluc[1] &~aluc[0] |
                 aluc[2] & a[31] &~b[31] &~r[31] &~aluc[1] &~aluc[0];
endmodule
```

The following Verilog HDL code implements the instruction memory. Instead of using general Verilog HDL statements, here we show how to use an LPM (library of parameterized modules), provided by Altera, to implement memories. lpm_rom is a read-only memory module and can be initialized with a memory initialization file (mif). Because ModelSim does not support mif, we converted mif to hex (hexadecimal) format. sci_intr.hex is such a file. We will list the contents of the initialization file in mif format in the next section.

```
module sci_intr (a,inst);                         // inst mem (rom)
    input  [31:0] a;                               // mem address
    output [31:0] inst;                            // mem data output
```

```
    lpm_rom rom (.address(a[7:2]),                      // word address
                 .q(inst),                              // mem data output
                 .inclock(),                            // no clock
                 .outclock(),                           // no clock
                 .memenab());                           // no write enable
    defparam rom.lpm_width           = 32,              // data: 32 bits
             rom.lpm_widthad         = 6,               // 2^6 = 64 words
             rom.lpm_file            = "sci_intr.hex",  // mem init file
             rom.lpm_outdata         = "UNREGISTERED",  // no reg (data)
             rom.lpm_address_control = "UNREGISTERED";  // no reg (addr)
endmodule
```

The following Verilog HDL code implements the data memory. We also use LPM. `lpm_ram_dq` is a synchronous random access memory module that needs a clock. The input signals of the memory, including address, data in, and write control, must be registered using `inclock`. The output signal can be either unregistered or registered using `outclock`.

```
module scd_intr (dataout,datain,addr,we,memclk);       // data mem (sram)
    input   [31:0] datain;                             // mem data input
    input   [31:0] addr;                               // mem address
    input          we;                                 // write enable
    input          memclk;                             // sync ram clock
    output  [31:0] dataout;                            // mem data output
    wire           inclk  = memclk;                    // in reg clock
    wire           outclk = memclk;                    // out reg clock
    lpm_ram_dq ram (.data(datain),                     // data in
                    .address(addr[6:2]),               // word address
                    .we(we),                           // write enable
                    .inclock(inclk),                   // in reg clock
                    .outclock(outclk),                 // out reg clock
                    .q(dataout));                      // mem data out
    defparam ram.lpm_width           = 32;             // data: 32 bits
    defparam ram.lpm_widthad         = 5;              // 2^5 = 32 words
    defparam ram.lpm_file            = "scd_intr.hex"; // mem init file
    defparam ram.lpm_indata          = "REGISTERED";   // in reg (data)
    defparam ram.lpm_outdata         = "REGISTERED";   // out reg (data)
    defparam ram.lpm_address_control = "REGISTERED";   // in reg (a, we)
endmodule
```

In our implementation, we also registered the output signal. Thus, in the single-cycle CPU, there must be two rising edges for the synchronous memory within a CPU cycle. We use `memclk` as both `inclock` and `outclock`. The memory module is initialized with a `scd_intr.hex` file. We will also list the contents of the memory initialization file in `mif` format in the next section.

6.3 The CPU Exception and Interrupt Tests

This section gives the test program and test data in `mif` format for verifying the interrupt/exception mechanism. The `mif` files need to be converted to `hex` format for the simulation with ModelSim. The simulation waveforms are also provided.

6.3.1 Test Program and Data

The contents of `sci_intr.mif` for the instruction memory initialization are listed below. We just show the entering to and leaving from the interrupt/exception handler; nothing is done inside the handler. The entry address of the handler is 0x00000008. Based on the `ExcCode` in the Cause register, the control is transferred to a corresponding address by using a jump table which is given in data memory. For the three exceptions, the EPC is incremented by 4 before returning from the exception handler.

```
DEPTH = 64;                 % Memory depth and width are required    %
WIDTH = 32;                 % Enter a decimal number                 %
ADDRESS_RADIX = HEX;        % Address and value radixes are optional  %
DATA_RADIX = HEX;           % Enter BIN, DEC, HEX, or OCT; unless     %
                            % otherwise specified, radixes = HEX      %
CONTENT
BEGIN
[0..3f] : 00000000;  % Range--Every address from 0 to 3f = 00000000  %
                %    reset:                                      %
  0: 0800001d; % (00)    j     start        # entry on reset         %
  1: 00000000; % (04)    nop                 #                        %
                %    exc_base:               # exception handler      %
  2: 401a6800; % (08)    mfc0  $26, c0_cause # read cp0 cause reg     %
  3: 335b000c; % (0c)    andi  $27, $26, 0xc # get exccode, 2 bits here %
  4: 8f7b0020; % (10)    lw    $27, j_table($27) # get address from table %
  5: 00000000; % (14)    nop                 #                        %
  6: 03600008; % (18)    jr    $27           # jump to that address   %
  7: 00000000; % (1c)    nop                 #                        %
                %    int_entry:              # 0. interrupt handler   %
  c: 00000000; % (30)    nop                 # deal with interrupt here %
  d: 42000018; % (34)    eret                # return from interrupt  %
  e: 00000000; % (38)    nop                 #                        %
                %    sys_entry:              # 1. syscall handler     %
  f: 00000000; % (3c)    nop                 # do something here      %
                %    epc_plus4:              #                        %
 10: 401a7000; % (40)    mfc0  $26, c0_epc   # get epc                %
 11: 235a0004; % (44)    addi  $26, $26, 4   # epc + 4                %
 12: 409a7000; % (48)    mtc0  $26, c0_epc   # epc <- epc + 4         %
 13: 42000018; % (4c)    eret                # return from exception  %
 14: 00000000; % (50)    nop                 #                        %
                %    uni_entry:              # 2. unimpl. inst. handler %
 15: 00000000; % (54)    nop                 # do something here      %
 16: 08000010; % (58)    j     epc_plus4     # return                 %
 17: 00000000; % (5c)    nop                 #                        %
```

```
            %        ovf_entry:                  # 3. overflow handler      %
1a: 00000000; % (68)     nop                     # do something here        %
1b: 08000010; % (6c)     j      epc_plus4        # return                   %
1c: 00000000; % (70)     nop                     #                          %
            %        start:                      #                          %
1d: 2008000f; % (74)     addi  $8, $0, 0xf       # im[3:0] <- 1111          %
1e: 40886000; % (78)     mtc0  $8, c0_status     # exc/intr enable          %
1f: 8c080048; % (7c)     lw    $8, 0x48($0)      # try overflow exception   %
20: 8c09004c; % (80)     lw    $9, 0x4c($0)      # caused by add            %
            %        ov:                          #                          %
21: 01094020; % (84)     add   $9, $9, $8        # overflow                 %
22: 00000000; % (88)     nop                     #                          %
            %        sys:                         #                          %
23: 0000000c; % (8c)     syscall                 # system call              %
24: 00000000; % (90)     nop                     #                          %
            %        unimpl:                      #                          %
25: 0128001a; % (94)     div   $9, $8            # div, but not implemented %
26: 00000000; % (98)     nop                     #                          %
            %        int:                         #                          %
27: 34040050; % (9c)     ori   $4, $1, 0x50      # address of data[0]       %
28: 20050004; % (a0)     addi  $5, $0,  4        # counter                  %
29: 00004020; % (a4)     add   $8, $0, $0        # sum <- 0                 %
            %        loop:                        #                          %
2a: 8c890000; % (a8)     lw    $9, 0($4)         # load data                %
2b: 20840004; % (ac)     addi  $4, $4,  4        # address + 4              %
2c: 01094020; % (b0)     add   $8, $8, $9        # sum                      %
2d: 20a5ffff; % (b4)     addi  $5, $5, -1        # counter - 1              %
2e: 14a0fffb; % (b8)     bne   $5, $0, loop      # finish?                  %
2f: 00000000; % (bc)     nop                     #                          %
            %        finish:                      #                          %
30: 08000030; % (c0)     j      finish           # dead loop                %
END ;
```

The contents of scd_intr.mif for the data memory initialization are listed below. An address jump table is given from the stating address of 0x00000020.

```
DEPTH = 32;             % Memory depth and width are required        %
WIDTH = 32;             % Enter a decimal number                     %
ADDRESS_RADIX = HEX;    % Address and value radixes are optional     %
DATA_RADIX = HEX;       % Enter BIN, DEC, HEX, or OCT; unless         %
                        % otherwise specified, radixes = HEX         %
CONTENT
BEGIN
[0..1f] : 00000000; % Range--Every address from 0 to 1f = 00000000    %
     % address table for internal exceptions and external interrupt   %
    8 : 00000030;       % (20) int_entry # 0. address for interrupt    %
    9 : 0000003c;       % (24) sys_entry # 1. address for system call  %
    a : 00000054;       % (28) uni_entry # 2. address for unimpl. instruction %
```

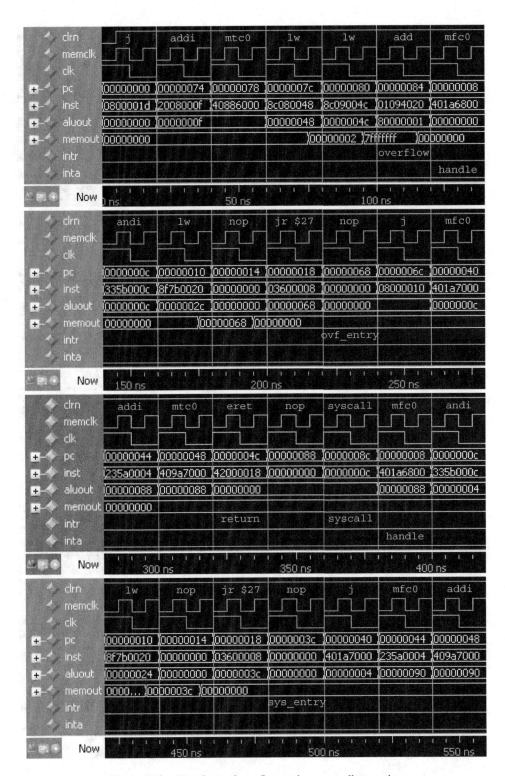

Figure 6.12 Waveform of overflow and system call exceptions

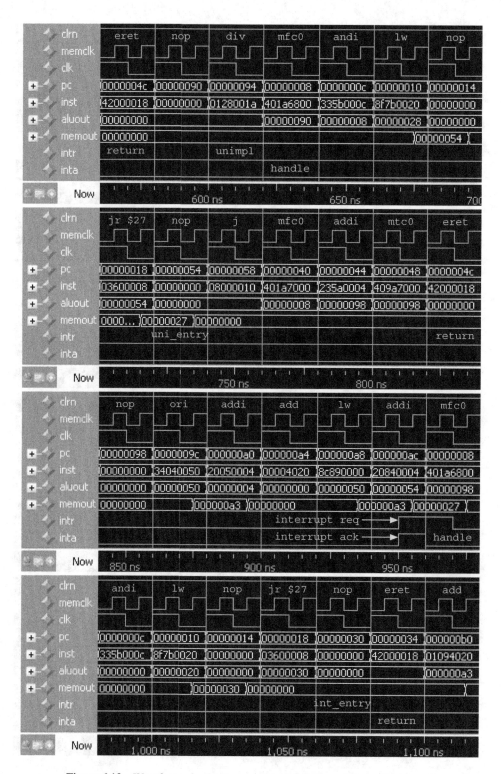

Figure 6.13   Waveform of unimplemented instruction exception and interrupt

```
 b : 00000068;    % (2c) ovf_entry # 3. address for signed overflow    %
12 : 00000002;    % (48) for testing overflow                          %
13 : 7fffffff;    % (4c) 2 + max_int -> overflow                       %
14 : 000000a3;    % (50) data[0]   0 +  a3 =  a3                        %
15 : 00000027;    % (54) data[1]  a3 +  27 =  ca                        %
16 : 00000079;    % (58) data[2]  ca +  79 = 143                        %
17 : 00000115;    % (5c) data[3] 143 + 115 = 258                        %
END ;
```

6.3.2   Test Waveform and Explanations

Figure 6.12 shows the waveform of overflow and system call exceptions. The add $9, $9, $8 instruction at 0x00000084 causes an overflow exception. The syscall instruction at 0x0000008c causes a system call exception.

Figure 6.13 shows the waveform of unimplemented instruction exception and interrupt. The div $9, $8 instruction at 0x00000094 causes an unimplemented instruction exception. An external interrupt occurs when the CPU is executing the addi $4, $4, 4 instruction at 0x000000ac.

Exercises

6.1 Design the interrupt controller shown in Figure 6.7 in Verilog HDL.

6.2 Design the Daisy chain circuit shown in Figure 6.8 in Verilog HDL.

6.3 Design a single-cycle CPU which can deal with multiple interrupts.

6.4 The hardware-unimplemented instructions can be implemented by software. If we can use lw instruction to load the unimplemented instruction, then we can emulate its execution in the exception handler. The following code illustrates the load of the instruction that causes the exception.

```
uni_entry:                              # unimpl. inst. exception handler
          mfc0   $26, c0_epc            # get epc
          lw     $26, 0($26)            # get the unimpl. instruction
emulator: ...                           # implement it here
finish:   mfc0   $26, c0_epc            # get epc
          addi   $26, $26, 4            # epc + 4
          mtc0   $26, c0_epc            # epc <-- epc + 4
          eret                          # return from exception
```

Unfortunately, the single-cycle CPU has separate instruction and data memories. The lw instruction cannot load data from the instruction memory. Consider how to implement the emulation of the unimplemented instructions in a single-cycle CPU computer.

6.5 Rewrite the interrupt/exception handler to implement nested interrupts. Note that some registers must be saved to the stack memory.

7

Multiple-Cycle CPU Design in Verilog HDL

The single-cycle (SC) CPU (central processing unit) executes each instruction in one clock cycle regardless of the complexity of the instructions. This chapter describes a method for designing a multiple-cycle (MC) CPU which uses less time to execute simple instructions.

7.1 Dividing Instruction Execution into Several Clock Cycles

The key design point of the MC CPU is to divide the execution of an instruction into several small steps. Each step takes a short clock cycle. Then we can use more cycles to execute complex instructions and use fewer cycles to execute simple instructions. The most complex instruction is `lw rt, offset(rs)` in our CPU. It takes five clock cycles:

1. Instruction fetch (IF)
2. Instruction decode and operand fetch (ID)
3. Execution (EXE)
4. Memory access (MEM)
5. Write back (WB).

The computational instructions take four cycles (no memory access cycle). The conditional instructions take three cycles. The branch target address can be calculated in the second cycle. Whether the branch is taken or not taken can be determined in the third cycle. The simplest instruction is `j address`. It takes two clock cycles. Instruction is fetched in the first cycle, and the program counter (PC) is written with the jump target address in the second cycle. Figure 7.1 compares the time requirements of the SC CPU and MC CPU when they execute the instructions of `lw`, `beq`, `add`, and `j`.

Table 7.1 lists the clock cycles taken by each instruction's execution. We also use PC+4 as the return address in our MC CPU design. The MIPS defines it as PC+8. We will use PC+8 in the design of a pipelined CPU in Chapter 8.

7.1.1 Instruction Fetch Cycle

In the IF cycle, an instruction is fetched and PC is incremented by 4:

```
IR <-- Memory[PC];
PC <-- PC + 4;
```

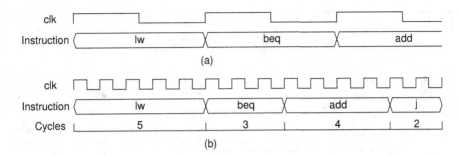

Figure 7.1 Timing comparison of (a) single-cycle and (b) multiple-cycle CPUs

Table 7.1 Clock cycles required by instructions

Instruction	Meaning	Cycles
add rd, rs, rt	Register add	4
sub rd, rs, rt	Register subtract	4
and rd, rs, rt	Register AND	4
or rd, rs, rt	Register OR	4
xor rd, rs, rt	Register XOR	4
sll rd, rt, sa	Shift left	4
srl rd, rt, sa	Logical shift right	4
sra rd, rt, sa	Arithmetic shift right	4
jr rs	Register jump	2
addi rt, rs, immediate	Immediate add	4
andi rt, rs, immediate	Immediate AND	4
ori rt, rs, immediate	Immediate OR	4
xori rt, rs, immediate	Immediate XOR	4
lw rt, offset(rs)	Load word	5
sw rt, offset(rs)	Store word	4
beq rs, rt, offset	Branch on equal	3
bne rs, rt, offset	Branch on not equal	3
lui rt, immediate	Load upper immediate	4
j address	Jump	2
jal address	Subroutine call	2

Therefore, we need two registers: IR (instruction register) and PC. Their write enable signals are `wir` (write IR) and `wpc` (write PC), respectively. The circuit is shown in Figure 7.2.

PC + 4 is done by the arithmetic logic unit (ALU). `aluc` should be x000 (performs addition, see Figure 4.13). There are three multiplexers. The `selpc` (select PC) signal should be 1 for selecting PC; `alusrcb` (ALU source B) should be 01 for selecting the constant 4; and `pcsrc` (PC source) should be 00 for selecting the ALU output (PC + 4).

7.1.2 Instruction Decode and Source Operand Read Cycle

In ID cycle, if the instruction in the IR is not a jump (`j`, `jal`, or `jr`), the CPU performs the following operations:

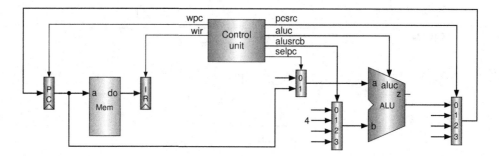

Figure 7.2   Block diagram of instruction fetch of an MC CPU

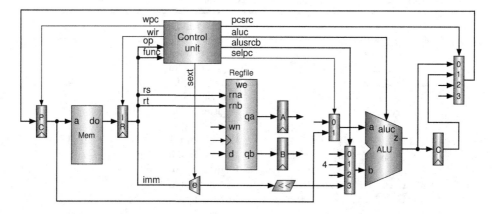

Figure 7.3   Block diagram of instruction decode of an MC CPU

```
A <-- RegisterFile[rs];
B <-- RegisterFile[rt];
C <-- PC + sign_extend(offset) << 2;
```

Three registers (A, B, and C) are used. Registers A and B store the contents of registers rs and rt, respectively. Register C stores the branch target address for beq or bne instruction. The circuit is shown in Figure 7.3.

For calculating the branch target address, the ALU performs addition (aluc = x000); PC (already added by 4 in the IF cycle) and sign-extended immediates (sext = 1) are selected (selpc = 1 and alusrcb = 11). The registers A, B, and C have no write enable signals; they are updated at every rising edge of the clock. Therefore, we must keep the IR unchanged during the execution of the current instruction.

If the instruction in the IR is a j, jal, or jr instruction, the CPU performs the following operations:

```
j:   PC <-- {PC[31:28],address,00};
jal: RegisterFile[31] <-- PC;
     PC <-- {PC[31:28],address,00};
jr:  PC <-- RegisterFile[rs];
```

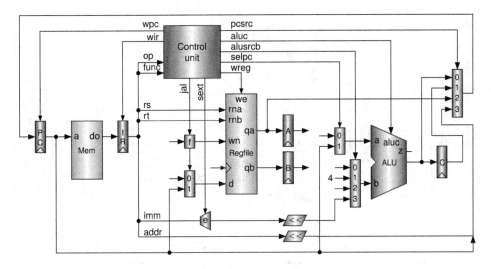

Figure 7.4   Block diagram for jump instructions in an MC CPU

The circuit is shown in Figure 7.4. wpc should be 1. For the j or jal instruction, the jump target address is selected (pcsrc = 11). For the jal instruction, the return address PC (already added by 4 in the IF cycle) should be saved to register $31 (register file write enable wreg = 1, jal = 1, and the component f generates a constant 31). For the jr instruction, let pcsrc = 10 to select the value in the register rs of the register file. The jump instructions complete their execution in ID cycle.

7.1.3   Execution Cycle

In the EXE cycle, if the instruction in the IR is a conditional branch instruction (beq or bne), the CPU performs the following operations:

```
beq: If (A == B) PC <-- C;
bne: If (A != B) PC <-- C;
```

The branch target address is in register C, which is calculated in the ID cycle. The circuit is shown in Figure 7.5. ALU performs subtraction (aluc = x100). The z flag is used to check the branch condition. If beq·z + bne·z̄ is true, then wpc is 1 (updates PC). pcsrc is 01 for selecting the branch target address. The conditional branch instructions finish their execution in this cycle.

For the other instructions, the CPU performs the following operations:

```
add/sub/and/or/xor: C <-- A op B;
sll:                C <-- B << sa;
srl:                C <-- B >> sa;
sra:                C <-- signed(B) >>> sa;
addi:               C <-- A +  sign_extend(immediate);
andi/ori/xori:      C <-- A op zero_extend(immediate);
lw/sw:              C <-- A +  sign_extend(offset);
lui:                C <-- immediate << 16;
```

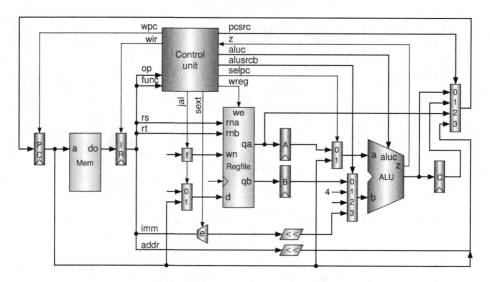

Figure 7.5   Block diagram for branch instructions in an MC CPU

All of these instructions perform calculations, and the result is saved to register C. The circuit is shown in Figure 7.6.

For add, sub, and, or, and xor instructions, except for aluc, their control signals are the same: shift = 0 and selpc = 0 (selects the content of register A) and alusrcb = 00 (selects the content of register B). For three shift instructions, the shift amount sa is selected for the ALU input A; therefore, shift = 1 and selpc = 0. The content of register B is selected, which will be shifted by ALU; therefore, alusrcb = 00. aluc of each of the shift instructions is different.

The rest are the I-format instructions. The extended immediate is selected (alusrcb = 10). For the addi, lw, or sw instruction, the immediate is sign-extended (sext = 1). For the andi, ori, or xori

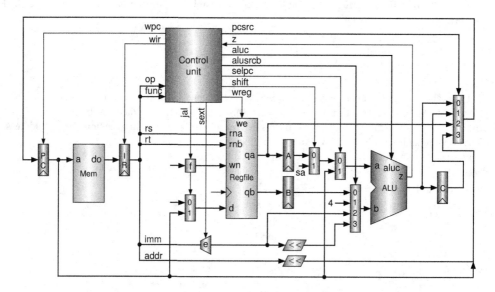

Figure 7.6   Block diagram for computational instructions in an MC CPU

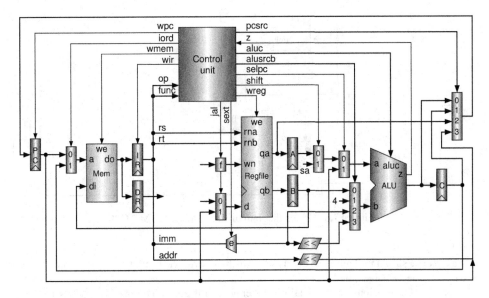

Figure 7.7 Block diagram for load and store instructions in an MC CPU

instruction, the immediate is zero-extended (sext = 0). The lui instruction does not care about the extension. The lw or sw instruction calculates the memory address in this cycle (aluc = x000).

Other control signals are as follows: wpc = 0 (do not update PC), pcsrc = xx (the selected address will not be used anymore), wir = 0 (do not update IR), and wreg = 0 (do not write register file).

7.1.4   Memory Access Cycle

The lw or sw instruction enters the MEM cycle and performs the following:

```
lw: DR <-- Memory[C];
sw: Memory[C] <-- B;
```

These two instructions access the memory. The circuit is shown in Figure 7.7. The memory address is in register C, which is calculated in the EXE cycle. Different from an SC CPU, in an MC CPU we use only one memory module for storing both instructions and data. Therefore, a 32-bit 2-to-1 multiplexer is used for selecting the instruction address or data address. The selection signal is iord (instruction or data). For lw and sw instructions, the content of register C will be selected as the memory address (iord = 1). For the lw instruction, DR is a data register, to which the data word read from memory is written. For the sw instruction, the content in register B is written to memory (wmem = 1). The sw instruction finishes the execution in this cycle.

7.1.5   Write Back Cycle

The following instructions enter the WB cycle and perform the following:

```
add/sub/and/or/xor/sll/srl/sra: RegisterFile[rd] <-- C;
addi/andi/ori/xori/lui:         RegisterFile[rt] <-- C;
lw:                             RegisterFile[rt] <-- DR;
```

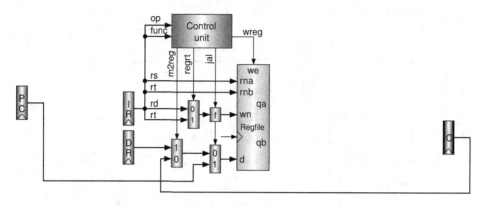

Figure 7.8   Block diagram of write back of an MC CPU

The circuit is shown in Figure 7.8. All these instructions write the result to the register file; therefore, wreg = 1. The jal signal will be 0.

For R-format instructions, the destination register number is rd; for I-format instructions, the destination register number is rt. Therefore, we use a 5-bit 2-to-1 multiplexer for selecting either rd or rt. The selection signal is regrt (register rt). The lw instruction writes the memory data in register DR to the register file, and other instructions write the ALU result in register C to the register file. We use a 32-bit 2-to-1 multiplexer for selecting the content in register C or DR. The selection signal is m2reg (memory data to register file).

All the other control signals will be zeroes (inactive). All the instructions finish their execution in this cycle. After the WB cycle, the CPU enters the IF cycle again for executing the next instruction.

7.2   Multiple-Cycle CPU Schematic and Verilog HDL Codes

This section shows the overall schematic circuit of the MC CPU and gives the Verilog HDL (hardware description language) code that implements the datapath of the CPU. The control unit design is little bit complex and will be described in the next section.

7.2.1   Multiple-Cycle CPU Structure

According to the discussions of the previous section, we can get the overall schematic circuit of the MC CPU, as shown in Figure 7.9. Except for the control unit, all the components in the figure have been described.

Logically, the CPU does not contain memory, so we redraw the circuit and show it in Figure 7.10. It can be called an MC computer.

7.2.2   Multiple-Cycle CPU Verilog HDL Codes

Below is the Verilog HDL code of the circuit shown in Figure 7.10. It invokes mccpu (CPU) and mcmem (memory) modules.

```
module mccomp (clk,clrn,q,a,b,alu,adr,tom,fromm,pc,ir,memclk);
    input       clk, clrn;                      // clock and reset
    input       memclk;                         // memory clk for sync ram
```

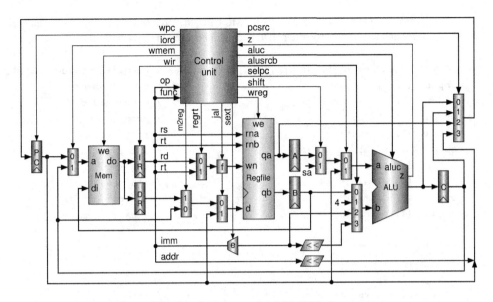

Figure 7.9   Block diagram of an MC CPU plus memory

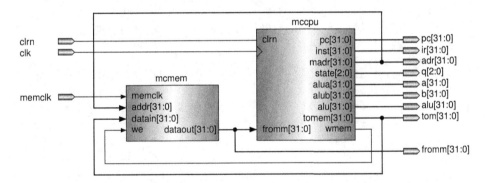

Figure 7.10   Block diagram of a MC computer

```
output [31:0] a;                    // alu input a
output [31:0] b;                    // alu input b
output [31:0] alu;                  // alu result
output [31:0] adr;                  // memory address
output [31:0] tom;                  // data to memory
output [31:0] fromm;                // data from memory
output [31:0] pc;                   // program counter
output [31:0] ir;                   // instruction register
output  [2:0] q;                    // state
wire          wmem;                 // memory write enable
mccpu mc_cpu (clk,clrn,fromm,pc,ir,a,b,alu,wmem,adr,tom,q);    // cpu
mcmem memory (fromm,tom,adr,wmem,memclk);                      // memory
endmodule
```

The following Verilog HDL code implements the MC CPU, the module of mccpu. It invokes the module of the control unit mccu whose Verilog HDL code will be given in the next section.

```verilog
module mccpu (clk,clrn,frommem,pc,inst,alua,alub,alu,wmem,madr,tomem,state);
    input   [31:0] frommem;                    // data from memory
    input          clk, clrn;                  // clock and reset
    output  [31:0] pc;                         // program counter
    output  [31:0] inst;                       // instruction
    output  [31:0] alua;                       // alu input a
    output  [31:0] alub;                       // alu input b
    output  [31:0] alu;                        // alu result
    output  [31:0] madr;                       // memory address
    output  [31:0] tomem;                      // data to memory
    output   [2:0] state;                      // state
    output         wmem;                       // memory write enable
    // instruction fields
    wire     [5:0] op   = inst[31:26];         // op
    wire     [4:0] rs   = inst[25:21];         // rs
    wire     [4:0] rt   = inst[20:16];         // rt
    wire     [4:0] rd   = inst[15:11];         // rd
    wire     [5:0] func = inst[05:00];         // func
    wire    [15:0] imm  = inst[15:00];         // immediate
    wire    [25:0] addr = inst[25:00];         // address
    // control signals
    wire     [3:0] aluc;                       // alu operation control
    wire     [1:0] pcsrc;                      // select pc source
    wire           wreg;                       // write regfile
    wire           regrt;                      // dest reg number is rt
    wire           m2reg;                      // instruction is an lw
    wire           shift;                      // instruction is a shift
    wire     [1:0] alusrcb;                    // alu input b selection
    wire           jal;                        // instruction is a jal
    wire           sext;                       // is sign extension
    wire           wpc;                        // write pc
    wire           wir;                        // write ir
    wire           iord;                       // select memory address
    wire           selpc;                      // select pc
    // datapath wires
    wire    [31:0] bpc;                        // branch target address
    wire    [31:0] npc;                        // next pc
    wire    [31:0] qa;                         // regfile output port a
    wire    [31:0] qb;                         // regfile output port b
    wire    [31:0] alua;                       // alu input a
    wire    [31:0] alub;                       // alu input b
    wire    [31:0] wd;                         // regfile write port data
    wire    [31:0] r;                          // alu out or mem
    wire    [31:0] sa   = {27'b0,inst[10:6]};  // shift amount
```

```
wire     [15:0] s16 = {16{sext & inst[15]}};      // 16-bit signs
wire     [31:0] i32 = {s16,imm};                  // 32-bit immediate
wire     [31:0] dis = {s16[13:0],imm,2'b00};      // word distance
wire     [31:0] jpc = {pc[31:28],addr,2'b00};     // jump target address
wire      [4:0] reg_dest;                          // rs or rt
wire      [4:0] wn  = reg_dest | {5{jal}};         // regfile write reg #
wire            z;                                 // alu, zero tag
wire     [31:0] rega;                              // register a
wire     [31:0] regb;                              // register b
wire     [31:0] regc;                              // register c
wire     [31:0] data;                              // output of dr
wire     [31:0] opa;                               // sa or output of reg a
// datapath
mccu control_unit (op,func,z,clk,clrn,            // control unit
                   wpc,wir,wmem,wreg,
                   iord,regrt,m2reg,aluc,
                   shift,selpc,alusrcb,
                   pcsrc,jal,sext,state);
dffe32  ip      (npc,clk,clrn,wpc,pc);            // pc register
dffe32  ir      (frommem,clk,clrn,wir,inst);     // instruction register
dff32   dr      (frommem,clk,clrn,data);         // data register
dff32   reg_a   (qa, clk,clrn,rega);             // register a
dff32   reg_b   (qb, clk,clrn,regb);             // register b
dff32   reg_c   (alu,clk,clrn,regc);             // register c
mux2x32 aorsa   (rega,sa,shift,opa);             // sa or output of reg a
mux2x32 alu_a   (opa,pc,selpc,alua);             // alu input a
mux4x32 alu_b   (regb,32'h4,i32,dis,alusrcb,alub); // alu input b
mux2x32 alu_m   (regc,data,m2reg,r);             // alu out or mem data
mux2x32 mem_a   (pc,regc,iord,madr);             // memory address
mux2x32 link    (r,pc,jal,wd);                   // r or pc
mux2x5  reg_wn  (rd,rt,regrt,reg_dest);          // rs or rt
mux4x32 nextpc  (alu,regc,qa,jpc,pcsrc,npc);     // next pc
regfile rf      (rs,rt,wd,wn,wreg,clk,clrn,qa,qb); // register file
alu alunit      (alua,alub,aluc,alu,z);          // alu
assign tomem = regb;                              // output of reg b
endmodule
```

7.3 Multiple-Cycle CPU Control Unit Design

In the MC CPU, different instructions take different clock cycles for execution. This can be implemented with a finite state machine, which is a typical sequential circuit. This section describes the design of the finite state machine for implementing the control unit of the MC CPU.

7.3.1 State Transition Diagram of the Control Unit

There are different state transition diagrams that can implement the control unit. Figure 7.11 shows just one of them. It uses the smallest number of states. A unique name and binary number are assigned to each

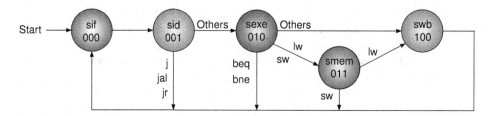

Figure 7.11   State transition diagram of an MC CPU

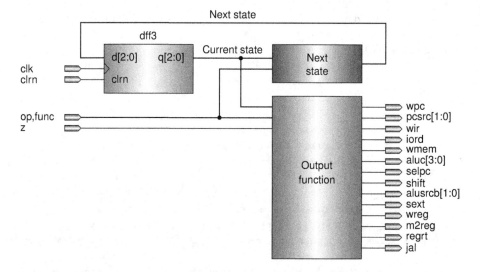

Figure 7.12   Block diagram of the control unit of an MC CPU

of the states. Each state takes one clock cycle. The arrowed lines and the corresponding labels (instruction names) show the transition from one state to another.

From the figure, we can see that the jump instructions take two cycles, the branch instructions take three cycles, the load instruction takes five cycles, and the others take four cycles. This satisfies the requirements listed in Table 7.1.

7.3.2   Circuit Structure of the Control Unit

As we described in Chapter 2, the sequential circuit consists of three parts: D flip-flops, next state block, and output function block. The last two blocks themselves are combinational circuits that can be designed by starting with truth tables.

Figure 7.12 shows the block diagram of the control unit. Because there are five states, we used three D flip-flops ($\lceil \log_2 5 \rceil = 3$). The output of the D flip-flops represents the current state, and the value of the next state will be written to the D flip-flops at the rising edge of the clock.

7.3.3   Next State Function

From the state transition diagram shown in Figure 7.11, we can draw the truth table for the next state block, as shown in Table 7.2.

Table 7.2 Truth table of next state function

Current state		Input	Next state	
State	q[2:0]	Instruction	State	d[2:0]
sif	0 0 0	x	sid	0 0 1
sid	0 0 1	i_j	sif	0 0 0
		i_jal	sif	0 0 0
		i_jr	sif	0 0 0
		Others	sexe	0 1 0
sexe	0 1 0	i_beq	sif	0 0 0
		i_bne	sif	0 0 0
		i_lw	smem	0 1 1
		i_sw	smem	0 1 1
		Others	swb	1 0 0
smem	0 1 1	i_lw	swb	1 0 0
		i_sw	sif	0 0 0
swb	1 0 0	x	sif	0 0 0

From the table, we can easily get the logic expressions of the output signals (d[2], d[1], and d[0]) of the next state block. The instruction can be decoded based on its op and func, which we have described in Chapter 5. Then the schematic circuit can be drawn based on the logic expressions. We do not show it here.

7.3.4 Output Function of Control Signal

The output function block generates the control signals of the MC CPU. According to the discussions of the previous section, we can prepare the truth table for the control signals, as shown in Table 7.3. Similarly, we can write the logic expression of each signal and design the circuit based on these expressions. As an exercise, complete the table.

7.3.5 Verilog HDL Codes of the Control Unit

We can write the Verilog HDL code for implementing the control unit, as given below. Only state is a real register; all others are nets generated by combinational circuits, although some of them are declared as reg type.

```
module mccu (op,func,z,clk,clrn,wpc,wir,wmem,wreg,iord,regrt,m2reg,aluc,
             shift,alusrca,alusrcb,pcsrc,jal,sext,state);    // control unit
    input      [5:0] op, func;                  // op, func
    input            clk, clrn;                 // clock and reset
    input            z;                         // alu zero flag
    output reg       wpc;                       // write pc
    output reg       wir;                       // write ir
    output reg       wmem;                      // memory write enable
    output reg       wreg;                      // write regfile
```

Table 7.3 Truth table of control signals

State	Inst.	z	wpc	pcsrc[1:0]	wir	iord	wmem	aluc[3:0]	selpc	shift	alusrcb[1:0]	sext	wreg	m2reg	regrt	jal
sif	x	x					0									
sid	i_j	x					0									
	i_jal	x					0									
	i_jr	x					0									
	others	x					0									
sexe	i_add	x	0	x x	0	x	0	x 0 0 0	0	0	0 0	x	0	x	x	0
	i_sub	x					0									
	i_and	x					0									
	i_or	x					0									
	i_xor	x					0									
	i_sll	x					0									
	i_srl	x					0									
	i_sra	x					0									
	i_addi	x					0									
	i_andi	x					0									
	i_ori	x					0									
	i_xori	x					0									
	i_lw	x					0									
	i_sw	x					0									
	i_beq	0 1					0									
	i_bne	0 1					0									
	i_lui	x					0									
smem	i_lw	x					0									
	i_sw	x					1									
swb	rtype	x					0									
	i_andi	x					0									
	i_ori	x					0									
	i_lui	x					0									
	i_lw	x					0									

```
output reg        iord;          // select memory address
output reg        regrt;         // dest reg number is rt
output reg        m2reg;         // instruction is an lw
output reg        shift;         // instruction is a shift
output reg        alusrca;       // select pc
output reg        jal;           // instruction is a jal
output reg        sext;          // is sign extension
output reg [3:0]  aluc;          // alu operation control
output reg [1:0]  alusrcb;       // alu input b selection
```

```verilog
output reg [1:0] pcsrc;                        // select pc source
output reg [2:0] state;                        // state
reg        [2:0] next_state;                   // next state
parameter  [2:0] sif  = 3'b000,                // IF   state
                 sid  = 3'b001,                // ID   state
                 sexe = 3'b010,                // EXE  state
                 smem = 3'b011,                // MEM  state
                 swb  = 3'b100;                // WB   state
wire rtype,i_add,i_sub,i_and,i_or,i_xor,i_sll,i_srl,i_sra,i_jr;
wire i_addi,i_andi,i_ori,i_xori,i_lw,i_sw,i_beq,i_bne,i_lui,i_j,i_jal;
// and (out, in1, in2, ...);                   // instruction decode
and (rtype,~op[5],~op[4],~op[3],~op[2],~op[1],~op[0]);
and (i_add,rtype, func[5],~func[4],~func[3],~func[2],~func[1],~func[0]);
and (i_sub,rtype, func[5],~func[4],~func[3],~func[2], func[1],~func[0]);
and (i_and,rtype, func[5],~func[4],~func[3], func[2],~func[1],~func[0]);
and (i_or, rtype, func[5],~func[4],~func[3], func[2],~func[1], func[0]);
and (i_xor,rtype, func[5],~func[4],~func[3], func[2], func[1],~func[0]);
and (i_sll,rtype,~func[5],~func[4],~func[3],~func[2],~func[1],~func[0]);
and (i_srl,rtype,~func[5],~func[4],~func[3],~func[2], func[1],~func[0]);
and (i_sra,rtype,~func[5],~func[4],~func[3],~func[2], func[1], func[0]);
and (i_jr, rtype,~func[5],~func[4], func[3],~func[2],~func[1],~func[0]);
and (i_addi,~op[5],~op[4], op[3],~op[2],~op[1],~op[0]);
and (i_andi,~op[5],~op[4], op[3], op[2],~op[1],~op[0]);
and (i_ori, ~op[5],~op[4], op[3], op[2],~op[1], op[0]);
and (i_xori,~op[5],~op[4], op[3], op[2], op[1],~op[0]);
and (i_lw,   op[5],~op[4],~op[3],~op[2], op[1], op[0]);
and (i_sw,   op[5],~op[4], op[3],~op[2], op[1], op[0]);
and (i_beq, ~op[5],~op[4],~op[3], op[2],~op[1],~op[0]);
and (i_bne, ~op[5],~op[4],~op[3], op[2],~op[1], op[0]);
and (i_lui, ~op[5],~op[4], op[3], op[2], op[1], op[0]);
and (i_j,   ~op[5],~op[4],~op[3],~op[2], op[1],~op[0]);
and (i_jal, ~op[5],~op[4],~op[3],~op[2], op[1], op[0]);
wire i_shift = i_sll | i_srl | i_sra;
always @* begin                                // default outputs:
    wpc     = 0;                               // do not write pc
    wir     = 0;                               // do not write ir
    wmem    = 0;                               // do not write memory
    wreg    = 0;                               // do not write regfile
    iord    = 0;                               // select pc as address
    aluc    = 4'bx000;                         // alu operation: add
    alusrca = 0;                               // alu a: reg a or sa
    alusrcb = 2'h0;                            // alu input b: reg b
    regrt   = 0;                               // reg dest no: rd
    m2reg   = 0;                               // select reg c
    shift   = 0;                               // select reg a
    pcsrc   = 2'h0;                            // select alu output
    jal     = 0;                               // not a jal
```

```
            sext    = 1;                              // sign extend
            case (state) //-------------------------------------------------- IF:
                sif: begin                            // IF state
                    wpc     = 1;                       // write PC
                    wir     = 1;                       // write IR
                    alusrca = 1;                       // PC
                    alusrcb = 2'h1;                    // 4
                    next_state = sid;                  // next state: ID
                end //-------------------------------------------------------- ID:
                sid: begin                            // ID state
                    if (i_j) begin                    // j instruction
                        pcsrc = 2'h3;                  // jump address
                        wpc   = 1;                     // write PC
                        next_state = sif;              // next state: IF
                    end else if (i_jal) begin          // jal instruction
                        pcsrc = 2'h3;                  // jump address
                        wpc   = 1;                     // write PC
                        jal   = 1;                     // reg no = 31
                        wreg  = 1;                     // save PC+4
                        next_state = sif;              // next state: IF
                    end else if (i_jr) begin           // jr instruction
                        pcsrc = 2'h2;                  // jump register
                        wpc   = 1;                     // write PC
                        next_state = sif;              // next state: IF
                    end else begin                     // other instructions
                        aluc    = 4'bx000;             // add
                        alusrca = 1;                   // PC
                        alusrcb = 2'h3;                // branch offset
                        next_state = sexe;             // next state: EXE
                    end
                end
            end //-------------------------------------------------------- EXE:
//           add sub and or xor sll srl sra lw sw beq bne addi andi ori xori lui
// aluc[3]   X   X   X   X   X   0   0   1  X  X   X   X   X    X    X   X    X
// aluc[2]   0   1   0   1   0   0   1   1  0  0   0   0   0    0    1   0    1
// aluc[1]   0   0   0   0   1   1   1   1  0  0   1   1   0    0    0   1    1
// aluc[0]   0   0   1   1   0   1   1   1  0  0   0   0   0    1    1   0    0
                sexe: begin                           // EXE state
                    aluc[3] = i_sra;
                    aluc[2] = i_sub | i_or  | i_srl | i_sra | i_ori  | i_lui;
                    aluc[1] = i_xor | i_sll | i_srl | i_sra | i_xori | i_beq |
                              i_bne | i_lui;
                    aluc[0] = i_and | i_or  | i_sll | i_srl | i_sra  | i_andi |
                              i_ori;
                    if (i_beq || i_bne) begin          // beq or bne inst
                        pcsrc = 2'h1;                  // branch address
```

```
                    wpc = i_beq & z | i_bne & ~z;   // write PC
                    next_state = sif;               // next state: IF
                end else begin                      // other instruction
                    if (i_lw || i_sw) begin         // lw or sw inst
                        alusrcb = 2'h2;             // select offset
                        next_state = smem;          // next state: MEM
                    end else begin                  // other instruction
                        if (i_shift) shift = 1;     // shift instruction
                        if (i_addi || i_andi || i_ori || i_xori || i_lui)
                            alusrcb = 2'h2;         // select immediate
                        if (i_andi || i_ori || i_xori) sext=0;   // 0 extend
                        next_state = swb;           // next state: WB
                    end
                end
            end //---------------------------------------------------- MEM:
        smem: begin                                 // MEM state
            iord = 1;                               // memory address = C
            if (i_lw) begin                         // load
                next_state = swb;                   // next state: WB
            end else begin                          // store
                wmem = 1;                           // write memory
                next_state = sif;                   // next state: IF
            end
        end //----------------------------------------------------- WB:
        swb: begin                                  // WB state
            if (i_lw) m2reg = 1;                    // select memory data
            if (i_lw || i_addi || i_andi || i_ori || i_xori || i_lui)
                regrt = 1;                          // reg dest no: rt
            wreg = 1;                               // write register file
            next_state = sif;                       // next state: IF
        end //----------------------------------------------------- END
        default: begin
            next_state = sif;                       // default state: IF
        end
    endcase
end
always @ (posedge clk or negedge clrn) begin
    if (!clrn) begin
        state <= sif;                               // reset state to IF
    end else begin
        state <= next_state;                        // state transition
    end
end
endmodule
```

7.4   Memory and Test Program

This section shows the memory, test program, and simulation waveforms.

7.4.1   Memory Design

Similar to the data memory described in Chapter 6, we use lpm_ram_dq of the Altera LPM (library of parameterized modules). The memory module is initialized with a mcmem.hex file.

```
module mcmem (dataout,datain,addr,we,memclk);       // memory (sram)
    input   [31:0] datain;                          // mem data input
    input   [31:0] addr;                            // mem address
    input          we;                              // write enable
    input          memclk;                          // sync ram clock
    output  [31:0] dataout;                         // mem data output
    lpm_ram_dq ram (.data(datain),                  // data in
                    .address(addr[7:2]),            // word address
                    .we(we),                        // write enable
                    .inclock(memclk),               // in reg clock
                    .outclock(memclk),              // out reg clock
                    .q(dataout));                   // mem data out
    defparam ram.lpm_width          = 32;           // data: 32 bits
    defparam ram.lpm_widthad        = 6;            // 2^6 = 64 words
    defparam ram.lpm_file           = "mcmem.hex";  // mem init file
    defparam ram.lpm_indata         = "REGISTERED"; // in reg (data)
    defparam ram.lpm_outdata        = "REGISTERED"; // out reg (data)
    defparam ram.lpm_address_control = "REGISTERED"; // in reg (a, we)
endmodule
```

7.4.2   Test Program

The following is the content of mcmem.mif used for testing the MC CPU. It must be converted to mcmem.hex for the use by ModelSim. Because there is only one memory module, the test code and data are put in the memory together.

```
DEPTH = 64;           % Memory depth and width are required            %
WIDTH = 32;           % Enter a decimal number                         %
ADDRESS_RADIX = HEX;  % Address and value radixes are optional         %
DATA_RADIX = HEX;     % Enter BIN, DEC, HEX, or OCT; unless            %
                      % otherwise specified, radixes = HEX             %
CONTENT
BEGIN
[0..3f] : 00000000; % Range--Every address from 0 to 3f = 00000000    %
 0 : 3c010000; % (00) main: lui  $1, 0         # address of data[0]    %
 1 : 34240080; % (04)       ori  $4, $1, 0x80  # address of data[0]    %
 2 : 20050004; % (08)       addi $5, $0,  4     # counter              %
 3 : 0c000018; % (0c) call: jal  sum           # call function         %
 4 : ac820000; % (10)       sw   $2, 0($4)      # store result         %
```

```
 5 : 8c890000; % (14)            lw   $9, 0($4)      # check sw               %
 6 : 01244022; % (18)            sub  $8, $9, $4     # sub: $8 <-- $9 - $4    %
 7 : 20050003; % (1c)            addi $5, $0, 3      # counter                %
 8 : 20a5ffff; % (20) loop2: addi $5, $5, -1         # counter - 1            %
 9 : 34a8ffff; % (24)            ori  $8, $5, 0xffff # zero-extend: 0000ffff  %
 a : 39085555; % (28)            xori $8, $8, 0x5555 # zero-extend: 0000aaaa  %
 b : 2009ffff; % (2c)            addi $9, $0, -1     # sign-extend: ffffffff  %
 c : 312affff; % (30)            andi $10, $9,0xffff # zero-extend: 0000ffff  %
 d : 01493025; % (34)            or   $6, $10, $9    # or:  ffffffff          %
 e : 01494026; % (38)            xor  $8, $10, $9    # xor: ffff0000          %
 f : 01463824; % (3c)            and  $7, $10, $6    # and: 0000ffff          %
10 : 10a00001; % (40)            beq  $5, $0, shift  # if $5 = 0, goto shift  %
11 : 08000008; % (44)            j    loop2          # jump loop2             %
12 : 2005ffff; % (48) shift: addi $5, $0, -1         # $5   = ffffffff        %
13 : 000543c0; % (4c)            sll  $8, $5, 15     # <<15 = ffff8000        %
14 : 00084400; % (50)            sll  $8, $8, 16     # <<16 = 80000000        %
15 : 00084403; % (54)            sra  $8, $8, 16     # >>16 = ffff8000(arith) %
16 : 000843c2; % (58)            srl  $8, $8, 15     # >>15 = 0001ffff(logic) %
17 : 08000017; % (5c) finish: j   finish             # dead loop              %
18 : 00004020; % (60) sum:   add  $8, $0, $0          # sum                    %
19 : 8c890000; % (64) loop:  lw   $9, 0($4)           # load data              %
1a : 20840004; % (68)            addi $4, $4, 4      # address + 4            %
1b : 01094020; % (6c)            add  $8, $8, $9     # sum                    %
1c : 20a5ffff; % (70)            addi $5, $5, -1     # counter - 1            %
1d : 14a0fffb; % (74)            bne  $5, $0, loop   # finish?                %
1e : 00081000; % (78)            sll  $2, $8, 0      # move result to v0      %
1f : 03e00008; % (7c)            jr   $31            # return                 %
20 : 000000a3; % (80) data[0]      0 +  a3 =  a3                              %
21 : 00000027; % (84) data[1]     a3 +  27 =  ca                              %
22 : 00000079; % (88) data[2]     ca +  79 = 143                             %
23 : 00000115; % (8c) data[3]    143 + 115 = 258                             %
24 : 00000000; % (90) sum, should be = 0x00000258, stored by sw             %
END ;
```

7.4.3 Multiple-Cycle CPU Simulation Waveforms

Some simulation waveforms are given in Figures 7.13–7.15. The state information is also shown in the figures. Figure 7.13 shows that the first instruction (lui) takes four clock cycles.

Figure 7.14 shows the waveform when the CPU executes the lw $9, 0($4) instruction (0x8c890000) at pc = 0x00000064. It takes five clock cycles.

Figure 7.15 shows the waveform when the CPU executes the jr $31 instruction (returning from subroutine) at pc = 0x0000007c. The return address is pc = 0x00000010. It takes two clock cycles. The next instruction, sw, takes four cycles.

It takes 1008 ns for the MC CPU to execute the test program. In contrast, the SC CPU takes 1320 ns for executing the same test program. The speedup is about 131%.

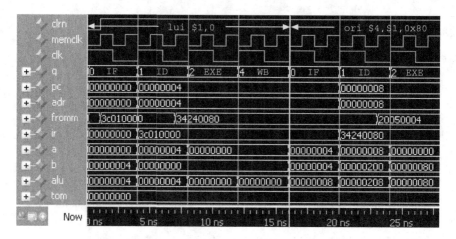

Figure 7.13   Waveform of an MC CPU (lui instruction)

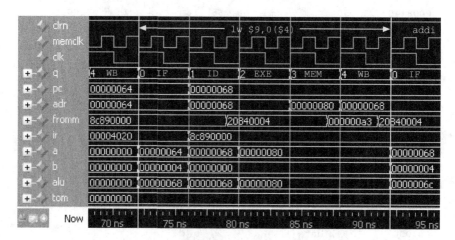

Figure 7.14   Waveform of an MC CPU (lw instruction)

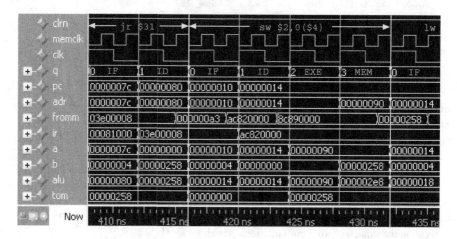

Figure 7.15   Waveform of an MC CPU (jr and sw instructions)

Exercises

7.1 Compare SC CPUs and MC CPUs. Indicate their advantages and disadvantages.

7.2 Can the registers A, B, and DR in Figure 7.9 be deleted?

7.3 Design the MC CPU with schematic capture.

7.4 We used the same program to test the SC CPU and MC CPU so that we knew the execution times taken by the two CPUs. Suppose the clock cycle of the SC CPU is little bit shorter than the five times of the clock cycle of the MC CPU, say 1 ns shorter. Then calculate the time taken by the SC CPU when executing the same test program and compare it with that of the MC CPU.

7.5 Design an MC CPU in Verilog HDL of the behavioral style. Requirement: the lui instruction takes only two clock cycles.

7.6 Design an MC CPU with the state transition diagram of Figure 7.16.

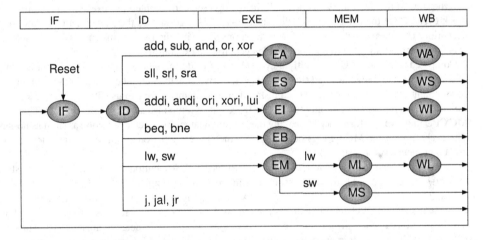

Figure 7.16   Another state transition diagram of an MC CPU

7.7 Referring to Chapter 6, design an MC CPU that can deal with interrupt and exceptions.

8

Design of Pipelined CPU with Precise Interrupt in Verilog HDL

The single-cycle (SC) CPU (central processing unit), described in Chapter 5, executes each instruction in one clock cycle, regardless of the complexity of instructions. The cycle time must be determined so that the slowest instruction, lw in our CPU's instruction set, can complete the execution correctly in one clock cycle. Once the cycle time is determined, every instruction uses the same length of time for the execution. This means that, although fast instructions, jump instructions for instance, can complete their executions early, they have to take one whole cycle for the executions.

Compared to the SC CPU, the multiple-cycle (MC) CPU, described in Chapter 7, divides the execution of an instruction into several small steps (short cycles). By using the finite state machine, we can use more or less number of cycles to execute each instruction. For example, we can use five cycles to execute the lw instruction, but we can use only two cycles to execute a jump instruction (jr, j, or jal). That is, the MC CPU improves performance by making the execution latency of each instruction as short as possible. The cycle time of the MC CPU is determined by the slowest functional unit, the arithmetic logic unit (ALU) or memory, for instance.

The common feature of both SC CPU and MC CPU is that the execution of an instruction can be started only after the execution of its predecessor (prior instruction) is completed. This chapter introduces how to design a five-stage pipelined CPU which allows overlapping execution of multiple instructions. In the five-stage pipelined CPU, although an instruction takes five clock cycles to pass through the pipeline, a new instruction can enter the pipeline during every clock cycle. Under ideal circumstances, the pipelined CPU can produce a result in every clock cycle. In contrast, the MC CPU produces a result in every 2–5 clock cycles. That is, the pipelined CPU improves the throughput, instead of shortening the execution latency of each instruction.

Because of the overlapping execution of multiple instructions, the pipelined CPU may encounter three types of hazards: structural hazards, control hazards, and data hazards, which will be explained starting in Section 8.2. We must find solutions to these problems. Furthermore, dealing with the exceptions and interrupts becomes more difficult in the design of the pipelined CPU than with the SC or MC CPU.

This chapter describes how to design a pipelined CPU in which a delayed branch technique is used to solve the control hazard problems; an internal forwarding technique and a pipeline stall mechanism are used to solve the data hazard problems. Then a precise interrupt and exception implementation on the pipelined CPU is described. The Verilog HDL (hardware description language) codes and simulation waveforms are also presented.

Computer Principles and Design in Verilog HDL, First Edition. Yamin Li.
© 2015 Tsinghua University Press. All rights reserved. Published 2015 by John Wiley & Sons Singapore Pte Ltd.
Companion Website: www.wiley.com/go/li/verilog

8.1   Pipelining

Pipelining is an implementation technique in which multiple instructions are overlapped in execution. Pipelining technique exists everywhere. A good example is the automobile assembly line. Suppose that a manufacturing plant starts to produce a next car after the current car is produced; how many cars can be produced per year in such a plant? Can it survive?

This section introduces the idea of the pipelining technique for building a fast CPU and describes a baseline structure, which is the starting point for constructing a pipelined CPU with the precise interrupt and exception mechanism.

8.1.1   The Concept of Pipelining

Consider an instruction as a car in the assembly line. We can use the pipelining technique to let a new instruction enter the pipeline while its predecessors are still in the pipeline. Similar to the MC CPU, we divide the execution of an instruction into five stages: instruction fetch (IF), instruction decode (ID), execution (EXE), memory access (MEM), and write back (WB).

Suppose that there is a four-element array x. The following assembly code fragment adds a scalar s, located in the $10 register, to the each element of x. There are 12 instructions in total. The memory address is shown in the left column.

```
# address   instruction              comment
   100:     lw   $2,  00($1)        # $2 <-- memory[$1+00];   load x[0]
   104:     lw   $3,  04($1)        # $3 <-- memory[$1+04];   load x[1]
   108:     lw   $4,  08($1)        # $4 <-- memory[$1+08];   load x[2]
   112:     lw   $5,  12($1)        # $5 <-- memory[$1+12];   load x[3]
   116:     add $6,  $2,  $10       # $6 <-- $2 + $10;      x[0] = x[0] + s
   120:     add $7,  $3,  $10       # $7 <-- $3 + $10;      x[1] = x[1] + s
   124:     add $8,  $4,  $10       # $8 <-- $4 + $10;      x[2] = x[2] + s
   128:     add $9,  $5,  $10       # $9 <-- $5 + $10;      x[3] = x[3] + s
   132:     sw   $6,  00($1)        # memory[$1+00] <-- $6; store x[0]
   136:     sw   $7,  04($1)        # memory[$1+04] <-- $7; store x[1]
   140:     sw   $8,  08($1)        # memory[$1+08] <-- $8; store x[2]
   144:     sw   $9,  12($1)        # memory[$1+12] <-- $9; store x[3]
```

Before introducing the pipelined CPU in detail, let's see how an SC CPU executes this code fragment. Suppose that there is an SC CPU that can execute only arithmetic instructions and load/store instructions. Figure 8.1 shows the manner of execution of the first `lw` instruction on the SC CPU.

An `lw rt, imm(rs)` instruction loads a data word from memory and writes it into register `rt`. The memory address is calculated by adding the sign-extended immediate (`imm`) to the data read from the register `rs`. The execution of the `lw $2,00($1)` instruction consists of the following five steps:

1. IF: The program counter (PC) contains 100; it is used as the instruction memory address. The instruction in the memory location 100 is fetched. It is `lw $2,00($1)`. Meanwhile, the PC + 4 is calculated by a dedicated adder.
2. ID: The content in register $1 (rs = 1) is read, and the immediate (`imm` = 0) is sign-extended. The control unit decodes the instruction. The outputs of the control signals are as follows: `aluimm = 1` (select the extended immediate as the ALU input B's data); `aluc = x000` (let ALU perform addition); `wmem = 0` (do not write data memory); `m2reg = 1` (select memory data out that will be written into register file); `regrt = 1` (select rt as the destination register number); and `wreg = 1` (write result into register $2 (rt = 2)).

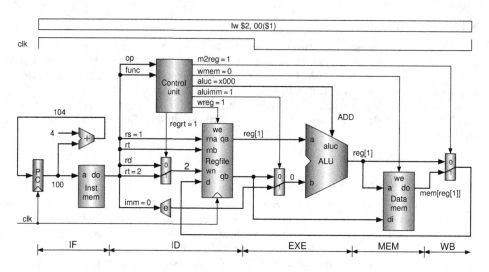

Figure 8.1   Single-cycle CPU executing a load instruction

3. EXE: The ALU performs addition to calculate the data memory address.
4. MEM: A 32-bit data word is read from the data memory.
5. WB: The word read from the data memory is written into the register file. Meanwhile, PC + 4 = 104 is written into the PC. Both writes happen at the rising edge of the clock signal (clk).

Assume that each step described above takes 1 ns (nanosecond). Then the time length of the clock cycle of the SC CPU is 5 ns. The total time for executing the 12 instructions is $5 \times 12 = 60$ ns.

In the MC CPU case, the cycle time is 1 ns. lw takes five cycles; add and sw instructions take four cycles. Then the total time of executing the 12 instructions on the MC CPU is $5 \times 4 + 4 \times 8 = 52$ ns.

On a pipelined CPU, the cycle time is also 1 ns, but a new instruction can enter the pipeline during every clock cycle. Referring to Figure 8.2, the pipeline is fulfilled from the fifth clock cycle. The total execution time is $12 + 5 - 1 = 16$ ns (actually is 15 ns because the sw instruction has no WB stage).

Figure 8.2 compares the timing chart of the three types of CPUs. Looking at the sixth clock cycle in Figure 8.2(c), there are five different operations that are performed simultaneously. Each of these operations belongs to a different instruction. This is the exact meaning of overlapping execution of multiple instructions. Also, looking at the EXE stage of every instruction, the job in this stage is done by the ALU. Under ideal conditions, the ALU is busy during every clock cycle.

Assume that a pipelined CPU has m stages and each stage takes exactly one clock cycle. Ignoring the delay time of the pipeline registers in between stages, executing n instructions on the pipelined CPU will take $n + m - 1$ cycles. In contrast, it takes $n \times m$ clock cycles on the SC CPU and $n \times k$ clock cycles on the MC CPU, where k is the average number of clock cycles an instruction takes for execution on the MC CPU with $1 < k \leq m$. That is, the CPI (average cycles per instruction) of an SC CPU is m; the CPI of an MC CPU is k; and the CPI of a pipelined CPU is $(n + m - 1)/n$. When n is sufficiently larger than m, the CPI of pipelined CPU is nearly 1. Therefore, the speedup of the pipelined CPU compared to the SC CPU is nearly m and it is k compared to the MC CPU.

8.1.2   The Baseline of Pipelined CPU

Because in a pipelined CPU there are multiple operations in each clock cycle, we must save the temporary results in each pipeline stage into pipeline registers for use in the follow-up stages. Remember that our pipelined CPU has five stages: IF, ID, EXE, MEM, and WB. The PC can be considered as the

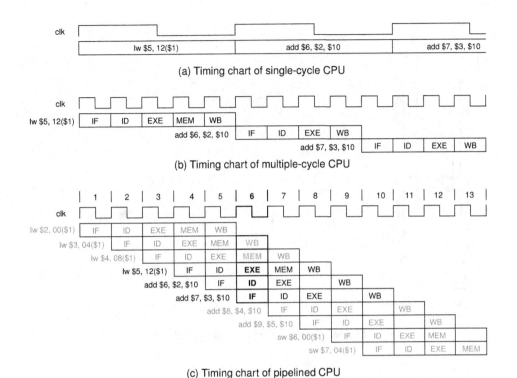

Figure 8.2 Timing comparison between the three types of CPUs

first pipeline register at the beginning of the first stage. We name the other pipeline registers as IF/ID, ID/EXE, EXE/MEM, and MEM/WB in sequence. Except for PC, the pipeline registers are transparent to programmers: they have no way to get the contents of these pipeline registers.

In order to understand in depth how the pipelined CPU works, we show the circuits that are required in each pipeline stage of a baseline CPU for executing the code fragment listed in the previous section.

8.1.2.1 Circuits of the Instruction Fetch Stage

The circuits in the IF stage are shown in Figure 8.3. Also, looking at the first clock cycle in Figure 8.2(c), the first `lw` instruction is being fetched. This is something like a newly built 5-year university in which there are only first-year students.

In the IF stage, there is an instruction memory module and an adder between two pipeline registers. The leftmost pipeline register is the PC; it holds 100. In the end of the first cycle (at the rising edge of `clk`), the instruction fetched from instruction memory is written into the IF/ID register (the first-year students are promoted to second year). Meanwhile, the output of the adder (PC + 4, the next PC) is written into PC (for recruiting new first-year students).

The pipelined CPU we will design can execute the branch and jump instructions. The next PC may be a branch or jump target address, not always PC + 4. We will describe it later.

8.1.2.2 Circuits of the Instruction Decode Stage

Referring to Figure 8.4, in the second cycle, the first instruction entered the ID stage. There are two jobs in the second cycle: to decode the first instruction in the ID stage, and to fetch the second instruction

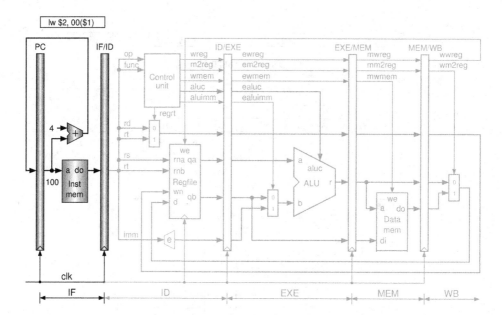

Figure 8.3   Pipeline instruction fetch (IF) stage

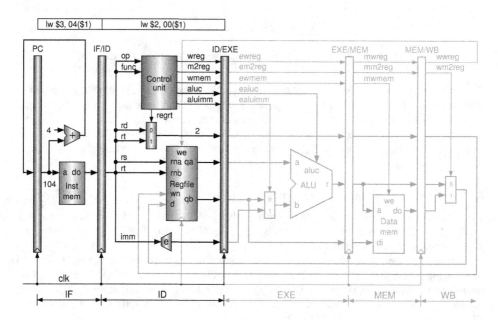

Figure 8.4   Pipeline instruction decode (ID) stage

in the IF stage. The two instructions are shown on the top of the figure: the first instruction is in the ID stage, and the second instruction is in the IF stage.

The first instruction in the ID stage comes from the IF/ID register. Two operands are read from the register file (Regfile in the figure) based on rs and rt, although the lw instruction does not use the operand in the register rt. The immediate (imm) is sign-extended into 32 bits. The regrt signal is used in the ID stage that selects the destination register number; all others must be written into the ID/EXE register for later use. At the end of the second clock cycle, all the data and control signals, except for regrt, in the ID stage are written into the ID/EXE register. At the same time, the PC and the IF/ID register are also updated.

8.1.2.3 Circuits of the Execution Stage

Referring to Figure 8.5, in the third cycle the first instruction entered the EXE stage. The ALU performs addition, and the multiplexer selects the immediate. A letter "e" is prefixed to each control signal in order to distinguish it from that in the ID stage. The second instruction is being decoded in the ID stage and the third instruction is being fetched in the IF stage. All the four pipeline registers are updated at the end of the cycle.

8.1.2.4 Circuits of the Memory Access Stage

Referring to Figure 8.6, in the fourth cycle the first instruction entered the MEM stage. The only task in this stage is to read data memory. All the control signals have a prefix "m." The second instruction entered the EXE stage; the third instruction is being decoded in the ID stage; and the fourth instruction is being fetched in the IF stage. All the five pipeline registers are updated at the end of the cycle.

8.1.2.5 Circuits of the Write Back Stage

Referring to Figure 8.7, in the fifth cycle the first instruction entered the WB stage. The memory data is selected and will be written into the register file at the end of the cycle. All the control signals have a prefix "w." The second instruction entered the MEM stage; the third instruction entered the EXE stage; the fourth instruction is being decoded in the ID stage; and the fifth instruction is being fetched in the IF stage. All the six pipeline registers are updated at the end of the cycle (the destination register is

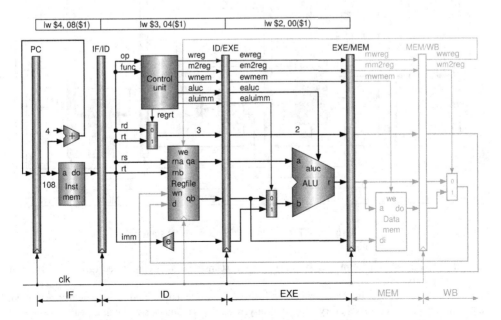

Figure 8.5 Pipeline execution (EXE) stage

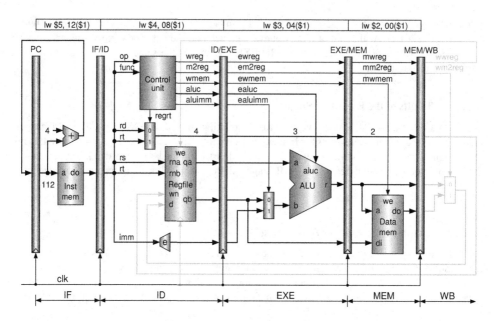

Figure 8.6 Pipeline memory access (MEM) stage

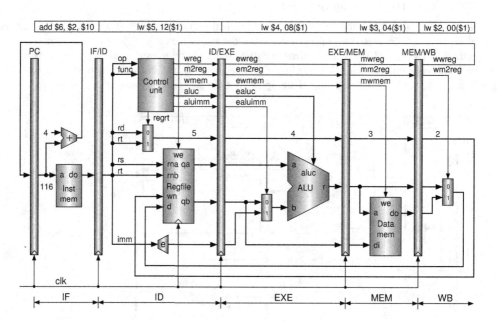

Figure 8.7 Pipeline write back (WB) stage

considered as the sixth pipeline register). Then the first instruction is committed (graduated from the 5-year university).

In each of the forthcoming clock cycles, an instruction will be committed and a new instruction will enter the pipeline. This is like, every year the fifth-year students graduate and the first-year students enter the 5-year university.

We use the structure shown in Figure 8.7 as a baseline for the design of our pipelined CPU. The branch and jump instructions cannot be executed by the baseline CPU. We will add some circuits to it to make the branch and jump instructions executable later.

8.2 Pipeline Hazards and Solutions

Under ideal conditions, the pipelined CPU can execute a useful instruction during every clock cycle. Because of the overlapping execution of multiple instructions, the pipelined CPU may encounter some situations where an instruction cannot be processed or it will not execute correctly. These situations are called hazards.

There are three types of hazards in a pipelined CPU: structural hazards, control hazards, and data hazards. This section explains these hazards and gives possible solutions to these problems.

8.2.1 Structural Hazards and Solutions

Structural hazards occur when two or more instructions attempt to use a hardware component at the same time but the component can serve only one request. An example of structural hazards is the memory access conflict. Suppose that there is only one memory module; the memory data load/store will conflict with IF because IF takes place during every clock cycle.

As shown in Figure 8.8, another example of the structural hazards is that, if an integer divider is designed in a low-cost iterative style that may take tens of clock cycles to get the quotient, the successive divide instruction cannot go through the pipeline until the current division is finished.

This integer division operation may cause another structural hazard: there may be two writes to the integer register file in a clock cycle.

Generally, any lack of sufficient hardware resources that results in pipeline stalls is a structural hazard. Because the current integrated circuit (IC) technology allows makers to fabricate a large number of transistors on a chip, most of the structural hazards can be solved. For example, there are dedicated instruction caches and data caches in modern CPUs. We can also design a register file with two write ports easily, which we will show in Chapter 10.

We can even design a fully pipelined divider that can accept a new divide instruction in every clock cycle. The problem is the utilization of such expensive functional units. A trade-off must be made between the hardware cost and the performance (the inverse of program execution time).

8.2.2 Data Hazards and Internal Forwarding

A data hazard is also known as data dependency. It means that an instruction i uses the execution result of its predecessor (instruction $i - 1$). Because the execution result of instruction $i - 1$ is written into register

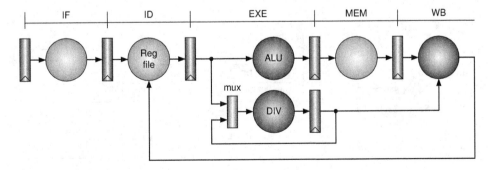

Figure 8.8 Structural hazards caused by an iterative divider

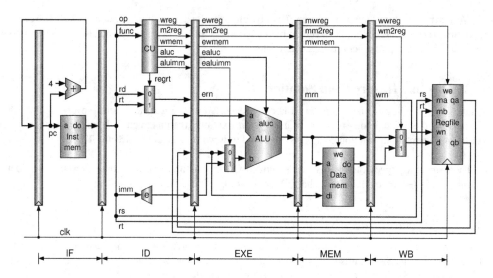

Figure 8.9   Writing result to the register file in the write back stage

file at the end of its WB stage, while the instruction *i* reads the register contents in its ID stage, a data word read from the register file is not the state-of-the-art result produced by instruction *i* − 1. Such situations are called data hazards.

In order to focus our attention on the WB stage easily, we redraw the baseline CPU by putting the register file on the WB stage where the execution result of an instruction is written, as shown in Figure 8.9. The state-of-the-art content can be read correctly in the ID stage after it is written at the end of its WB stage.

Let's see the code example shown in Figure 8.10. The first instruction (add $3, $1, $2) reads two register operands from registers $1 and $2 in the ID stage (the second clock cycle), adds them in the EXE stage (the third clock cycle), and writes the sum into the register $3 at the end of the WB stage (the fifth clock cycle). The last instruction (and) reads $3 in the sixth clock cycle without any problem. But the second instruction (sub), the third instruction (or), and the fourth instruction (xor) want to read the $3 register in the third, fourth, and fifth clock cycles, respectively. Data hazards occur in these three cases. If these three instructions use the data read from the register file, their execution results will be wrong. Now, the question is how to deal with these hazards.

One solution is to force these instructions to wait until the register $3 is updated by the first instruction. This waiting is called a pipeline stall. Pipeline stalls decrease the CPU performance. Is there any way to not stall the pipeline and use the state-of-the-art data to calculate the correct results?

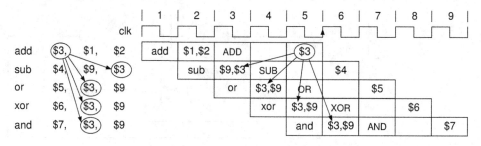

Figure 8.10   Data hazard examples

The answer is "yes." The operation of the sub instruction is subtraction, which is the job of the ALU. When ALU starts to do subtraction, it would have already finished the addition for the add instruction. The result (sum) is in an EXE/MEM pipeline register. Then it becomes possible to design a special path to forward the result from the MEM stage of the add instruction to the EXE stage of the sub instruction, as shown in the fourth clock cycle in Figure 8.11.

This technique is called an internal forwarding or bypass. Referring to Figure 8.11 again, the execution result in the WB stage of the add instruction can be forwarded to the EXE stage of the or instruction. Because writing the result to the register file in the WB stage is a simple operation, we can perform it in the first half of a clock cycle so that the state-of-the-art data can be read in the second half of the clock cycle. Then the data hazard between xor and add will also be solved.

Using the forwarding style of Figure 8.11, the sub instruction uses the data stored in the EXE/MEM pipeline register to perform the subtraction in its EXE stage. Our design adopts another forwarding style in which the output of the ALU in the EXE stage is forwarded directly to the ID stage (for sub instruction) and will be stored into the ID/EXE pipeline register, as shown in the third clock cycle in Figure 8.12. Similarly, the result in the MEM stage is also forwarded to the ID stage (for the or instruction). The details of why we adopt this forwarding style will be explained later.

By using the internal forwarding technique, all the data dependencies between two instructions of the arithmetic and logic calculation types can be solved without stalling the pipeline. But the data dependencies with a load instruction are different.

Referring to Figure 8.13, the output of the ALU in the EXE stage of the lw instruction cannot be forwarded because it is a memory address. What we can forward is the output of the data memory in the MEM stage. In this case, the pipeline must be stalled for one clock cycle for waiting for the memory data.

Figure 8.14 illustrates an implementation for dealing with data hazards. A NOT gate inverts the clock signal so that in WB stage the result can be written to register file in the falling edge of the clock. Two extra 32-bit 4-to-1 multiplexers in the ID stage are used to forward the ALU output in the EXE stage, the ALU output in the MEM stage, or the memory output in the MEM stage, to the ID stage.

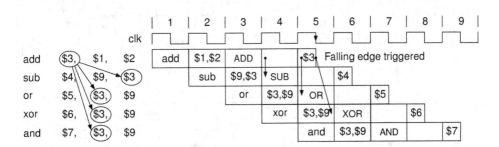

Figure 8.11   Internal forwarding to the EXE stage

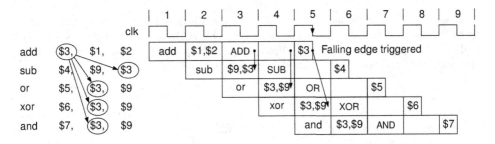

Figure 8.12   Internal forwarding to the ID stage

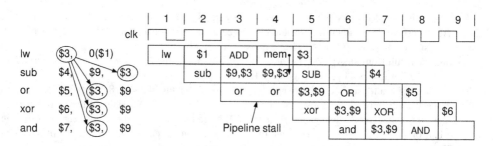

Figure 8.13 Data hazard with load instruction

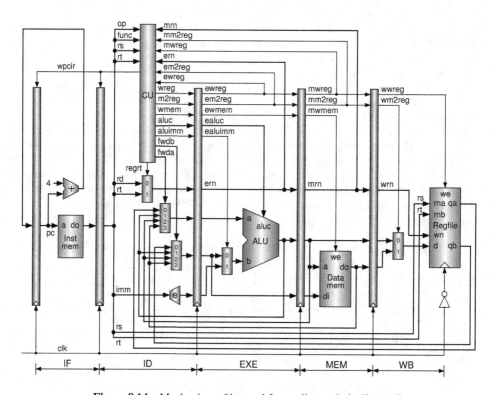

Figure 8.14 Mechanism of internal forwarding and pipeline stall

The selection signal of the multiplexer for ALU's input A is named `fwda` (forward A) and the other, for ALU's input B, is named `fwdb` (forward B). If there is no data hazard, the multiplexer selects the data read from the register file. The following Verilog HDL codes generate the `fwda` and `fwdb` signals. All the signal names in the codes can be found in Figure 8.14.

```
// forward control signal for alu input a
fwda = 2'b00;                                          // default: no hazards
if (ewreg & (ern != 0) & (ern == rs) & ~em2reg) begin
    fwda = 2'b01;                                      // select exe_alu
end else begin
```

```
    if (mwreg & (mrn != 0) & (mrn == rs) & ~mm2reg) begin
        fwda = 2'b10;                                // select mem_alu
    end else begin
        if (mwreg & (mrn != 0) & (mrn == rs) & mm2reg) begin
            fwda = 2'b11;                            // select mem_lw
        end
    end
end
```

```
// forward control signal for alu input b
fwdb = 2'b00;                                        // default: no hazards
if (ewreg & (ern != 0) & (ern == rt) & ~em2reg) begin
    fwdb = 2'b01;                                    // select exe_alu
end else begin
    if (mwreg & (mrn != 0) & (mrn == rt) & ~mm2reg) begin
        fwdb = 2'b10;                                // select mem_alu
    end else begin
        if (mwreg & (mrn != 0) & (mrn == rt) & mm2reg) begin
            fwdb = 2'b11;                            // select mem_lw
        end
    end
end
```

Figure 8.15 shows the timing chart of the pipeline `stall` signal. The inverse of the signal is used as the write enable for the PC and the IF/ID pipeline register (`wpcir`).

The `stall` signal becomes true when an instruction in the ID stage uses the result of an `lw` instruction which is in the EXE stage. Thus the `stall` signal can be generated by the following Verilog HDL code.

```
stall = ewreg & em2reg & (ern != 0) & (i_rs & (ern == rs) |
                                        i_rt & (ern == rt));
```

where `i_rs` and `i_rt` indicate that an instruction uses the contents of the `rs` register and the `rt` register, respectively.

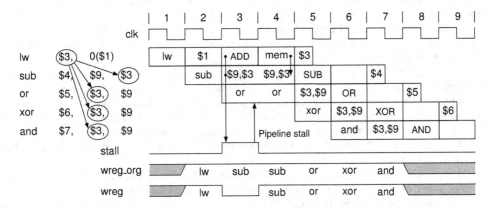

Figure 8.15 Implementing pipeline stall and instruction cancellation

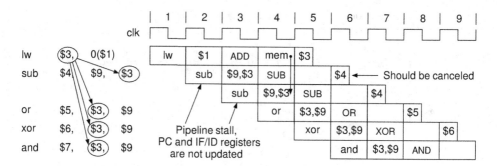

Figure 8.16   Pipeline stall causing an instruction to be executed twice

There is an important thing we must not forget. The pipeline stall is implemented by prohibiting the updates of the PC and the IF/ID pipeline register. But the instruction that is already in the IF/ID register (sub instruction in our example) will be decoded and fed to the next pipeline stage. This will result in an instruction being executed twice, as shown in Figure 8.16.

To prevent an instruction from being executed twice, we must cancel the first instruction. Canceling an instruction is easy: prevent it from updating the states of the CPU and memory. In our case, we just need to disable the register file write signal (wreg) and the memory write signal (wmem) with the following codes.

```
wreg = wreg_org & ~stall;          // prevent from executing twice
wmem = wmem_org & ~stall;          // prevent from executing twice
```

where wreg_org is the original signal of the register file write enable and wmem_org is the original signal of the memory write enable which did not care about the pipeline stalls. The timing of the new wreg signal is also shown in Figure 8.15, from which we can see that the first sub instruction was canceled.

8.2.3   Control Hazards and Delayed Branch

The control hazard occurs when a pipelined CPU executes a branch or jump instruction. The jump target address of a jump instruction (jr, j, or jal) can be determined in the ID stage and it will be written into PC at the end of the ID stage. But because the pipelined CPU fetches an instruction during every clock cycle, the next instruction is being fetched during the ID stage of the jump instruction. How to deal with this next instruction is a problem.

The control hazard caused by a conditional branch instruction (beq or bne) becomes more serious than that of a jump instruction because the condition must be evaluated in addition to the calculation of the branch target address.

We can calculate the branch target address in the ID stage. But if the condition (equal or not equal) is checked in the EXE stage, after transferring control to the branch target address, two follow-up instructions to the branch instruction are already in the pipeline, as shown in Figure 8.17(a), where the two follow-up instructions to the branch instruction are represented with question marks.

If the condition can be checked also in the ID stage, then the situation becomes the same as that of a jump instruction, as shown in Figure 8.17(b). In either case, we must do something on the fetched instruction(s) next to the branch or jump instruction.

There are mainly two methods to deal with the instruction(s) next to a branch or jump instruction. One method is to cancel it (them). The other is to let it (them) be executed. The second method is called a delayed branch. The positions in between the location of a jump or branch instruction and the jump or branch target address are called delay slots. MIPS (microprocessor without interlocked pipeline stages)

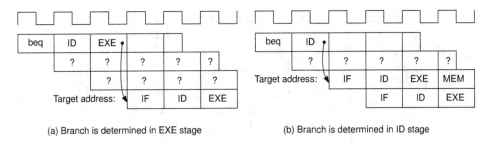

(a) Branch is determined in EXE stage   (b) Branch is determined in ID stage

Figure 8.17 Determining whether a branch is taken or not taken

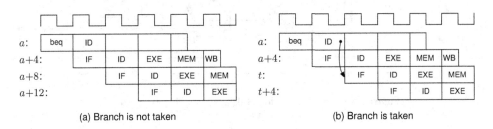

(a) Branch is not taken   (b) Branch is taken

Figure 8.18 Delayed branch

ISA (instruction set architecture) adopts a one-delay-slot mechanism: the instruction located in delay slot is always executed no matter whether the branch is taken or not, as shown in Figure 8.18. Figure 8.18(a) shows the case where the branch is not taken. Figure 8.18(b) shows the case where the branch is taken; t is the branch target address. In both cases, the instruction located in $a + 4$ (delay slot) is always executed, no matter whether the branch is taken or not.

In order to implement the delayed branch with one delay slot, we must let the conditional branch instructions finish the executions in the ID stage. There should be no problem for calculating the branch target address within the ID stage. For checking the condition, we can perform an exclusive OR (XOR) on the two source operands:

```
rsrtequ  = ~|(da^db);                    // (da == db)
```

where the `rsrtequ` signal indicates whether `da` and `db` are equal or not. Both `da` and `db` should be the state-of-the-art data. Referring to Figure 8.19, we use the outputs of the multiplexers for internal forwarding as `da` and `db`. This is the reason why we put the forwarding to the ID stage instead of to the EXE stage.

Because of the delayed branch, the return address of the MIPS `jal` instruction is PC + 8. Figure 8.20 illustrates the execution of the `jal` instruction.

The instruction located in delay slot (PC + 4) was already executed before transferring control to a function (or a subroutine). The return address should be PC + 8, which is written into $31 register in the WB stage by the `jal` instruction. The return from subroutine can be done by the instruction of `jr $31`. The `jr rs` instruction reads the content of register `rs` and writes it into the PC.

8.3   The Circuit of the Pipelined CPU and Verilog HDL Codes

The section describes the entire circuit of the pipelined CPU in detail and gives its Verilog HDL codes and the simulation waveforms.

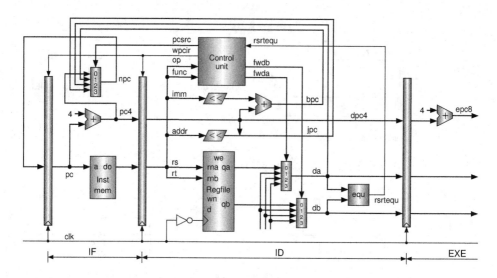

Figure 8.19   Implementation of delayed branch with one delay slot

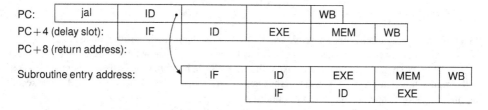

Figure 8.20   Return address of the function call instruction

8.3.1   The Circuit of the Pipelined CPU

The pipelined CPU can execute 20 instructions. We have described the meanings of the instructions in Chapter 4. Note that the jal instruction saves PC + 8 to register $31 because of the delayed branch.

Figure 8.21 illustrates the detailed circuit of the pipelined CPU, plus instruction memory and data memory. As described before, there are five pipeline stages: IF, ID, EXE, MEM, and WB. Four pipeline registers are inserted in between the five stages, namely IF/ID, ID/EXE, EXE/MEM, and MEM/WB registers. These four registers are transparent to programmers. The PC can be considered as the first pipeline register located in front of the IF stage, and a register of the register file can be considered as the sixth (last) pipeline register at the end of the WB stage. These two registers are accessible to programmers.

In the IF stage, an instruction is fetched from instruction memory, and the PC is incremented by 4 if the instruction in the ID stage is neither a branch nor a jump instruction, and there is no pipeline stall. There are four sources for the next PC:

1. pc4: PC + 4,
2. bpc: branch target address of a beq or bne instruction,
3. da: target address in register of a jr instruction, and
4. jpc: jump target address of a j or jal instruction.

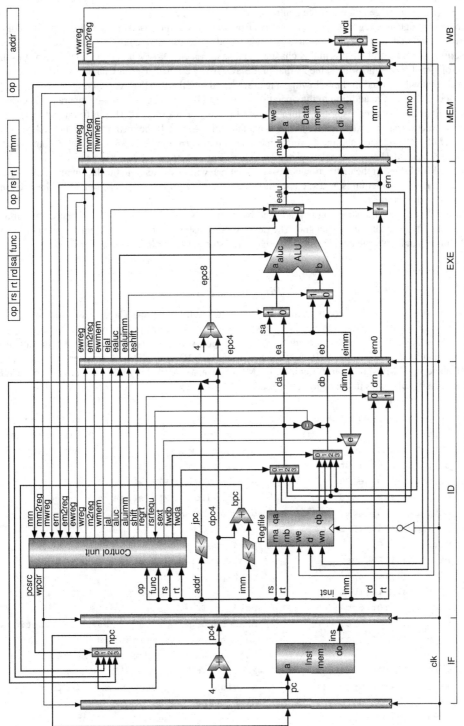

Figure 8.21 Detailed circuit of the pipelined CPU

The selection for the next PC (npc) is done by a 32-bit 4-to-1 multiplexer whose selection signal is pcsrc (PC source), generated by the control unit in the ID stage.

In the ID stage, two register operands are read from the register file based on rs and rt; the immediate (imm) is extended; and the instruction is decoded based on op (and func) by the control unit. We have described some control signals in this chapter, such as pcsrc, fwda, fwdb, and wpcir. The rest of the control signals are almost the same as that in the SC CPU we introduced in Chapter 5. All the control signals that will be used in the following stages are saved into the ID/EXE registers.

In the EXE stage, in addition to the operation performed by the ALU, the PC + 8 operation is carried out by an adder for generating the return address for the jal instruction. The shift amount (sa) for a shift instruction can be extracted from the immediate field (eimm). If the instruction in the EXE stage is a jal, PC + 8 is selected and the destination register number (ern) is set to 31 (done by f component). Otherwise, the ALU output is selected and let ern = ern0 (rd or rt in the EXE stage).

In the MEM stage, if the instruction is an sw, the data mb will be written into the data memory addressed by malu. If the instruction is an lw, the memory data addressed by malu is read out. Other instructions do nothing in this stage.

In the WB stage, an instruction is graduated by writing the result, either the ALU result or memory data, into a register of the register file. The destination register number is wrn (register number in the WB stage). And the write enable signal is wwreg (register write enable in WB stage).

8.3.2   Verilog HDL Codes of the Pipelined CPU

The complete Verilog HDL codes of the pipelined CPU are listed below. The codes of some modules, such as multiplexers, the register file, and the ALU, were already given in the previous chapters.

Below is the codes of the top module, pipelinedcpu, which consists of the modules in sequence of PC, circuits in the IF stage, IF/ID registers, circuits in the ID stage, ID/EXE registers, circuits in the EXE stage, EXE/MEM registers, circuits in the MEM stage, MEM/WB registers, and circuits in the WB stage. The signal names in the code can be found in Figure 8.21.

```
module pipelinedcpu (clk,clrn,pc,inst,ealu,malu,wdi);        // pipelined cpu
    input           clk, clrn;      // clock and reset         // plus inst mem
    output [31:0] pc;               // program counter         // and   data mem
    output [31:0] inst;             // instruction in ID stage
    output [31:0] ealu;             // alu result in EXE stage
    output [31:0] malu;             // alu result in MEM stage
    output [31:0] wdi;              // data to be written into register file
    // signals in IF stage
    wire   [31:0] pc4;              // pc+4 in IF stage
    wire   [31:0] ins;              // instruction in IF stage
    wire   [31:0] npc;              // next pc in IF stage
    // signals in ID stage
    wire   [31:0] dpc4;             // pc+4 in ID stage
    wire   [31:0] bpc;              // branch target of beq and bne instructions
    wire   [31:0] jpc;              // jump target of jr instruction
    wire   [31:0] da,db;            // two operands a and b in ID stage
    wire   [31:0] dimm;             // 32-bit extended immediate in ID stage
    wire   [4:0]  drn;              // destination register number in ID stage
    wire   [3:0]  daluc;            // alu control in ID stage
    wire   [1:0]  pcsrc;            // next pc (npc) select in ID stage
    wire          wpcir;            // pipepc and pipeir write enable
    wire          dwreg;            // register file write enable in ID stage
```

```verilog
wire          dm2reg;        // memory to register in ID stage
wire          dwmem;         // memory write in ID stage
wire          daluimm;       // alu input b is an immediate in ID stage
wire          dshift;        // shift in ID stage
wire          djal;          // jal in ID stage
// signals in EXE stage
wire   [31:0] epc4;          // pc+4 in EXE stage
wire   [31:0] ea,eb;         // two operands a and b in EXE stage
wire   [31:0] eimm;          // 32-bit extended immediate in EXE stage
wire    [4:0] ern0;          // temporary register number in WB stage
wire    [4:0] ern;           // destination register number in EXE stage
wire    [3:0] ealuc;         // alu control in EXE stage
wire          ewreg;         // register file write enable in EXE stage
wire          em2reg;        // memory to register in EXE stage
wire          ewmem;         // memory write in EXE stage
wire          ealuimm;       // alu input b is an immediate in EXE stage
wire          eshift;        // shift in EXE stage
wire          ejal;          // jal in EXE stage
// signals in MEM stage
wire   [31:0] mb;            // operand b in MEM stage
wire   [31:0] mmo;           // memory data out in MEM stage
wire    [4:0] mrn;           // destination register number in MEM stage
wire          mwreg;         // register file write enable in MEM stage
wire          mm2reg;        // memory to register in MEM stage
wire          mwmem;         // memory write in MEM stage
// signals in WB stage
wire   [31:0] wmo;           // memory data out in WB stage
wire   [31:0] walu;          // alu result in WB stage
wire    [4:0] wrn;           // destination register number in WB stage
wire          wwreg;         // register file write enable in WB stage
wire          wm2reg;        // memory to register in WB stage
// program counter
pipepc   prog_cnt (npc,wpcir,clk,clrn,pc);
pipeif   if_stage (pcsrc,pc,bpc,da,jpc,npc,pc4,ins);        // IF stage
// IF/ID pipeline register
pipeir    fd_reg (pc4,ins,wpcir,clk,clrn,dpc4,inst);
pipeid   id_stage (mwreg,mrn,ern,ewreg,em2reg,mm2reg,dpc4,inst,wrn,wdi,
                   ealu,malu,mmo,wwreg,clk,clrn,bpc,jpc,pcsrc,wpcir,
                   dwreg,dm2reg,dwmem,daluc,daluimm,da,db,dimm,drn,
                   dshift,djal);                            // ID stage
// ID/EXE pipeline register
pipedereg  de_reg (dwreg,dm2reg,dwmem,daluc,daluimm,da,db,dimm,drn,
                   dshift,djal,dpc4,clk,clrn,ewreg,em2reg,ewmem,
                   ealuc,ealuimm,ea,eb,eimm,ern0,eshift,ejal,epc4);
pipeexe exe_stage (ealuc,ealuimm,ea,eb,eimm,eshift,ern0,epc4,ejal,
                   ern,ealu);                               // EXE stage
// EXE/MEM pipeline register
```

```
    pipeemreg   em_reg  (ewreg,em2reg,ewmem,ealu,eb,ern,clk,clrn,mwreg,
                         mm2reg,mwmem,malu,mb,mrn);
    pipemem mem_stage (mwmem,malu,mb,clk,mmo);                    // MEM stage
    // MEM/WB pipeline register
    pipemwreg   mw_reg  (mwreg,mm2reg,mmo,malu,mrn,clk,clrn,wwreg,wm2reg,
                         wmo,walu,wrn);
    pipewb   wb_stage (walu,wmo,wm2reg,wdi);                      // WB stage
endmodule
```

Next, we give each module's Verilog HDL code. Below is the Verilog HDL codes of a PC which is a simple 32-bit D flip-flop with a write enable. The write enable is used for pipeline stall.

```
module pipepc (npc,wpc,clk,clrn,pc);                    // program counter
    input          clk, clrn;                           // clock and reset
    input          wpc;                                 // pc write enable
    input  [31:0] npc;                                  // next pc
    output [31:0] pc;                                    // program counter
    // dffe32         (d,   clk,clrn,e,   q);
    dffe32 prog_cntr (npc,clk,clrn,wpc,pc);             // program counter
endmodule
```

Below is the Verilog HDL codes of the circuits in the IF stage. There are three modules. A 32-bit 4-to-1 multiplexer (mux4x32) is used for selecting an address (npc) from pc4 (PC + 4), bpc (branch target address), rpc (target address in register), and jpc (jump target address). A 32-bit carry-lookahead adder (cla32) is used for calculating PC + 4. The instruction memory (pl_inst_mem) stores the program. The Verilog HDL codes of the instruction memory will be given in the next section.

```
module pipeif (pcsrc,pc,bpc,rpc,jpc,npc,pc4,ins);     // IF stage
    input  [31:0] pc;                                  // program counter
    input  [31:0] bpc;                                 // branch target
    input  [31:0] rpc;                                 // jump target of jr
    input  [31:0] jpc;                                 // jump target of j/jal
    input   [1:0] pcsrc;                               // next pc (npc) select
    output [31:0] npc;                                 // next pc
    output [31:0] pc4;                                 // pc + 4
    output [31:0] ins;                                 // inst from inst mem
    mux4x32 next_pc (pc4,bpc,rpc,jpc,pcsrc,npc);       // npc select
    cla32   pc_plus4 (pc,32'h4,1'b0,pc4);              // pc + 4
    pl_inst_mem inst_mem (pc,ins);                     // inst mem
endmodule
```

Below is the Verilog HDL codes of the IF/ID pipeline register. We used two 32-bit D flip-flops with write enable (dffe32) to store PC + 4 and the instruction fetched from instruction memory. The write enable is used for pipeline stall.

```
module pipeir (pc4,ins,wir,clk,clrn,dpc4,inst);  // IF/ID pipeline register
    input          clk, clrn;                         // clock and reset
    input          wir;                               // write enable
```

```
    input   [31:0] pc4;                        // pc + 4 in IF stage
    input   [31:0] ins;                        // instruction in IF stage
    output  [31:0] dpc4;                       // pc + 4 in ID stage
    output  [31:0] inst;                       // instruction in ID stage
    // dffe32              (d,   clk,clrn,e,   q);
    dffe32 pc_plus4     (pc4,clk,clrn,wir,dpc4);   // pc+4 register
    dffe32 instruction (ins,clk,clrn,wir,inst);    // inst register
endmodule
```

Below is the Verilog HDL codes of the circuits in the ID stage. The modules that are invoked in this stage include the following: (i) the control unit (pipeidcu); (ii) the register file (regfile); (iii) a 5-bit 2-to-1 multiplexer (mux2x5) for selecting the destination register number; (iv) two 32-bit 4-to-1 multiplexers for internal forwarding; and (v) a carry-lookahead adder (cla32) for calculating the branch target address. The other jobs are to extend the immediate to 32 bits and to evaluate the condition (rsrtequ) for conditional branch instructions.

```
module pipeid (mwreg,mrn,ern,ewreg,em2reg,mm2reg,dpc4,inst,wrn,wdi,ealu,
               malu,mmo,wwreg,clk,clrn,bpc,jpc,pcsrc,nostall,wreg,m2reg,
               wmem,aluc,aluimm,a,b,dimm,rn,shift,jal);// ID stage
    input          clk, clrn;                  // clock and reset
    input   [31:0] dpc4;                       // pc+4 in ID
    input   [31:0] inst;                       // inst in ID
    input   [31:0] wdi;                        // data in WB
    input   [31:0] ealu;                       // alu res in EXE
    input   [31:0] malu;                       // alu res in MEM
    input   [31:0] mmo;                        // mem out in MEM
    input    [4:0] ern;                        // dest reg # in EXE
    input    [4:0] mrn;                        // dest reg # in MEM
    input    [4:0] wrn;                        // dest reg # in WB
    input          ewreg;                      // wreg in EXE
    input          em2reg;                     // m2reg in EXE
    input          mwreg;                      // wreg in MEM
    input          mm2reg;                     // m2reg in MEM
    input          wwreg;                      // wreg in MEM
    output  [31:0] bpc;                        // branch target
    output  [31:0] jpc;                        // jump target
    output  [31:0] a, b;                       // operands a and b
    output  [31:0] dimm;                       // 32-bit immediate
    output   [4:0] rn;                         // dest reg #
    output   [3:0] aluc;                       // alu control
    output   [1:0] pcsrc;                      // next pc select
    output         nostall;                    // no pipeline stall
    output         wreg;                       // write regfile
    output         m2reg;                      // mem to reg
    output         wmem;                       // write memory
    output         aluimm;                     // alu input b is imm
    output         shift;                      // inst is a shift
    output         jal;                        // inst is jal
```

```
wire      [5:0] op   = inst[31:26];                     // op
wire      [4:0] rs   = inst[25:21];                     // rs
wire      [4:0] rt   = inst[20:16];                     // rt
wire      [4:0] rd   = inst[15:11];                     // rd
wire      [5:0] func = inst[05:00];                     // func
wire      [15:0] imm = inst[15:00];                     // immediate
wire      [25:0] addr = inst[25:00];                    // address
wire            regrt;                                  // dest reg # is rt
wire            sext;                                   // sign extend
wire      [31:0] qa, qb;                                // regfile outputs
wire      [1:0] fwda, fwdb;                             // forward a and b
wire      [15:0] s16 = {16{sext & inst[15]}};           // 16-bit signs
wire      [31:0] dis = {dimm[29:0],2'b00};              // branch offset
wire            rsrtequ = ~|(a^b);                      // reg[rs] == reg[rt]
pipeidcu cu (mwreg,mrn,ern,ewreg,em2reg,mm2reg,         // control unit
            rsrtequ,func,op,rs,rt,wreg,m2reg,
            wmem,aluc,regrt,aluimm,fwda,fwdb,
            nostall,sext,pcsrc,shift,jal);
regfile r_f (rs,rt,wdi,wrn,wwreg,~clk,clrn,qa,qb);      // register file
mux2x5  d_r (rd,rt,regrt,rn);                           // select dest reg #
mux4x32 s_a (qa,ealu,malu,mmo,fwda,a);                  // forward for a
mux4x32 s_b (qb,ealu,malu,mmo,fwdb,b);                  // forward for b
cla32 b_adr (dpc4,dis,1'b0,bpc);                        // branch target
assign dimm = {s16,imm};                               // 32-bit imm
assign jpc  = {dpc4[31:28],addr,2'b00};                // jump target
endmodule
```

The Verilog HDL codes of the control unit (pipeidcu) are listed below. All the control signals are generated here, which can be categorized to three types: (i) selection signals of multiplexers; (ii) register and memory write enables (wpcir, wreg, and wmem); and (iii) ALU operation control codes (aluc). The control unit of the pipelined CPU is a combinational circuit. It is almost the same as that of the SC CPU, except for the part of data hazard detection. There are three signals for dealing with data hazards: fwda, fwdb, and nostall (used as wpcir).

```
module pipeidcu (mwreg,mrn,ern,ewreg,em2reg,mm2reg,rsrtequ,func,op,rs,rt,
                wreg,m2reg,wmem,aluc,regrt,aluimm,fwda,fwdb,nostall,sext,
                pcsrc,shift,jal); // control unit in ID stage
    input   [5:0] op,func;   // op and func fields in instruction
    input   [4:0] rs,rt;     // rs and rt fields in instruction
    input   [4:0] ern;       // destination register number in EXE stage
    input   [4:0] mrn;       // destination register number in MEM stage
    input         ewreg;     // register file write enable in EXE stage
    input         em2reg;    // memory to register in EXE stage
    input         mwreg;     // register file write enable in MEM stage
    input         mm2reg;    // memory to register in MEM stage
    input         rsrtequ;   // reg[rs] == reg[rt]
    output  [3:0] aluc;      // alu control
    output  [1:0] pcsrc;     // next pc (npc) select
```

```
output    [1:0] fwda;      // forward a: 00:qa; 01:exe; 10:mem; 11:mem_mem
output    [1:0] fwdb;      // forward b: 00:qb; 01:exe; 10:mem; 11:mem_mem
output          wreg;      // register file write enable
output          m2reg;     // memory to register
output          wmem;      // memory write
output          aluimm;    // alu input b is an immediate
output          shift;     // instruction in ID stage is a shift
output          jal;       // instruction in ID stage is jal
output          regrt;     // destination register number is rt
output          sext;      // sign extend
output          nostall;   // no stall (pipepc and pipeir write enable)
// instruction decode
wire rtype,i_add,i_sub,i_and,i_or,i_xor,i_sll,i_srl,i_sra,i_jr;
wire i_addi,i_andi,i_ori,i_xori,i_lw,i_sw,i_beq,i_bne,i_lui,i_j,i_jal;
and (rtype,~op[5],~op[4],~op[3],~op[2],~op[1],~op[0]);        // r format
and (i_add,rtype, func[5],~func[4],~func[3],~func[2],~func[1],~func[0]);
and (i_sub,rtype, func[5],~func[4],~func[3],~func[2], func[1],~func[0]);
and (i_and,rtype, func[5],~func[4],~func[3], func[2],~func[1],~func[0]);
and (i_or, rtype, func[5],~func[4],~func[3], func[2],~func[1], func[0]);
and (i_xor,rtype, func[5],~func[4],~func[3], func[2], func[1],~func[0]);
and (i_sll,rtype,~func[5],~func[4],~func[3],~func[2],~func[1],~func[0]);
and (i_srl,rtype,~func[5],~func[4],~func[3],~func[2], func[1],~func[0]);
and (i_sra,rtype,~func[5],~func[4],~func[3],~func[2], func[1], func[0]);
and (i_jr, rtype,~func[5],~func[4], func[3],~func[2],~func[1],~func[0]);
and (i_addi,~op[5],~op[4], op[3],~op[2],~op[1],~op[0]);        // i format
and (i_andi,~op[5],~op[4], op[3], op[2],~op[1],~op[0]);
and (i_ori, ~op[5],~op[4], op[3], op[2],~op[1], op[0]);
and (i_xori,~op[5],~op[4], op[3], op[2], op[1],~op[0]);
and (i_lw,   op[5],~op[4],~op[3],~op[2], op[1], op[0]);
and (i_sw,   op[5],~op[4], op[3],~op[2], op[1], op[0]);
and (i_beq, ~op[5],~op[4],~op[3], op[2],~op[1],~op[0]);
and (i_bne, ~op[5],~op[4],~op[3], op[2],~op[1], op[0]);
and (i_lui, ~op[5],~op[4], op[3], op[2], op[1], op[0]);
and (i_j,   ~op[5],~op[4],~op[3],~op[2], op[1],~op[0]);        // j format
and (i_jal, ~op[5],~op[4],~op[3],~op[2], op[1], op[0]);
// instructions that use rs
wire i_rs = i_add  | i_sub  | i_and  | i_or  | i_xor | i_jr  | i_addi |
            i_andi | i_ori  | i_xori | i_lw  | i_sw  | i_beq | i_bne;
// instructions that use rt
wire i_rt = i_add  | i_sub  | i_and  | i_or  | i_xor | i_sll | i_srl  |
            i_sra  | i_sw   | i_beq  | i_bne;
// pipeline stall caused by data dependency with lw instruction
assign nostall = ~(ewreg & em2reg & (ern != 0) & (i_rs & (ern == rs) |
                                         i_rt & (ern == rt)));
reg [1:0] fwda, fwdb;  // forwarding, multiplexer's select signals
always @ (ewreg, mwreg, ern, mrn, em2reg, mm2reg, rs, rt) begin
    // forward control signal for alu input a
```

```
        fwda = 2'b00;                                    // default: no hazards
    if (ewreg & (ern != 0) & (ern == rs) & ~em2reg) begin
        fwda = 2'b01;                                    // select exe_alu
    end else begin
        if (mwreg & (mrn != 0) & (mrn == rs) & ~mm2reg) begin
            fwda = 2'b10;                                // select mem_alu
        end else begin
            if (mwreg & (mrn != 0) & (mrn == rs) & mm2reg) begin
                fwda = 2'b11;                            // select mem_lw
            end
        end
    end
    // forward control signal for alu input b
    fwdb = 2'b00;                                        // default: no hazards
    if (ewreg & (ern != 0) & (ern == rt) & ~em2reg) begin
        fwdb = 2'b01;                                    // select exe_alu
    end else begin
        if (mwreg & (mrn != 0) & (mrn == rt) & ~mm2reg) begin
            fwdb = 2'b10;                                // select mem_alu
        end else begin
            if (mwreg & (mrn != 0) & (mrn == rt) & mm2reg) begin
                fwdb = 2'b11;                            // select mem_lw
            end
        end
    end
end
// control signals
assign wreg     = (i_add |i_sub |i_and |i_or  |i_xor |i_sll |i_srl |
                   i_sra |i_addi|i_andi|i_ori |i_xori|i_lw  |i_lui |
                   i_jal)& nostall;         // prevent from executing twice
assign regrt    = i_addi|i_andi|i_ori |i_xori|i_lw  |i_lui;
assign jal      = i_jal;
assign m2reg    = i_lw;
assign shift    = i_sll |i_srl |i_sra;
assign aluimm   = i_addi|i_andi|i_ori |i_xori|i_lw  |i_lui |i_sw;
assign sext     = i_addi|i_lw  |i_sw  |i_beq |i_bne;
assign aluc[3]  = i_sra;
assign aluc[2]  = i_sub |i_or  |i_srl |i_sra |i_ori |i_lui;
assign aluc[1]  = i_xor |i_sll |i_srl |i_sra |i_xori|i_beq |i_bne|i_lui;
assign aluc[0]  = i_and |i_or  |i_sll |i_srl |i_sra |i_andi|i_ori;
assign wmem     = i_sw  & nostall;          // prevent from executing twice
assign pcsrc[1] = i_jr  |i_j   |i_jal;
assign pcsrc[0] = i_beq & rsrtequ | i_bne & ~rsrtequ | i_j | i_jal;
endmodule
```

Below is the Verilog HDL codes of the ID/EXE pipeline register. Four 32-bit data (two register operands, an immediate, a PC + 4), a 5-bit destination register number, a 4-bit ALU control, and six 1-bit control signals are stored in the pipeline registers.

```verilog
module pipedereg (dwreg,dm2reg,dwmem,daluc,daluimm,da,db,dimm,drn,dshift,
                  djal,dpc4,clk,clrn,ewreg,em2reg,ewmem,ealuc,ealuimm,ea,
                  eb,eimm,ern,eshift,ejal,epc4); // ID/EXE pipeline register
    input          clk, clrn;            // clock and reset
    input   [31:0] da, db;               // a and b in ID stage
    input   [31:0] dimm;                 // immediate in ID stage
    input   [31:0] dpc4;                 // pc+4 in ID stage
    input    [4:0] drn;                  // register number in ID stage
    input    [3:0] daluc;                // alu control in ID stage
    input          dwreg,dm2reg,dwmem,daluimm,dshift,djal;    // in ID stage
    output  [31:0] ea, eb;               // a and b in EXE stage
    output  [31:0] eimm;                 // immediate in EXE stage
    output  [31:0] epc4;                 // pc+4 in EXE stage
    output   [4:0] ern;                  // register number in EXE stage
    output   [3:0] ealuc;                // alu control in EXE stage
    output         ewreg,em2reg,ewmem,ealuimm,eshift,ejal;    // in EXE stage
    reg     [31:0] ea, eb, eimm, epc4;
    reg      [4:0] ern;
    reg      [3:0] ealuc;
    reg            ewreg,em2reg,ewmem,ealuimm,eshift,ejal;
    always @(negedge clrn or posedge clk)
        if (!clrn) begin                 // clear
            ewreg    <= 0;           em2reg  <= 0;
            ewmem    <= 0;           ealuc   <= 0;
            ealuimm  <= 0;           ea      <= 0;
            eb       <= 0;           eimm    <= 0;
            ern      <= 0;           eshift  <= 0;
            ejal     <= 0;           epc4    <= 0;
        end else begin                   // register
            ewreg    <= dwreg;       em2reg  <= dm2reg;
            ewmem    <= dwmem;       ealuc   <= daluc;
            ealuimm  <= daluimm;     ea      <= da;
            eb       <= db;          eimm    <= dimm;
            ern      <= drn;         eshift  <= dshift;
            ejal     <= djal;        epc4    <= dpc4;
        end
endmodule
```

Below is the Verilog HDL codes of the circuits in the EXE stage. The main job is calculation, done by the ALU. The others include PC + 8 (epc8), selecting data for the ALU inputs (ealua and ealub), preparing a constant 31 for the jal instruction or bypassing the destination register number (ern), and selecting ALU output or PC + 8 for the output of the EXE stage (ealu).

```verilog
module pipeexe (ealuc,ealuimm,ea,eb,eimm,eshift,ern0,epc4,ejal,ern,ealu);
    input   [31:0] ea, eb;                       // all in EXE stage
    input   [31:0] eimm;                         // imm
    input   [31:0] epc4;                         // pc+4
    input    [4:0] ern0;                         // temporary dest reg #
    input    [3:0] ealuc;                        // aluc
```

```
    input          ealuimm;                      // aluimm
    input          eshift;                       // shift
    input          ejal;                         // jal
    output [31:0] ealu;                          // EXE stage result
    output  [4:0] ern;                           // dest reg #
    wire   [31:0] alua;                          // alu input a
    wire   [31:0] alub;                          // alu input b
    wire   [31:0] ealu0;                         // alu result
    wire   [31:0] epc8;                          // pc+8
    wire          z;                             // alu z flag, not used
    wire   [31:0] esa = {eimm[5:0],eimm[31:6]};  // shift amount
    cla32  ret_addr (epc4,32'h4,1'b0,epc8);      // pc+8
    mux2x32 alu_in_a (ea,esa,eshift,alua);       // alu input a
    mux2x32 alu_in_b (eb,eimm,ealuimm,alub);     // alu input b
    mux2x32 save_pc8 (ealu0,epc8,ejal,ealu);     // alu result or pc+8
    assign ern = ern0 | {5{ejal}};               // dest reg #, jal: 31
    alu al_unit (alua,alub,ealuc,ealu0,z);       // alu result, z flag
endmodule
```

Below is the Verilog HDL codes of the EXE/MEM pipeline register. The following signals are registered: a 32-bit result of the EXE stage (ealu), a 32-bit data of the register rt in the EXE stage (eb) for sw instruction, a 5-bit destination register number (ern), a 1-bit register file write enable (ewreg), a 1-bit memory-to-register control signal (em2reg), and a 1-bit memory write enable (ewmem).

```
module pipeemreg (ewreg,em2reg,ewmem,ealu,eb,ern,clk,clrn,mwreg,mm2reg,
                  mwmem,malu,mb,mrn);             // EXE/MEM pipeline register
    input          clk, clrn;                     // clock and reset
    input  [31:0] ealu;                           // alu control in EXE stage
    input  [31:0] eb;                             // b in EXE stage
    input   [4:0] ern;                            // register number in EXE stage
    input          ewreg,em2reg,ewmem;            // in EXE stage
    output [31:0] malu;                           // alu control in MEM stage
    output [31:0] mb;                             // b in MEM stage
    output  [4:0] mrn;                            // register number in MEM stage
    output         mwreg,mm2reg,mwmem;            // in MEM stage
    reg    [31:0] malu,mb;
    reg     [4:0] mrn;
    reg            mwreg,mm2reg,mwmem;
    always @(negedge clrn or posedge clk)
        if (!clrn) begin                          // clear
            mwreg   <= 0;         mm2reg <= 0;
            mwmem   <= 0;         malu   <= 0;
            mb      <= 0;         mrn    <= 0;
        end else begin                            // register
            mwreg   <= ewreg;     mm2reg <= em2reg;
            mwmem   <= ewmem;     malu   <= ealu;
            mb      <= eb;        mrn    <= ern;
        end
endmodule
```

There is only one component in the MEM stage, data memory, for memory access instructions (lw and sw). The Verilog HDL codes of the data memory (pl_data_mem) will be given in the next section.

```
module pipemem (we,addr,datain,clk,dataout);        // MEM stage
    input           clk;                            // clock
    input   [31:0] addr;                            // address
    input   [31:0] datain;                          // data in (to mem)
    input           we;                             // memory write
    output [31:0] dataout;                          // data out (from mem)
    pl_data_mem dmem (clk,dataout,datain,addr,we);  // data memory
endmodule
```

Below is the Verilog HDL codes of the MEM/WB pipeline register. The following signals are registered: a 32-bit ALU result in the MEM stage (malu), a 32-bit memory data (mmo), a 5-bit destination register number (mrn), a 1-bit register file write enable (mwreg), and a 1-bit memory-to-register control signal (mm2reg).

```
module pipemwreg (mwreg,mm2reg,mmo,malu,mrn,clk,clrn,wwreg,wm2reg,wmo,walu,
                  wrn);                             // MEM/WB pipeline register
    input           clk, clrn;                      // clock and reset
    input   [31:0] mmo;                             // memory data out in MEM stage
    input   [31:0] malu;                            // alu control in MEM stage
    input   [4:0] mrn;                              // register number in MEM stage
    input           mwreg, mm2reg;                  // in MEM stage
    output [31:0] wmo;                              // memory data out in WB stage
    output [31:0] walu;                             // alu control in WB stage
    output   [4:0] wrn;                             // register number in WB stage
    output          wwreg, wm2reg;                  // in WB stage
    reg     [31:0] wmo, walu;
    reg     [4:0] wrn;
    reg             wwreg,wm2reg;
    always @(negedge clrn or posedge clk)
      if (!clrn) begin                              // clear
          wwreg <= 0;              wm2reg <= 0;
          wmo   <= 0;              walu   <= 0;
          wrn   <= 0;
      end else begin                                // register
          wwreg <= mwreg;          wm2reg <= mm2reg;
          wmo   <= mmo;            walu   <= malu;
          wrn   <= mrn;
      end
endmodule
```

There is only one component in the WB stage: a 32-bit 2-to-1 multiplexer, for selecting a data word from walu and wmo that will be written into the register file. The register indexed by wrn in the register file will be updated with the selected data word if the write enable wwreg is true. Writing the register file happens on the falling edge of the clock. An instruction is graduated at this stage.

```
module pipewb (walu,wmo,wm2reg,wdi);        // WB stage
    input   [31:0] walu;                    // alu result or pc+8 in WB stage
    input   [31:0] wmo;                     // data out (from mem) in WB stage
    input         wm2reg;                   // memory to register in WB stage
    output  [31:0] wdi;                     // data to be written into regfile
    mux2x32 wb (walu,wmo,wm2reg,wdi);       // select for wdi
endmodule
```

8.3.3 Test Program and Simulation Waveforms

We give the test code and test data for verifying the correctness of our pipelined CPU and show the simulation waveforms, generated with ModelSim.

Below is the test program that is stored in a ROM (read-only memory). We used the general Verilog HDL codes to implement the ROM. Although a ROM is said to be a memory, it actually is a combinational circuit. In the test program, we checked all the 20 instructions. The main part of the test program is a subroutine in which four 32-bit memory words are summed by a for loop. After returning from the subroutine, the sum is stored in the data memory by an sw instruction. A code pattern that causes pipeline stall is also prepared within the loop. We used word address to assign the content of each word (a 32-bit instruction). The parenthesized hexadecimal number in the center of each line is the byte address (PC).

```
module pl_inst_mem (a,inst);              // instruction memory, rom
    input   [31:0] a;                     // rom address
    output  [31:0] inst;                  // rom content = rom[a]
    wire    [31:0] rom [0:63];            // rom cells: 64 words * 32 bits
    // rom[word_addr] = instruction      // (pc) label    instruction
    assign rom[6'h00] = 32'h3c010000;     // (00) main:    lui  $1, 0
    assign rom[6'h01] = 32'h34240050;     // (04)          ori  $4, $1, 80
    assign rom[6'h02] = 32'h0c00001b;     // (08) call:    jal  sum
    assign rom[6'h03] = 32'h20050004;     // (0c) dslot1:  addi $5, $0,  4
    assign rom[6'h04] = 32'hac820000;     // (10) return:  sw   $2, 0($4)
    assign rom[6'h05] = 32'h8c890000;     // (14)          lw   $9, 0($4)
    assign rom[6'h06] = 32'h01244022;     // (18)          sub  $8, $9, $4
    assign rom[6'h07] = 32'h20050003;     // (1c)          addi $5, $0,  3
    assign rom[6'h08] = 32'h20a5ffff;     // (20) loop2:   addi $5, $5, -1
    assign rom[6'h09] = 32'h34a8ffff;     // (24)          ori  $8, $5, 0xffff
    assign rom[6'h0a] = 32'h39085555;     // (28)          xori $8, $8, 0x5555
    assign rom[6'h0b] = 32'h2009ffff;     // (2c)          addi $9, $0, -1
    assign rom[6'h0c] = 32'h312affff;     // (30)          andi $10,$9,0xffff
    assign rom[6'h0d] = 32'h01493025;     // (34)          or   $6, $10, $9
    assign rom[6'h0e] = 32'h01494026;     // (38)          xor  $8, $10, $9
    assign rom[6'h0f] = 32'h01463824;     // (3c)          and  $7, $10, $6
    assign rom[6'h10] = 32'h10a00003;     // (40)          beq  $5, $0, shift
    assign rom[6'h11] = 32'h00000000;     // (44) dslot2:  nop
    assign rom[6'h12] = 32'h08000008;     // (48)          j    loop2
    assign rom[6'h13] = 32'h00000000;     // (4c) dslot3:  nop
    assign rom[6'h14] = 32'h2005ffff;     // (50) shift:   addi $5, $0, -1
    assign rom[6'h15] = 32'h000543c0;     // (54)          sll  $8, $5, 15
    assign rom[6'h16] = 32'h00084400;     // (58)          sll  $8, $8, 16
```

```
    assign rom[6'h17] = 32'h00084403;  // (5c)          sra  $8, $8, 16
    assign rom[6'h18] = 32'h000843c2;  // (60)          srl  $8, $8, 15
    assign rom[6'h19] = 32'h08000019;  // (64) finish: j    finish
    assign rom[6'h1a] = 32'h00000000;  // (68) dslot4: nop
    assign rom[6'h1b] = 32'h00004020;  // (6c) sum:    add  $8, $0, $0
    assign rom[6'h1c] = 32'h8c890000;  // (70) loop:   lw   $9, 0($4)
    assign rom[6'h1d] = 32'h01094020;  // (74) stall:  add  $8, $8, $9
    assign rom[6'h1e] = 32'h20a5ffff;  // (78)         addi $5, $5, -1
    assign rom[6'h1f] = 32'h14a0fffc;  // (7c)         bne  $5, $0, loop
    assign rom[6'h20] = 32'h20840004;  // (80) dslot5: addi $4, $4,  4
    assign rom[6'h21] = 32'h03e00008;  // (84)         jr   $31
    assign rom[6'h22] = 32'h00081000;  // (88) dslot6: sll  $2, $8, 0
    assign inst = rom[a[7:2]];         // use 6-bit word address to read rom
endmodule
```

Below is the test data that are stored in a RAM (random-access memory). Four 32-bit words in the RAM will be read by `lw` instructions. The test program will store a word in the location next to the four words. We also used the general Verilog HDL codes to implement the RAM.

```
module pl_data_mem (clk,dataout,datain,addr,we); // data memory, ram
    input           clk;              // clock
    input   [31:0] addr;              // ram address
    input   [31:0] datain;            // data in (to memory)
    input           we;               // write enable
    output  [31:0] dataout;           // data out (from memory)
    reg     [31:0] ram [0:31];        // ram cells: 32 words * 32 bits
    assign dataout = ram[addr[6:2]];  // use 5-bit word address
    always @ (posedge clk) begin
        if (we) ram[addr[6:2]] = datain;  // write ram
    end
    integer i;
    initial begin                     // ram initialization
        for (i = 0; i < 32; i = i + 1)
            ram[i] = 0;
        // ram[word_addr] = data      // (byte_addr) item in data array
        ram[5'h14] = 32'h000000a3;    // (50) data[0]   0 +  a3 =   a3
        ram[5'h15] = 32'h00000027;    // (54) data[1]  a3 +  27 =   ca
        ram[5'h16] = 32'h00000079;    // (58) data[2]  ca +  79 =  143
        ram[5'h17] = 32'h00000115;    // (5c) data[3] 143 + 115 =  258
        // ram[5'h18] should be 0x00000258, the sum stored by sw instruction
    end
endmodule
```

Now, it is the time to show the simulation waveforms. Figure 8.22 illustrates the waveforms when the pipelined CPU executes the `jal` instruction (PC = 0x00000008). The instruction in the delay slot (PC = 0x0000000c) is executed also. The target address of the `jal` instruction is 0x0000006c, the entry of a subroutine (`sum`). We can see that the result at the EXE stage of the `jal` instruction is 0x00000010, which is the return address (from the subroutine).

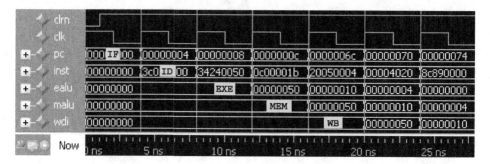

Figure 8.22 Waveforms of the pipelined CPU (call subroutine)

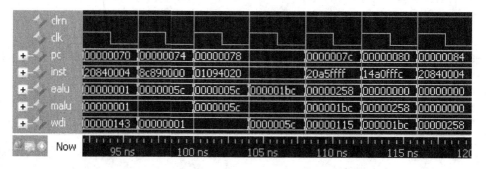

Figure 8.23 Waveforms of the pipelined CPU (pipeline stall)

We have added the stage names in the figure. `pc`, `inst`, `ealu`, `malu`, and `wdi` are signals in the IF, ID, EXE, MEM, and WB stages, respectively.

Figure 8.23 illustrates the waveforms when the pipelined CPU executes the `lw $9, 0($4)` instruction (PC = 0x00000070) and its follow-up, the `add $8, $8, $9` instruction (PC = 0x00000074) in the fourth (last) round of the `for` loop. Because the `add` instruction has a data dependency with the `lw` instruction, the pipeline is stalled for one clock cycle. The last data word of the four-element array (0x00000115) in the memory location of 0x0000005c is loaded by the `lw` instruction. We can also see that the sum of the four elements (0x00000258) is calculated correctly in the EXE stage of the `add` instruction.

Figure 8.24 illustrates the waveforms when the pipelined CPU executes the `jr` instruction (PC = 0x00000084) to return from the subroutine. The instruction in the delay slot (PC = 0x00000088) is executed also. The return address is 0x00000010. The figure also shows that the sum (0x00000258) is stored in the memory in the location 0x00000060 and it is loaded again correctly from the same memory location by an `lw` instruction. The `sub` instruction (PC = 0x00000018) has a data dependency with the `lw` instruction, and the pipeline is stalled again for one clock cycle.

The waveforms shown in Figures 8.22–8.24 are the same as what we expected. It seems that our pipelined CPU works correctly logically.

8.4 Precise Interrupts/Exceptions in Pipelined CPU

In Chapter 6, we described how to deal with interrupts and exceptions in the SC CPU. The SC CPU acknowledges an interrupt request by transferring control to an interrupt handler after completely

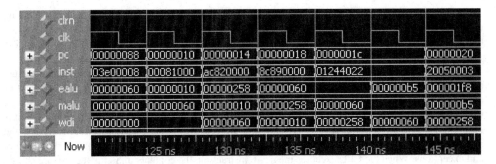

Figure 8.24 Waveforms of the pipelined CPU (return from subroutine)

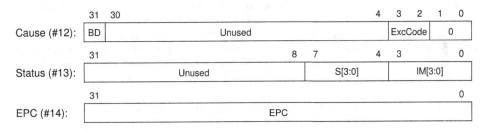

Figure 8.25 Three registers for implementing precise interrupt

finishing the execution of the current instruction. Different from SC or MC CPUs, in a pipelined CPU there are multiple instructions that are overlapped in execution during every clock cycle. We cannot find a point of time at which we can say that all the instructions in the pipeline have finished their executions. Therefore it is more difficult to deal with interrupts in pipelined CPUs than in SC and MC CPUs. Moreover, the pipelined CPU adopts the delayed branch. This makes it much more difficult.

If a pipelined CPU can deal with interrupts in the same beautiful way as an SC CPU does, this pipelined CPU is said to have a precise interrupt mechanism. More specifically, when an interrupt occurs, the instruction that is staying at the ID stage and its predecessors must be executed safely, and its follow-up instructions must be canceled. After the interrupt request is serviced, the control is returned to the place of the first canceled instruction: the program then continues as if nothing happened.

This section introduces an implementation of precise interrupts/exceptions on the pipelined CPU, gives the Verilog HDL codes of the implementation, and shows the simulation waveforms.

8.4.1 Exception and Interrupt Types and Related Registers

The pipelined CPU introduced in this chapter deals with only one external interrupt request and three exceptions: arithmetic overflow, unimplemented instruction, and system call. Figure 8.25 shows the formats of the CP0 registers, which we will use in the design of the precise interrupt/exception mechanism.

The EPC (exception program counter) register (register 14 of CP0) is used for saving the interrupt/exception return address. When the CPU executes an `eret` (return from exception) instruction, the content in the EPC is written into the PC. When an interrupt occurs, the address of the instruction in the IF stage will be saved into the EPC (the instruction in the ID stage will be executed). However, if the instruction in the ID stage is a branch or a jump instruction when an interrupt occurs, the branch or jump instruction should be canceled and its address saved in the EPC. When an instruction generates an exception, the address of the instruction is saved into the EPC so that the instruction is capable of being executed again. If there is no need to execute it again, the return address must be updated by adding a 4 to the EPC before executing the eret instruction. Note that the eret instruction itself has no delay slot.

The Status register (register 13 of CP0) is used for masking the interrupt and exceptions. The bits 0, 1, 2, and 3 in the IM field are the mask control bits of interrupt, system call, unimplemented instruction, and arithmetic overflow, respectively. A 1 in the IM field indicates that the corresponding interrupt or exception is enabled; a 0 indicates that it is disabled. The S field is used for saving the contents in the IM field. In our implementation, when an interrupt or an exception happens and its corresponding IM bit is enabled, the contents in Status register are shifted to the left by 4 bits before transferring control to the interrupt/exception handler. This shift has two functions: to save the IM and to disable all the interrupt and exceptions. When the CPU executes the eret instruction, the contents in Status register are shifted to the right by 4 bits, restoring the original mask settings.

The ExcCode field in the Cause register (register 12 of CP0) indicates the source of the interrupt or exception. Its encoding in our design is listed in Table 8.1. The BD bit in the Cause register is set to 1 when the instruction that generated an exception is located in a delay slot. The BD bit is also set to 1 when an interrupt occurs, and the instruction in the ID stage is located in a delay slot.

The system call (Sys) and unimplemented instruction (Unimpl) exceptions can be determined in the ID stage, and arithmetic overflow (Ov) may happen when the ALU performs addition or subtraction in the EXE stage, but an external interrupt (Int) may occur in any stage. In order to implement the precise interrupt mechanism, we classify the time when an interrupt occurs into three cases.

1. When an interrupt occurs, the instruction in the ID stage is a jump or a branch;
2. When an interrupt occurs, the ID stage is in a delay slot;
3. Other cases (ordinary cases) than cases 1 and 2.

8.4.1.1 Interrupt at Ordinary Case

In an ordinary case, when an interrupt occurs, the instruction in the ID stage is executed; its follow-up instruction which is being fetched using the PC is canceled. Therefore, the value of the PC should be saved into the EPC. At the end of the clock cycle, the entry address of the interrupt handler (BASE) is written into the PC. These actions are illustrated in Figure 8.26.

In addition to saving the return address and transferring control to the interrupt handler, the contents in Status register are shifted to the left by 4 bits and the encoding of the interrupt is written into the ExcCode field of the Cause register.

Canceling an instruction is realized as follows: when an interrupt occurs in a clock cycle, a signal, called a cancel, is generated in the ID stage. This cancel signal cannot be used immediately to cancel the instruction which is being fetched in the IF stage. An instruction can be canceled in the ID stage. Therefore, we store the cancel signal into the ID/EXE pipeline register for the use in the next clock cycle when the instruction that should be canceled enters the ID stage. We use the name e_cancel for the cancel signal in the EXE stage. This signal is used to prohibit all the write signals generated in the ID stage so that the instruction is canceled.

Table 8.1 Stage where exception or interrupt may occur

ExcCode	Mnemonic	Mask	Description	Occur in
0	Int	IM[0]	External interrupt	Any stage
1	Sys	IM[1]	System call	ID stage
2	Unimpl	IM[2]	Unimplemented instruction	ID stage
3	Ov	IM[3]	Arithmetic overflow	EXE stage

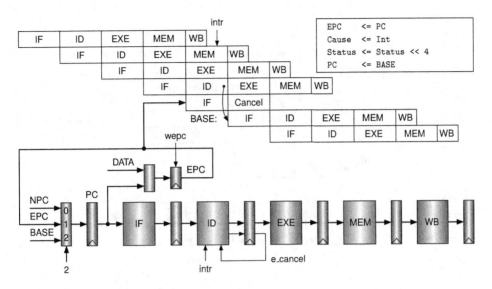

Figure 8.26 Interrupt occurring in an ordinary case

8.4.1.2 Interrupt at Jump or Branch

Figure 8.27 shows the case in which, when an interrupt occurs, the instruction in the ID stage is a branch (or a jump). The main idea is to cancel the branch instruction as well as the instruction that is being fetched (the instruction located in the delay slot).

Different from the ordinary case, in which the instruction in the ID stage when an interrupt occurs is executed, in the case shown in Figure 8.27 the branch instruction itself is canceled. Therefore, the address of the branch instruction (PCD in the figure) should be saved into the EPC. The cancel signal generated

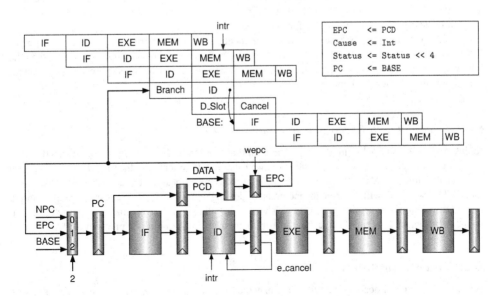

Figure 8.27 Interrupt occurring when decoding the branch instruction

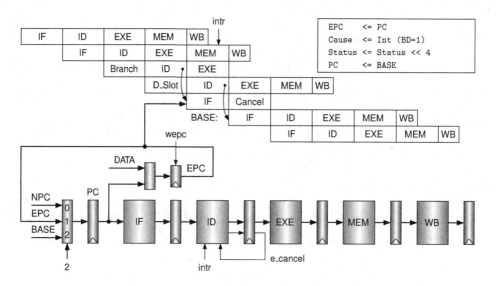

Figure 8.28   Interrupt occurring in the ID stage of delay slot

in the ID stage is used for canceling the branch immediately, and its registered signal `e_cancel` is used for canceling the instruction in the delay slot.

8.4.1.3   Interrupt at Delay Slot

Figure 8.28 shows the case in which, when an interrupt occurs, the ID stage is in a delay slot. We let the instruction in the ID stage be executed and cancel the instruction that is being fetched in the IF stage. (The PC contains the branch target address if the branch is taken.) Therefore, the PC should be saved to the EPC. This case is almost the same as the ordinary case except for setting the BD bit in the Cause register. Basically, the BD bit is useless for interrupt.

Up to now, we have described the method of implementing a precise interrupt mechanism on a pipelined CPU. There are two key points in the implementation: one is how to cancel an instruction, and the other is how to save the address of the canceled instruction into the EPC which stores the restart address when returning from the interrupt handler.

8.4.2   Dealing with Exceptions in Pipelined CPU

An interrupt is an external request that has no temporal relation to the program being executed. But an exception results directly from the execution of the program. It is better to use the following two phrases to mention the interrupt and exception: external asynchronous interrupt and internal synchronous exception.

In the following, we will describe the methods for dealing with the exceptions of system call, unimplemented instruction, and arithmetic overflow. The system call and the unimplemented instruction exceptions can be detected in the ID stage, but the arithmetic overflow exception happens in the EXE stage.

8.4.2.1   Exception of System Call

System call exception occurs when the CPU executes a `syscall` instruction. This instruction is used to request the operating system kernel to provide a service. It is almost like the `jal` instruction. The

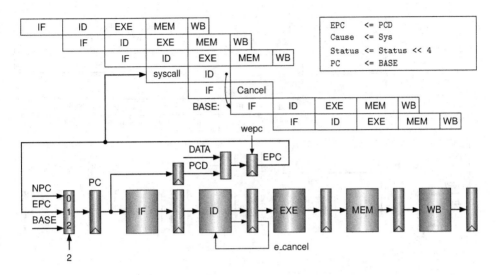

Figure 8.29   Execution of system call instruction

differences between these two instructions are that the `jal` instruction saves PC + 8 into register $31 of the register file and is followed by a delay slot in PC + 4, but the `syscall` instruction saves its own address into the EPC and has no delay slot: its follow-up instruction is always canceled.

To avoid executing the `syscall` instruction again on return from the exception handler, the return address saved in the EPC must be incremented by 4, as shown in the codes below; thus the canceled instruction will be executed.

```
epc_plus4:
    mfc0  $k0, c0_epc      # load the content of epc to register 26
    addiu $k0, $k0, 4      # increment by four
    mtc0  $k0, c0_epc      # save it back to epc
    eret                   # pc <-- epc (return from the exception handler)
    nop                    # will be canceled
```

Figure 8.29 illustrates the case of executing the `syscall` instruction. The PCD (the address of the `syscall` instruction) is saved into the EPC. The signal `cancel` generated in the ID stage is stored into an ID/EXE pipeline register whose output `e_cancel` is used for canceling the next instruction to `syscall`, as described in the previous section. At the end of the ID stage of `syscall`, the entry address of the exception handler (`BASE`) is written into the PC. In our implementation, it is not allowed to put the `syscall` instruction in a delay slot.

8.4.2.2   Exception of Unimplemented Instruction

An exception of the unimplemented instruction occurs when the CPU executes a legal instruction in the ISA but the CPU hardware does not implement it. This exception is different from the exception of an undefined instruction. The undefined instruction is not a legal instruction in the ISA.

By using the unimplemented instruction exception, we can emulate the execution of an instruction that is not supported by hardware. For example, we can write two code fragments to perform the division and square root operations for the emulation of a division and a square root instruction, respectively.

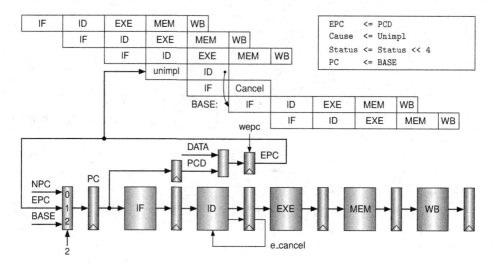

Figure 8.30   Unimplemented instruction exception

Figure 8.30 illustrates the execution of an unimplemented instruction. Dealing with this exception is mostly the same as that with the `syscall` instruction. The only difference is in the `ExcCode`. Also, in our implementation, it is not allowed to put an unimplemented instruction in a delay slot.

8.4.2.3   Exception of Arithmetic Overflow

All the interrupt and exceptions we have described up to now are checked in the ID stage. But overflow happens in the EXE stage of an arithmetic instruction. This one clock cycle delay makes dealing with the overflow exceptions a little more difficult. Figure 8.31 illustrates how to deal with an overflow exception in an ordinary case (not in a delay slot). The signal of an overflow exception (`exc_ovr`) generated by an

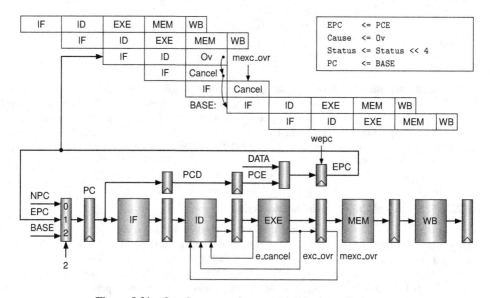

Figure 8.31   Overflow exception occurring in the ordinary case

arithmetic instruction in the EXE stage is sent to the control unit which is located in the ID stage, and the entry address of the exception handler (BASE) is selected for the next PC.

There are three instructions that must be canceled. One is the instruction that has generated the overflow exception, and the others are the two follow-up instructions. Canceling the last two instructions can be done inside the control unit by using the signals exc_ovr and mexc_ovr (exc_ovr in the MEM stage), but canceling the first instruction that generated the overflow exception has to be done in the EXE stage (outside the control unit). The PCE (PC in the EXE stage) that points to the instruction which generated the overflow exception must be saved to the EPC.

Figure 8.32 illustrates how to deal with an overflow exception generated by an arithmetic instruction that is located in a delay slot. In this case, the address of the branch or jump instruction must be saved into the EPC and BD must be set. The PCM register (PC in the MEM stage) contains this address. The method of canceling three instructions is similar to that in the ordinary case.

The exception handler can read the EPC and BD bit in the Cause register to check which instruction generated the overflow exception: if BD is zero, EPC points to the instruction that generated the exception; otherwise, EPC + 4 points to that instruction. However, there is no way to resume the execution of the program because the overflow exception will occur again once its execution is resumed. Usually, when such exceptions happen, the exception handler displays some messages based on the instruction pointed by the EPC or EPC + 4, and stops the program.

Because an exception is generated in the ID or EXE stage, two exceptions may occur at the same time, as shown in Figure 8.33, where an overflow exception and a system call exception occur in the same clock cycle.

We defined the ExcCode of overflow as 11 and that of system call as 01. In the case above, an ExcCode of 11 (overflow) will be generated. We say that an overflow exception has a higher priority than a system call exception. Similarly, any exception has a higher priority than an interrupt.

We have described how to deal with an external asynchronous interrupt and three internal synchronous exceptions. The most important job is to save an address into the EPC. We summarize it in Table 8.2.

An overflow exception only occurs in the EXE stage of an instruction execution. An interrupt can occur in any pipeline stage. The system call and unimplemented instructions can be determined in the ID stage. They are neither branch nor jump instructions; thus we mark them as "Impossible" in the table. Branch or jump instructions do not care about their overflow when calculating target addresses in the ID stage,

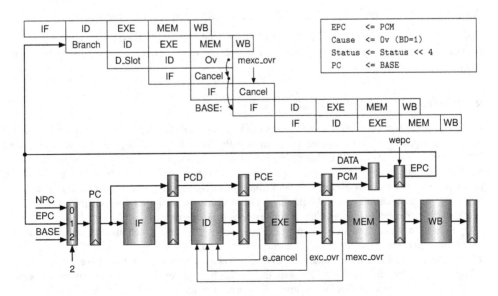

Figure 8.32   Overflow exception occurring in the delay slot

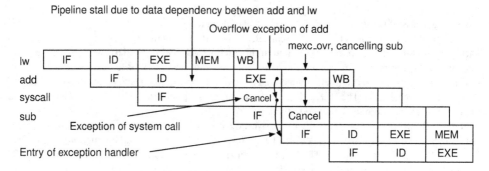

Figure 8.33 Two exceptions that might be occurring in the same clock cycle

Table 8.2 Writing EPC on exception or interrupt

Mnemonic	Branch/jump in ID stage	Occurs in delay slot	Others
Int	EPC ⇐ PCD	EPC ⇐ PC	EPC ⇐ PC
Sys	Impossible	Not allowed	EPC ⇐ PCD
Unimpl	Impossible	Not allowed	EPC ⇐ PCD
Ov	Not occur	EPC ⇐ PCM	EPC ⇐ PCE

so we designate them as "Not occur" in the table. In our design of the pipelined CPU, the system call and unimplemented instructions are not allowed to be put in the place of a delay slot.

From the table we know that we must use a multiplexer to select one address from the PC, PCD (PC in the ID stage), PCE (PC in the EXE stage), and PCM (PC in the MEM stage), for the input of EPC register.

8.5   Design of Pipelined CPU with Precise Interrupt/Exception

Now we are ready to design the pipelined CPU with a precise interrupt/exception mechanism. This section shows the hardware structure of the CPU, lists its Verilog HDL codes and the test program, and gives the simulation waveforms.

8.5.1   The Structure of the Pipelined CPU

In addition to the basic instructions, the instructions that move data between the general-purpose register file and the registers of CP0 are implemented in the CPU. We put the CP0 registers (Cause, Status, and EPC) in between the ID and EXE stages. The mfc0 (move to CP0) instruction reads the data word from a CP0 register and writes it into a register of the general-purpose register file in the WB stage. The mtc0 (move to CP0) instruction reads the data word from a general-purpose register in the ID stage and writes it into a CP0 register at the end of the ID stage. The pipeline stages of these two instructions are shown in Figure 8.34. The internal forwarding is applicable to these instructions.

Figure 8.35 shows the formats of mfc0, mtc0, syscall, and eret instructions. The field rt is the register number in the general-purpose register file, and the field rd is the register number of CP0 registers. The register numbers of Cause, Status, and EPC registers are 12, 13, and 14, respectively.

Figure 8.36 shows the pipelined CPU with the precise interrupt/exception mechanism. It is designed by adding the necessary circuits for dealing with the interrupt and exceptions to the basic pipelined CPU shown in Figure 8.21. Because of space limitation, some circuits are not drawn in Figure 8.36.

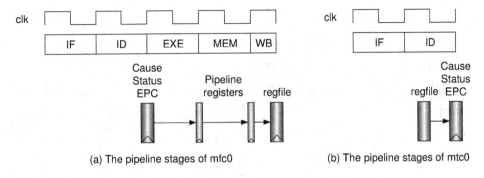

(a) The pipeline stages of mfc0   (b) The pipeline stages of mtc0

Figure 8.34 Pipeline stages of mfc0 and mtc0 instructions

	31 30 29 28 27 26	25 24 23 22 21	20 19 18 17 16	15 14 13 12 11	10 9 8 7 6 5 4 3 2 1 0
mfc0	0 1 0 0 0 0	0 0 0 0 0	rt	rd	0 0 0 0 0 0 0 0 0 0 0
mtc0	0 1 0 0 0 0	0 0 1 0 0	rt	rd	0 0 0 0 0 0 0 0 0 0 0
syscall	0 0 0 0 0 0	0 0 0 0 0	0 0 0 0 0	0 0 0 0 0	0 0 0 0 0 0 0 1 1 0 0
eret	0 1 0 0 0 0	1 0 0 0 0	0 0 0 0 0	0 0 0 0 0	0 0 0 0 0 0 0 1 1 0 0 0

Figure 8.35 Instruction formats of mfc0, mtc0, syscall, and eret

The main parts of the extra circuits are the three CP0 registers: Cause, Status, and EPC. Each of them can be read with the mfc0 instruction (see the OR gate and two multiplexers in the EXE stage in Figure 8.36) and can be written with the mtc0 instruction (the db is selected for the writing). The write enable signals of Cause, Status, and EPC registers are wcau, wsta, and wepc, respectively. These three write enable signals are drawn at the center of the figure.

The three CP0 registers (Cause, Status, and EPC) are updated not only by the mtc0 instruction but also by entering or return from the interrupt/exception handler. When the CPU acknowledges an interrupt or an exception, the Cause register will be updated with the source encoding of the interrupt or exception (cause in the figure); the Status register will be shifted left by 4 bits (sta_left in the figure); and the EPC register will be updated with the restart address, selected from the PC, PCD, PCE, and PCM by the selection signal sepc. The Status register is shifted right back by 4 bits (sta_right in the figure) at the return from the interrupt/exception handler.

An instruction in the ID stage can be canceled immediately by the cancel signal. An instruction in the IF stage can be canceled once it enters the ID stage by the e_cancel signal in the EXE stage which was originally generated at the ID stage and then saved into an ID/EXE pipeline register. However, the instruction that caused the overflow exception must be canceled in the EXE stage. The signal exc_ovr in Figure 8.36 is used to cancel the instruction in the EXE stage (to disable the register write signal ewreg0) and the instruction in the ID stage immediately. It must also be saved into an EXE/MEM pipeline register in order to cancel the instruction that is being fetched from instruction memory (canceled once it enters the ID stage).

8.5.2 The Verilog HDL Codes of the Pipelined CPU

Based on the pipelined CPU structure shown in Figure 8.36, we wrote the Verilog HDL codes to implement the pipelined CPU with a precise interrupt/exception mechanism. The codes are given

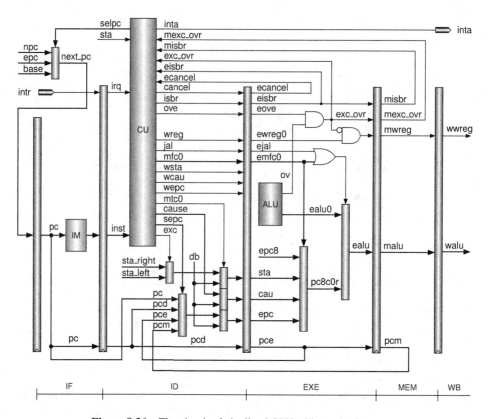

Figure 8.36   The circuit of pipelined CPU with precise interrupt

in two modules: the module `cu_exc_int.v` contains the codes of the control unit, and the module `pipelined_cpu_exc_int.v` contains all the remaining codes.

Below is the codes of the module `pipelined_cpu_exc_int.v`. We put all the pipeline registers and the circuits of all the pipeline stages in this module (we recommend that readers do not use such a style to write codes). First, we define all of the signals including the input and output signals and the internal signals that are used inside the module. The comments are mainly given to the signals that contribute to the implementation of the precise interrupt and exceptions. Then the codes are given in the order of the PC, circuits of the IF stage, IF/ID registers, circuits of the ID stage, ID/EXE registers, circuits of the EXE stage, EXE/MEM registers, circuits of the MEM stage, MEM/WB registers, and circuits of the WB stage. At the end of the WB stage, the result (`wdi` in the codes) is written into the destination register (`wrn`) in the register file if the write enable (`wwreg`) is active. The instruction memory and data memory modules are invoked in IF and MEM stages, respectively.

```
module pipelined_cpu_exc_int (clk,clrn,pc,inst,ealu,malu,wdi,intr,inta);
    input           clk, clrn;          // clock and reset
    input           intr;               // interrupt request
    output [31:0]   pc;                 // program counter
    output [31:0]   inst;               // instruction in ID stage
    output [31:0]   ealu;               // result in EXE stage
    output [31:0]   malu;               // result in MEM stage
```

```
output  [31:0] wdi;                       // result in WB stage
output         inta;                      // interrupt acknowledgement
parameter      exc_base = 32'h00000008; // exception handler entry
// signals in IF stage
wire    [31:0] pc4,ins,npc;
wire    [31:0] next_pc;                   // next pc
// signals in ID stage
wire    [31:0] dpc4,bpc,jpc,da,db,imm,qa,qb;
wire    [5:0]  op,func;
wire    [4:0]  rs,rt,rd,rn;
wire    [3:0]  aluc;
wire    [1:0]  pcsrc,fwda,fwdb;
wire           wreg,m2reg,wmem,aluimm,shift,jal,sext,regrt,rsrtequ,wpcir;
wire    [31:0] pcd;                       // pc in ID stage
wire    [31:0] cause;                     // cause content
wire    [31:0] sta_in;                    // status register, data in
wire    [31:0] cau_in;                    // cause  register, data in
wire    [31:0] epc_in;                    // epc    register, data in
wire    [31:0] epcin;                     // pc, pcd, pce, or pcm
wire    [31:0] stalr;                     // state shift left or right
wire    [1:0]  mfc0;                      // select pc+8, sta, cau, or epc
wire    [1:0]  selpc;                     // select for next_pc
wire    [1:0]  sepc;                      // select for epcin
wire           isbr;                      // is a branch or a jump
wire           ove;                       // ov enable = arith & sta[3]
wire           cancel;                    // cancel next instruction
wire           exc;                       // exc or int occurs
wire           mtc0;                      // move to c0 instruction
wire           wsta;                      // status register write enable
wire           wcau;                      // cause  register write enable
wire           wepc;                      // epc    register write enable
wire           irq;                       // latched intr
// signals in EXE stage
wire    [31:0] ealua,ealub,esa,ealu0,epc8;
reg     [31:0] ea,eb,eimm,epc4;
reg     [4:0]  ern0;
wire    [4:0]  ern;
reg     [3:0]  ealuc;
reg            ewreg0,em2reg,ewmem,ealuimm,eshift,ejal;
wire           ewreg,zero;
wire           exc_ovr;                   // overflow exc in EXE stage
reg     [31:0] pce;                       // pc in EXE stage
wire    [31:0] sta;                       // status register, data out
wire    [31:0] cau;                       // cause  register, data out
wire    [31:0] epc;                       // epc    register, data out
wire    [31:0] pc8c0r;                    // epc8, sta, cau, or epc
reg     [1:0]  emfc0;                     // mfc0   in EXE stage
```

```
reg              eisbr;                    // isbr  in EXE stage
reg              eove;                     // ove   in EXE stage
reg              ecancel;                  // cancel in EXE stage
wire             ov;                       // overflow flag
// signals in MEM stage
wire    [31:0]  mmo;
reg     [31:0]  malu,mb;
reg      [4:0]  mrn;
reg              mwreg,mm2reg,mwmem;
reg     [31:0]  pcm;                       // pc      in MEM stage
reg              misbr;                    // isbr    in MEM stage
reg              mexc_ovr;                 // exc_ovr in MEM stage
// signals in WB stage
reg     [31:0]  wmo,walu;
reg      [4:0]  wrn;
reg              wwreg,wm2reg;
// program counter
dffe32 prog_cnt (next_pc,clk,clrn,wpcir,pc);          // pc
// IF stage
cla32 pc_plus4 (pc,32'h4,1'b0,pc4);                   // pc+4
mux4x32 nextpc (pc4,bpc,da,jpc,pcsrc,npc);            // next pc
pl_exc_i_mem inst_mem (pc,ins);                       // inst mem
// IF/ID pipeline register
dffe32 pc_4_r (pc4,clk,clrn,wpcir,dpc4);              // pc+4 reg
dffe32 inst_r (ins,clk,clrn,wpcir,inst);              // ir
dffe32 pcd_r  ( pc,clk,clrn,wpcir,pcd);               // pcd reg
dff    intr_r (intr,clk,clrn,irq);                    // interrupt req reg
// ID stage
assign op   = inst[31:26];                            // op
assign rs   = inst[25:21];                            // rs
assign rt   = inst[20:16];                            // rt
assign rd   = inst[15:11];                            // rd
assign func = inst[05:00];                            // func
assign imm  = {{16{sext&inst[15]}},inst[15:0]};
assign jpc  = {dpc4[31:28],inst[25:0],2'b00};         // jump target
regfile rf (rs,rt,wdi,wrn,wwreg,~clk,clrn,qa,qb);     // reg file
mux2x5  des_reg_no (rd,rt,regrt,rn);                  // destination reg
mux4x32 operand_a (qa,ealu,malu,mmo,fwda,da);         // forward a
mux4x32 operand_b (qb,ealu,malu,mmo,fwdb,db);         // forward b
assign  rsrtequ = ~|(da^db);                          // rsrtequ = (da==db)
cla32 br_addr (dpc4,{imm[29:0],2'b00},1'b0,bpc);      // branch target
cu_exc_int cu (mwreg,mrn,ern,ewreg,em2reg,mm2reg,rsrtequ,func,op,rs,
               rt,rd,rs,wreg,m2reg,wmem,aluc,regrt,aluimm,fwda,fwdb,
               wpcir,sext,pcsrc,shift,jal,irq,sta,ecancel,eisbr,misbr,
               inta,selpc,exc,sepc,cause,mtc0,wepc,wcau,wsta,mfc0,isbr,
               ove,cancel,exc_ovr,mexc_ovr);
dffe32  c0_status (sta_in,clk,clrn,wsta,sta);         // status register
```

```verilog
  dffe32  c0_cause  (cau_in,clk,clrn,wcau,cau);    // cause register
  dffe32  c0_epc    (epc_in,clk,clrn,wepc,epc);    // epc register
  mux2x32 sta_mx (stalr,db,mtc0,sta_in);           // mux for status reg
  mux2x32 cau_mx (cause,db,mtc0,cau_in);           // mux for cause reg
  mux2x32 epc_mx (epcin,db,mtc0,epc_in);           // mux for epc reg
  mux2x32 sta_lr ({4'h0,sta[31:4]},{sta[27:0],4'h0},exc,stalr);
  mux4x32 epc_10 (pc,pcd,pce,pcm,sepc,epcin);      // select epc source
  mux4x32 irq_pc (npc,epc,exc_base,32'h0,selpc,next_pc); // for pc
  mux4x32 fromc0 (epc8,sta,cau,epc,emfc0,pc8c0r);  // for mfc0
  // ID/EXE pipeline register
  always @(negedge clrn or posedge clk)
    if (!clrn) begin
        ewreg0 <= 0;          em2reg  <= 0;          ewmem <= 0;
        ealuc  <= 0;          ealuimm <= 0;          ea    <= 0;
        eb     <= 0;          eimm    <= 0;          ern0  <= 0;
        eshift <= 0;          ejal    <= 0;          epc4  <= 0;
        eove   <= 0;          ecancel <= 0;          eisbr <= 0;
        emfc0  <= 0;          pce     <= 0;
    end else begin
        ewreg0 <= wreg;       em2reg  <= m2reg;      ewmem <= wmem;
        ealuc  <= aluc;       ealuimm <= aluimm;     ea    <= da;
        eb     <= db;         eimm    <= imm;         ern0  <= rn;
        eshift <= shift;      ejal    <= jal;         epc4  <= dpc4;
        eove   <= ove;        ecancel <= cancel;     eisbr <= isbr;
        emfc0  <= mfc0;       pce     <= pcd;
    end
  // EXE stage
  assign     esa = {eimm[5:0],eimm[31:6]};
  cla32   ret_addr (epc4,32'h4,1'b0,epc8);
  mux2x32 alu_ina  (ea,esa,eshift,ealua);
  mux2x32 alu_inb  (eb,eimm,ealuimm,ealub);
  mux2x32 save_pc8 (ealu0,pc8c0r,ejal|emfc0[1]|emfc0[0],ealu); // c0 regs
  assign  ern = ern0 | {5{ejal}};
  alu_ov  al_unit (ealua,ealub,ealuc,ealu0,zero,ov);
  assign  exc_ovr = ov & eove;                      // overflow exception
  assign  ewreg   = ewreg0 & ~exc_ovr;              // cancel overflow inst
  // EXE/MEM pipeline register
  always @(negedge clrn or posedge clk)
    if (!clrn) begin
        mwreg  <= 0;          mm2reg  <= 0;          mwmem <= 0;
        malu   <= 0;          mb      <= 0;          mrn   <= 0;
        misbr  <= 0;          pcm     <= 0;          mexc_ovr <= 0;
    end else begin
        mwreg  <= ewreg;      mm2reg  <= em2reg;     mwmem <= ewmem;
        malu   <= ealu;       mb      <= eb;          mrn   <= ern;
        misbr  <= eisbr;      pcm     <= pce;         mexc_ovr <= exc_ovr;
    end
```

```
    // MEM stage
    p1_exc_d_mem data_mem (clk,mmo,mb,malu,mwmem);    // data mem
    // MEM/WB pipeline register
    always @(negedge clrn or posedge clk)
      if (!clrn) begin
          wwreg  <= 0;          wm2reg <= 0;           wmo  <= 0;
          walu   <= 0;          wrn    <= 0;
      end else begin
          wwreg  <= mwreg;      wm2reg <= mm2reg;      wmo  <= mmo;
          walu   <= malu;       wrn    <= mrn;
      end
    // WB stage
    mux2x32 wb_stage (walu,wmo,wm2reg,wdi);            // alu res or mem data
endmodule
```

Below is the Verilog HDL codes of the control unit that is invoked in the ID stage by the module pipelined_cpu_exc_int.v. For the integrity of the module, some codes that are already given in the control unit module of the basic pipelined CPU are also listed here. The control unit is also a combinational circuit, although some signals are declared as reg type.

```
module cu_exc_int (mwreg,mrn,ern,ewreg,em2reg,mm2reg,rsrtequ,func,op,rs,rt,
                   rd,op1,wreg,m2reg,wmem,aluc,regrt,aluimm,fwda,fwdb,wpcir,
                   sext,pcsrc,shift,jal,irq,sta,ecancel,eis_branch,
                   mis_branch,inta,selpc,exc,sepc,cause,mtc0,wepc,wcau,wsta,
                   mfc0,is_branch,ove,cancel,exc_ovr,mexc_ovr); // ctrl unit
    input    [31:0] sta;               // status: IM[3:0]: ov,unimpl,sys,int
    input    [5:0]  op,func;
    input    [4:0]  mrn,ern,rs,rt,rd;
    input    [4:0]  op1;               // for decode mfc0, mtc0, and eret
    input           mwreg,ewreg,em2reg,mm2reg,rsrtequ;
    input           irq;               // interrupt request
    input           ecancel;           // cancel in EXE stage
    input           eis_branch;        // is_branch in EXE stage
    input           mis_branch;        // is_branch in MEM stage
    input           exc_ovr;           // overflow exception occurs
    input           mexc_ovr;          // exc_ovr in MEM stage
    output   [31:0] cause;             // cause content
    output   [3:0]  aluc;
    output   [1:0]  pcsrc,fwda,fwdb;
    output   [1:0]  selpc;             // 00: npc;  01: epc; 10: exc_base
    output   [1:0]  mfc0;              // 00: epc8; 01: sta; 10: cau; 11: epc
    output   [1:0]  sepc;              // 00: pc;   01: pcd; 10: pce; 11: pcm
    output          wpcir,wreg,m2reg,wmem,regrt,aluimm,sext,shift,jal;
    output          inta;              // interrupt acknowledgement
    output          exc;               // any int or exc happened
    output          mtc0;              // is mtc0 instruction
    output          wsta;              // status register write enable
    output          wcau;              // cause  register write enable
```

```
output          wepc;               // epc   register write enable
output          is_branch;          // is a branch or a jump
output          ove;                // ov enable = arith & sta[3]
output          cancel;             // exception cancels next instruction
reg     [1:0] fwda,fwdb;
wire    [1:0] exccode;              // exccode
wire          rtype,i_add,i_sub,i_and,i_or,i_xor,i_sll,i_srl,i_sra;
wire          i_jr,i_addi,i_andi,i_ori,i_xori,i_lw,i_sw,i_beq,i_bne;
wire          i_lui,i_j,i_jal,i_rs,i_rt;
wire          exc_int;              // exception of interrupt
wire          exc_sys;              // exception of system call
wire          exc_uni;              // exception of unimplemented inst
wire          c0_type;              // cp0 instructions
wire          i_syscall;            // is syscall instruction
wire          i_mfc0;               // is mfc0 instruction
wire          i_mtc0;               // is mtc0 instruction
wire          i_eret;               // is eret instruction
wire          unimplemented_inst;   // is an unimplemented inst
wire          rd_is_status;         // rd is status
wire          rd_is_cause;          // rd is cause
wire          rd_is_epc;            // rd is epc
wire    arith   = i_add | i_sub | i_addi;           // for overflow
assign is_branch = i_beq | i_bne | i_jr | i_j | i_jal; // has delay slot
assign exc_int   = sta[0] & irq;                 // 0. exc_int
assign exc_sys   = sta[1] & i_syscall;           // 1. exc_sys
assign exc_uni   = sta[2] & unimplemented_inst;  // 2. exc_uni
assign ove       = sta[3] & arith;               // 3. exc_ovr enable
assign inta      = exc_int;                       // ack immediately
assign exc       = exc_int | exc_sys | exc_uni | exc_ovr; // all int_exc
assign cancel    = exc | i_eret;    // always cancel next inst, eret also
// sel epc:     id is_branch  eis_branch   mis_branch      others
// exc_int      PCD (01)      PC  (00)     PC  (00)        PC  (00)
// exc_sys      x             x            PCD (01)        PCD (01)
// exc_uni      x             x            PCD (01)        PCD (01)
// exc_ovr      x             x            PCM (11)        PCE (10)
assign sepc[0] = exc_int & is_branch | exc_sys | exc_uni |
                 exc_ovr & mis_branch;
assign sepc[1] = exc_ovr;
// exccode:   0 0 : irq
//            0 1 : i_syscall
//            1 0 : unimplemented_inst
//            1 1 : exc_ovr
assign exccode[0]    = i_syscall            | exc_ovr;
assign exccode[1]    = unimplemented_inst | exc_ovr;
assign cause         = {eis_branch,27'h0,exccode,2'b00}; // BD
assign mtc0          = i_mtc0;
assign wsta          = exc | mtc0 & rd_is_status | i_eret;
```

```
assign wcau          = exc | mtc0 & rd_is_cause;
assign wepc          = exc | mtc0 & rd_is_epc;
assign rd_is_status = (rd == 5'd12);              // cp0 status register
assign rd_is_cause  = (rd == 5'd13);              // cp0 cause register
assign rd_is_epc    = (rd == 5'd14);              // cp0 epc register
// mfc0:     0 0 : epc8
//           0 1 : sta
//           1 0 : cau
//           1 1 : epc
assign mfc0[0] = i_mfc0 & rd_is_status | i_mfc0 & rd_is_epc;
assign mfc0[1] = i_mfc0 & rd_is_cause  | i_mfc0 & rd_is_epc;
// selpc:    0 0 : npc
//           0 1 : epc
//           1 0 : exc_base
//           1 1 : x
assign selpc[0] = i_eret;
assign selpc[1] = exc;
assign c0_type  = ~op[5]  & op[4]  &~op[3] &~op[2] &~op[1] &~op[0];
assign i_mfc0   = c0_type &~op1[4] &~op1[3] &~op1[2] &~op1[1] &~op1[0];
assign i_mtc0   = c0_type &~op1[4] &~op1[3] & op1[2] &~op1[1] &~op1[0];
assign i_eret   = c0_type & op1[4] &~op1[3] &~op1[2] &~op1[1] &~op1[0] &
                  ~func[5] & func[4] & func[3] &~func[2] &~func[1] &~func[0];
assign i_syscall = rtype  & ~func[5] & ~func[4] & func[3] & func[2] &
                   ~func[1] & ~func[0];
assign unimplemented_inst = ~(i_mfc0 | i_mtc0 | i_eret | i_syscall |
      i_add | i_sub | i_and | i_or | i_xor | i_sll | i_srl | i_sra |
      i_jr | i_addi | i_andi | i_ori | i_xori | i_lw | i_sw | i_beq |
      i_bne | i_lui | i_j   | i_jal); // except for implemented insts
and (rtype,~op[5],~op[4],~op[3],~op[2],~op[1],~op[0]);          // r format
and (i_add,rtype, func[5],~func[4],~func[3],~func[2],~func[1],~func[0]);
and (i_sub,rtype, func[5],~func[4],~func[3],~func[2], func[1],~func[0]);
and (i_and,rtype, func[5],~func[4],~func[3], func[2],~func[1],~func[0]);
and (i_or, rtype, func[5],~func[4],~func[3], func[2],~func[1], func[0]);
and (i_xor,rtype, func[5],~func[4],~func[3], func[2], func[1],~func[0]);
and (i_sll,rtype,~func[5],~func[4],~func[3],~func[2],~func[1],~func[0]);
and (i_srl,rtype,~func[5],~func[4],~func[3],~func[2], func[1],~func[0]);
and (i_sra,rtype,~func[5],~func[4],~func[3],~func[2], func[1], func[0]);
and (i_jr, rtype,~func[5],~func[4], func[3],~func[2],~func[1],~func[0]);
and (i_addi,~op[5],~op[4], op[3],~op[2],~op[1],~op[0]);          // i format
and (i_andi,~op[5],~op[4], op[3], op[2],~op[1],~op[0]);
and (i_ori, ~op[5],~op[4], op[3], op[2],~op[1], op[0]);
and (i_xori,~op[5],~op[4], op[3], op[2], op[1],~op[0]);
and (i_lw,   op[5],~op[4],~op[3],~op[2], op[1], op[0]);
and (i_sw,   op[5],~op[4], op[3],~op[2], op[1], op[0]);
and (i_beq, ~op[5],~op[4],~op[3], op[2],~op[1],~op[0]);
and (i_bne, ~op[5],~op[4],~op[3], op[2],~op[1], op[0]);
and (i_lui, ~op[5],~op[4], op[3], op[2], op[1], op[0]);
```

```
and (i_j,    ~op[5],~op[4],~op[3],~op[2], op[1],~op[0]);        // i format
and (i_jal, ~op[5],~op[4],~op[3],~op[2], op[1], op[0]);
assign i_rs = i_add | i_sub | i_and | i_or | i_xor | i_jr | i_addi |
              i_andi | i_ori | i_xori | i_lw | i_sw | i_beq | i_bne;
assign i_rt = i_add | i_sub | i_and | i_or | i_xor | i_sll | i_srl |
              i_sra | i_sw | i_beq | i_bne | i_mtc0;   // mtc0 added
assign wpcir = ~(ewreg & em2reg & (ern != 0) & (i_rs & (ern == rs) |
                                                 i_rt & (ern == rt)));
always @ (ewreg or mwreg or ern or mrn or em2reg or mm2reg or rs or rt)
    begin
        fwda = 2'b00;                                  // default: no hazards
        if (ewreg & (ern != 0) & (ern == rs) & ~em2reg) begin
            fwda = 2'b01;                              // select exe_alu
        end else begin
            if (mwreg & (mrn != 0) & (mrn == rs) & ~mm2reg) begin
                fwda = 2'b10;                          // select mem_alu
            end else begin
                if (mwreg & (mrn != 0) & (mrn == rs) & mm2reg) begin
                    fwda = 2'b11;                      // select mem_lw
                end
            end
        end
        fwdb = 2'b00;                                  // default: no hazards
        if (ewreg & (ern != 0) & (ern == rt) & ~em2reg) begin
            fwdb = 2'b01;                              // select exe_alu
        end else begin
            if (mwreg & (mrn != 0) & (mrn == rt) & ~mm2reg) begin
                fwdb = 2'b10;                          // select mem_alu
            end else begin
                if (mwreg & (mrn != 0) & (mrn == rt) & mm2reg) begin
                    fwdb = 2'b11;                      // select mem_lw
                end
            end
        end
    end
assign wmem    = i_sw & wpcir & ~ecancel & ~exc_ovr & ~mexc_ovr;
assign regrt   = i_addi|i_andi|i_ori |i_xori|i_lw |i_lui |i_mfc0;
assign jal     = i_jal;
assign m2reg   = i_lw;
assign shift   = i_sll |i_srl |i_sra;
assign aluimm  = i_addi|i_andi|i_ori |i_xori|i_lw |i_lui |i_sw;
assign sext    = i_addi|i_lw |i_sw  |i_beq |i_bne;
assign aluc[3] = i_sra;
assign aluc[2] = i_sub |i_or  |i_srl |i_sra |i_ori |i_lui;
assign aluc[1] = i_xor |i_sll |i_srl |i_sra |i_xori|i_beq |i_bne|i_lui;
assign aluc[0] = i_and |i_or  |i_sll |i_srl |i_sra |i_andi|i_ori;
assign pcsrc[1] = i_jr  |i_j   |i_jal;
```

```
    assign pcsrc[0] = i_beq & rsrtequ |i_bne & ~rsrtequ | i_j | i_jal;
    assign wreg     =(i_add |i_sub |i_and |i_or  |i_xor|i_sll  |
                      i_srl |i_sra |i_addi|i_andi|i_ori|i_xori |
                      i_lw  |i_lui |i_jal |i_mfc0) & // mfc0 added
                      wpcir & ~ecancel & ~exc_ovr & ~mexc_ovr;
endmodule
```

8.5.3   Test Program for Exception and Interrupt

Below is the general Verilog HDL codes of the instruction memory in which the test program is stored.
The program first enables all the exceptions and interrupts. Then, it executes a div instruction, which will
cause an unimplemented instruction exception. Next, there is a syscall instruction that will generate
a system call exception. During the execution of the for loop, the interrupts will be tested. Finally, an
overflow exception in the delay slot is tested, and it will cause the program to exit because it is not a
restartable exception.

The interrupt and exception handler is located in the memory starting at 0x00000008. It gets the Exc-
Code from the Cause register and jumps to the entry address of the corresponding exception handler
based on an address table in data memory.

```
module p1_exc_i_mem (a,inst);        // instruction memory, rom
    input   [31:0] a;                // address
    output  [31:0] inst;             // instruction
    wire    [31:0] rom [0:63];       // rom cells: 64 words * 32 bits
    // rom[word_addr] = instruction  // (pc) label        instruction
    assign rom[6'h00] = 32'h0800001d; // (00) main:        j    start
    assign rom[6'h01] = 32'h00000000; // (04)              nop
    // common entry of exc and intr
    assign rom[6'h02] = 32'h401a6800; // (08) exc_base:   mfc0 $26, c0_cause
    assign rom[6'h03] = 32'h335b000c; // (0c)             andi $27, $26, 0xc
    assign rom[6'h04] = 32'h8f7b0020; // (10)             lw $27,j_table($27)
    assign rom[6'h05] = 32'h00000000; // (14)             nop
    assign rom[6'h06] = 32'h03600008; // (18)             jr   $27
    assign rom[6'h07] = 32'h00000000; // (1c)             nop
    // 0x00000030: intr handler
    assign rom[6'h0c] = 32'h00000000; // (30) int_entry: nop
    assign rom[6'h0d] = 32'h42000018; // (34)             eret
    assign rom[6'h0e] = 32'h00000000; // (38)             nop
    // 0x0000003c: syscall handler
    assign rom[6'h0f] = 32'h00000000; // (3c) sys_entry: nop
    assign rom[6'h10] = 32'h401a7000; // (40) epc_plus4: mfc0 $26, c0_epc
    assign rom[6'h11] = 32'h235a0004; // (44)             addi $26, $26, 4
    assign rom[6'h12] = 32'h409a7000; // (48)             mtc0 $26, c0_EPC
    assign rom[6'h13] = 32'h42000018; // (4c) e_return:  eret
    assign rom[6'h14] = 32'h00000000; // (50)             nop
    // 0x00000054: unimpl handler
    assign rom[6'h15] = 32'h00000000; // (54) uni_entry: nop
    assign rom[6'h16] = 32'h08000010; // (58)             j       epc_plus4
    assign rom[6'h17] = 32'h00000000; // (5c)             nop
```

```
    // 0x00000068: overflow handler
    assign rom[6'h1a] = 32'h00000000;  // (68) ovf_entry: nop
    assign rom[6'h1b] = 32'h0800002f;  // (6c)           j     exit
    assign rom[6'h1c] = 32'h00000000;  // (70)           nop
    // start: enable exc and intr
    assign rom[6'h1d] = 32'h2008000f;  // (74) start:    addi $8, $0, 0xf
    assign rom[6'h1e] = 32'h40886000;  // (78) exc_ena:  mtc0 $8, c0_status
    // unimplemented instruction
    assign rom[6'h1f] = 32'h0128001a;  // (7c) unimpl:   div  $9, $8
    assign rom[6'h20] = 32'h00000000;  // (80)           nop
    // system call
    assign rom[6'h21] = 32'h0000000c;  // (84) sys:      syscall
    assign rom[6'h22] = 32'h00000000;  // (88)           nop
    // loop code for testing intr
    assign rom[6'h23] = 32'h34040050;  // (8c) int:      ori  $4, $1, 0x50
    assign rom[6'h24] = 32'h20050004;  // (90)           addi $5, $0, 4
    assign rom[6'h25] = 32'h00004020;  // (94)           add  $8, $0, $0
    assign rom[6'h26] = 32'h8c890000;  // (98) loop:     lw   $9, 0($4)
    assign rom[6'h27] = 32'h01094020;  // (9c)           add  $8, $8, $9
    assign rom[6'h28] = 32'h20a5ffff;  // (a0)           addi $5, $5, -1
    assign rom[6'h29] = 32'h14a0fffc;  // (a4)           bne  $5, $0, loop
    assign rom[6'h2a] = 32'h20840004;  // (a8)           addi $4, $4, 4 # DS
    assign rom[6'h2b] = 32'h8c080048;  // (ac) ov:       lw   $8, 0x48($0)
    assign rom[6'h2c] = 32'h8c09004c;  // (b0)           lw   $9, 0x4c($0)
    // jump to start forever
    assign rom[6'h2d] = 32'h0800001d;  // (b4) forever:  j     start
    // overflow in delay slot
    assign rom[6'h2e] = 32'h01094020;  // (b8)           add  $9, $9, $8 #ov
    // if not overflow, go to start
    // exit, should be jal $31 to os
    assign rom[6'h2f] = 32'h0800002f;  // (bc) exit:     j     exit
    assign rom[6'h30] = 32'h00000000;  // (c0)           nop
    assign inst = rom[a[7:2]];         // use 6-bit word address to read rom
endmodule
```

Below is the general Verilog HDL codes of the data memory in which the test data are stored. Starting from the location 0x00000020, an address table is prepared for jumping to a particular exception handler. Then two 32-bit words are stored for testing the arithmetic overflow. Finally, there are four words for executing a loop. The interrupts are tested during the execution of the loop.

```
module pl_exc_d_mem (clk,dataout,datain,addr,we); // data memory, ram
    input         clk;                  // clock
    input         we;                   // write enable
    input  [31:0] datain;               // data in (to memory)
    input  [31:0] addr;                 // ram address
    output [31:0] dataout;              // data out (from memory)
    reg    [31:0] ram [0:31];           // ram cells: 32 words * 32 bits
    assign dataout = ram[addr[6:2]];    // use 6-bit word address
```

```
    always @ (posedge clk) begin
        if (we) ram[addr[6:2]] = datain;    // write ram
    end
    integer i;
    initial begin                              // ram initialization
        for (i = 0; i < 32; i = i + 1)
            ram[i] = 0;
        // ram[word_addr] = data              // (byte_addr) item in data array
        ram[5'h08] = 32'h00000030;            // (20) 0. int_entry
        ram[5'h09] = 32'h0000003c;            // (24) 1. sys_entry
        ram[5'h0a] = 32'h00000054;            // (28) 2. uni_entry
        ram[5'h0b] = 32'h00000068;            // (2c) 3. ovr_entry
        ram[5'h12] = 32'h00000002;            // (48) for testing overflow
        ram[5'h13] = 32'h7fffffff;            // (4c) 2 + max_int -> overflow
        ram[5'h14] = 32'h000000a3;            // (50) data[0]   0 +  a3 =  a3
        ram[5'h15] = 32'h00000027;            // (54) data[1]  a3 +  27 =  ca
        ram[5'h16] = 32'h00000079;            // (58) data[2]  ca +  79 = 143
        ram[5'h17] = 32'h00000115;            // (5c) data[3] 143 + 115 = 258
    end
endmodule
```

8.5.4   Simulation Waveforms of Exception and Interrupt

In sequence, this section shows the simulation waveforms of (i) unimplemented instruction exception; (ii) system call exception; (iii) interrupt that occurs when a branch instruction is in ID stage; and (iv) arithmetic overflow exception.

Figure 8.37 shows the waveforms when an unimplemented instruction exception occurs. In the memory location 0x0000007c, there is a div instruction which is not implemented in the CPU. The exception is detected in the ID stage. The ExcCode is 2, which is stored in the Cause register by the hardware. Then the control is transferred to 0x00000008, the common entry of the exception handler.

In the exception handler, the ExcCode in the Cause register is checked, and it jump to 0x00000054 based on a jump table. Our example code adds a 4 to the EPC and returns to 0x00000080, the next location to the div instruction. The execution sequence of the instructions in the figure is listed below for easy checking.

```
00: 0800001d; // main:      j     start      # jump to start
04: 00000000; //            nop              # delay slot
74: 2008000f; // start:     addi  $8, $0, 0xf # im[3:0] = 1111
78: 40886000; // exc_ena:   mtc0  $8, c0_status # enable exc and intr
7c: 0128001a; // unimpl:    div   $9, $8      # unimplemented instruction
80: 00000000; //            nop              # canceled, go to handler
08: 401a6800; // exc_base:  mfc0  $26, c0_cause # read cause register
0c: 335b000c; //            andi  $27, $26, 0xc # get exccode
10: 8f7b0020; //            lw    $27, j_table($27) # address table
14: 00000000; //            nop              # wait for load result
18: 03600008; //            jr    $27        # jump to uni_entry
1c: 00000000; //            nop              # delay slot
54: 00000000; // uni_entry: nop              # do nothing
58: 08000010; //            j     epc_plus4  # jump to epc_plus4
```

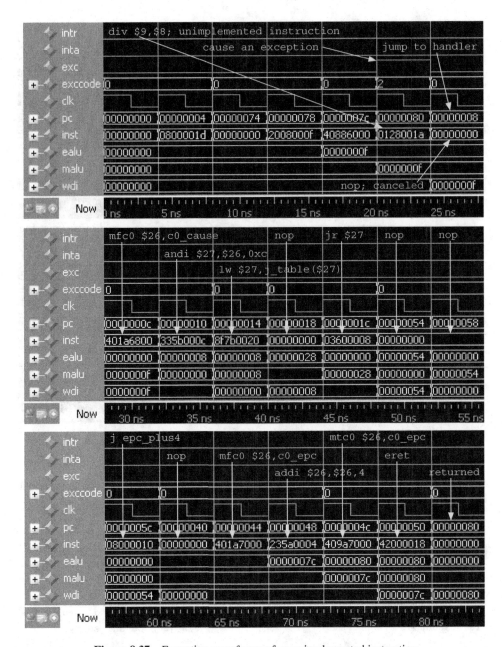

Figure 8.37   Exception waveforms of an unimplemented instruction

```
5c: 00000000; //                  nop           # delay slot
40: 401a7000; // epc_plus4: mfc0 $26, c0_epc     # read epc register
44: 235a0004; //           addi $26, $26, 4      # epc + 4
48: 409a7000; //           mtc0 $26, c0_epc      # write epc
4c: 42000018; // e_return:  eret                 # return from exception
50: 00000000; //                  nop            # will be canceled
```

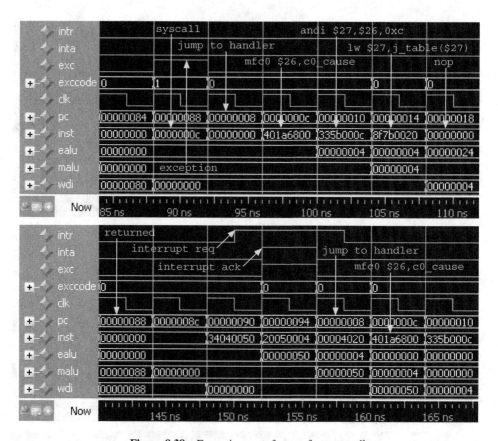

Figure 8.38 Exception waveforms of system call

Figure 8.38 shows the waveforms when a system call exception occurs. The execution sequence of the instructions in the figure is listed below.

```
84: 0000000c; // sys:      syscall           # system call instruction
88: 00000000; //           nop               # canceled, go to handler
08: 401a6800; // exc_base: mfc0 $26, c0_cause # read cause register
0c: 335b000c; //           andi $27, $26, 0xc # get exccode
10: 8f7b0020; //           lw   $27, j_table($27) # address table
14: 00000000; //           nop               # wait for load result
18: 03600008; //           jr   $27          # jump to uni_entry
```

We have not shown all the waveforms in the figure. Referring to the second part of the figure, the canceled instruction at 0x00000088 was restarted at the return from the exception handler (see the execution sequence of the instructions listed below). Figure 8.38 also shows the waveforms when an interrupt occurs.

```
88: 00000000; //           nop               # restart canceled inst
8c: 34040050; // int:      ori $4, $1, 0x50  #
90: 20050004; //           addi $5, $0, 4    # interrupt occurs
```

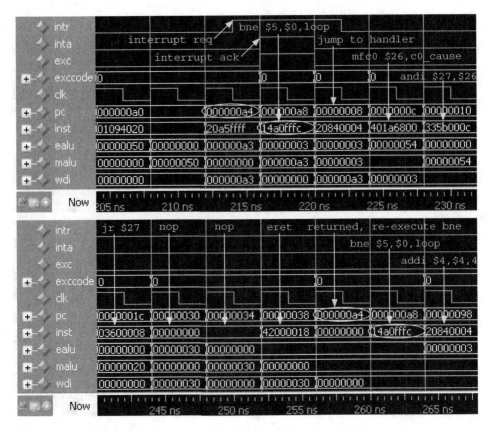

Figure 8.39 Interrupt waveforms (a branch instruction is in the ID stage)

```
94: 00004020; //           add  $8, $0, $0   # interrupt ack
08: 401a6800; // exc_base: mfc0 $26, c0_cause # read cause register
0c: 335b000c; //           andi $27, $26, 0xc # get exccode
10: 8f7b0020; //           lw   $27, j_table($27) # address table
```

Figure 8.39 shows the waveforms when an interrupt occurs but a branch instruction is being decoded during the same clock cycle. Which address should be written into the PC, the entry of the interrupt handler, or the branch target address? The answer is, of course, the entry of the interrupt handler. The address of the branch instruction is saved into the EPC. The execution sequence of the instructions in the figure is listed below.

```
a0: 20a5ffff; //           addi $5, $5, -1   #
a4: 14a0fffc; //           bne  $5, $0, loop # interrupt occurs
a8: 20840004; //           addi $4, $4, 4    # interrupt ack
08: 401a6800; // exc_base: mfc0 $26, c0_cause # read cause register
0c: 335b000c; //           andi $27, $26, 0xc # get exccode
10: 8f7b0020; //           lw   $27, j_table($27) # address table
    ... ...
18: 03600008; //           jr   $27          # jump to uni_entry
```

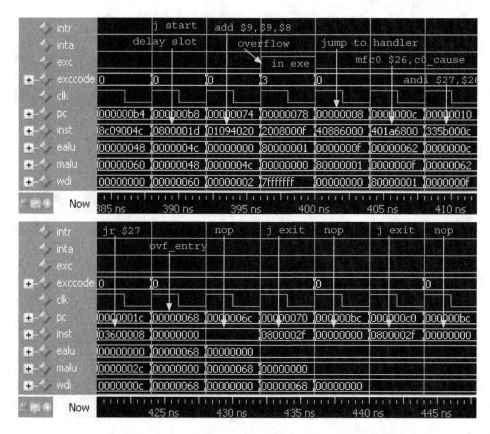

Figure 8.40   Exception waveforms of an arithmetic overflow

```
1c: 00000000; //                nop              # delay slot
30: 00000000; // int_entry:     nop              # do nothing
34: 42000018; //                eret             # return from exception
38: 00000000; //                nop              # delay slot
a4: 14a0fffc; //                bne  $5, $0, loop # restart canceled inst
a8: 20840004; //                addi $4, $4, 4   # delay slot
98: 8c890000; // loop:          lw   $9, 0($4)   # load memory word
```

Figure 8.40 shows the waveforms when an overflow exception occurs, which was generated by the add in the location 0x000000b8 – a delay slot of the j instruction. The address of j, 0x000000b4, is written into the EPC, and the BD field in the Cause register is set. The program was aborted by the exception handler. The execution sequence of the instructions in the figure is listed below.

```
b4: 0800001d; // forever:    j    start          # jump to start forever
b8: 01094020; //             add  $9, $9, $8      # overflow in delay slot
74: 2008000f; // start:      addi $8, $0, 0xf     # im[3:0] = 1111
78: 40886000; // exc_ena:    mtc0 $8, c0_status   # enable exc and intr
08: 401a6800; // exc_base:   mfc0 $26, c0_cause   # read cause register
```

```
0c: 335b000c; //              andi $27, $26, 0xc # get exccode
10: 8f7b0020; //              lw   $27, j_table($27) # address table
    ... ...
18: 03600008; //              jr   $27              # jump to uni_entry
1c: 00000000; //              nop                   # delay slot
68: 00000000; // ovf_entry: nop                     # do nothing
6c: 0800002f; //              j    exit             # jump to exit
70: 00000000; //              nop                   # delay slot
bc: 0800002f; // exit:        j    exit             # exit, should be jal $31
c0: 00000000; //              nop                   # delay slot
bc: 0800002f; // exit:        j    exit             # exit, should be jal $31
```

We have checked the simulation waveforms for external asynchronous interrupts and internal synchronous exceptions. All the waveforms show that our pipelined CPU with the precise interrupt/exception mechanism works correctly logically.

Exercises

8.1 Suppose in an ideal pipelined CPU the execution of each instruction takes m stages and each stage takes one clock cycle.
  (a) Give an equation for calculating the total number of clock cycles (N) when the CPU executes n instructions.
  (b) The term CPI stands for the average number of clock cycles each instruction takes to execute. Give a general equation for calculating the CPI of the CPU when the CPU executes n instructions.
  (c) Calculate $\lim_{n \to \infty}$ CPI.

8.2 In the pipelined CPU described in this chapter, if an sw instruction has a data dependency with its predecessor, an lw instruction, the pipeline will stall for one clock cycle. Redesign the CPU to eliminate this one cycle stall.

8.3 Redesign the interrupt mechanism to postpone the acknowledgment of interrupt until there is neither a branch nor a jump instruction in the ID stage.

8.4 Add one another interrupt request input for a timer; also design the timer itself and write a timer interrupt handler.

8.5 Implement the exception of the misaligned word address for the data memory accesses with lw and sw instructions.

8.6 Write an unimplemented instruction handler to emulate the executions of a square root instruction.

8.7 Consider what will happen if the system call instruction is located in a delay slot?

8.8 Investigate the techniques of the code optimization for eliminating the pipeline stalls caused by the data dependency with a load instruction and for replacing the nop instructions in delay slots with useful instructions.

8.9 Investigate the techniques of branch prediction and write a report.

9

Floating-Point Algorithms and FPU Design in Verilog HDL

Up to now, the central processing units (CPUs) we designed had only an integer unit (IU) which can operate on integers. In the modern CPUs, there is a floating-point unit (FPU) which can operate on floating-point numbers. This chapter introduces the IEEE formats of floating-point numbers, describes the circuit design of floating-point calculations, and gives the simulation waveforms of the circuits.

9.1   IEEE 754 Floating-Point Data Formats

The IEEE 754 Standard defines mainly two floating-point formats, `float` and `double`, which we commonly use in software development. Suppose that in a C program we declare three variables: k, s, and x, as shown below:

```
int      k = -1;
float    s = -1.75;
double   x =  1.75;
```

Then a compiler will convert them to the following binary patterns: (Ignore the underbars that divide a binary pattern into three parts for easy reading.)

Variable	Binary pattern	Bits
k:	11111111111111111111111111111111	32
s:	1_01111111_11000000000000000000000	32
x:	0_01111111111_1100000000000000000000000000000000000000000000000000	64

The `float s = -1.75` is converted to a 32-bit pattern in the IEEE 754 single-precision floating-point format. Referring to Figure 9.1, an IEEE 754 single-precision floating-point number consists of three parts: s, e, and f.

Computer Principles and Design in Verilog HDL, First Edition. Yamin Li.
© 2015 Tsinghua University Press. All rights reserved. Published 2015 by John Wiley & Sons Singapore Pte Ltd.
Companion Website: www.wiley.com/go/li/verilog

Figure 9.1   IEEE 754 format of a single-precision floating-point number

The 1-bit s (sign) indicates the sign of the number: $s = 1$ for a negative number and $s = 0$ for a nonnegative number. The 8-bit e (exponent) is the biased exponent with the bias of 127. The 23-bit f (fraction) is the significand (mantissa). The value (V) of a single-precision floating-point number is defined by IEEE as follows:

- Normalized: If $0 < e < 255$, then $V = (-1)^s \times 1.f \times 2^{e-127}$;
- $+0$, -0: If $e = 0$ and $f = 0$, then $V = (-1)^s \times 0$;
- Denormalized: If $e = 0$ and $f \neq 0$, then $V = (-1)^s \times 0.f \times 2^{-126}$;
- $+\infty$, $-\infty$: If $e = 255$ and $f = 0$, then $V = (-1)^s \infty$; and
- NaN (not a number): If $e = 255$ and $f \neq 0$, then $V = $ NaN.

For a normalized number, the 1 of $1.f$ is not stored in the 32-bit data; it is called a hidden bit. For a denormalized number, the hidden bit is a 0. Some examples of the single-precision floating-point numbers are shown below.

- Normalized numbers ($V = (-1)^s \times 1.f \times 2^{e-127}$):
  $1\_01111110\_10000000000000000000000 = -1.1 \times 2^{126-127} = -0.75$
  $0\_00000001\_00000000000000000000000 = +1.0 \times 2^{1-127} = +2^{-126}$
  $0\_11111110\_00000000000000000000000 = +1.0 \times 2^{254-127} = +2^{127}$
  $0\_11111110\_11111111111111111111111 = +(2 - 2^{-23}) \times 2^{254-127}$
- Zero, infinity, and NaN:
  $0\_00000000\_00000000000000000000000 = +0$
  $1\_00000000\_00000000000000000000000 = -0$
  $0\_11111111\_00000000000000000000000 = +\infty$
  $1\_11111111\_00000000000000000000000 = -\infty$
  $0_11111111_10000000000000000000000 = $ NaN
  $1_11111111_00000100001100000000000 = $ NaN
- Denormalized numbers ($V = (-1)^s \times 0.f \times 2^{-126}$):
  $0\_00000000\_10000000000000000000000 = +0.1 \times 2^{-126} = +2^{-127}$
  $0\_00000000\_00000000000000000000001 = +2^{-23} \times 2^{-126} = +2^{-149}$

The following steps show how to convert a real number in decimal format to the IEEE 754 single-precision floating-point format.

$$V = -6.25_{10}$$
$$= -(6_{10} + 0.25_{10})$$
$$= -(110_2 + 0.01_2)$$
$$= -110.01_2$$
$$= -1.1001_2 \times 2^2$$
$$= -1.1001_2 \times 2^{129-127} = (-1)^s \times 1.f \times 2^{e-127}$$

Figure 9.2   IEEE 754 format of a double-precision floating-point number

That is, $s = 1$, $e = 10000001$, and $f = 10010000000000000000000$, so in IEEE 754 single-precision floating-point format it is 1_10000001_10010000000000000000000.

Figure 9.2 shows the IEEE 754 format of a double-precision floating-point number. It has 64 bits in total. The exponent has 11 bits and the bias is 1023. The value (V) of a double-precision floating-point number is defined by IEEE as follows:

- Normalized: If $0 < e < 2047$, then $V = (-1)^s \times 1.f \times 2^{e-1023}$;
- +0, −0: If $e = 0$ and $f = 0$, then $V = (-1)^s \times 0$;
- Denormalized: If $e = 0$ and $f \neq 0$, then $V = (-1)^s \times 0.f \times 2^{-1022}$;
- +∞, −∞: If $e = 2047$ and $f = 0$, then $V = (-1)^s \infty$; and
- NaN: If $e = 2047$ and $f \neq 0$, then $V = $ NaN.

9.2   Converting between Floating-Point Number and Integer

The conversions between floating-point numbers and integers are common operations in software development. This section presents the converter design in Verilog HDL (hardware description language) and gives the simulation waveforms.

9.2.1   Converting Floating-Point Number to Integer

We know that a 32-bit integer d represented in 2's complement has a range of $-2^{31} \leq d \leq +2^{31} - 1$. A 32-bit single-precision floating-point number has a much wider range than the integer. This means that many floating-point numbers cannot be converted to integer correctly.

Let's take an example to show how to convert a floating-point number a into an integer d. Suppose $a = 4\text{effffff}_{16} = 0\_10011101\_11111111111111111111111_2$. That is, the sign $s_a = 0$, the exponent $e_a = 127 + 30$, and the significand $1.f_a = 1.11111111111111111111111_2$. By shifting $1.f_a$ to the left by 30 bits, we get the final result $d = 01111111111111111111111110000000_2 = 7\text{ffff80}_{16} = 2147483520_{10}$. Actually, $a = 4\text{effffff}_{16}$ is the maximum of the floating-point numbers that can be converted to an integer correctly.

Here is another example. Suppose $a = 1\_10011110\_00000000000000000000000_2$. That is, the sign $s_a = 1$, the exponent $e_a = 127 + 31$, and the significand $1.f_a = 1.00000000000000000000000_2$. By shifting $1.f_a$ to the left by 31 bits, we get the absolute of d. Because a is negative, we invert the absolute and add a 1 to it, and then we get the final result $d = 80000000_{16} = -2147483648_{10}$. This is the minimum negative floating-point number that can be converted to an integer correctly.

From these examples, we give the algorithm for converting a single-precision floating-point number $a = \{s_a, e_a, f_a\}$ to an integer, which as follows:

1. Attach an 8-bit 0 to $1f_a$ (the point of $1.f_a$ is shifted to the right by 31 bits).
2. Shift it to the right by $127 + 31 - e_a$ bits.
3. If $s_a = 1$, invert and add 1.

   Below is the Verilog HDL code that converts an IEEE 754 single-precision floating-point number a to a 32-bit integer d. If a exceeds the range $-2^{31} \le d \le +2^{31} - 1$, an output signal, invalid, will be a 1. Because a can represent a decimal fraction but d cannot, we add an output signal, p_lost, to indicate "precision lost." If a is a denormalized number, the denorm signal outputs a 1.

```verilog
module f2i (a,d,p_lost,denorm,invalid);            // convert float to integer
    input   [31:0] a;                              // float
    output  [31:0] d;                              // integer
    output         p_lost;                         // precision lost
    output         denorm;                         // denormalized
    output         invalid;                        // inf,nan,out_of_range
    reg     [31:0] d;                              // will be combinational
    reg            p_lost;                         // will be combinational
    reg            invalid;                        // will be combinational
    wire           hidden_bit = |a[30:23];         // hidden bit
    wire           frac_is_not_0 = |a[22:0];
    assign         denorm  = ~hidden_bit & frac_is_not_0;
    wire           is_zero = ~hidden_bit & ~frac_is_not_0;
    wire           sign = a[31];                   // sign
    wire    [8:0]  shift_right_bits = 9'd158 - {1'b0,a[30:23]};   // 127 + 31
    wire    [55:0] frac0 = {hidden_bit,a[22:0],32'h0};   // 32 + 24 = 56 bits
    wire    [55:0] f_abs = ($signed(shift_right_bits) > 9'd32)?    // shift
                       frac0 >> 6'd32 : frac0 >> shift_right_bits;
    wire           lost_bits = |f_abs[23:0];       // if != 0, p_lost = 1
    wire    [31:0] int32 = sign? ~f_abs[55:24] + 32'd1 : f_abs[55:24];
    always @ * begin
        if (denorm) begin                          //denormalized
            p_lost = 1;
            invalid = 0;
            d = 32'h00000000;
        end else begin                             // not denormalized
            if (shift_right_bits[8]) begin         // too big
                p_lost = 0;
                invalid = 1;
                d = 32'h80000000;
            end else begin                         // shift right
                if (shift_right_bits[7:0] > 8'h1f) begin // too small
                    if (is_zero) p_lost = 0;
                    else         p_lost = 1;
                    invalid = 0;
                    d = 32'h00000000;
                end else begin
                    if (sign != int32[31]) begin // out of range
                        p_lost = 0;
                        invalid = 1;
                        d = 32'h80000000;
                    end else begin               // normal case
```

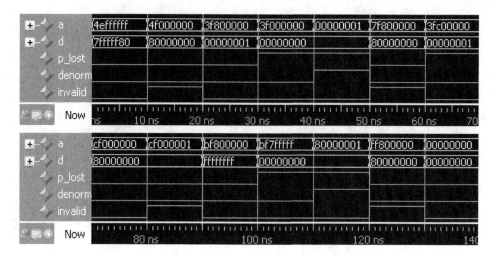

Figure 9.3   Waveform of f2i converter

```
                    if (lost_bits) p_lost = 1;
                    else           p_lost = 0;
                    invalid = 0;
                    d = int32;
                end
            end
        end
      end
    end
endmodule
```

By following the Intel x86 conventions, the floating-point numbers that are out of the integer-representable range are converted to 0x80000000 (the most negative number). The simulation waveforms of the code are shown in Figure 9.3.

In order to verify the correctness of the Verilog HDL code, we wrote a C program that does the conversions on x86 FPU, as listed below. It also shows how to use assembly program (particular CPU instructions) in a C program.

```
#include <stdio.h>                            // f2i.c, x86 FPU float to int test
#define Chop_disable_exception \
asm volatile("fldcw _RoundChop_disable_exception")    // write control word
int _RoundChop_disable_exception = 0x1c3f;            // mask exceptions
#define Read_fp_state_word \
asm volatile("fstsw _fp_state_word")                  // read state word
int _fp_state_word;                                   // FPU state
#define Clear_exceptions_of_sw \
asm volatile("fclex")                                 // clear state
int main (void) {
```

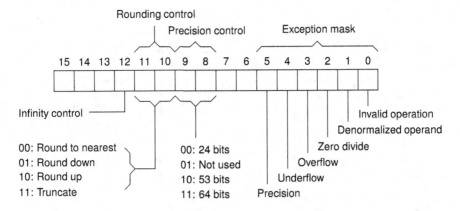

Figure 9.4 Intel x86 FPU control word

```
union {
    int intword;
    float floatword;
} u;
int d;
while (1) {
    fprintf (stderr,"input a float number in hex format: ");
    fscanf (stdin,"%x",&u.intword);              // read an fp number
    Chop_disable_exception;                      // mask exceptions
    Clear_exceptions_of_sw;                      // clear state
    d = u.floatword;                             // f2i
    fprintf (stderr,"f = %08x = %0.6f\n",  u.intword, u.floatword);
    fprintf (stderr,"d = %08x = %08d\n", d, d);  // display integer
    Read_fp_state_word;                          // read FPU state
    fprintf (stderr,"fp_state_word = %04X\n",_fp_state_word);
}
}
```

The program first reads a float number in hexadecimal format from keyboard and saves it as an integer to a union u. Then the exceptions of the x86 FPU are disabled by writing a 16-bit word to the x86 FPU's control register with the x86 fldcw instruction. Next, the x86 FPU's state register is cleared with the x86 fclex instruction. Then the float number is converted to an integer d by reading the data in the union u as a float number and assigning it to the integer d. After this conversion, the x86 FPU's state register is read with the x86 fstsw instruction and the state is shown on display.

Figure 9.4 shows the format of the x86 FPU control word. The program writes 0x1c3f, or 0001110000111111 in binary format, to the control register, allows positive and negative infinity, rounds toward zero (truncate), uses 24-bit significand, and disables all the exceptions. If the exceptions are not disabled, the program will be aborted when an exception occurs.

Figure 9.5 shows the format of the x86 FPU state word. The states we are interested in are "invalid operation," "denormalized operand," and "precision (lost)."

The C program can be compiled and executed as shown below. Table 9.1 lists some execution results. The conversion results shown in Figure 9.3 are the same as those in Table 9.1. Therefore, we can say that our Verilog HDL code is correct, assuming that x86 FPU works correctly.

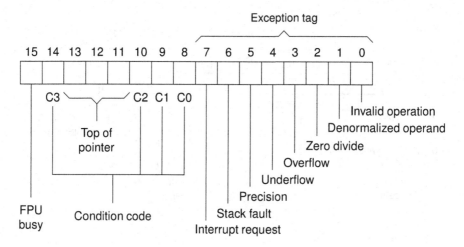

Figure 9.5   Intel x86 FPU state word

Table 9.1   Examples of converting a float to an integer

Float	4effffff	4f000000	3f800000	3f000000	00000001	7f800000	3fc00000
Integer	7fffff80	80000000	00000001	00000000	00000000	80000000	00000001
p_lost	0	0	0	1	1	0	1
denorm	0	0	0	0	1	0	0
invalid	0	1	0	0	0	1	0
Float	cf000000	cf000001	bf800000	bf7fffff	80000001	ff800000	00000000
Integer	80000000	80000000	ffffffff	00000000	00000000	80000000	00000000
p_lost	0	0	0	1	1	0	0
denorm	0	0	0	0	1	0	0
invalid	0	1	0	0	0	1	0

```
[cpu_verilog]$ gcc f2i.c -o f2i
[cpu_verilog]$ ./f2i
input a float number in hex format: 4effffff
f = 4effffff = 2147483520.000000
d = 7fffff80 = 2147483520
fp_state_word = 0000
```

9.2.2   Converting an Integer to a Floating-Point Number

Any integer can be converted to an IEEE 754 single-precision floating-point number, but integers have more significant bits than float numbers, meaning that the precision may be lost after the conversion.

Let's take an example to show how to convert an integer d to a float number a. Suppose $d = $ 1fffffff$_{16}$. The conversion steps are given below.

$$d = 0001111111111111111111111111111._2 \qquad \text{(to shift "." to left by 31 bits)}$$
$$= 0.001111111111111111111111111111111_2 \times 2^{31} \qquad \text{(to } 1.f_a \text{ format)}$$
$$= 1.111111111111111111111111111000_2 \times 2^{31-3} \qquad \text{(3 bits shifted)}$$
$$\approx 1.11111111111111111111111_2 \times 2^{(127+31-3)-127} \qquad \text{(IEEE float format)}$$
$$= (-1)^s \times 1.f \times 2^{e-127} \qquad \text{(IEEE float format)}$$

Then $s = 0$, $e = 127 + 31 - 3 = 10011011_2$, $f = 11111111111111111111111_2$. The final result is $a = 0_10011011_11111111111111111111111_2 = $ 4dffffff$_{16}$. Because the value of the last 8 bits that were thrown away is not 0, the precision is lost.

If d is negative, it should be converted to positive first. The following is the Verilog HDL code that converts an integer to an IEEE 754 single-precision floating-point number. The key point is to determine the sa (shift amount), by which the significand is shifted to the left so that the shifted significand (f0 in the code) has the format $1. f$. The idea to do this is to determine each bit of sa, from the most significant bit to the least significant bit.

```
module i2f (d,a,p_lost);             // convert integer to float
   input   [31:0]  d;                // integer
   output  [31:0]  a;                // float
   output          p_lost;           // precision lost
   wire            sign = d[31];     // sign
   wire    [31:0]  f5 = sign? -d : d; // absolute
   wire    [31:0]  f4,f3,f2,f1,f0;
   wire    [4:0]   sa;               // shift amount (to 1.f)
   assign          sa[4] = ~|f5[31:16];  // 16-bit 0
   assign          f4 = sa[4]? {f5[15:0],16'b0} : f5;
   assign          sa[3] = ~|f4[31:24];  // 8-bit 0
   assign          f3 = sa[3]? {f4[23:0], 8'b0} : f4;
   assign          sa[2] = ~|f3[31:28];  // 4-bit 0
   assign          f2 = sa[2]? {f3[27:0], 4'b0} : f3;
   assign          sa[1] = ~|f2[31:30];  // 2-bit 0
   assign          f1 = sa[1]? {f2[29:0], 2'b0} : f2;
   assign          sa[0] = ~f1[31];      // 1-bit 0
   assign          f0 = sa[0]? {f1[30:0], 1'b0} : f1;
   assign          p_lost = |f0[7:0];    // not 0
   wire    [22:0]  fraction = f0[30:8];  // f0[31] = 1, hidden bit
   wire    [7:0]   exponent = 8'h9e - {3'h0,sa};  // 0x9e = 158 = 127 + 31
   assign          a = (d == 0)? 0 : {sign,exponent,fraction};
endmodule
```

The simulation waveforms of the code are shown in Figure 9.6. Similarly, we can write a C program, i2f.c, to check the correctness of the Verilog HDL code.

9.3 Floating-Point Adder (FADD) Design

This section describes the design of an adder that can perform addition or subtraction on IEEE 754 single-precision floating-point numbers.

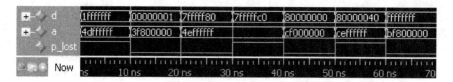

Figure 9.6   Waveform of an i2f converter

9.3.1   Floating-Point Addition Algorithm

Let's take an example to see the addition $s = a + b$ on the two float numbers a and b. Suppose a and b have the values in IEEE 754 float format as shown below.

$$a:\ 0\_01111000\_11000000000000000010001\ (0x3c600011)$$
$$b:\ 1\_01111101\_00000100000000000000000\ (0xbe820000)$$

That is, $s_a = 0$, $e_a = 01111000 = 120$, and $1.f_a = 1.11000000000000000010001$; $s_b = 1$, $e_b = 01111101 = 125$, and $1.f_b = 1.00000100000000000000000$. The calculation will take three steps: alignment, calculation, and normalization, which we describe below.

9.3.1.1   Alignment

a and b in the example above are two normalized float numbers. Although we calculate $a + b$, the signs of a and b are different, so we must perform subtraction on their absolutes. By checking their exponents, we know that the absolute of b is larger than that of a, so we will perform $|b| - |a|$. The sign of s is negative, the same as the number that has a larger absolute. And the exponent of s is 125 temporarily.

The subtraction is performed on the significands (hidden bits and fractions). To do it, we must make a and b to have the same exponent. Because $e_b - e_a = 125 - 120 = 5$, we shift $1.f_a$ to the right by 5 bits. That is

$$a = 2^{120-127} \times 1.11000000000000000010001$$

$$= 2^{125-127} \times 0.0000111000000000000000000\_10001$$

The least five significant bits of the shifted significand of a, 10001, will be treated as shown below in the IEEE 754 standard: the leftmost bit, 1, is called a guard bit. The next bit, 0, is called a round bit. And the rest of the bits, 001, are reduced to one bit, called a sticky bit, by performing a reduction OR on these bits, that is, the sticky bit is set whenever there are nonzero bits to the right of the round bit. We got three bits: 101. We use grs to denote these three bits. grs will participate in the calculation in order to get a more precise result.

9.3.1.2   Calculation

Now we can perform the subtraction operation on the two significands as below (subtracting the smaller significand from the larger significant).

```
                                              grs
      01.00000100000000000000000  000  (significand of b)
   −  00.00001110000000000000000  101  (significand of a)
      00.11110101111111111111111  011  (significand of s)
```

Because the result of an addition on two significands may be larger than 2 (but <4), we use an additional bit to the most significant bit position during the calculation. The total number of bits is $1 + 24 + 3$, or 28.

9.3.1.3 Normalization

The calculated result of the significand must be converted to the format $1.f$. Thus we search the first "1" from the left to determine by how many bits the significand must be shifted to the left. In the above example, we shift the significand to the left by 1 bit. Meanwhile, we must subtract a 1 from the exponent in order to keep the value unchanged. That is

$$s = -2^{125-127} \times 0.11110101111111111111111\_011$$

$$= -2^{124-127} \times 1.1110101111111111111110\_110$$

The question now we have is how to deal with the rightmost three bits 110. The answer is "rounding." The IEEE 754 standard mainly defines the following four rounding modes as shown below.

1. Round to nearest (even if $grs = 100$).
2. Round toward minus infinity $(-\infty)$, also called round down.
3. Round toward plus infinity $(+\infty)$, also called round up.
4. Round toward zero, also called truncate or chop.

If we use round to nearest mode, we get the final result of s as below.

$$s:\ 1\_01111100\_11101011111111111111111 \quad (\text{0xbe75ffff})$$

We may get different results by using different rounding modes. The following C program tests the float addition and subtraction results under the four rounding modes. The meanings of control word and state word of the Intel x86 FPU are shown in Figures 9.4 and 9.5.

```c
#include <stdio.h>                     // fadd_test.c, rounding test
#define Near asm volatile("fldcw _RoundNear")
#define Down asm volatile("fldcw _RoundDown")
#define Up   asm volatile("fldcw _RoundUp")
#define Chop asm volatile("fldcw _RoundChop")
int _RoundNear  = 0x103f;              // round code = 00 round to nearest
int _RoundDown  = 0x143f;              // round code = 01 round toward -infinity
int _RoundUp    = 0x183f;              // round code = 10 round toward +infinity
int _RoundChop  = 0x1c3f;              // round code = 11 round toward 0
int main(void) {
    union {
        int intword;
        float floatword;
    } u, v, s, t;
    while (1) {
        fprintf (stderr,"input 1st fp number in hex format: ");
        fscanf (stdin,"%x",&u.intword);
        fprintf (stderr,"input 2nd fp number in hex format: ");
        fscanf (stdin,"%x",&v.intword);
```

```
        Near;                                           // round to nearest
        s.floatword = u.floatword + v.floatword;
        t.floatword = u.floatword - v.floatword;
        fprintf (stderr,"the sum of 2 fp numbers is (near): "
                  "%08x\t%08x\n",s.intword,t.intword);
        Down;                                           // round toward -infinity
        s.floatword = u.floatword + v.floatword;
        t.floatword = u.floatword - v.floatword;
        fprintf (stderr,"the sum of 2 fp numbers is (down): "
                  "%08x\t%08x\n",s.intword,t.intword);
        Up;                                             // round toward +infinity
        s.floatword = u.floatword + v.floatword;
        t.floatword = u.floatword - v.floatword;
        fprintf (stderr,"the sum of 2 fp numbers is ( up ): "
                  "%08x\t%08x\n",s.intword,t.intword);
        Chop;                                           // round toward 0
        s.floatword = u.floatword + v.floatword;
        t.floatword = u.floatword - v.floatword;
        fprintf (stderr,"the sum of 2 fp numbers is (chop): "
                  "%08x\t%08x\n",s.intword,t.intword);
    }
}
```

The compilation command and the execution example are listed below. The two input float numbers, 0x3c600011 and 0xbe820000, are the numbers that we used in this section. In addition to the addition, we also calculated the subtraction results, which are shown on the right side.

```
[cpu_verilog]$ gcc fadd_test.c -o fadd_test
[cpu_verilog]$ ./fadd_test
input 1st fp number in hex format: 3c600011
input 2nd fp number in hex format: be820000
the sum of 2 fp numbers is (near): be75ffff      3e890001
the sum of 2 fp numbers is (down): be75ffff      3e890000
the sum of 2 fp numbers is ( up ): be75fffe      3e890001
the sum of 2 fp numbers is (chop): be75fffe      3e890000
```

9.3.1.4 Special Cases

If a or b is an NaN, whether or not an addition or a subtraction, the result s will be an NaN. If the two numbers are infinities, the results are shown in Table 9.2, where sub = 1 means $s = a - b$, or $s = a + b$ otherwise.

If a is a 0, a normalized number, or a denormalized number, then $a + (\pm\infty) = \pm\infty$ and $a - (\pm\infty) = \mp\infty$. If b is such a number, then $\pm\infty \pm b = \pm\infty$.

Table 9.2 Results of operations on two infinity numbers

sub	a	b	s	Comment
0	$+\infty$	$+\infty$	$+\infty$	$(+\infty) + (+\infty)$
0	$-\infty$	$-\infty$	$-\infty$	$(-\infty) + (-\infty)$
1	$+\infty$	$-\infty$	$+\infty$	$(+\infty) - (-\infty)$
1	$-\infty$	$+\infty$	$-\infty$	$(-\infty) - (+\infty)$
0	$+\infty$	$-\infty$	NaN	$(+\infty) + (-\infty)$
0	$-\infty$	$+\infty$	NaN	$(-\infty) + (+\infty)$
1	$+\infty$	$+\infty$	NaN	$(+\infty) - (+\infty)$
1	$-\infty$	$-\infty$	NaN	$(-\infty) - (-\infty)$

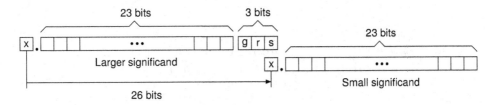

Figure 9.7 Shifting to the right by at most 26 bits is enough

Table 9.3 FP rounding examples

Sign	Fraction	g	r	s	Nearest	Truncate	Up	Down
+	1001	0	0	0	1001	1001	1001	1001
	1001	1	0	0	1010	1001	1010	1001
	1000	1	0	0	1000	1000	1001	1000
	1001	0	1	1	1001	1001	1010	1001
	1001	1	1	0	1010	1001	1010	1001
−	1001	0	0	0	1001	1001	1001	1001
	1001	1	0	0	1010	1001	1001	1010
	1000	1	0	0	1000	1000	1000	1001
	1001	0	1	1	1001	1001	1001	1010
	1001	1	1	0	1010	1001	1001	1010

9.3.1.5 Implementation Details of FADD

As we explained before, three additional bits, *grs*, are used for the significand calculation, and $1.f$ or $0.f$ of the number whose absolute is smaller than another will be shifted to the right by at most 26 bits, as shown in Figure 9.7.

Table 9.3 lists some rounding examples in which the column "fraction" gives only the four least significant bits before rounding. The 4-bit results after rounding are listed in the right part of the table.

Table 9.4 summarizes all the cases of `frac_plus_1 = 1`, that is, a 1 will be added to the least significant bit of the fraction (Frac in the table). The `rm[1:0]` signals are the encodings of the rounding modes.

Table 9.4 Fraction increment when rounding

rm[1:0]	Frac	g	r	s	Sign	frac_plus_1	Rounding mode
00	1	1	0	0	x	1	Round to even
00	x	1	{not 00}		x	1	Round to nearest
01	x	{not 0 0 0}			1	1	Round toward $-\infty$
10	x	{not 0 0 0}			0	1	Round toward $+\infty$

In the normalization step, the result exponent is adjusted based on the number of shifted bits to make the significand have the format $1.f$. Furthermore, the result exponent may need to be adjusted again because of the rounding. See the following example. In the rounding step, adding a 1 to the least significant bit of the significand results in the loss of the $1.f$ format. In such a case, we must set the significand to 1.0 and add a 1 to the exponent.

$$
\begin{aligned}
& 0\,1.1\,1\,1\,1\,1\,1\,1\,1\,1\,1\,1\,1\,1\,1\,1\,1\,1\,1\,1\,1\,1\,1\,1 \;(1.f) \\
+ \;& 0\,0.0\,0\,0\,0\,0\,0\,0\,0\,0\,0\,0\,0\,0\,0\,0\,0\,0\,0\,0\,0\,0\,0\,1 \;(\text{frac\_plus\_1}) \\
\hline
= \;& 1\,0.0\,0\,0\,0\,0\,0\,0\,0\,0\,0\,0\,0\,0\,0\,0\,0\,0\,0\,0\,0\,0\,0\,0 \\
\rightarrow \;& 0\,1.0\,0\,0\,0\,0\,0\,0\,0\,0\,0\,0\,0\,0\,0\,0\,0\,0\,0\,0\,0\,0\,0\,0 \;(1.f)
\end{aligned}
$$

The result of the float addition or subtraction may be overflow. For example, given two normalized float numbers a and b as below, calculate $s = a + b$.

$$
\begin{aligned}
a\text{:}\;& 0\_11111110\_11111111111111111111111 \;(\text{0x7f7fffff}) \\
b\text{:}\;& 0\_11111011\_11111111111111111111111 \;(\text{0x7dffffff})
\end{aligned}
$$

After the alignment by shifting $1.f_b$ to the right by 3 bits ($e_a - e_b = 11111110 - 11111011 = 00000011$), we perform addition on the two significands as below.

$$
\begin{array}{r}
grs \\
0\,1.1\,1\,1\,1\,1\,1\,1\,1\,1\,1\,1\,1\,1\,1\,1\,1\,1\,1\,1\,1\,1\,1\,1\,1 \;\;000 \\
+ \;\;0\,0.0\,0\,1\,1\,1\,1\,1\,1\,1\,1\,1\,1\,1\,1\,1\,1\,1\,1\,1\,1\,1\,1\,1 \;\;111 \\
\hline
= \;\;1\,0.0\,0\,1\,1\,1\,1\,1\,1\,1\,1\,1\,1\,1\,1\,1\,1\,1\,1\,1\,1\,1\,1\,0 \;\;111 \\
\rightarrow \;\;0\,1.0\,0\,0\,1\,1\,1\,1\,1\,1\,1\,1\,1\,1\,1\,1\,1\,1\,1\,1\,1\,1\,1 \;\;011
\end{array}
$$

In the normalization step, the result significand is converted to the format $1.f$ by shifting the significand to the right by 1 bit, and the exponent is incremented by 1. Then the exponent becomes 11111111, which is not a valid exponent of a normalized float number. We say that an overflow has occurred. The IEEE 754 Standard treads the overflow with the following rules, corresponding to the rounding modes.

1. Round to nearest mode carries all overflows to infinity (∞) with the sign of the intermediate result.
2. Round toward zero mode carries all overflows to the largest normalized number with the sign of the intermediate result.
3. Round toward $-\infty$ mode carries positive overflows to the largest normalized number, and carries negative overflows to $-\infty$.
4. Round toward $+\infty$ mode carries negative overflows to the most negative normalized number, and carries positive overflows to $+\infty$.

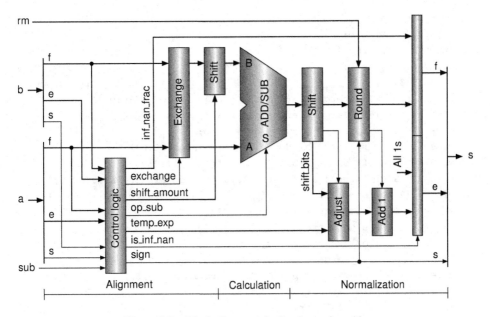

Figure 9.8   Block diagram of a floating-point adder

Then the addition result of the example above will be either infinity or the maximum (the calculated significand is not used), as shown below.

> s: 0_11111111_00000000000000000000000 (Round to nearest or round up)
> s: 0_11111110_11111111111111111111111 (Round down or truncate)

9.3.2   Circuit and Verilog HDL Codes of FADD

As described previously, a float addition or subtraction can be done in three steps: alignment, calculation, and normalization. Figure 9.8 shows the block diagram of the single-precision floating-point adder. The input signals include two float numbers a and b represented in IEEE 754 single-precision format, 2-bit round modes rm, and 1-bit sub. If sub is a 1, the adder performs subtraction; it performs addition otherwise. The output y is the calculation result, also represented in IEEE 754 single-precision format.

In the alignment step, the hidden bit and the fraction of the float number whose absolute is smaller than the other is shifted to the right so that the two numbers will have the same exponent. This is done by the blocks exchange and shift whose control signals (whether exchange and shift amount) are generated by the block control logic. The control logic block also generates the sign and intermediate exponent of the result number, the operation control of the calculation (addition or subtraction), and a signal that indicates whether there is an NaN or infinity number in the inputs a and b.

An addition or subtraction is performed by the block add/sub in the calculation step. In the normalization step, the intermediate significand (the output of the add/sub block) is shifted so that it has the format 1.f, and the intermediate exponent is adjusted based on the shift amount (shift_bits). The rounding is done according to the input rm, and it may cause the exponent to be adjusted again as we described previously. Finally, we get the float result by dealing with the overflow and the special cases. The Verilog HDL code of the single-precision floating-point adder is given below.

```verilog
module fadder (a,b,sub,rm,s);                           // fadder
    input    [31:0] a,b;                                // fp a and b
    input    [1:0] rm;                                  // round mode
    input          sub;                                 // 1: sub; 0: add
    output   [31:0] s;                                  // fp output
    wire           exchange = ({1'b0,b[30:0]} > {1'b0,a[30:0]});
    wire    [31:0] fp_large = exchange? b : a;
    wire    [31:0] fp_small = exchange? a : b;
    wire           fp_large_hidden_bit = |fp_large[30:23];
    wire           fp_small_hidden_bit = |fp_small[30:23];
    wire    [23:0] large_frac24 = {fp_large_hidden_bit,fp_large[22:0]};
    wire    [23:0] small_frac24 = {fp_small_hidden_bit,fp_small[22:0]};
    wire    [7:0]  temp_exp = fp_large[30:23];
    wire           sign = exchange? sub ^ b[31] : a[31];
    wire           op_sub = sub ^ fp_large[31] ^ fp_small[31];
    wire           fp_large_expo_is_ff = &fp_large[30:23]; // exp == 0xff
    wire           fp_small_expo_is_ff = &fp_small[30:23];
    wire           fp_large_frac_is_00 = ~|fp_large[22:0];  // frac == 0x0
    wire           fp_small_frac_is_00 = ~|fp_small[22:0];
    wire           fp_large_is_inf=fp_large_expo_is_ff & fp_large_frac_is_00;
    wire           fp_small_is_inf=fp_small_expo_is_ff & fp_small_frac_is_00;
    wire           fp_large_is_nan=fp_large_expo_is_ff & ~fp_large_frac_is_00;
    wire           fp_small_is_nan=fp_small_expo_is_ff & ~fp_small_frac_is_00;
    wire           s_is_inf = fp_large_is_inf | fp_small_is_inf;
    wire           s_is_nan = fp_large_is_nan | fp_small_is_nan |
                            ((sub ^ fp_small[31] ^ fp_large[31]) &
                            fp_large_is_inf & fp_small_is_inf);
    wire    [22:0] nan_frac = ({1'b0,a[22:0]} > {1'b0,b[22:0]}) ?
                            {1'b1,a[21:0]} : {1'b1,b[21:0]};
    wire    [22:0] inf_nan_frac = s_is_nan? nan_frac : 23'h0;
    wire    [7:0]  exp_diff = fp_large[30:23] - fp_small[30:23];
    wire           small_den_only = (fp_large[30:23] != 0) &
                            (fp_small[30:23] == 0);
    wire    [7:0]  shift_amount = small_den_only? exp_diff - 8'h1 : exp_diff;
    wire    [49:0] small_frac50 = (shift_amount >= 26)?
                    {26'h0,small_frac24} :
                    {small_frac24,26'h0} >> shift_amount;
    wire    [26:0] small_frac27 = {small_frac50[49:24],|small_frac50[23:0]};
    wire    [27:0] aligned_large_frac = {1'b0,large_frac24,3'b000};
    wire    [27:0] aligned_small_frac = {1'b0,small_frac27};
    wire    [27:0] cal_frac = op_sub?
                    aligned_large_frac - aligned_small_frac :
                    aligned_large_frac + aligned_small_frac;
    wire    [26:0] f4,f3,f2,f1,f0;
    wire    [4:0]  zeros;
    assign         zeros[4] = ~|cal_frac[26:11];            // 16-bit 0
    assign         f4 = zeros[4]? {cal_frac[10:0],16'b0} : cal_frac[26:0];
```

```
assign        zeros[3] = ~|f4[26:19];              //  8-bit 0
assign        f3 = zeros[3]? {f4[18:0], 8'b0} : f4;
assign        zeros[2] = ~|f3[26:23];              //  4-bit 0
assign        f2 = zeros[2]? {f3[22:0], 4'b0} : f3;
assign        zeros[1] = ~|f2[26:25];              //  2-bit 0
assign        f1 = zeros[1]? {f2[24:0], 2'b0} : f2;
assign        zeros[0] = ~f1[26];                  //  1-bit 0
assign        f0 = zeros[0]? {f1[25:0], 1'b0} : f1;
reg    [7:0] exp0;
reg    [26:0] frac0;
always @ * begin
   if (cal_frac[27]) begin           // 1x.xxxxxxxxxxxxxxxxxxxxxxx xxx
      frac0 = cal_frac[27:1];        //  1.xxxxxxxxxxxxxxxxxxxxxxx xxx
      exp0 = temp_exp + 8'h1;
   end else begin
      if ((temp_exp > zeros) && (f0[26])) begin // a normalized number
         exp0 = temp_exp - zeros;
         frac0 = f0;                 //  1.xxxxxxxxxxxxxxxxxxxxxxx xxx
      end else begin                 // is a denormalized number or 0
         exp0 = 0;
         if (temp_exp != 0)          // (e - 127) = ((e - 1) - 126)
            frac0 = cal_frac[26:0] << (temp_exp - 8'h1);
         else frac0 = cal_frac[26:0];
      end
   end
end
wire frac_plus_1 =                   // for rounding
    ~rm[1] & ~rm[0] & frac0[2] & (frac0[1] | frac0[0]) |
    ~rm[1] & ~rm[0] & frac0[2] & ~frac0[1] & ~frac0[0]  & frac0[3] |
    ~rm[1] & rm[0] & (frac0[2] | frac0[1] | frac0[0]) & sign |
    rm[1] & ~rm[0] & (frac0[2] | frac0[1] | frac0[0]) & ~sign;
wire   [24:0] frac_round = {1'b0,frac0[26:3]} + frac_plus_1;
wire   [7:0] exponent = frac_round[24]? exp0 + 8'h1 : exp0;
wire          overflow = &exp0 | &exponent;
assign s = final_result(overflow,rm,sign,s_is_nan,s_is_inf,exponent,
                        frac_round[22:0],inf_nan_frac);
function  [31:0] final_result;
   input         overflow;
   input  [1:0] rm;
   input         sign;
   input         is_nan;
   input         is_inf;
   input  [7:0] exponent;
   input  [22:0] fraction, inf_nan_frac;
   casex ({overflow,rm,sign,s_is_nan,s_is_inf})
      6'b1_00_x_0_x : final_result = {sign,8'hff,23'h000000};   // inf
      6'b1_01_0_0_x : final_result = {sign,8'hfe,23'h7fffff};   // max
```

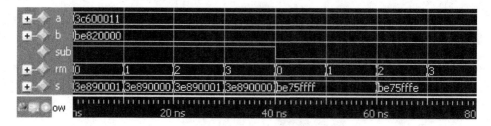

Figure 9.9   Waveform of the floating-point adder

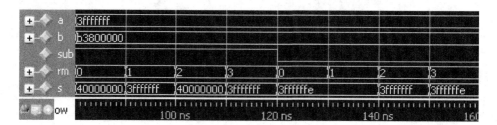

Figure 9.10   Increasing exponent caused by rounding

```
        6'b1_01_1_0_x : final_result = {sign,8'hff,23'h000000};   // inf
        6'b1_10_0_0_x : final_result = {sign,8'hff,23'h000000};   // inf
        6'b1_10_1_0_x : final_result = {sign,8'hfe,23'h7fffff};   // max
        6'b1_11_x_0_x : final_result = {sign,8'hfe,23'h7fffff};   // max
        6'b0_xx_x_0_0 : final_result = {sign,exponent,fraction};  // nor
        6'bx_xx_x_1_x : final_result = {1'b1,8'hff,inf_nan_frac}; // nan
        6'bx_xx_x_0_1 : final_result = {sign,8'hff,inf_nan_frac}; // inf
        default       : final_result = {sign,8'h00,23'h000000};   // 0
      endcase
    endfunction
endmodule
```

Figure 9.9 shows the simulation results with the two input float numbers that were used as the example in the beginning of this section. The results are the same as those generated by the Intel x86 FPU.

Figure 9.10 shows the simulation waveforms of the case in which the result exponents are different under the different rounding modes when performing subtraction on the following two float numbers.

a: 0_01111111_11111111111111111111111 (0x3fffffff)
b: 1_01100111_00000000000000000000000 (0xb3800000)

Because $e_a - e_b = 24$, we shift $1.f_b$ to the right by 24 bits; and because s_a and s_b are different, we perform addition on the two significands of a and b:

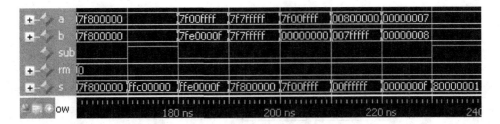

Figure 9.11   Special cases of floating-point additions/subtractions

$$
\begin{array}{rll}
& grs & \\
01.11111111111111111111111\ 111 & 000 & \text{(significand of } a) \\
+\quad 00.00000000000000000000000\ 000 & 100 & \text{(significand of } b) \\
\hline
=\quad 01.11111111111111111111111\ 111 & 100 & \text{(significand of } s) \\
\rightarrow\quad 01.11111111111111111111111\ 111 & & \text{(truncate or round down)} \\
\rightarrow\quad 10.00000000000000000000000\ 000 & & \text{(round to nearest or round up)} \\
\rightarrow\quad 01.00000000000000000000000\ 000 & & \text{(to } 1.f \text{ and exponent } + 1)
\end{array}
$$

The result is 0x40000000 with the round to nearest or round up modes, and it is 0x3fffffff with truncate or round down modes. The exponents of the two results are different.

Figure 9.11 shows the simulation waveforms with the inputs of the eight special cases. In sequence, the waveforms show

1. $(+\infty) + (+\infty) = +\infty$;
2. $(+\infty) - (+\infty) =$ NaN;
3. a normalized number + NaN = NaN;
4. the sum of the two largest numbers is $+\infty$ (round to nearest);
5. a normalized number + 0;
6. the smallest normalized number + the largest denormalized number;
7. a denormalized number + a denormalized number; and
8. a denormalized number − a denormalized number.

9.3.3   Pipelined FADD Design

Figure 9.12 shows the block diagram of a pipelined single-precision floating-point adder. We simply divide the operations into three stages: alignment, calculation, and normalization. The pipeline registers are inserted in between the stages and have a write enable signal which will be used in CPU design to stall the pipeline.

Such pipelined float adder/subtracter can perform a float addition or subtraction during every clock cycle. In our implementation, the calculation stage takes shorter times than the other two stages. The dividing should make each stage taking nearly equal times and make the cycle time as short as possible.

Below is the top module of the pipelined floating-point adder. It invokes five modules, corresponding to the five blocks shown in Figure 9.12. The letters "a" (alignment), "c" (calculation), and "n" (normalization) are prefixed to signals in the alignment, calculation, and normalization stages, respectively.

```
module pipelined_fadder (a,b,sub,rm,s,clk,clrn,e);    // pipelined fp adder
    input          clk, clrn;                          // clock and reset
    input   [31:0] a, b;                               // fp a and b
```

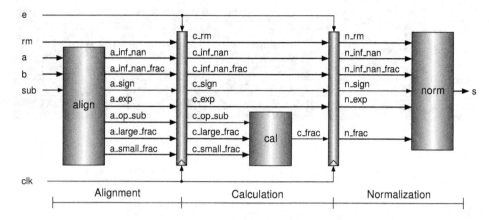

Figure 9.12   Block diagram of a pipelined floating-point adder

```
input    [1:0] rm;                              // round mode
input          sub;                             // 1: sub; 0: add
input          e;                               // enable
output [31:0] s;                                // fp output
wire    [26:0] a_small_frac;
wire    [23:0] a_large_frac;
wire    [22:0] a_inf_nan_frac;
wire     [7:0] a_exp;
wire           a_is_nan,a_is_inf;
wire           a_sign;
wire           a_op_sub;
// exe1: alignment stage
fadd_align alignment (a,b,sub,a_is_nan,a_is_inf,a_inf_nan_frac,a_sign,
                      a_exp,a_op_sub,a_large_frac,a_small_frac);
wire    [26:0] c_small_frac;
wire    [23:0] c_large_frac;
wire    [22:0] c_inf_nan_frac;
wire     [7:0] c_exp;
wire     [1:0] c_rm;
wire           c_is_nan,c_is_inf;
wire           c_sign;
wire           c_op_sub;
// pipelined registers
reg_align_cal reg_ac (rm,a_is_nan,a_is_inf,a_inf_nan_frac,a_sign,a_exp,
                      a_op_sub,a_large_frac,a_small_frac,clk,clrn,e,
                      c_rm,c_is_nan,c_is_inf,c_inf_nan_frac,c_sign,
                      c_exp,c_op_sub,c_large_frac,c_small_frac);
wire    [27:0] c_frac;
// exe2: calculation stage
fadd_cal calculation (c_op_sub,c_large_frac,c_small_frac,c_frac);
wire    [27:0] n_frac;
```

```
    wire    [22:0] n_inf_nan_frac;
    wire    [7:0] n_exp;
    wire    [1:0] n_rm;
    wire          n_is_nan,n_is_inf;
    wire          n_sign;
    // pipelined registers
    reg_cal_norm reg_cn (c_rm,c_is_nan,c_is_inf,c_inf_nan_frac,c_sign,c_exp,
                         c_frac,clk,clrn,e,n_rm,n_is_nan,n_is_inf,
                         n_inf_nan_frac,n_sign,n_exp,n_frac);
    // exe3: normalization stage
    fadd_norm normalization (n_rm,n_is_nan,n_is_inf,n_inf_nan_frac,n_sign,
                             n_exp,n_frac,s);
endmodule
```

Below is the Verilog HDL code of the alignment stage.

```
module fadd_align (a,b,sub,s_is_nan,s_is_inf,inf_nan_frac,sign,temp_exp,
                   op_sub,large_frac24,small_frac27);        //alignment stage
    input   [31:0] a,b;
    input          sub;
    output  [26:0] small_frac27;
    output  [23:0] large_frac24;
    output  [22:0] inf_nan_frac;
    output   [7:0] temp_exp;
    output         s_is_nan;
    output         s_is_inf;
    output         sign;
    output         op_sub;
    wire           exchange = (b[30:0] > a[30:0]);
    wire    [31:0] fp_large = exchange? b : a;
    wire    [31:0] fp_small = exchange? a : b;
    wire           fp_large_hidden_bit = |fp_large[30:23];
    wire           fp_small_hidden_bit = |fp_small[30:23];
    wire    [23:0] large_frac24 = {fp_large_hidden_bit,fp_large[22:0]};
    wire    [23:0] small_frac24 = {fp_small_hidden_bit,fp_small[22:0]};
    assign         temp_exp = fp_large[30:23];
    assign         sign = exchange? sub ^ b[31] : a[31];
    assign         op_sub = sub ^ fp_large[31] ^ fp_small[31];
    wire           fp_large_expo_is_ff = &fp_large[30:23]; // exp == 0xff
    wire           fp_small_expo_is_ff = &fp_small[30:23];
    wire           fp_large_frac_is_00 = ~|fp_large[22:0];  // frac == 0x0
    wire           fp_small_frac_is_00 = ~|fp_small[22:0];
    wire           fp_large_is_inf=fp_large_expo_is_ff & fp_large_frac_is_00;
    wire           fp_small_is_inf=fp_small_expo_is_ff & fp_small_frac_is_00;
    wire           fp_large_is_nan=fp_large_expo_is_ff & ~fp_large_frac_is_00;
    wire           fp_small_is_nan=fp_small_expo_is_ff & ~fp_small_frac_is_00;
    assign         s_is_inf = fp_large_is_inf | fp_small_is_inf;
```

```
    wire            s_is_nan = fp_large_is_nan | fp_small_is_nan |
                             ((sub ^ fp_small[31] ^ fp_large[31]) &
                             fp_large_is_inf & fp_small_is_inf);
    wire    [22:0] nan_frac = (a[21:0] > b[21:0])?
                             {1'b1,a[21:0]} : {1'b1,b[21:0]};
    assign          inf_nan_frac = s_is_nan? nan_frac : 23'h0;
    wire    [7:0]  exp_diff = fp_large[30:23] - fp_small[30:23];
    wire            small_den_only = (fp_large[30:23] != 0) &
                             (fp_small[30:23] == 0);
    wire    [7:0]  shift_amount = small_den_only? exp_diff - 8'h1 : exp_diff;
    wire    [49:0] small_frac50 = (shift_amount >= 26)?
                             {26'h0,small_frac24} :
                             {small_frac24,26'h0} >> shift_amount;
    assign          small_frac27 = {small_frac50[49:24],|small_frac50[23:0]};
endmodule
```

Below is the Verilog HDL code of the pipeline registers in between the alignment and calculation stages.

```
module reg_align_cal (a_rm,a_is_nan,a_is_inf,a_inf_nan_frac,a_sign,a_exp,
                      a_op_sub,a_large_frac,a_small_frac,clk,clrn,e,c_rm,
                      c_is_nan,c_is_inf,c_inf_nan_frac,c_sign,c_exp,
                      c_op_sub,c_large_frac,c_small_frac);  // pipeline regs
    input       [26:0] a_small_frac;
    input       [23:0] a_large_frac;
    input       [22:0] a_inf_nan_frac;
    input        [7:0] a_exp;
    input        [1:0] a_rm;
    input              a_is_nan, a_is_inf, a_sign, a_op_sub;
    input              e;                                // e: enable
    input              clk, clrn;                        // clock and reset
    output reg  [26:0] c_small_frac;
    output reg  [23:0] c_large_frac;
    output reg  [22:0] c_inf_nan_frac;
    output reg   [7:0] c_exp;
    output reg   [1:0] c_rm;
    output reg         c_is_nan,c_is_inf,c_sign,c_op_sub;
    always @ (posedge clk or negedge clrn) begin
        if (!clrn) begin
            c_rm            <= 0;
            c_is_nan        <= 0;
            c_is_inf        <= 0;
            c_inf_nan_frac  <= 0;
            c_sign          <= 0;
            c_exp           <= 0;
            c_op_sub        <= 0;
            c_large_frac    <= 0;
```

```
                    c_small_frac    <= 0;
            end else if (e) begin
                c_rm            <= a_rm;
                c_is_nan        <= a_is_nan;
                c_is_inf        <= a_is_inf;
                c_inf_nan_frac  <= a_inf_nan_frac;
                c_sign          <= a_sign;
                c_exp           <= a_exp;
                c_op_sub        <= a_op_sub;
                c_large_frac    <= a_large_frac;
                c_small_frac    <= a_small_frac;
            end
        end
    end
endmodule
```

Below is the Verilog HDL code of the calculation stage.

```
module fadd_cal (op_sub,large_frac24,small_frac27, cal_frac); // calculation
    input   [23:0] large_frac24;
    input          op_sub;
    input   [26:0] small_frac27;
    output  [27:0] cal_frac;
    wire    [27:0] aligned_large_frac = {1'b0,large_frac24,3'b000};
    wire    [27:0] aligned_small_frac = {1'b0,small_frac27};
    assign         cal_frac = op_sub?
                            aligned_large_frac - aligned_small_frac :
                            aligned_large_frac + aligned_small_frac;
endmodule
```

Below is the Verilog HDL code of the pipeline registers in between the calculation and normalization stages.

```
module reg_cal_norm (c_rm,c_is_nan,c_is_inf,c_inf_nan_frac,c_sign,c_exp,
                    c_frac,clk,clrn,e,n_rm,n_is_nan,n_is_inf,
                    n_inf_nan_frac,n_sign,n_exp,n_frac);    // pipeline regs
    input       [27:0] c_frac;
    input       [22:0] c_inf_nan_frac;
    input       [7:0] c_exp;
    input       [1:0] c_rm;
    input             c_is_nan, c_is_inf, c_sign;
    input             e;                              // e: enable
    input             clk, clrn;                      // clock and reset
    output reg  [27:0] n_frac;
    output reg  [22:0] n_inf_nan_frac;
    output reg  [7:0] n_exp;
    output reg  [1:0] n_rm;
    output reg        n_is_nan,n_is_inf,n_sign;
    always @ (posedge clk or negedge clrn) begin
```

```
      if (!clrn) begin
          n_rm              <= 0;
          n_is_nan          <= 0;
          n_is_inf          <= 0;
          n_inf_nan_frac <= 0;
          n_sign            <= 0;
          n_exp             <= 0;
          n_frac            <= 0;
      end else if (e) begin
          n_rm              <= c_rm;
          n_is_nan          <= c_is_nan;
          n_is_inf          <= c_is_inf;
          n_inf_nan_frac <= c_inf_nan_frac;
          n_sign            <= c_sign;
          n_exp             <= c_exp;
          n_frac            <= c_frac;
      end
   end
endmodule
```

Below is the Verilog HDL code of the normalization stage.

```
module fadd_norm (rm,is_nan,is_inf,inf_nan_frac,sign,temp_exp,cal_frac,s);
    input   [27:0] cal_frac;
    input   [22:0] inf_nan_frac;
    input    [7:0] temp_exp;
    input    [1:0] rm;
    input          is_nan,is_inf;
    input          sign;
    output [31:0] s;
    wire    [26:0] f4,f3,f2,f1,f0;
    wire     [4:0] zeros;
    assign         zeros[4] = ~|cal_frac[26:11];                    // 16-bit 0
    assign         f4 = zeros[4]? {cal_frac[10:0],16'b0} : cal_frac[26:0];
    assign         zeros[3] = ~|f4[26:19];                          //  8-bit 0
    assign         f3 = zeros[3]? {f4[18:0], 8'b0} : f4;
    assign         zeros[2] = ~|f3[26:23];                          //  4-bit 0
    assign         f2 = zeros[2]? {f3[22:0], 4'b0} : f3;
    assign         zeros[1] = ~|f2[26:25];                          //  2-bit 0
    assign         f1 = zeros[1]? {f2[24:0], 2'b0} : f2;
    assign         zeros[0] = ~f1[26];                              //  1-bit 0
    assign         f0 = zeros[0]? {f1[25:0], 1'b0} : f1;
    reg     [26:0] frac0;
    reg      [7:0] exp0;
    always @ * begin
       if (cal_frac[27]) begin
           frac0 = cal_frac[27:1];      // 1x.xxxxxxxxxxxxxxxxxxxxxxxx xxx
```

```
                exp0 = temp_exp + 8'h1;          // 1.xxxxxxxxxxxxxxxxxxxxxxx xxx
        end else begin
            if ((temp_exp > zeros) && (f0[26])) begin // a normalized number
                exp0 = temp_exp - zeros;
                frac0 = f0;                       // 01.xxxxxxxxxxxxxxxxxxxxxx xxx
            end else begin                        // is a denormalized number or 0
                exp0 = 0;
                if (temp_exp != 0)                // (e - 127) = ((e - 1) - 126)
                    frac0 = cal_frac[26:0] << (temp_exp - 8'h1);
                else frac0 = cal_frac[26:0];
            end
        end
    end
    wire frac_plus_1 =                       // for rounding
        ~rm[1] & ~rm[0] & frac0[2] & (frac0[1] | frac0[0]) |
        ~rm[1] & ~rm[0] & frac0[2] & ~frac0[1] & ~frac0[0] & frac0[3] |
        ~rm[1] & rm[0] & (frac0[2] | frac0[1] | frac0[0]) & sign |
        rm[1] & ~rm[0] & (frac0[2] | frac0[1] | frac0[0]) & ~sign;
    wire    [24:0] frac_round = {1'b0,frac0[26:3]} + frac_plus_1;
    wire    [7:0] exponent = frac_round[24]? exp0 + 8'h1 : exp0;
    wire          overflow = &exp0 | &exponent;
    assign s = final_result(overflow, rm, sign, is_nan, is_inf, exponent,
                            frac_round[22:0], inf_nan_frac);
    function [31:0] final_result;
        input          overflow;
        input   [1:0] rm;
        input          sign, is_nan, is_inf;
        input   [7:0] exponent;
        input   [22:0] fraction, inf_nan_frac;
        casex ({overflow, rm, sign, is_nan, is_inf})
            6'b1_00_x_0_x : final_result = {sign,8'hff,23'h000000};   // inf
            6'b1_01_0_0_x : final_result = {sign,8'hfe,23'h7fffff};   // max
            6'b1_01_1_0_x : final_result = {sign,8'hff,23'h000000};   // inf
            6'b1_10_0_0_x : final_result = {sign,8'hff,23'h000000};   // inf
            6'b1_10_1_0_x : final_result = {sign,8'hfe,23'h7fffff};   // max
            6'b1_11_x_0_x : final_result = {sign,8'hfe,23'h7fffff};   // max
            6'b0_xx_x_0_0 : final_result = {sign,exponent,fraction};  // nor
            6'bx_xx_x_1_x : final_result = {1'b1,8'hff,inf_nan_frac}; // nan
            6'bx_xx_x_0_1 : final_result = {sign,8'hff,inf_nan_frac}; // inf
            default        : final_result = {sign,8'h00,23'h000000};  // 0
        endcase
    endfunction
endmodule
```

Figure 9.13 shows the simulation waveforms of the pipelined floating-point adder. The first result is available in the third clock cycle.

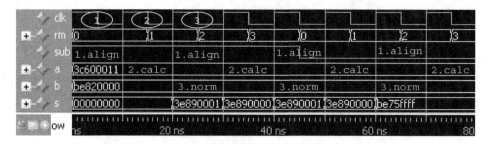

Figure 9.13 Waveform of the pipelined floating-point adder

9.4   Floating-Point Multiplier (FMUL) Design

This section describes the algorithm of floating-point multiplications, the design of a float multiplier with Wallace tree, and the design of a pipelined float multiplier.

9.4.1   Floating-Point Multiplication Algorithm

Because there is no alignment step, the algorithm of float multiplication is simpler than that of float addition/subtraction.

First, consider the float multiplication on two normalized float numbers: given $a = \{s_a, e_a, f_a\}$ and $b = \{s_b, e_b, f_b\}$ in the IEEE float format, calculate the product $c = \{s_c, e_c, f_c\} = a \times b$. The sign of c is $s_c = s_a \oplus s_b$; the absolute of c is $|c| = |a| \times |b| = (2^{e_a-127} \times 1.f_a) \times (2^{e_b-127} \times 1.f_b) = 2^{(e_a+e_b-127)-127} \times (1.f_a \times 1.f_b)$. We have $1.0 \le (1.f_a \times 1.f_b) < 4.0$. If $1.f_a \times 1.f_b < 2.0$, $e_c = e_a + e_b - 127$ and $1.f_c = 1.f_a \times 1.f_b$. Otherwise, $e_c = e_a + e_b - 127 + 1$ and $1.f_c = (1.f_a \times 1.f_b) >> 1$ (shift to the right by one bit). Because the exponent e of a normalized float number satisfies $1 \le e \le 254$, we have $-125 \le (e_a + e_b - 127) \le 381$ or $-124 \le (e_a + e_b - 127 + 1) \le 382$. That is, e_c may exceed the range 1–254. If $1 \le e_c \le 254$, the product is a normalized float number. If $e_c > 254$, the result is an ∞ ($e_c = 255$ and $f_c = 0$). When $e_c < 1$, if the result is larger than or equal to $2^{-126} \times 0.00000000000000000000001 = 2^{-149}$, the product can be represented with a denormalized float number; otherwise, it is represented with a 0.

Then, consider the float multiplication on a normalized float number and a denormalized float number. Because $1 \le e \le 254$ for a normalized number, the real exponent e' (i.e., $e - 127$) satisfies $-126 \le e' \le 127$. Suppose $a = \{s_a, e_a, f_a\}$ is a normalized float number and $b = \{s_b, e_b, f_b\}$ is a denormalized float number ($e_b = 0$, $f_b \ne 0$). The absolute of $c = a \times b$ is $|c| = |a| \times |b| = (2^{e_a-127} \times 1.f_a) \times (2^{-126} \times 0.f_b) = 2^{(e_a-253)} \times (1.f_a \times 0.f_b)$. The largest absolute is $2^{254-253} \times (2 - 2^{-23}) \times (1 - 2^{-23}) = 2 \times (2 - 3 \times 2^{-23} + 2^{-46})$, which is a normalized float number. The smallest absolute is $2^{1-253} \times 1.0 \times 2^{-23} = 2^{-275}$, it exceeds the range the e' should be in, and, therefore, the product may be represented by a denormalized float number or a 0.

Next, consider the float multiplication on two denormalized float numbers. Suppose both $a = \{s_a, e_a, f_a\}$ and $b = \{s_b, e_b, f_b\}$ are denormalized float numbers. The absolute of $c = a \times b$ is $|c| = |a| \times |b| = (2^{-126} \times 0.f_a) \times (2^{-126} \times 0.f_b) = 2^{-252} \times (0.f_a \times 0.f_b)$. It is smaller than the smallest number a denormalized float number can represent; the result is represented by a 0.

Finally, consider some special calculations: (i) if $b \ne 0$ and $b \ne$ NaN, $\infty \times b = \infty$; (ii) NaN $\times b =$ NaN; and (iii) $\infty \times 0 =$ NaN.

The following C program is helpful for the design of a float multiplier.

```
#include<stdio.h>                    // fmul_test.c, rounding test
#define Near asm volatile("fldcw _RoundNear")
#define Down asm volatile("fldcw _RoundDown")
```

```
#define Up   asm volatile("fldcw _RoundUp")
#define Chop asm volatile("fldcw _RoundChop")
int _RoundNear = 0x103f;           // round code = 00 round to nearest
int _RoundDown = 0x143f;           // round code = 01 round toward -infinity
int _RoundUp   = 0x183f;           // round code = 10 round toward +infinity
int _RoundChop = 0x1c3f;           // round code = 11 round toward 0
int main(void){
    union {
        int intword;
        float floatword;
    } u, v, s;
    while (1) {
        fprintf (stderr,"input 1st f_p number in hex format: ");
        fscanf (stdin,"%x",&u.intword);
        fprintf (stderr,"input 2nd f_p number in hex format: ");
        fscanf (stdin,"%x",&v.intword);
        Near;                              // round to nearest
        s.floatword = u.floatword * v.floatword;
        fprintf (stderr,"the prod of 2 fp numbers is (near): "
                 "%08x\n",s.intword);
        Down;                              // round toward -infinity
        s.floatword = u.floatword * v.floatword;
        fprintf (stderr,"the prod of 2 fp numbers is (down): "
                 "%08x\n",s.intword);
        Up;                                // round toward +infinity
        s.floatword = u.floatword * v.floatword;
        fprintf (stderr,"the prod of 2 fp numbers is ( up ): "
                 "%08x\n",s.intword);
        Chop;                              // round toward 0
        s.floatword = u.floatword * v.floatword;
        fprintf (stderr,"the prod of 2 fp numbers is (chop): "
                 "%08x\n",s.intword);
    }
}
```

We can compile it with the command "gcc fmul.c -o fmul" and execute it with the command "./fmul." Table 9.5 lists some execution results. In sequence, the results show the multiplications on

Table 9.5 Example results of floating-point multiplications

	1	2	3	4	5	6	7	8
a	3fc00000	00800000	7f7fffff	00800000	003fffff	7f800000	7f800000	7ff000ff
b	3fc00000	00800000	7f7fffff	3f000000	40000000	00ffffff	00000000	3f80ff00
near	40100000	00000000	7f800000	00400000	007ffffe	7f800000	ffc00000	7ff000ff
down	40100000	00000000	7f7fffff	00400000	007ffffe	7f800000	ffc00000	7ff000ff
up	40100000	00000001	7f800000	00400000	007ffffe	7f800000	ffc00000	7ff000ff
chop	40100000	00000000	7f7fffff	00400000	007ffffe	7f800000	ffc00000	7ff000ff

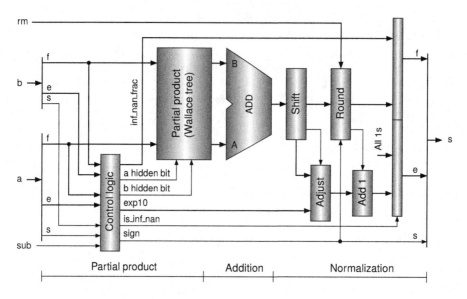

Figure 9.14   Block diagram of the floating-point multiplier

1. two normalized numbers: $1.5 \times 1.5 = 2.25$ (exponent adds 1);
2. two smallest normalized numbers, note the result with round up;
3. two largest normalized numbers, the product is $+\infty$ or the largest number;
4. two normalized numbers, the product is a denormalized float number;
5. a denormalized number and a normalized number;
6. $+\infty$ and a normalized number, the result is $+\infty$;
7. $+\infty$ and 0, the result is an NaN; and
8. an NaN and a normalized float number, the result is an NaN.

These results will be used to verify our single-precision floating-point multiplier.

9.4.2   Circuit and Verilog HDL Codes of FMUL

Figure 9.14 shows the block diagram of the single-precision floating-point multiplier. It consists of three parts: a Wallace tree for producing the partial product (sum and carry); an adder for generating the product by adding the sum and carry; and the circuits for normalization.

An 8-bit Wallace tree has been introduced in Chapter 3. Figure 9.15 shows the structure of the 24-bit Wallace tree. The AND matrix is not shown in the figure.

The following is the Verilog HDL code of a float multiplier. An input number may be a normalized number, a denormalized number, an infinity, a zero, or an NaN. The multiplication is done by a 24-bit Wallace tree. Because the Verilog HDL code of the Wallace tree is too long to be listed here, we just show its structure in Figure 9.15. Note that the result of zero times infinity is indeterminate (NaN), which is represented with 0xffc00000 in Intel x86 FPU.

```
module fmul (a,b,rm,s);                                    // fmul
    input   [31:0] a,b;                                    // fp a and b
    input   [1:0] rm;                                      // round mode
    output [31:0] s;                                       // fp output
```

```
wire          a_expo_is_00 = ~|a[30:23];                    // exp = 00
wire          b_expo_is_00 = ~|b[30:23];
wire          a_expo_is_ff =  &a[30:23];                    // exp = ff
wire          b_expo_is_ff =  &b[30:23];
wire          a_frac_is_00 = ~|a[22:0];                     // frac = 0
wire          b_frac_is_00 = ~|b[22:0];
wire          a_is_inf = a_expo_is_ff & a_frac_is_00;
wire          b_is_inf = b_expo_is_ff & b_frac_is_00;
wire          a_is_nan = a_expo_is_ff & ~a_frac_is_00;
wire          b_is_nan = b_expo_is_ff & ~b_frac_is_00;
wire          a_is_0   = a_expo_is_00 & a_frac_is_00;
wire          b_is_0   = b_expo_is_00 & b_frac_is_00;
wire          s_is_inf = a_is_inf | b_is_inf;
wire          s_is_nan = a_is_nan | (a_is_inf & b_is_0) |
                         b_is_nan | (b_is_inf & a_is_0);
wire    [22:0] nan_frac = (a[21:0] > b[21:0])?
                          {1'b1,a[21:0]} : {1'b1,b[21:0]};
wire    [22:0] inf_nan_frac = s_is_nan? nan_frac : 23'h0;
wire          sign = a[31] ^ b[31];
wire    [9:0] exp10 = {2'h0,a[30:23]} + {2'h0,b[30:23]} - 10'h7f +
                      a_expo_is_00 + b_expo_is_00;          // -126
wire    [23:0] a_frac24 = {~a_expo_is_00,a[22:0]};
wire    [23:0] b_frac24 = {~b_expo_is_00,b[22:0]};
wire    [47:0] z;
wire    [47:8] z_sum;
wire    [47:8] z_carry;
// multiplication 1 (carry and sum generated by wallace tree)
wallace_24x24 wt24 (a_frac24,b_frac24,z_sum,z_carry,z[7:0]);
// multiplication 2 (product by adding carry and sum)
assign        z[47:8] = {1'b0,z_sum} + z_carry;   // xx.xxxxxxxxxxxx...
// normalization
wire    [46:0] z5,z4,z3,z2,z1,z0;     //  x.xxxxxxxxxxxxxxxxxxxxxxxx...
wire    [5:0] zeros;
assign        zeros[5] = ~|z[46:15];                        // 32-bit 0
assign        z5 = zeros[5]? {z[14:0],32'b0} : z[46:0];
assign        zeros[4] = ~|z5[46:31];                       // 16-bit 0
assign        z4 = zeros[4]? {z5[30:0],16'b0} : z5;
assign        zeros[3] = ~|z4[46:39];                       //  8-bit 0
assign        z3 = zeros[3]? {z4[38:0], 8'b0} : z4;
assign        zeros[2] = ~|z3[46:43];                       //  4-bit 0
assign        z2 = zeros[2]? {z3[42:0], 4'b0} : z3;
assign        zeros[1] = ~|z2[46:45];                       //  2-bit 0
assign        z1 = zeros[1]? {z2[44:0], 2'b0} : z2;
assign        zeros[0] = ~z1[46];                           //  1-bit 0
assign        z0 = zeros[0]? {z1[45:0], 1'b0} : z1;
reg     [46:0] frac0; // temporary fraction
reg     [9:0] exp0;  // temporary exponent
```

```verilog
    always @ * begin
        if (z[47]) begin                    // 1x.xxxxxxxxxxxxxxxxxxxxxxxx...
            exp0 = exp10 + 10'h1;
            frac0 = z[47:1];                // 1.xxxxxxxxxxxxxxxxxxxxxxxx...
        end else begin
            if (!exp10[9] && (exp10[8:0] > zeros) && z0[46]) begin
                exp0 = exp10 - zeros;
                frac0 = z0;                 // 1.xxxxxxxxxxxxxxxxxxxxxxxx...
            end else begin                  // is a denormalized number or 0
                exp0 = 0;
                if (!exp10[9] && (exp10 != 0))            // e > 0
                    frac0 = z[46:0] << (exp10 - 10'h1);   // e-127 -> -126
                else frac0 = z[46:0] >> (10'h1 - exp10);  // e = 0 or neg
            end
        end
    end
    wire    [26:0] frac = {frac0[46:21],|frac0[20:0]};     // x.xx...xx grs
    wire          frac_plus_1 =                            // for rounding
        ~rm[1] & ~rm[0] & frac0[2] & (frac0[1] | frac0[0]) |
        ~rm[1] & ~rm[0] & frac0[2] & ~frac0[1] & ~frac0[0]  & frac0[3] |
        ~rm[1] &  rm[0] & (frac0[2] | frac0[1] | frac0[0]) & sign |
         rm[1] & ~rm[0] & (frac0[2] | frac0[1] | frac0[0]) & ~sign;
    wire    [24:0] frac_round = {1'b0,frac[26:3]} + frac_plus_1;
    wire    [9:0] exp1 = frac_round[24]? exp0 + 10'h1 : exp0;
    wire          overflow = (exp0 >= 10'h0ff) | (exp1 >= 10'h0ff);
    assign s = final_result(overflow, rm, sign, s_is_inf, s_is_nan,
                        exp1[7:0], frac_round[22:0], inf_nan_frac);
    function [31:0] final_result;
        input         overflow;
        input   [1:0] rm;
        input         sign, s_is_inf, s_is_nan;
        input   [7:0] exponent;
        input   [22:0] fraction, inf_nan_frac;
        casex ({overflow, rm, sign, s_is_nan, s_is_inf})
            6'b1_00_x_0_x : final_result = {sign,8'hff,23'h000000};    // inf
            6'b1_01_0_0_x : final_result = {sign,8'hfe,23'h7fffff};    // max
            6'b1_01_1_0_x : final_result = {sign,8'hff,23'h000000};    // inf
            6'b1_10_0_0_x : final_result = {sign,8'hff,23'h000000};    // inf
            6'b1_10_1_0_x : final_result = {sign,8'hfe,23'h7fffff};    // max
            6'b1_11_x_0_x : final_result = {sign,8'hfe,23'h7fffff};    // max
            6'b0_xx_x_0_0 : final_result = {sign,exponent,fraction};   // nor
            6'bx_xx_x_1_x : final_result = {1'b1,8'hff,inf_nan_frac};  // nan
            6'bx_xx_x_0_1 : final_result = {sign,8'hff,inf_nan_frac};  // inf
            default       : final_result = {sign,8'h00,23'h000000};    // 0
        endcase
    endfunction
endmodule
```

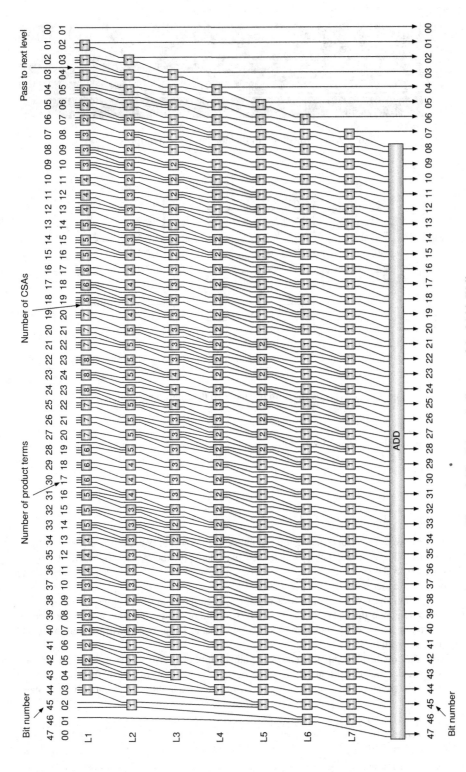

Figure 9.15 Schematic diagram of the 24-bit Wallace tree

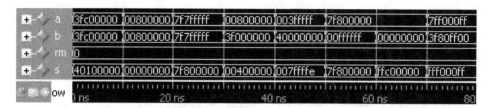

Figure 9.16   Waveform of the floating-point multiplier

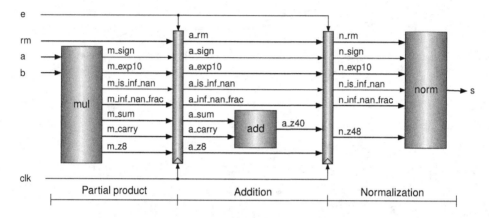

Figure 9.17   Block diagram of pipelined floating-point multiplier

Figure 9.16 shows the simulation waveforms of the float multiplier. We only show the results with the round to the nearest mode. The simulation results are the same as those in Table 9.5, which are the outputs of Intel x86 FPU.

9.4.3   Pipelined Wallace Tree FMUL Design

We use two clock cycles to perform the significand multiplication with a Wallace tree. In order to be able to execute a float multiplication on every clock cycle, we add a pipeline register to the Wallace tree, as shown in Figure 9.15. This register is inserted in between the array of the carry-save adders (CSAs) and the carry-propagate adder.

Figure 9.17 shows the block diagram of a pipelined single-precision floating-point multiplier. We divide the operations into three stages: partial product, addition, and normalization. The pipeline registers are inserted in between the stages and have a write enable signal which will be used in the CPU design to stall the pipeline.

Below is the top module of the pipelined floating-point multiplier. It invokes five modules, corresponding to the five blocks shown in Figure 9.17. The letters "m" (multiplication), "a" (addition), and "n" (normalization) are prefixed to the signal names in partial product, addition, and normalization stages, respectively.

```
module pipelined_fmul (a,b,rm,s,clk,clrn,e);        // pipelined fp mul
    input   [31:0] a, b;                            // fp a and b
    input   [1:0] rm;                               // round mode
```

```verilog
input           e;                              // enable
input           clk, clrn;                      // clock and reset
output [31:0] s;                                // fp output
wire            m_sign;
wire    [9:0] m_exp10;
wire            m_is_nan;
wire            m_is_inf;
wire    [22:0] m_inf_nan_frac;
wire    [39:0] m_sum;
wire    [39:0] m_carry;
wire    [7:0] m_z8;
// exe1: partial product stage (carry and sum generated by wallace tree)
fmul_mul mul1 (a,b,m_sign,m_exp10,m_is_nan,m_is_inf,m_inf_nan_frac,
               m_sum,m_carry,m_z8);
wire    [1:0] a_rm;
wire            a_sign;
wire    [9:0] a_exp10;
wire            a_is_nan;
wire            a_is_inf;
wire    [22:0] a_inf_nan_frac;
wire    [39:0] a_sum;
wire    [39:0] a_carry;
wire    [7:0] a_z8;
// pipeline register
reg_mul_add reg_ma(rm,m_sign,m_exp10,m_is_nan,m_is_inf,m_inf_nan_frac,
                   m_sum,m_carry,m_z8,clk,clrn,e, a_rm,a_sign,a_exp10,
                   a_is_nan,a_is_inf,a_inf_nan_frac,a_sum,a_carry,a_z8);
wire    [47:8] a_z40;
// exe2: addition stage (product by adding carry and sum)
fmul_add mul2 (a_sum,a_carry,a_z40);
wire    [47:0] a_z48 = {a_z40,a_z8};
wire    [1:0] n_rm;
wire            n_sign;
wire    [9:0] n_exp10;
wire            n_is_nan;
wire            n_is_inf;
wire    [22:0] n_inf_nan_frac;
wire    [47:0] n_z48;
// pipeline register
reg_add_norm reg_an (a_rm,a_sign,a_exp10,a_is_nan,a_is_inf,
                     a_inf_nan_frac,a_z48,clk,clrn,e, n_rm,n_sign,
                     n_exp10,n_is_nan,n_is_inf,n_inf_nan_frac,n_z48);
// exe3: normalization stage
fmul_norm mul3 (n_rm,n_sign,n_exp10,n_is_nan,n_is_inf,n_inf_nan_frac,
                n_z48,s);
endmodule
```

Below is the Verilog HDL code of the partial product stage.

```
module fmul_mul (a,b,sign,exp10,s_is_nan,s_is_inf,inf_nan_frac,z_sum,
                 z_carry,z8);                                    // mul stage
    input   [31:0] a, b;
    output  [39:0] z_sum;
    output  [39:0] z_carry;
    output  [22:0] inf_nan_frac;
    output   [9:0] exp10;
    output   [7:0] z8;
    output         sign;
    output         s_is_nan;
    output         s_is_inf;
    wire           a_expo_is_00 = ~|a[30:23];                    // exp = 00
    wire           b_expo_is_00 = ~|b[30:23];
    wire           a_expo_is_ff =  &a[30:23];                    // exp = ff
    wire           b_expo_is_ff =  &b[30:23];
    wire           a_frac_is_00 = ~|a[22:0];                     // frac = 0
    wire           b_frac_is_00 = ~|b[22:0];
    wire           a_is_inf = a_expo_is_ff & a_frac_is_00;
    wire           b_is_inf = b_expo_is_ff & b_frac_is_00;
    wire           a_is_nan = a_expo_is_ff & ~a_frac_is_00;
    wire           b_is_nan = b_expo_is_ff & ~b_frac_is_00;
    wire           a_is_0   = a_expo_is_00 & a_frac_is_00;
    wire           b_is_0   = b_expo_is_00 & b_frac_is_00;
    assign         s_is_inf = a_is_inf |  b_is_inf;
    assign         s_is_nan = a_is_nan | (a_is_inf & b_is_0) |
                              b_is_nan | (b_is_inf & a_is_0);
    wire    [22:0] nan_frac = (a[21:0] > b[21:0])?
                              {1'b1,a[21:0]} : {1'b1,b[21:0]};
    assign         inf_nan_frac = s_is_nan? nan_frac : 23'h0;
    assign         sign = a[31] ^ b[31];
    assign         exp10 = {2'h0,a[30:23]} + {2'h0,b[30:23]} - 10'h7f +
                           a_expo_is_00 + b_expo_is_00;          // -126
    wire    [23:0] a_frac24 = {~a_expo_is_00,a[22:0]};
    wire    [23:0] b_frac24 = {~b_expo_is_00,b[22:0]};
    wallace_24x24 wt24 (a_frac24,b_frac24,z_sum,z_carry,z8);
endmodule
```

Below is the Verilog HDL code of the pipeline registers in between the partial product and addition stages.

```
module reg_mul_add (m_rm,m_sign,m_exp10,m_is_nan,m_is_inf,m_inf_nan_frac,
                    m_sum,m_carry,m_z8,clk,clrn,e,a_rm,a_sign,a_exp10,
                    a_is_nan,a_is_inf,a_inf_nan_frac,a_sum,a_carry,a_z8);
    input      [39:0] m_sum;                                // partial mul stage
    input      [39:0] m_carry;
```

```
    input       [22:0] m_inf_nan_frac;
    input        [9:0] m_exp10;
    input        [7:0] m_z8;
    input        [1:0] m_rm;
    input              m_sign;
    input              m_is_nan;
    input              m_is_inf;
    input              e;                           // enable
    input              clk, clrn;                   // clock and reset
    output reg [39:0] a_sum;                        // addition stage
    output reg [39:0] a_carry;
    output reg [22:0] a_inf_nan_frac;
    output reg  [9:0] a_exp10;
    output reg  [7:0] a_z8;
    output reg  [1:0] a_rm;
    output reg         a_sign;
    output reg         a_is_nan;
    output reg         a_is_inf;
    always @ (posedge clk or negedge clrn) begin
        if (!clrn) begin
            a_rm            <= 0;
            a_sign          <= 0;
            a_exp10         <= 0;
            a_is_nan        <= 0;
            a_is_inf        <= 0;
            a_inf_nan_frac  <= 0;
            a_sum           <= 0;
            a_carry         <= 0;
            a_z8            <= 0;
        end else if (e) begin
            a_rm            <= m_rm;
            a_sign          <= m_sign;
            a_exp10         <= m_exp10;
            a_is_nan        <= m_is_nan;
            a_is_inf        <= m_is_inf;
            a_inf_nan_frac  <= m_inf_nan_frac;
            a_sum           <= m_sum;
            a_carry         <= m_carry;
            a_z8            <= m_z8;
        end
    end
endmodule
```

Below is the Verilog HDL code of the addition stage.

```
module fmul_add (z_sum, z_carry, z);                        // fmul add
    input   [39:0] z_sum;
```

```
    input    [39:0] z_carry;
    output   [47:8] z;
    assign          z = z_sum + z_carry;
endmodule
```

Below is the Verilog HDL code of the pipeline registers in between the addition and normalization stages.

```
module reg_add_norm (a_rm,a_sign,a_exp10,a_is_nan,a_is_inf,a_inf_nan_frac,
                     a_z48,clk,clrn,e,n_rm,n_sign,n_exp10,n_is_nan,
                     n_is_inf,n_inf_nan_frac,n_z48);  // pipeline register
    input       [47:0] a_z48;                         // addition stage
    input       [22:0] a_inf_nan_frac;
    input        [9:0] a_exp10;
    input        [1:0] a_rm;
    input              a_sign;
    input              a_is_nan;
    input              a_is_inf;
    input              e;                              // e: enable
    input              clk, clrn;                      // clock and reset
    output reg  [47:0] n_z48;                          // normalization stage
    output reg  [22:0] n_inf_nan_frac;
    output reg   [9:0] n_exp10;
    output reg   [1:0] n_rm;
    output reg         n_sign;
    output reg         n_is_nan;
    output reg         n_is_inf;
    always @ (posedge clk or negedge clrn) begin
        if (!clrn) begin
            n_rm           <= 0;
            n_sign         <= 0;
            n_exp10        <= 0;
            n_is_nan       <= 0;
            n_is_inf       <= 0;
            n_inf_nan_frac <= 0;
            n_z48          <= 0;
        end else if (e) begin
            n_rm           <= a_rm;
            n_sign         <= a_sign;
            n_exp10        <= a_exp10;
            n_is_nan       <= a_is_nan;
            n_is_inf       <= a_is_inf;
            n_inf_nan_frac <= a_inf_nan_frac;
            n_z48          <= a_z48;
        end
    end
endmodule
```

Below is the Verilog HDL code of the normalization stage.

```verilog
module fmul_norm (rm,sign,exp10,is_nan,is_inf,inf_nan_frac,z,s);// fmul norm
    input   [47:0]  z;                        // xx.xxxxxxxxxxxxxxxxxxxxxxxxxx...
    input   [22:0]  inf_nan_frac;
    input   [9:0]   exp10;
    input   [1:0]   rm;
    input           sign;
    input           is_nan;
    input           is_inf;
    output  [31:0]  s;
    wire    [46:0]  z5,z4,z3,z2,z1,z0;        // x.xxxxxxxxxxxxxxxxxxxxxxxxxx...
    wire    [5:0]   zeros;
    assign          zeros[5] = ~|z[46:15];                        // 32-bit 0
    assign          z5 = zeros[5]? {z[14:0],32'b0} : z[46:0];
    assign          zeros[4] = ~|z5[46:31];                       // 16-bit 0
    assign          z4 = zeros[4]? {z5[30:0],16'b0} : z5;
    assign          zeros[3] = ~|z4[46:39];                       //  8-bit 0
    assign          z3 = zeros[3]? {z4[38:0], 8'b0} : z4;
    assign          zeros[2] = ~|z3[46:43];                       //  4-bit 0
    assign          z2 = zeros[2]? {z3[42:0], 4'b0} : z3;
    assign          zeros[1] = ~|z2[46:45];                       //  2-bit 0
    assign          z1 = zeros[1]? {z2[44:0], 2'b0} : z2;
    assign          zeros[0] = ~z1[46];                           //  1-bit 0
    assign          z0 = zeros[0]? {z1[45:0], 1'b0} : z1;
    reg     [46:0]  frac0;                           // temporary fraction
    reg     [9:0]   exp0;                            // temporary exponent
    always @ * begin
        if (z[47]) begin                      // 1x.xxxxxxxxxxxxxxxxxxxxxxxxxx...
            exp0 = exp10 + 10'h1;
            frac0 = z[47:1];                  //  1.xxxxxxxxxxxxxxxxxxxxxxxxxx...
        end else begin
            if (!exp10[9] && (exp10[8:0] > zeros) && z0[46]) begin
                exp0 = exp10 - zeros;
                frac0 = z0;                   //  1.xxxxxxxxxxxxxxxxxxxxxxxxxx...
            end else begin                    // is a denormalized number or 0
                exp0 = 0;
                if (!exp10[9] && (exp10 != 0))        // e > 0
                    frac0 = z[46:0] << (exp10 - 10'h1);   // e-127 -> -126
                else frac0 = z[46:0] >> (10'h1 - exp10);  // e = 0 or neg
            end
        end
    end
    wire    [26:0]  frac = {frac0[46:21],|frac0[20:0]};   // x.xx...xx grs
    wire            frac_plus_1 =                        // for rounding
        ~rm[1] & ~rm[0] &  frac0[2] & (frac0[1] |  frac0[0]) |
        ~rm[1] & ~rm[0] &  frac0[2] & ~frac0[1] & ~frac0[0] & frac0[3] |
```

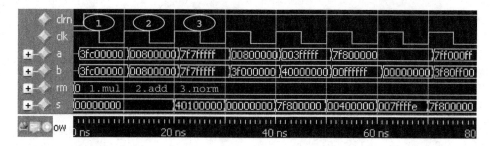

Figure 9.18   Waveform of the pipelined floating-point multiplier

```
        ~rm[1] & rm[0] & (frac0[2] | frac0[1] |   frac0[0]) & sign |
         rm[1] & ~rm[0] & (frac0[2] | frac0[1] |   frac0[0]) & ~sign;
wire   [24:0] frac_round = {1'b0,frac[26:3]} + frac_plus_1;
wire   [9:0] exp1 = frac_round[24]? exp0 + 10'h1 : exp0;
wire        overflow = (exp0 >= 10'h0ff) | (exp1 >= 10'h0ff);
assign s = final_result(overflow, rm, sign, is_nan, is_inf, exp1[7:0],
                        frac_round[22:0], inf_nan_frac);
function  [31:0] final_result;
    input        overflow;
    input  [1:0] rm;
    input        sign, is_nan, is_inf;
    input  [7:0] exponent;
    input  [22:0] fraction,inf_nan_frac;
    casex ({overflow,rm,sign,is_nan,is_inf})
        6'b1_00_x_0_x : final_result = {sign,8'hff,23'h000000};   // inf
        6'b1_01_0_0_x : final_result = {sign,8'hfe,23'h7fffff};   // max
        6'b1_01_1_0_x : final_result = {sign,8'hff,23'h000000};   // inf
        6'b1_10_0_0_x : final_result = {sign,8'hff,23'h000000};   // inf
        6'b1_10_1_0_x : final_result = {sign,8'hfe,23'h7fffff};   // max
        6'b1_11_x_0_x : final_result = {sign,8'hfe,23'h7fffff};   // max
        6'b0_xx_x_0_0 : final_result = {sign,exponent,fraction};  // nor
        6'bx_xx_x_1_x : final_result = {1'b1,8'hff,inf_nan_frac}; // nan
        6'bx_xx_x_0_1 : final_result = {sign,8'hff,inf_nan_frac}; // inf
        default       : final_result = {sign,8'h00,23'h000000};   // 0
    endcase
    endfunction
endmodule
```

Figure 9.18 shows the simulation waveforms of the pipelined floating-point multiplier. The first result is available in the third clock cycle. The partial product and addition are performed in the first cycle and the second cycle, respectively.

9.5   Floating-Point Divider (FDIV) Design

This section describes the float division algorithm and the design of the float divider which uses the Newton–Raphson algorithm to implement the significand division.

9.5.1   Floating-Point Division Algorithm

The float division algorithm is similar to that of float multiplication with two different points: one is calculating the result exponent with a subtraction, and the other is calculating the result fraction with a division.

First, consider the float division on two normalized float numbers: given $a = \{s_a, e_a, f_a\}$ and $b = \{s_b, e_b, f_b\}$ in the IEEE 754 float format, calculate $c = \{s_c, e_c, f_c\} = a/b$. The sign of c is $s_c = s_a \oplus s_b$; the absolute of c is $|c| = |a|/|b| = (2^{e_a-127} \times 1.f_a)/(2^{e_b-127} \times 1.f_b) = 2^{(e_a-e_b+127)-127} \times (1.f_a/1.f_b)$. We have $0.5 < (1.f_a/1.f_b) < 2.0$. If $1.f_a/1.f_b \geq 1.0$, $e_c = e_a - e_b + 127$ and $1.f_c = 1.f_a/1.f_b$. Otherwise, $e_c = e_a - e_b + 127 - 1$ and $1.f_c = (1.f_a/1.f_b) \ll 1$ (shift to the left by one bit). Because the exponent e of a normalized float number satisfies $1 \leq e \leq 254$, we have $-126 \leq (e_a - e_b + 127) \leq 380$ or $-127 \leq (e_a - e_b + 127 - 1) \leq 379$. That is, e_c may exceed the range 1–254. If $1 \leq e_c \leq 254$, the quotient is a normalized float number. If $e_c > 254$, the result is an ∞ ($e_c = 255$ and $f_c = 0$). When $e_c < 1$, if the result is larger than or equal to $2^{-126} \times 0.00000000000000000000001 = 2^{-149}$, the quotient can be represented with a denormalized float number; otherwise, it is represented with a 0.

Second, consider dividing a normalized float number by a denormalized float number. Because $1 \leq e \leq 254$ for a normalized number, the real exponent e' (i.e., $e - 127$) satisfies $-126 \leq e' \leq 127$. Suppose $a = \{s_a, e_a, f_a\}$ is a normalized float number and $b = \{s_b, e_b, f_b\}$ is a denormalized number ($e_b = 0$, $f_b \neq 0$). The absolute of $c = a/b$ is $|c| = |a|/|b| = (2^{e_a-127} \times 1.f_a)/(2^{-126} \times 0.f_b) = 2^{e_a-1} \times (1.f_a/0.f_b)$. The largest absolute is $2^{254-1} \times (2 - 2^{-23})/2^{-23} = 2^{276} \times (2 - 2^{-23})$; it is larger than the largest representable number. The result is represented with an ∞. The smallest absolute is $2^{1-1} \times 1.0/(1 - 2^{-23})$; it is a normalized number.

Then, consider dividing a denormalized float number by a normalized float number. Suppose $a = \{s_a, e_a, f_a\}$ is a denormalized float number ($e_a = 0$, $f_a \neq 0$) and $b = \{s_b, e_b, f_b\}$ is a normalized number. The largest absolute of $c = a/b$ is $|c| = |a|/|b| = (2^{-126} \times (1 - 2^{-23}))/(2^{1-127} \times 1.0) = 1 - 2^{-23}$; it is a normalized number. The smallest absolute is $(2^{-126} \times 2^{-23}/(2^{254-127} \times (2 - 2^{-23})) = 2^{-276}/(2 - 2^{-23})$; the result is represented with a 0. That is, the quotient in this case may be a normalized number, a denormalized number, or a 0.

Next, consider the float division on two denormalized float numbers. Suppose both $a = \{s_a, e_a, f_a\}$ and $b = \{s_b, e_b, f_b\}$ are denormalized float numbers. The absolute of $c = a/b$ is $|c| = |a|/|b| = (2^{-126} \times 0.f_a)/(2^{-126} \times 0.f_b) = 0.f_a/0.f_b$. No matter what the values f_a and f_b are, the result is a normalized float number.

Finally, consider some special calculations: (i) if $a \neq \infty$ and $a \neq$ NaN, $a/\infty = 0$; (ii) if $a \neq 0$ and $a \neq$ NaN, $a/0 = \infty$; (iii) $0/0 =$ NaN; (iv) $\infty/\infty =$ NaN; and (v) NaN$/b =$ NaN.

9.5.2   Circuit and Verilog HDL Codes of FDIV

We use Newton–Raphson algorithm to perform the significand division, which was already described in Chapter 3. Figure 9.19 shows the block diagram of the single-precision floating-point divider.

The Newton–Raphson division algorithm consists of two parts: (i) iterative calculation: $x_{i+1} = x_i(2 - x_i b)$ and (ii) quotient calculation: $q = a \times x_n$. The second part performs a multiplication, which can be implemented with a pipelined Wallace tree. Therefore, the operation of the float division takes the following four steps:

1. Newton iteration, which calculates x_n;
2. partial product, which calculates carry and sum with Wallace CSA array;
3. addition, which calculates the product by adding the carry and sum; and
4. normalization, which generates the final result in the IEEE float format.

Steps 2–4 are done in the pipeline manner, and step 1 is done in an iterative manner. Figure 9.20 shows the pipeline stages of the floating-point divider. During the Newton iteration in the ID (instruction decode)

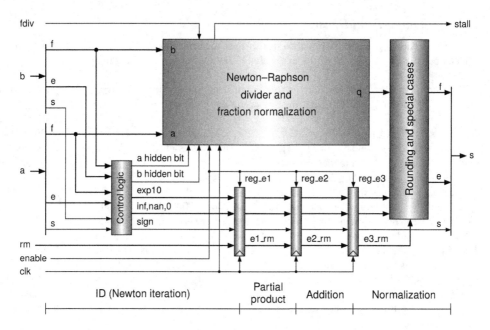

Figure 9.19 Block diagram of the floating-point divider

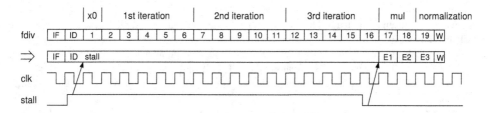

Figure 9.20 Pipeline stages of the floating-point divider

stage, we stall the pipeline. Three iterations are carried out. Each iteration takes five clock cycles to calculate $x_{i+1} = x_i(2 - x_i b)$, which consists of two multiplications and one subtraction. The multiplication is done with the Wallace tree, which takes two clock cycles and the subtraction takes one cycle.

The following Verilog HDL code implements the float divider. It invokes a module of new-ton24, which implements Newton–Raphson algorithm for the significand division. Because the Newton–Raphson algorithm requires the normalized dividend and divisor (the most significant bit must be a 1) and our divider allows denormalized inputs, we use the module shift_to_msb_equ_1 to shift fractions to the normalized format for the inputs to the newton24 module.

```
module fdiv_newton (a,b,rm,fdiv,ena,clk,clrn,  s,busy,stall,count,reg_x);
    input   [31:0] a,b;                          // fp s = a / b
    input   [1:0]  rm;                           // round mode
    input          fdiv;                         // ID stage: fdiv = i_fdiv
    input          ena;                          // enable
    input          clk, clrn;                    // clock and reset
```

```verilog
output [31:0] s;                            // fp output
output [25:0] reg_x;                        // x_i
output  [4:0] count;                        // for iteration control
output        busy;                         // for generating stall
output        stall;                        // for pipeline stall
parameter ZERO = 31'h00000000;
parameter INF  = 31'h7f800000;
parameter NaN  = 31'h7fc00000;
parameter MAX  = 31'h7f7fffff;
wire          a_expo_is_00 = ~|a[30:23];    // a_expo = 00
wire          b_expo_is_00 = ~|b[30:23];    // b_expo = 00
wire          a_expo_is_ff = &a[30:23];     // a_expo = ff
wire          b_expo_is_ff = &b[30:23];     // b_expo = ff
wire          a_frac_is_00 = ~|a[22:0];     // a_frac = 00
wire          b_frac_is_00 = ~|b[22:0];     // b_frac = 00
wire          sign = a[31] ^ b[31];
wire    [9:0] exp_10 = {2'h0,a[30:23]} - {2'h0,b[30:23]} + 10'h7f;
wire   [23:0] a_temp24 = a_expo_is_00? {a[22:0],1'b0} : {1'b1,a[22:0]};
wire   [23:0] b_temp24 = b_expo_is_00? {b[22:0],1'b0} : {1'b1,b[22:0]};
wire   [23:0] a_frac24,b_frac24;   // to 1xx...x for denormalized number
wire    [4:0] shamt_a,shamt_b;     // how many bits shifted
shift_to_msb_equ_1 shift_a (a_temp24,a_frac24,shamt_a);   // to 1xx...xx
shift_to_msb_equ_1 shift_b (b_temp24,b_frac24,shamt_b);   // to 1xx...xx
wire    [9:0] exp10 = exp_10 - shamt_a + shamt_b;
reg           e1_sign,e1_ae00,e1_aeff,e1_af00,e1_be00,e1_beff,e1_bf00;
reg           e2_sign,e2_ae00,e2_aeff,e2_af00,e2_be00,e2_beff,e2_bf00;
reg           e3_sign,e3_ae00,e3_aeff,e3_af00,e3_be00,e3_beff,e3_bf00;
reg     [1:0] e1_rm,e2_rm,e3_rm;
reg     [9:0] e1_exp10,e2_exp10,e3_exp10;
always @ (negedge clrn or posedge clk)
  if (!clrn) begin   // 3 pipeline registers: reg_e1, reg_e2, and reg_e3
      // reg_e1                   // reg_e2                 // reg_e3
      e1_sign <= 0;               e2_sign <= 0;           e3_sign <= 0;
      e1_rm   <= 0;               e2_rm   <= 0;           e3_rm   <= 0;
      e1_exp10<= 0;               e2_exp10<= 0;           e3_exp10<= 0;
      e1_ae00 <= 0;               e2_ae00 <= 0;           e3_ae00 <= 0;
      e1_aeff <= 0;               e2_aeff <= 0;           e3_aeff <= 0;
      e1_af00 <= 0;               e2_af00 <= 0;           e3_af00 <= 0;
      e1_be00 <= 0;               e2_be00 <= 0;           e3_be00 <= 0;
      e1_beff <= 0;               e2_beff <= 0;           e3_beff <= 0;
      e1_bf00 <= 0;               e2_bf00 <= 0;           e3_bf00 <= 0;
  end else if (ena) begin
      e1_sign <= sign;            e2_sign <= e1_sign;  e3_sign <= e2_sign;
      e1_rm   <= rm;              e2_rm   <= e1_rm;    e3_rm   <= e2_rm;
      e1_exp10<= exp10;           e2_exp10<= e1_exp10; e3_exp10<= e2_exp10;
      e1_ae00 <= a_expo_is_00;    e2_ae00 <= e1_ae00;  e3_ae00 <= e2_ae00;
      e1_aeff <= a_expo_is_ff;    e2_aeff <= e1_aeff;  e3_aeff <= e2_aeff;
```

```
        e1_af00 <= a_frac_is_00; e2_af00 <= e1_af00;  e3_af00 <= e2_af00;
        e1_be00 <= b_expo_is_00; e2_be00 <= e1_be00;  e3_be00 <= e2_be00;
        e1_beff <= b_expo_is_ff; e2_beff <= e1_beff;  e3_beff <= e2_beff;
        e1_bf00 <= b_frac_is_00; e2_bf00 <= e1_bf00;  e3_bf00 <= e2_bf00;
    end
wire    [31:0] q;                      // af24 / bf24 = 1.xxxxx...x or 0.1xxxx...x
newton24 frac_newton (a_frac24,b_frac24,fdiv,ena,clk,clrn,
                      q,busy,count,reg_x,stall);
wire    [31:0] z0 = q[31] ? q : {q[30:0],1'b0};      // 1.xxxxx...x
wire    [9:0] exp_adj = q[31] ? e3_exp10 : e3_exp10 - 10'b1;
reg     [9:0] exp0;
reg     [31:0] frac0;
always @ * begin
    if (exp_adj[9]) begin                            // exp is negative
        exp0 = 0;
        if (z0[31])                                  // 1.xx...x
          frac0 = z0 >> (10'b1 - exp_adj);           // denormalized (-126)
        else frac0 = 0;
    end else if (exp_adj == 0) begin                 // exp is 0
        exp0 = 0;
        frac0 = {1'b0,z0[31:2],|z0[1:0]};            // denormalized (-126)
    end else begin                                   // exp > 0
        if (exp_adj > 254) begin                     // inf
            exp0 = 10'hff;
            frac0 = 0;
        end else begin                               // normalized
            exp0 = exp_adj;
            frac0 = z0;
        end
    end
end
wire    [26:0] frac = {frac0[31:6],|frac0[5:0]};     // sticky
wire           frac_plus_1 =
    ~e3_rm[1] & ~e3_rm[0] &  frac[3] &  frac[2] & ~frac[1] & ~frac[0] |
    ~e3_rm[1] & ~e3_rm[0] &  frac[2] & (frac[1] |  frac[0]) |
    ~e3_rm[1] &  e3_rm[0] & (frac[2] |  frac[1] |  frac[0]) & e3_sign |
     e3_rm[1] & ~e3_rm[0] & (frac[2] |  frac[1] |  frac[0]) & ~e3_sign;
wire    [24:0] frac_round = {1'b0,frac[26:3]} + frac_plus_1;
wire    [9:0] exp1 = frac_round[24]? exp0 + 10'h1 : exp0;
wire           overflow = (exp1 >= 10'h0ff);         // overflow
wire    [7:0] exponent;
wire    [22:0] fraction;
assign {exponent,fraction} = final_result(overflow,e3_rm,e3_sign,
                             e3_ae00,e3_aeff,e3_af00,e3_be00,e3_beff,
                             e3_bf00,{exp1[7:0],frac_round[22:0]});
assign         s = {e3_sign,exponent,fraction};
function [30:0] final_result;
```

```verilog
    input         overflow;
    input   [1:0] e3_rm;
    input         e3_sign;
    input         a_e00,a_eff,a_f00, b_e00,b_eff,b_f00;
    input [30:0] calc;
    casex ({overflow,e3_rm,e3_sign,a_e00,a_eff,a_f00,b_e00,b_eff,b_f00})
        10'b100x_xxx_xxx : final_result = INF;    // overflow
        10'b1010_xxx_xxx : final_result = MAX;    // overflow
        10'b1011_xxx_xxx : final_result = INF;    // overflow
        10'b1100_xxx_xxx : final_result = INF;    // overflow
        10'b1101_xxx_xxx : final_result = MAX;    // overflow
        10'b111x_xxx_xxx : final_result = MAX;    // overflow
        10'b0xxx_010_xxx : final_result = NaN;    // NaN / any
        10'b0xxx_011_010 : final_result = NaN;    // inf / NaN
        10'b0xxx_100_010 : final_result = NaN;    // den / NaN
        10'b0xxx_101_010 : final_result = NaN;    //   0 / NaN
        10'b0xxx_00x_010 : final_result = NaN;    // nor / NaN
        10'b0xxx_011_011 : final_result = NaN;    // inf / inf
        10'b0xxx_100_011 : final_result = ZERO;   // den / inf
        10'b0xxx_101_011 : final_result = ZERO;   //   0 / inf
        10'b0xxx_00x_011 : final_result = ZERO;   // nor / inf
        10'b0xxx_011_101 : final_result = INF;    // inf / 0
        10'b0xxx_100_101 : final_result = INF;    // den / 0
        10'b0xxx_101_101 : final_result = NaN;    //   0 / 0
        10'b0xxx_00x_101 : final_result = INF;    // nor / 0
        10'b0xxx_011_100 : final_result = INF;    // inf / den
        10'b0xxx_100_100 : final_result = calc;   // den / den
        10'b0xxx_101_100 : final_result = ZERO;   //   0 / den
        10'b0xxx_00x_100 : final_result = calc;   // nor / den
        10'b0xxx_011_00x : final_result = INF;    // inf / nor
        10'b0xxx_100_00x : final_result = calc;   // den / nor
        10'b0xxx_101_00x : final_result = ZERO;   //   0 / nor
        10'b0xxx_00x_00x : final_result = calc;   // nor / nor
        default          : final_result = ZERO;
    endcase
  endfunction
endmodule
```

The following is the Verilog HDL code of shift_to_msb_equ_1 module that shifts the input a to the left to generate the output b whose most significant bit is a 1. The output signal sa indicates the shift amount that is used for the exponent adjusting.

```verilog
module shift_to_msb_equ_1 (a,b,sa);      // shift left until msb = 1
    input   [23:0] a;                    // shift a = xxx...xx
    output  [23:0] b;                    // to    b = 1xx...xx
    output   [4:0] sa;                   // how many bits shifted
    wire    [23:0] a5,a4,a3,a2,a1,a0;
    assign a5 = a;
```

```
    assign sa[4]    = ~|a5[23:08];                  // 16-bit 0
    assign a4 = sa[4]? {a5[07:00],16'b0}  : a5;
    assign sa[3]    = ~|a4[23:16];                  //  8-bit 0
    assign a3 = sa[3]? {a4[15:00],  8'b0}  : a4;
    assign sa[2]    = ~|a3[23:20];                  //  4-bit 0
    assign a2 = sa[2]? {a3[19:00],  4'b0}  : a3;
    assign sa[1]    = ~|a2[23:22];                  //  2-bit 0
    assign a1 = sa[1]? {a2[21:00],  2'b0}  : a2;
    assign sa[0]    = ~a1[23];                      //  1-bit 0
    assign a0 = sa[0]? {a1[22:00],  1'b0}  : a1;
    assign b = a0;
endmodule
```

Figure 9.21 shows the schematic diagram of the Newton–Raphson divider. The circuit is similar to the 32-bit Newton–Raphson divider that was described in Chapter 3 and shown in Figure 3.26, but here we used the system clock and added the pipeline registers to calculate $q = a \times x_n$.

We use a counter to control the operation of the divider. Referring to Figure 9.22, initially the counter contains a 0. On receiving a float division (fdiv) instruction, a signal busy is set to active to indicate that the circuit is busy for the Newton–Raphson iterations. It will become inactive whenever the iterations finish. The stall signal is used to stall the pipeline.

Below is the Verilog HDL code of the newton24 module that implements the circuit in Figure 9.21. The two multiplications for the iteration of $x_{i+1} = x_i(2 - x_i b)$ are done by the modules wallace_26x24_product and wallace_26x26_product, which we give later. The calculation of $q = a \times x_n$ is done with a two-stage pipeline. The partial product is done by the module wallace_24x26 in the first stage. The quotient q is calculated by an adder in the second stage. Rounding is not necessary here because in FPU, there is a normalization stage to generate a final result in the IEEE float format.

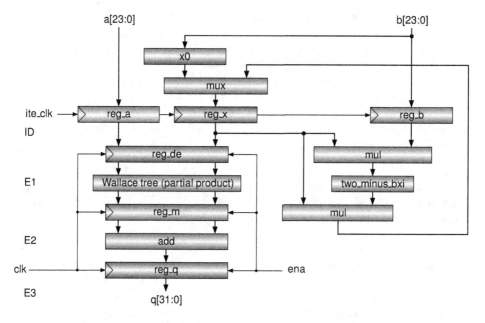

Figure 9.21   Schematic diagram of the 24-bit Newton–Raphson divider

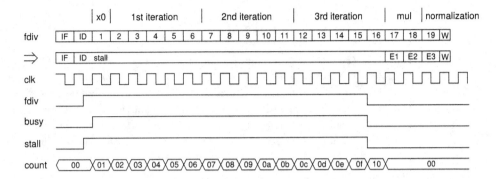

Figure 9.22   Stall caused by the division instruction

```
module newton24 (a,b,fdiv,ena,clk,clrn,q,busy,count,reg_x,stall);
    input   [23:0] a;                        // dividend: .1xxx...x
    input   [23:0] b;                        // divisor:  .1xxx...x
    input          fdiv;                     // ID stage: i_fdiv
    input          clk, clrn;                // clock and reset
    input          ena;                      // enable, save partial product
    output  [31:0] q;                        // quotient: x.xxxxx...x
    output reg     busy;                     // cannot receive new div
    output  [4:0]  count;                    // counter
    output  [25:0] reg_x;                    // for sim test only 01.xx...x
    output         stall;                    // for pipeline stall
    reg     [31:0] q;                        // 32-bit:  x.xxxxx...x
    reg     [25:0] reg_x;                    // 26-bit: xx.xxxxx...x
    reg     [23:0] reg_a;                    // 24-bit:   .1xxxx...x
    reg     [23:0] reg_b;                    // 24-bit:   .1xxxx...x
    reg     [4:0]  count;                    // 3 iterations * 5 cycles
    wire    [49:0] bxi;                      //   xx.xxxxx...x
    wire    [51:0] x52;                      //  xxx.xxxxx...x
    wire    [49:0] d_x;                      //   0x.xxxxx...x
    wire    [31:0] e2p;                      // sticky
    wire    [7:0]  x0 = rom(b[22:19]);       // x0: from rom table
    always @ (posedge clk or negedge clrn) begin
        if (!clrn) begin
            busy  <= 0;
            count <= 0;
            reg_x <= 0;
        end else begin
            if (fdiv & (count == 0)) begin   // do once only
                count <= 5'b1;               // set count
                busy  <= 1'b1;               // set to busy
            end else begin                   // 3 iterations
                if (count == 5'h01) begin
                    reg_a <= a;              //   .1xxxx...x
```

```
                    reg_b <= b;                                  //   .1xxxx...x
                    reg_x <= {2'b1,x0,16'b0};                    // 01.xxxx0...0
                end
            if  (count != 0) count <= count + 5'b1;   // count++
            if  (count == 5'h0f) busy <= 0;           // ready for next
            if  (count == 5'h10) count <= 5'b0;       // reset count
            if ((count == 5'h06) ||    // save result of 1st iteration
                (count == 5'h0b) ||    // save result of 2nd iteration
                (count == 5'h10))      // no need to save here actually
                    reg_x <= x52[50:25];   // xx.xxxxx...x
        end
    end
end
assign stall = fdiv & (count == 0) | busy;
// wallace_26x24_product    (a,    b,    z);
wallace_26x24_product bxxi (reg_b,reg_x,bxi);        //          xi * b
wire    [25:0] b26 = ~bxi[48:23] + 1'b1;             //         2 - xi * b
wallace_26x26_product xip1 (reg_x,b26,x52);          // xi * (2 - xi * b)
reg     [25:0] reg_de_x; // pipeline register in between id and e1, x
reg     [23:0] reg_de_a; // pipeline register in between id and e1, a
wire    [49:0] m_s;        // sum
wire    [49:8] m_c;        // carry
// wallace_24x26 (a,         b,          x,          y,  z);
wallace_24x26 wt (reg_de_a,reg_de_x,m_s[49:8],m_c,m_s[7:0]); // a * xn
reg     [49:0] a_s;      // pipeline register in between e1 and e2, sum
reg     [49:8] a_c;      // pipeline register in between e1 and e2, carry
assign d_x = {1'b0,a_s} + {a_c,8'b0};                // 0x.xxxxx...x
assign e2p = {d_x[48:18],|d_x[17:0]};                // sticky
always @ (negedge clrn or posedge clk)
  if (!clrn) begin                              // pipeline registers
      reg_de_x <= 0;                  reg_de_a <= 0;        // id-e1
      a_s       <= 0;                  a_c      <= 0;        // e1-e2
      q         <= 0;                                       // e2-e3
  end else if (ena) begin     // x52[50:25]: the result of 3rd iteration
      reg_de_x <= x52[50:25];          reg_de_a <= reg_a;   // id-e1
      a_s       <= m_s;                a_c      <= m_c;      // e1-e2
      q         <= e2p;                                     // e2-e3
  end
function [7:0] rom;                                    // a rom table
    input [3:0] b;
    case (b)
        4'h0: rom = 8'hff;           4'h1: rom = 8'hdf;
        4'h2: rom = 8'hc3;           4'h3: rom = 8'haa;
        4'h4: rom = 8'h93;           4'h5: rom = 8'h7f;
        4'h6: rom = 8'h6d;           4'h7: rom = 8'h5c;
        4'h8: rom = 8'h4d;           4'h9: rom = 8'h3f;
        4'ha: rom = 8'h33;           4'hb: rom = 8'h27;
```

```
              4'hc: rom = 8'h1c;           4'hd: rom = 8'h12;
              4'he: rom = 8'h08;           4'hf: rom = 8'h00;
        endcase
    endfunction
endmodule
```

The following lists the Verilog HDL code of the `wallace_26x24_product` module. It calculates the product of a 26-bit number and a 24-bit number by using the Wallace tree module `wallace_26x24` and an adder.

```
module wallace_26x24_product (a,b,z);             // 26*24 wt product
    input   [25:00] b;                            // 26 bits
    input   [23:00] a;                            // 24 bits
    output  [49:00] z;                            // product
    wire    [49:08] x;                            // sum high
    wire    [49:08] y;                            // carry high
    wire    [49:08] z_high;                       // product high
    wire    [07:00] z_low;                        // product low
    wallace_26x24 wt_partial (a, b, x, y, z_low); // partial product
    assign z_high = x + y;
    assign z = {z_high,z_low};                    // product
endmodule
```

The following lists the Verilog HDL code of the `wallace_26x26_product` module. It calculates the product of two 26-bit numbers by using the Wallace tree module `wallace_26x26` and an adder.

```
module wallace_26x26_product (a,b,z);             // 26*26 wt product
    input   [25:00] a;                            // 26 bits
    input   [25:00] b;                            // 26 bits
    output  [51:00] z;                            // product
    wire    [51:08] x;                            // sum high
    wire    [51:08] y;                            // carry high
    wire    [51:08] z_high;                       // product high
    wire    [07:00] z_low;                        // product low
    wallace_26x26 wt_partial (a, b, x, y, z_low); // partial product
    assign z_high = x + y;
    assign z = {z_high,z_low};                    // product
endmodule
```

The three modules, namely `wallace_24x26`, `wallace_26x24`, and `wallace_26x26`, mentioned above, generate the partial product (carry and sum). The Verilog HDL codes of these modules are not listed in this book because the codes are too long. The reader can refer to Figure 9.15, the structure of a 24-bit by 24-bit Wallace tree, to write the codes of the three modules.

Figure 9.23 shows the simulation waveforms of two float division examples. The first example performs 8.0/4.0 (0x41000000 / 0x40800000 in float format); the result is 2.0 (0x40000000 in float format). The second example performs division on two denormalized float numbers, 0x0000fe01 / 0x000000ff; the result is a normalized float number $0x37f0000 = 0\_10000110\_11111110000000000000000_2$ in the IEEE float format. It is $2^{128+6-127} \times 1.1111111_2 = 11111111_2 = 0xff = 255_{10}$.

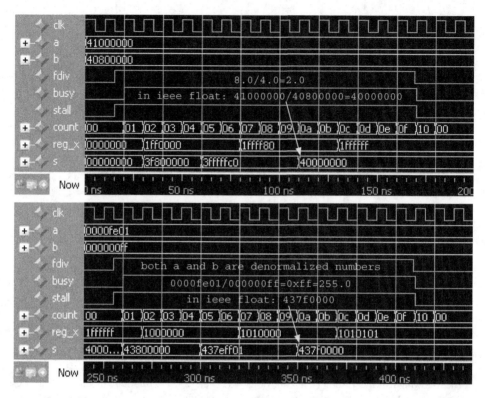

Figure 9.23   Waveform of the floating-point divider

The float divider takes 19 clock cycles to perform its operation: 1 cycle for reading the ROM table, 15 (3 × 5) for Newton iterations, 2 for multiplication, and 1 for normalization. It can accept a float division instruction during every 16 clock cycles because the last three stages are pipelined.

9.6   Floating-Point Square Root (FSQRT) Design

This section describes the algorithm and design of the float square rooter which uses the Newton–Raphson algorithm to implement the significand square root.

9.6.1   Floating-Point Square Root Algorithm

The float square root algorithm is simpler than the float division algorithm because there is only one input radicand and its square root is closer toward 1 than the radicand, for example, $\sqrt{0.25 \times 10^{-6}} = 0.5 \times 10^{-3}$ and $\sqrt{2.25 \times 10^6} = 1.5 \times 10^3$.

First, consider the float square root of a normalized float number: given a radicand $d = \{s_d, e_d, f_d\}$ in the IEEE format, calculate $q = \{s_q, e_q, f_q\} = \sqrt{d}$. If $s_d = 1$ (negative), the result is an NaN. Otherwise, the result should be $\sqrt{2^{e_d-127} \times 1.f_d}$. We use the Newton–Raphson algorithm to calculate the square root. Therefore, we must change $1.f_d$ to the format $0.1xx\cdots x$ (shift to the right by 1 bit) or $0.01x\cdots x$ (shift to the right by 2 bits), and correspondingly adjust the exponent by adding a 1 or 2 to it, so that the adjusted exponent is divisible by 2 (even). Then the root's significand has the format $0.1xx\cdots\cdots x$. We shift the root to the left by 1 bit to the format $1.f_q$ and decrease the root's exponent by 1. We summarize the square root algorithm for a normalized float number as follows:

1. If $e_d - 127$ is even, that is, e_d is odd, shift $1.f_d$ to the right by 2 bits: $1.f_q = \sqrt{1.f_d \gg 2} \ll 1$ and $e_q = (e_d - 127 + 2)/2 + 127 - 1 = (e_d \gg 1) + 63 + e_d\%2$.
2. If $e_d - 127$ is odd, that is, e_d is even, shift $1.f_d$ to the right by 1 bit: $1.f_q = \sqrt{1.f_d \gg 1} \ll 1$ and $e_q = (e_d - 127 + 1)/2 + 127 - 1 = (e_d \gg 1) + 63 + e_d\%2$.

Therefore, no matter whether e_d is even or odd, we always perform $e_q = (e_d \gg 1) + 63 + e_d\%2$. Below we give an example to show the calculation of the square root of a normalized float number. Suppose we have a float number

$$d = 41100000_{16} = 0\_10000010\_00100000000000000000000_2$$

That is, $s_d = 0$, $e_d = 10000010_2 = 130_{10}$, and $f_d = 00100000000000000000000_2$. Its value is $2^{130-127} \times 1.00100000000000000000000_2 = 2^3 \times 1.001_2 = 1001_2 = 9_{10}$. We know that the root is $q = 3_{10} = 11_2 = 1.1_2 \times 2^1 = 1.1_2 \times 2^{128-127}$, that is, $s_q = 0$, $e_q = 10000000_2$, and $f_q = 10000000000000000000000_2$, or 40400000_{16} in the IEEE float format. Below we show the calculations for getting this result by the algorithm.

Because $e_d = 10000010_2 = 130_{10}$ is even, shift $1.f_d$ to the right by 1 bit; we get 0.1001_2. By using the Newton–Raphson algorithm, we get the root of 0.11_2. Shift it to the left by 1 bit, and we get $1.1_2 = 1.f_q$. $e_q = e_d \gg 1 + 63 + e_d\% 2 = 65 + 63 + 0 = 128 = 10000000_2$. Then we get $q = 40400000_{16}$.

Next, consider the float square root of a denormalized float number. The value of a denormalized number $d = \{0, 0, f_d\}$ in the IEEE format is $d = 2^{-126} \times 0.f_d$. The smallest representable number is $d = 2^{-126} \times 2^{-23}$; it is larger than $2^{-126} \times 2^{-24}$. Even if $d = 2^{-126} \times 2^{-24}$, $q = \sqrt{d} = 2^{-63} \times 2^{-12} = 2^{-75} = 2^{52-127}$, which is a normalized float number; therefore we conclude that the square root of a denormalized float number is a normalized float number.

To calculate $q = \{0, e_q, f_q\} = \sqrt{d}$, where $d = \{0, 0, f_d\}$ is a denormalized float number, we must change $0.f_d$ to the format $0.1xx \cdots x$ or $0.01x \cdots x$. The shift amount must be even, so that it is divisible by 2. Then the root's significand has the format $0.1xx \cdots x$. We shift the root to the left by 1 bit to the format $1.f_q$ and decrease the root's exponent by 1. We use t to denote the shift amount; then the root's exponent $e_q = (-126 - t)/2 - 1 + 127 = 63 - t/2$. We summarize the square root algorithm for a denormalized float number as follows:

1. Shift $0.f_d$ by even bits t to get $0.f'_d$ with a format of $0.1xx \cdots x$ or $0.01x \cdots x$.
2. $1.f_q = \sqrt{0.f'_d} \ll 1$.
3. $e_q = 63 - t/2$.

Below we give an example to show the calculation of the square root of a denormalized float number. Suppose we have a denormalized float number

$$d = 00003200_{16} = 0\_00000000\_00000000011001000000000_2$$

That is, $s_d = 0$, $e_d = 00000000_2 = 0_{10}$, and $f_d = 00000000011001000000000_2$. We shift $0.f_d$ to the left by 8 bits; it becomes $0.01100100000000000000000_2$. By using the Newton–Raphson algorithm, we get the root 0.101_2. Shift it to the left by 1 bit, and we get $1.01_2 = 1.f_q$. $e_q = 63 - 8/2 = 59 = 00111011_2$. Then $q = 1da00000_{16}$.

The value of d is $2^{-126} \times 0.00000000011001000000000_2 = 2^{-126} \times 11001_2 \times 2^{-14} = 2^{-140} \times 25_{10}$. We know that its root is $q = 2^{-70} \times 5_{10} = 2^{-70} \times 101_2 = 2^{-68} \times 1.01_2 = 2^{59-127} \times 1.01_2$, that is, $s_q = 0$, $e_q = 00111011_2$, and $f_q = 01000000000000000000000_2$, or $1da00000_{16}$, showing that the result calculated by the algorithm above is correct.

Finally, consider some special calculations: (i) if $s_d = 1$, that is, d is negative, the result is an NaN; (ii) $\sqrt{0} = 0$; (iii) $\sqrt{+\infty} = +\infty$; and (iv) $\sqrt{\text{NaN}} = \text{NaN}$.

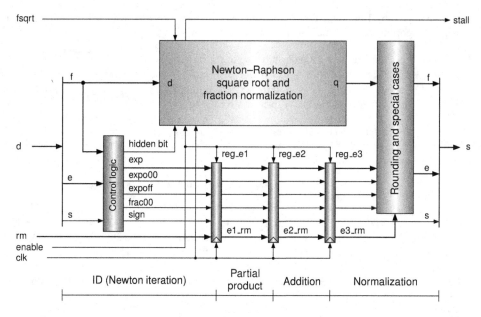

Figure 9.24   Block diagram of the floating-point square root

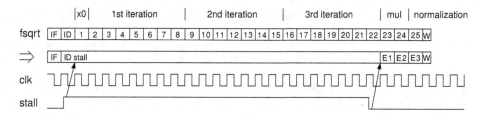

Figure 9.25   Pipeline stages of the floating-point square root

9.6.2   Circuit and Verilog HDL Codes of FSQRT

We use the Newton–Raphson algorithm to perform the significand square root, which was already described in Chapter 3. Figure 9.24 shows the block diagram of the single-precision floating-point square rooter.

The Newton–Raphson square root algorithm consists of two parts: (i) iterative calculation: $x_{i+1} = x_i(3 - x_i^2 d)/2$ and (ii) square root calculation: $q = d \times x_n$. The second part performs a multiplication, which can be implemented with a pipelined Wallace tree. Therefore, the operation of the float square root takes four steps:

1. Newton iteration, which calculates x_n;
2. partial product, which calculates carry and sum with the Wallace CSA array;
3. addition, which calculates the product by adding the carry and sum; and
4. normalization, which generates the final result in the IEEE float format.

Steps 2–4 are done in a pipeline manner and step 1 is done in an iterative manner. Figure 9.25 shows the pipeline stages of the floating-point square rooter. During the Newton iteration in the ID stage, we stall the pipeline. Three iterations are carried out. Each iteration takes seven clock cycles to calculate $x_{i+1} = x_i(3 - x_i^2 d)/2$, which consists of three multiplications and one subtraction.

The following Verilog HDL code implements the float square rooter. It invokes the module `root_newton24`, which implements the Newton–Raphson algorithm for the significand square root. Because the Newton–Raphson algorithm requires the normalized radicand (the two most significant bits must be 1x or 01) and our square rooter allows denormalized inputs, we use the module `shift_even_bits` to shift fractions to the normalized format.

```
module fsqrt_newton (d,rm,fsqrt,ena,clk,clrn, s,busy,stall,count,reg_x);
    input           clk, clrn;              // clock and reset
    input   [31:0]  d;                      // fp s = root(d)
    input   [1:0]   rm;                     // round mode
    input           fsqrt;                  // ID stage: fsqrt = i_fsqrt
    input           ena;                    // enable
    output  [31:0]  s;                      // fp output
    output  [25:0]  reg_x;                  // x_i
    output  [4:0]   count;                  // for iteration control
    output          busy;                   // for generating stall
    output          stall;                  // for pipeline stall
    parameter       ZERO = 32'h00000000;
    parameter       INF  = 32'h7f800000;
    parameter       NaN  = 32'h7fc00000;
    wire            d_expo_is_00 = ~|d[30:23];   // d_expo = 00
    wire            d_expo_is_ff =  &d[30:23];    // d_expo = ff
    wire            d_frac_is_00 = ~|d[22:0];     // d_frac = 00
    wire            sign = d[31];
    //              e_q  = (e_d >> 1)      +    63 + (e_d % 2)
    wire    [7:0]   exp_8 = {1'b0,d[30:24]} + 8'd63 + d[23];  // normalized
    //                    = 0              +    63 + 0 = 63   // denormalized
    //              d_f24 = denormalized? .f_d,0 :     .1,f_d // shifted 1 bit
    wire    [23:0]  d_f24 = d_expo_is_00? {d[22:0],1'b0} : {1'b1,d[22:0]};
    //              tmp = e_d is odd? shift one more bit : 1 bit, no change
    wire    [23:0]  d_temp24 = d[23]? {1'b0,d_f24[23:1]} : d_f24;
    wire    [23:0]  d_frac24;  // .1xx...x or .01x...x for denormalized number
    wire    [4:0]   shamt_d;   // shift amount, even number
    shift_even_bits shift_d (d_temp24,d_frac24,shamt_d);
    // denormalized: e_q = 63 - shamt_d / 2
    // normalized:   e_q = exp_8 - 0
    wire    [7:0]   exp0 = exp_8 - {4'h0,shamt_d[4:1]};
    reg             e1_sign,e1_e00,e1_eff,e1_f00;
    reg             e2_sign,e2_e00,e2_eff,e2_f00;
    reg             e3_sign,e3_e00,e3_eff,e3_f00;
    reg     [1:0]   e1_rm,e2_rm,e3_rm;
    reg     [7:0]   e1_exp,e2_exp,e3_exp;
    always @ (negedge clrn or posedge clk)
        if (!clrn) begin  // 3 pipeline registers: reg_e1, reg_e2, and reg_e3
            // reg_e1              reg_e2                 reg_e3
            e1_sign <= 0;          e2_sign <= 0;          e3_sign <= 0;
```

```
        e1_rm   <= 0;          e2_rm   <= 0;       e3_rm   <= 0;
        e1_exp  <= 0;          e2_exp  <= 0;       e3_exp  <= 0;
        e1_e00  <= 0;          e2_e00  <= 0;       e3_e00  <= 0;
        e1_eff  <= 0;          e2_eff  <= 0;       e3_eff  <= 0;
        e1_f00  <= 0;          e2_f00  <= 0;       e3_f00  <= 0;
    end else if (ena) begin
        e1_sign <= sign;       e2_sign <= e1_sign; e3_sign <= e2_sign;
        e1_rm   <= rm;         e2_rm   <= e1_rm;   e3_rm   <= e2_rm;
        e1_exp  <= exp0;       e2_exp  <= e1_exp;  e3_exp  <= e2_exp;
        e1_e00  <= d_expo_is_00; e2_e00 <= e1_e00; e3_e00  <= e2_e00;
        e1_eff  <= d_expo_is_ff; e2_eff <= e1_eff; e3_eff  <= e2_eff;
        e1_f00  <= d_frac_is_00; e2_f00 <= e1_f00; e3_f00  <= e2_f00;
    end
    wire    [31:0] frac0;                          // root = 1.xxxx...x
    root_newton24 frac_newton (d_frac24,fsqrt,ena,clk,clrn,
                        frac0,busy,count,reg_x,stall);
    wire    [26:0] frac = {frac0[31:6],|frac0[5:0]};    // sticky
    wire           frac_plus_1 =
        ~e3_rm[1] & ~e3_rm[0] & frac[3] & frac[2] & ~frac[1] & ~frac[0] |
        ~e3_rm[1] & ~e3_rm[0] & frac[2] & (frac[1] | frac[0]) |
        ~e3_rm[1] & e3_rm[0] & (frac[2] | frac[1] | frac[0]) & e3_sign |
        e3_rm[1] & ~e3_rm[0] & (frac[2] | frac[1] | frac[0]) & ~e3_sign;
    wire    [24:0] frac_rnd = {1'b0,frac[26:3]} + frac_plus_1;
    wire    [7:0] expo_new = frac_rnd[24]? e3_exp + 8'h1 : e3_exp;
    wire    [22:0] frac_new = frac_rnd[24]? frac_rnd[23:1]: frac_rnd[22:0];
    assign s = final_result(e3_sign,e3_e00,e3_eff,e3_f00,
                        {e3_sign,expo_new,frac_new});
    function [31:0] final_result;
        input       d_sign,d_e00,d_eff,d_f00;
        input [31:0] calc;
        casex ({d_sign,d_e00,d_eff,d_f00})
            4'b1xxx : final_result = NaN;       // -
            4'b000x : final_result = calc;      // nor
            4'b0100 : final_result = calc;      // den
            4'b0010 : final_result = NaN;       // nan
            4'b0011 : final_result = INF;       // inf
            default : final_result = ZERO;      // 0
        endcase
    endfunction
endmodule
```

The following is the Verilog HDL code of the shift_even_bits module, which shifts the input a to the left by even bits to generate the output b whose two most significant bits are 1x or 01. The output signal sa indicates the shift amount that is used for the exponent adjustment.

```
module shift_even_bits (a,b,sa);   // shift even bits until msb is 1x or 01
    input   [23:0] a;              // shift a = xxx...x by even bits
    output  [23:0] b;              // to    b = 1xx...x or 01x...x
```

```
output    [4:0] sa;              // shift amount, even number
wire     [23:0] a5,a4,a3,a2,a1;
assign a5 = a;
assign sa[4]    = ~|a5[23:08];                              // 16-bit 0
assign a4 = sa[4]? {a5[07:00],16'b0}  : a5;
assign sa[3]    = ~|a4[23:16];                              //  8-bit 0
assign a3 = sa[3]? {a4[15:00],  8'b0}  : a4;
assign sa[2]    = ~|a3[23:20];                              //  4-bit 0
assign a2 = sa[2]? {a3[19:00],  4'b0}  : a3;
assign sa[1]    = ~|a2[23:22];                              //  2-bit 0
assign a1 = sa[1]? {a2[21:00],  2'b0}  : a2;
assign sa[0] = 0;
assign b = a1;
endmodule
```

Figure 9.26 shows the schematic diagram of the Newton–Raphson square rooter. The circuit is similar to the 32-bit Newton–Raphson square rooter, which was described in Chapter 3 and shown in Figure 3.34, but here we used the system clock and added the pipeline registers to calculate $q = d \times x_n$.

We use a counter to control the operation of the square rooter. Referring to Figure 9.27, initially the counter contains a 0. On receiving a float square root (fsqrt) instruction, a signal busy is set to active to indicate that the circuit is busy for the Newton iterations. It will become inactive whenever the iterations finish. The stall signal is used to stall the pipeline.

Below is the Verilog HDL code of the root_newton24 module that implements the circuit in Figure 9.26. x_0 is read from a 32×8-bit ROM table. The three multiplications for the iteration of $x_{i+1} = x_i(3 - x_i^2 d)/2$ are done by the modules wallace_26x26_product, wallace_24x28_product, and

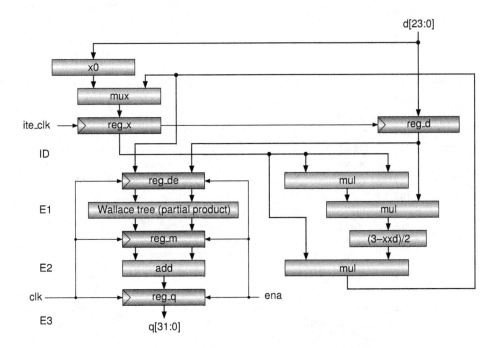

Figure 9.26   Block diagram of the Newton–Raphson square rooter

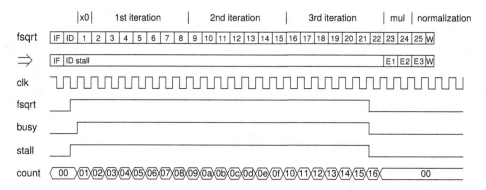

Figure 9.27   Stall caused by the square root instruction

`wallace_26x26_product`. The calculation of $q = d \times x_n$ is done with a two-stage pipeline. The partial product is done by the module `wallace_24x26` in the first stage. The quotient q is calculated by an adder in the second stage. Rounding is not necessary here because in FPU, there is a normalization stage to generate the final result in the IEEE float format.

```
module root_newton24 (d,fsqrt,ena,clk,clrn,q,busy,count,reg_x,stall);
    input    [23:0] d;                          // radicand: .1xx...x  .01x...x
    input           fsqrt;                      // ID stage: fsqrt = i_fsqrt
    input           clk, clrn;                  // clock and reset
    input           ena;                        // enable, save partial product
    output   [31:0] q;                          // root: .1xxx...x
    output          busy;                       // cannot receive new div
    output          stall;                      // stall to save result
    output   [4:0]  count;                      // for sim test only
    output   [25:0] reg_x;                      // for sim test only 01.xx...x
    reg      [31:0] q;                          // root:      .1xxx...x
    reg      [23:0] reg_d;                      // 24-bit:    .xxxx...xx
    reg      [25:0] reg_x;                      // 26-bit: xx.1xxx...xx
    reg      [4:0]  count;                      // 3 iterations * 7 cycles
    reg             busy;                       // cannot receive new fsqrt
    wire     [7:0]  x0 = rom(d[23:19]);         // x0: from rom table
    wire     [51:0] x_2,x2d,x52;                // xxxx.xxxxx...x
    always @ (posedge clk or negedge clrn) begin
        if (!clrn) begin
            count <= 0;
            busy  <= 0;
            reg_x <= 0;
        end else begin
            if (fsqrt & (count == 0)) begin     // do once only
                count <= 5'b1;                  // set count
                busy  <= 1'b1;                  // set to busy
            end else begin                      // 3 iterations
```

```verilog
            if  (count == 5'h01) begin
                reg_x <= {2'b1,x0,16'b0};                // 01.xxxx0...0
                reg_d <= d;                              //   .1xxxx...x
            end
            if  (count != 0) count <= count + 5'b1;      // count++
            if  (count == 5'h15) busy  <= 0;             // ready for next
            if  (count == 5'h16) count <= 0;             // reset count
            if ((count == 5'h08) ||    // save result of 1st iteration
                (count == 5'h0f) ||    // save result of 2nd iteration
                (count == 5'h16))      // no need to save here actually
                reg_x <= x52[50:25];   // /2 = xx.xxxxx...x
        end
    end
end
assign stall = fsqrt & (count == 0) | busy;
// wallace_26x26_product (a,     b,     z);
wallace_26x26_product x2 (reg_x,reg_x,x_2);              // xi(3-xi*xi*d)/2
wallace_24x28_product xd (reg_d,x_2[51:24],x2d);
wire    [25:0] b26 = 26'h3000000 - x2d[49:24];          // xx.xxxxx...x
wallace_26x26_product xip1 (reg_x,b26,x52);
reg     [25:0] reg_de_x; // pipeline register in between id and e1, x
reg     [23:0] reg_de_d; // pipeline register in between id and e1, d
wire    [49:0] m_s;       // sum:  41 + 8 = 49-bit
wire    [49:8] m_c;       // carry: 42-bit
// wallace_24x26 (a,          b,          x,          y,  z);
wallace_24x26 wt (reg_de_d,reg_de_x,m_s[49:8],m_c,m_s[7:0]);    // d * x
reg     [49:0] a_s;       // pipeline register in between e1 and e2, sum
reg     [49:8] a_c;       // pipeline register in between e1 and e2, carry
wire    [49:0] d_x = {1'b0,a_s} + {a_c,8'b0};           // 0x.xxxxx...x
wire    [31:0] e2p = {d_x[47:17],|d_x[16:0]};           // sticky
always @ (negedge clrn or posedge clk)
    if (!clrn) begin                              // pipeline registers
        reg_de_x <= 0;         reg_de_d <= 0;           // id-e1
        a_s      <= 0;         a_c      <= 0;           // e1-e2
        q        <= 0;                                  // e2-e3
    end else if (ena) begin     // x52[50:25]: the result of 3rd iteration
        reg_de_x <= x52[50:25]; reg_de_d <= reg_d;      // id-e1
        a_s      <= m_s;        a_c      <= m_c;         // e1-e2
        q        <= e2p;                                // e2-e3
    end
function  [7:0] rom;                          // a rom table: 1/d^{1/2}
    input [4:0] d;
    case (d)
        5'h08: rom = 8'hff;          5'h09: rom = 8'he1;
        5'h0a: rom = 8'hc7;          5'h0b: rom = 8'hb1;
        5'h0c: rom = 8'h9e;          5'h0d: rom = 8'h9e;
        5'h0e: rom = 8'h7f;          5'h0f: rom = 8'h72;
```

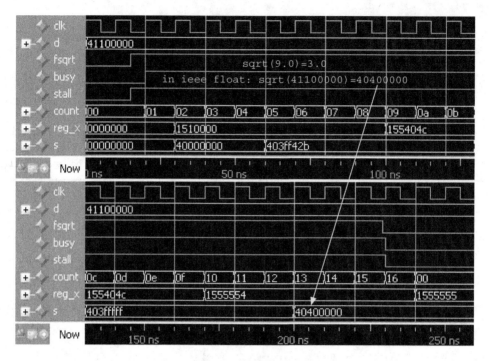

Figure 9.28 Waveform of the square root of a normalized float number

```
        5'h10: rom = 8'h66;              5'h11: rom = 8'h5b;
        5'h12: rom = 8'h51;              5'h13: rom = 8'h48;
        5'h14: rom = 8'h3f;              5'h15: rom = 8'h37;
        5'h16: rom = 8'h30;              5'h17: rom = 8'h29;
        5'h18: rom = 8'h23;              5'h19: rom = 8'h1d;
        5'h1a: rom = 8'h17;              5'h1b: rom = 8'h12;
        5'h1c: rom = 8'h0d;              5'h1d: rom = 8'h08;
        5'h1e: rom = 8'h04;              5'h1f: rom = 8'h00;
        default: rom = 8'hff;                      // 0 - 7: not be accessed
      endcase
    endfunction
endmodule
```

The following list is the Verilog HDL code of the `wallace_24x28_product` module. It calculates the product of a 24-bit number and a 28-bit number by using the Wallace tree module `wallace_24x28` and an adder.

```
module wallace_24x28_product (a,b,z);            // 24*28 wt product
    input   [23:00] a;                           // 24 bits
    input   [27:00] b;                           // 28 bits
    output  [51:00] z;                           // product
    wire    [51:08] x;                           // sum high
    wire    [51:08] y;                           // carry high
```

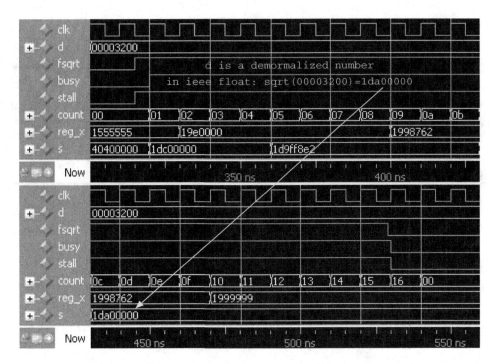

Figure 9.29   Waveform of the square root of a denormalized float number

```
wire    [51:08] z_high;                          // product high
wire    [07:00] z_low;                           // product low
wallace_24x28 wt_partial (a, b, x, y, z_low);    // partial product
assign z_high = x + y;
assign z = {z_high, z_low};                      // product
endmodule
```

Figure 9.28 shows the simulation waveforms of calculating the square root of a normalized number, which was explained at the beginning of this section.

Figure 9.29 shows the simulation waveforms of calculating the square root of a denormalized number, which was also explained at the beginning of this section.

The float square rooter takes 25 clock cycles to perform its operation: 1 cycle for reading ROM table, 21 (3×7) cycles for Newton iterations, 2 cycles for multiplication, and 1 cycle for normalization. It can accept a float division instruction in every 22 clock cycles because the last three stages are pipelined.

Exercises

9.1 The conversion circuits between float and integer given in this chapter implement only the truncate rounding mode. Redesign the conversion circuits that can deal with all the four rounding modes.

9.2 Design a pipelined float adder with five stages: two for alignment, one for calculation, and two for normalization.

9.3   Develop a Verilog HDL code that implements a 26-bit by 26-bit Wallace tree.

9.4   Writing Verilog HDL code for a Wallace tree by hand is a tedious work. Try to develop a program in any programming language that can generate the Verilog HDL code for an m-bit by n-bit Wallace tree automatically.

9.5   The Newton–Raphson division and square root algorithms use a ROM table to get x_0. Because the ROM table is quite small, we can put the access to the ROM table into the ID stage such that the latency of the division or square root can be reduced by one cycle. Redesign the float divider and square rooter to do so.

9.6   Implement a float divider with the Goldschmidt algorithm.

9.7   Implement a float square rooter with the Goldschmidt algorithm.

9.8   Consider the possibility of designing a fully pipelined float divider and a fully pipelined float square rooter that can accept a new float division or square root instruction in every clock cycle.

10

Design of Pipelined CPU with FPU in Verilog HDL

This chapter describes the design of a pipelined CPU (central processing unit) which has a floating-point unit (FPU). The instructions executed by the FPU include float addition, subtraction, multiplication, division, and square root. These instructions fetch operands from a floating-point register file, rather than from the general-purpose register file which is for use by the integer unit (IU). Therefore, it needs to implement the float load and store instructions which transfer data between the memory and the floating-point register file.

10.1 CPU/FPU Pipeline Model

This section introduces the instructions related to the FPU and their pipeline models.

10.1.1 FPU Instructions

The CPU that will be designed in this chapter can execute 27 instructions, 20 instructions of which were explained and implemented in the pipelined CPU described in Chapter 8. The seven new instructions are explained below.

```
add.s/sub.s/mul.s/div.s fd, fs, ft   # fd <-- fs op ft;
```

These four instructions perform float addition, subtraction, multiplication, and division, respectively. They have the same format: fs and ft are two source register numbers and fd is a destination register number of the floating-point register file.

```
sqrt.s fd, fs                        # fd <-- root (fs);
```

This instruction performs the float square root. It requires only one source number (fs) and a destination register number (fd).

```
lwc1 ft, offset(rs)                  # ft <-- memory[rs + offset];
swc1 ft, offset(rs)                  # memory[rs + offset] <-- ft;
```

Computer Principles and Design in Verilog HDL, First Edition. Yamin Li.
© 2015 Tsinghua University Press. All rights reserved. Published 2015 by John Wiley & Sons Singapore Pte Ltd.
Companion Website: www.wiley.com/go/li/verilog

Table 10.1   Seven instructions related to the floating-point unit

Inst.	[31:26]	[25:21]	[20:16]	[15:11]	[10:6]	[5:0]	Description
lwc1	110001	rs	ft		offset		Load FP word
swc1	111001	rs	ft		offset		Store FP word
add.s	010001	10000	ft	fs	fd	000000	FP add
sub.s	010001	10000	ft	fs	fd	000001	FP subtract
mul.s	010001	10000	ft	fs	fd	000010	FP multiplication
div.s	010001	10000	ft	fs	fd	000011	FP division
sqrt.s	010001	10000	00000	fs	fd	000100	FP square root

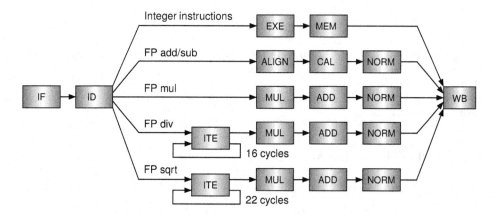

Figure 10.1   Pipeline models of the IU and the floating-point unit

These two instructions transfer data between the memory and the floating-point register file. MIPS calls the FPU as Coprocessor 1; therefore c1 is used in the instruction mnemonic. The lwc1 instruction loads a word from the memory and writes the word to the register ft of the floating-point register file. The memory address is the sum of the content in the register rs of the general-purpose register file and offset, which is a 16-bit immediate in the instruction and sign-extended to 32 bits. That is, the memory address calculation method is the same as that of the lw instruction. The swc1 instruction stores a word of the register ft of the floating-point register file in the memory. The memory address calculation of swc1 is the same as that of the lwc1 instruction.

Table 10.1 lists the format and the encodings of these seven instructions.

10.1.2   Basic Model of CPU/FPU Pipeline

The pipeline models of lwc1 and swc1 are similar to those of the lw and sw, respectively. The pipeline models of the computational floating-point instructions are a little more complex due to the iterations for the float division and square root instructions. Figure 10.1 shows the pipeline models of these instructions.

The pipeline model of the integer instructions, including lwc1 and swc1, has five stages, which is the same as for the pipelined CPU described in Chapter 8. There are six pipeline stages for the float addition and subtraction instructions. Fraction alignment is done in the ALIGN stage; the CAL stage is for the significand calculation; and the NORM stage is used to normalize the calculation result to the IEEE

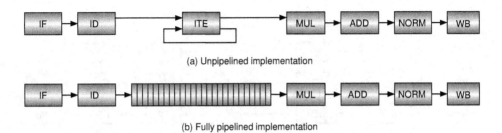

(a) Unpipelined implementation

(b) Fully pipelined implementation

Figure 10.2 Unpipelined and fully pipelined implementations

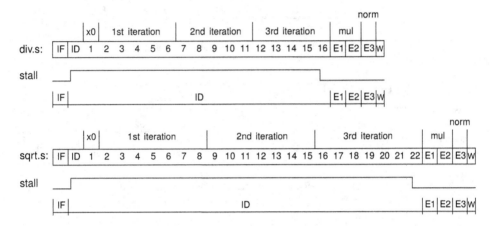

Figure 10.3 Stall signal for floating-point division and square root instructions

float format. The float multiplication instruction also adopts six pipeline stages. The partial product is generated by a Wallace tree in the MUL stage; and the product is calculated in the ADD stage. For the float division and square root instructions, we use the Newton–Raphson iterations, which are done in the ITE stage. Then, a multiplication is needed, as we described in the previous chapter. This multiplication can be done similar to the float multiplication instruction.

The Newton–Raphson division and square root algorithms take long time for the execution. There are two methods of implementing such long time execution: unpipelined implementation and fully pipelined implementation. As shown in Figure 10.2, the unpipelined implementation uses the same circuit for the iterative execution; it has low cost but cannot accept a new float division or square root instruction in every clock cycle. The fully pipelined implementation uses dedicated circuits for the pipelined execution. It can accept a new float division or square root instruction in every clock cycle but its cost is high. Note that the instruction latencies of the two implementations are the same[1].

We adopt the unpipelined implementation in our CPU design. To do this, we use a stall signal to stall the pipeline during the execution of the Newton–Raphson iterations and combine it with the instruction decode (ID) stage, as shown in Figure 10.3. $1/b$ and $1/\sqrt{d}$ are calculated during the new ID stage for div.s and sqrt.s instructions, respectively. Three iterations are needed because we used a small ROM (read only memory) table for getting the initial value of x_0. Then a multiplication is performed for

---

[1] The latency here is defined as the clock cycles the next instruction can use the result.

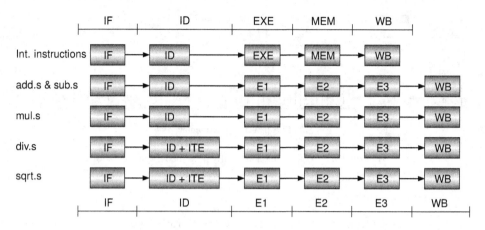

Figure 10.4   Unified floating-point pipeline model

Figure 10.5   Register file requiring two write ports

getting the quotient $q = a \times (1/b) = a/b$ or the square root $q = d \times (1/\sqrt{d}) = \sqrt{d}$. This multiplication is implemented in the pipeline manner.

Thus we have a unified six-stage pipeline model for all the float computational instructions: IF, ID, E1, E2, E3, and WB, as shown in Figure 10.4. E1, E2, and E3 stand for the execution stages 1, 2, and 3, respectively. In the following discussions, we use this six-stage pipeline model and do not mention whether there is a Newton–Raphson iteration.

10.2   Design of Register File with Two Write Ports

There are thirty-two 32-bit registers in the floating-point register file. Different from the general-purpose (integer) register file where register 0 always contains a constant 0, the register 0 in the floating-point register file is a regular register.

In our floating-point pipeline model, there is a case where two writes may appear at a same clock cycle, as shown in Figure 10.5. One is the float computational instruction writing the result to the floating-point register file; the other is the lwc1 instruction writing the memory data to the register file, due to the two instructions having different pipeline stages.

Although we can postpone the second write by hardware or optimize the code sequence by software (compiler), here we design a floating-point register file with two write ports: if the destination register numbers of the two writes are different, these two writes can be carried out in parallel. Otherwise, the result of the float computational instruction will be discarded and the lwc1 instruction will write its data to the destination register. Practical programs can avoid a such case because the first write is meaningless.

We have discussed how to design a register file of a one write port in Chapter 5. The symbol of the register file with two write ports is shown in the top left of Figure 10.6. rn0 and rn1 are two register

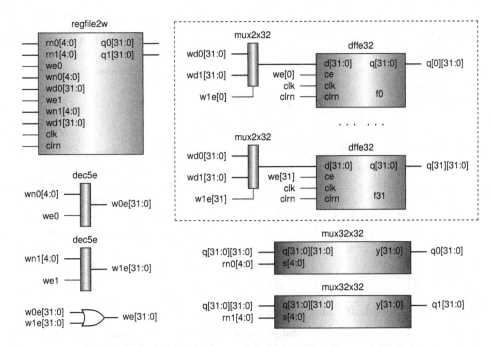

Figure 10.6 Schematic diagram of a register file with two write ports

numbers for readings; their corresponding contents are outputted at the read ports q0 and q1, respectively. Write port 0 has signals we0, wn0, and wd0. Write port 1 has signals we1, wn1, and wd1. we is a write enable signal; wn is the destination register number; and wd is the data word that will be written to the register file.

The two read ports can be easily implemented with two 32-to-1 multiplexers. For enabling the two writes, we use multiplexers in front of the 32 registers. The output of a multiplexer is connected to the data input of a register (dffe32). The selection signal, w1e[i], comes from write port 1, where $i = 0, 1, 2, \cdots, 31$. This means that write port 1 has a higher priority than port 0, because whenever the selection signal is a 1, the data wd1 will be selected, no matter whether there is a write in port 0. The w1e signals are generated by a 5-32 decoder (dec5e), which decodes wn1 and generates 32 write enable signals of which at most one signal is active. Another decoder decodes wn0 and generates w0e. Each dffe32 has a write enable we; it is connected to an ORed result of w0e and w1e, generated by the two decoders, respectively.

We can write the Verilog HDL (hardware description language) codes in dataflow style to implement the register file based on Figure 10.6 and the descriptions above, but the code may be too long and too hard to write and check. Below we give a rather shorter (maybe shortest) Verilog HDL code in behavioral style to implement the register file of two write ports. Instead of using 0 and 1 for the port numbers, we use a and b for the read port numbers, and use x and y for the write port numbers. The key part is the last if statement, which should not be too difficult to understand.

```
module regfile2w (rna,rnb,dx,wnx,wex,dy,wny,wey,clk,clrn,qa,qb);
    input   [31:0] dx, dy;              // write data
    input   [4:0] rna, rnb, wnx, wny;   // reg numbers
    input          wex, wey;            // write enables
    input          clk, clrn;           // clock and reset
```

```
output  [31:0] qa, qb;                                  // read data
reg     [31:0] register [0:31];                         // 32 32-bit registers
assign         qa = register[rna];                      // read port a
assign         qb = register[rnb];                      // read port b
integer i;
always @(posedge clk or negedge clrn)                   // write port
    if (!clrn) begin
        for (i=0; i<32; i=i+1)
            register[i] <= 0;                           // reset
    end else begin
        if (wey)              // write port y has a higher priority than x
            register[wny] <= dy;                        // write port y
        if (wex && (!wey || (wnx != wny)))
            register[wnx] <= dx;                        // write port x
    end
endmodule
```

10.3   Data Dependency and Pipeline Stalls

This section describes the data hazards and pipeline stalls caused by the instructions related to the FPU. The contents include (i) the data dependencies among float computational instructions; (ii) the data dependencies between lwc1/swc1 instructions and float computational instructions; and (iii) the pipeline stalls for the execution of the div.s and sqrt.s instructions.

Figure 10.7 shows the basic structure of the pipelined CPU, which mainly consists of an IU (lower part of the figure) and an FPU (upper part). The pipeline registers are inserted in between the pipeline stages. The circuit for Newton–Raphson iterations is not drawn in the figure; that should be included in the ID stage. In addition to the integer register file, there is a floating-point register file. The data word loaded by the lwc1 instruction from the data memory will be written to the floating-point register file through the write port y, which has a higher priority than the write port x. The write port x is used for writing the float result calculated by the instruction add.s, sub.s, mul.s, div.s, or sqrt.s. The data word read from the read port b of the floating-point register file will be written to the data memory by

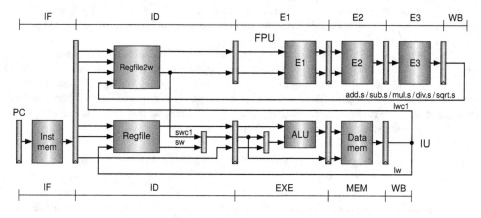

Figure 10.7   Baseline diagram of IU and FPU

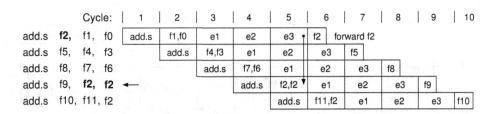

Figure 10.8 No-stall case for floating-point data hazard

the `swc1` instruction. Because the integer `sw` instruction also writes data to the data memory, we used a multiplexer in the ID stage of the IU.

In this baseline circuit, there is no circuit for internal forwarding and pipeline stalls. We will add the necessary circuits to it for building a complete CPU.

10.3.1   Internal Forwarding

The problem of the data dependency between the float computational instructions (`add.s`, `sub.s`, `mul.s`, `div.s`, and `sqrt.s`) is more serious than that between the integer instructions, because the calculation of the float numbers takes three execution stages (E1, E2, and E3), which is two more than it takes for the execution of integer instructions.

We start from the simplest case to show the float data dependency. Referring to Figure 10.8, the first `add.s` instruction writes the result of float addition to register `f2` in the WB stage in the first half of the clock cycle 6. The fifth (last) `add.s` instruction can read the result at the second half of the clock cycle 6 without any problem. But the fourth `add.s` instruction (pointed by an arrowed line) wants to read it in clock cycle 5 when the result was not yet written to the register `f2`. This causes the problem of float data dependency.

Fortunately, we can use the internal forwarding mechanism to bypass the result in the E3 stage of the first `add.s` instruction to the ID stage of the fourth `add.s` instruction, and there is no need to stall the pipeline. The circuit of internal forwarding is shown in Figure 10.9. The data `e3d` in the E3 stage is sent back to the inputs of two multiplexers in the ID stage. The selection signals of the multiplexers are named `fwdfa` (forward float to A) and `fwdfb` (forward float to B).

To generate these two selection signals, we need to have information about the instruction in the E3 stage, such as the float destination register number (`e3n`) and whether it writes the float register file (`e3w`). Then, we can easily write the logic expressions of the two selection signals in Verilog HDL format as shown below.

```
fwdfa = e3w & (e3n == fs);          // forward fpu e3d to fp a
fwdfb = e3w & (e3n == ft);          // forward fpu e3d to fp b
```

where `fs` and `ft` are source float register numbers of the instruction in the ID stage.

Next, consider a more serious case of the float data dependency, as shown in Figure 10.10. The third `add.s` instruction uses the result of the first `add.s` instruction. We must stall the pipeline for one clock cycle to wait for the data ready and bypass the result from the E3 stage to the ID stage for the third `add.s` instruction. We cannot bypass the intermediate result in the E2 stage that was not yet normalized.

Simultaneously, let us see the most serious case of the float data dependency, as shown in Figure 10.11, where the second `add.s` instruction wants to use the result of its prior instruction. In this case, we must stall the pipeline for two clock cycles.

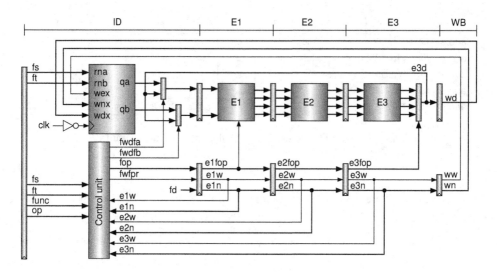

Figure 10.9   Block diagram of forwarding floating-point data

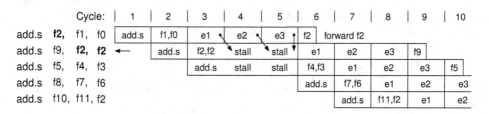

Figure 10.10   One-cycle stall case for floating-point data hazard

Figure 10.11   Two-cycle stall case for floating-point data hazard

In the last two cases, it needs to generate a stall signal, which is used to stall the pipeline. According to the discussions above, we can write the logic expression of the stall signal, `stall_fp`, as follows in Verilog HDL format.

```
i_fs = i_fadd | i_fsub | i_fmul | i_fdiv | i_fsqrt;          // use fs
i_ft = i_fadd | i_fsub | i_fmul | i_fdiv;                    // use ft
stall_fp = (e1w & (i_fs & (e1n == fs) | i_ft & (e1n == ft))) |
           (e2w & (i_fs & (e2n == fs) | i_ft & (e2n == ft)));
```

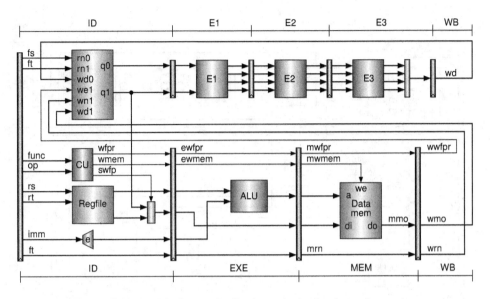

Figure 10.12   Block diagram for floating-point load and store instructions

Cycle:	1	2	3	4	5	6	7	8	9	10
lwc1 f0, 100($1)	lwc1	$1	addr	mem	f0					
lwc1 **f1**, 100($2)		lwc1	$2	addr	mem	f1	forward f1 (mmo)			
sub.s f3, f5, f4			sub.s	f5,f4	e1	e2	e3	f3		
add.s f2, **f1**, f0 ←				add.s	f1,f0	e1	e2	e3	f2	
lwc1 f4 100($3)					lwc1	$3	addr	mem	f4	

Figure 10.13   Data hazard with floating-point load

where i_fs and i_ft indicate the instruction in the ID stage using the float registers fs and ft, respectively. In either case, the pipeline will stall; therefore, we use an OR operator in the expression of the stall_fp signal. The first line of the expression is for the data dependency between the instruction in the ID stage and that in the E1 stage; the second line is for the data dependency between the instruction in the ID stage and that in the E2 stage. For the signals e1w, e1n, e2w, and e2n, refer to Figure 10.9.

10.3.2   Pipeline Stall Caused by lwc1 and swc1 Instructions

Figure 10.12 shows the execution datapath for the instructions lwc1 and swc1.

The lwc1 instruction loads a float data word from the data memory. Similar to the lw instruction, lwc1 may also cause a pipeline stall in case of the follow-up instruction using the memory data loaded by lwc1. The swc1 instruction stores a float data word to the data memory. If the data to be stored by swc1 is the result of the prior float computational instruction, we must also stall the pipeline.

First, consider the case of lwc1. Referring to Figure 10.13, the fourth instruction, add.s, uses the content in the float register f1, which is the data loaded from the data memory by the second lwc1 instruction. The data can be read out from the data memory in the MEM stage of the lwc1 instruction and can be bypassed to the ID stage of the add.s instruction. There is no need to stall the pipeline.

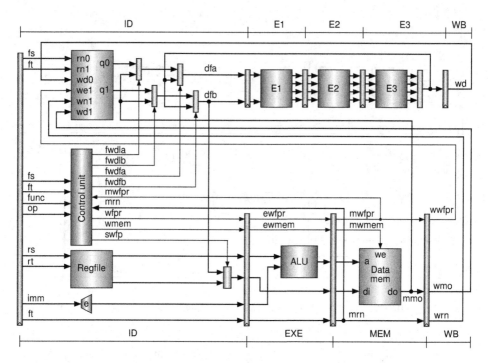

Figure 10.14   Block diagram for forwarding memory data

Figure 10.15   One-cycle stall caused by floating-point load

Figure 10.14 shows the datapath for bypassing the memory data mmo to FPU.

We used two multiplexers in the ID stage of the FPU. Their selection signals are fwdla (forward load to A) and fwdlb (forward load to B), respectively. Then

```
fwdla = mwfpr & (mrn == fs);                    // forward mmo to fp a
fwdlb = mwfpr & (mrn == ft);                    // forward mmo to fp b
```

where mwfpr (write FP register in the MEM stage) indicates that the instruction in the MEM stage is an lwc1 and that mrn is the destination register number in the MEM stage.

Referring to Figure 10.15, if the loaded memory data word is used immediately by the follow-up instruction, by one clock cycle the pipeline must be stalled.

Similarly, we have the following expression of a stall signal for lwc1. ewfpr indicates that the instruction in the EXE stage is a lwc1 and that ern is the destination register number in the EXE stage.

```
stall_lwc1 = ewfpr & (i_fs & (ern == fs) | i_ft & (ern == ft));
```

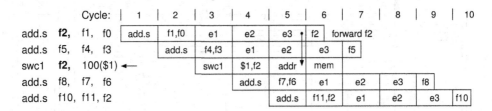

	Cycle:	1	2	3	4	5	6	7	8	9	10
add.s	f2, f1, f0	add.s	f1,f0	e1	e2	e3	f2	forward f2			
add.s	f5, f4, f3		add.s	f4,f3	e1	e2	e3	f5			
add.s	f8, f7, f6			add.s	f7,f6	e1	e2	e3	f8		
swc1	f2, 100($1) ←				swc1	$1,f2	addr	mem			
add.s	f10, f11, f2					add.s	f11,f2	e1	e2	e3	f10

Figure 10.16 Forwarding floating-point data to the instruction decode stage

	Cycle:	1	2	3	4	5	6	7	8	9	10
add.s	f2, f1, f0	add.s	f1,f0	e1	e2	e3	f2	forward f2			
add.s	f5, f4, f3		add.s	f4,f3	e1	e2	e3	f5			
swc1	f2, 100($1) ←		swc1	$1,f2	addr	mem					
add.s	f8, f7, f6			add.s	f7,f6	e1	e2	e3	f8		
add.s	f10, f11, f2				add.s	f11,f2	e1	e2	e3	f10	

Figure 10.17 Forwarding floating-point data to the execution stage

	Cycle:	1	2	3	4	5	6	7	8	9	10
add.s	f2, f1, f0	add.s	f1,f0	e1	e2	e3	f2	forward f2			
swc1	f2, 100($1) ←		swc1	$1,f2	stall	addr	mem				
add.s	f5, f4, f3			add.s	stall	f4,f3	e1	e2	e3	f5	
add.s	f8, f7, f6					add.s	f7,f6	e1	e2	e3	f8
add.s	f10, f11, f2						add.s	f11,f2	e1	e2	e3

Figure 10.18 One-cycle stall caused by floating-point store

Next, consider the case of the swc1 instruction. Figure 10.16 shows the case where swc1 (the fourth instruction) stores a float data word which is calculated by the first add.s instruction. This data can be bypassed from the E3 stage to the ID stage.

Figure 10.17 shows the case where swc1 (the third instruction) stores a float data word which is calculated by the first add.s instruction. This data can be bypassed from the E3 stage to the EXE stage (the data will be written to memory in the MEM stage).

Again, we use two multiplexers to bypass the float data in the E3 stage to the ID and EXE stages, respectively. Their selection signals can be generated by the following expressions: the swfp (store FP data) signal indicates that the instruction in the ID stage is swc1. fwdfe will be written to the pipeline register in between the ID and EXE stages, and is used by the multiplexer in the EXE stage.

```
fwdf  = swfp & e3w & (ft == e3n);               // forward to id  stage
fwdfe = swfp & e2w & (ft == e2n);               // forward to exe stage
```

In both the cases discussed above, there is no need to stall the pipeline. But, referring to Figure 10.18, if an swc1 instruction stores a float data word which is calculated by the prior instruction, we must stall the pipeline by one cycle.

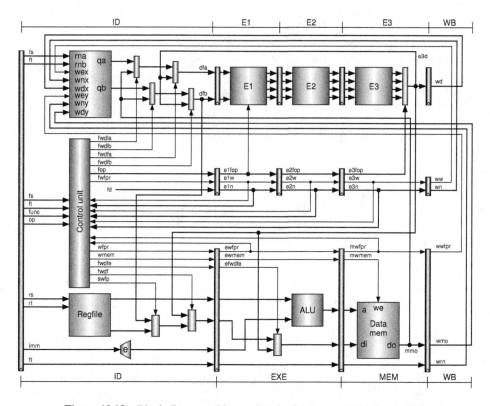

Figure 10.19   Block diagram of forwarding for floating-point load and store

This stall signal can be generated by the following expression: e1w indicates that the instruction in the E1 stage will write the float register file and that e1n is the destination register number.

```
stall_swc1 = swfp & e1w & (ft == e1n);              // stall caused by swc1
```

We have discussed all the bypass and stall cases for lwc1 and swc1 instructions. We give all the corresponding circuits in Figure 10.19.

10.3.3   Pipeline Stall Caused by Division and Square Root

The div.s and sqrt.s instructions also stall the pipeline, as shown in Figure 10.20.

The stall signal for div.s and sqrt.s instructions can be easily generated with the following expression. The second signal, w_fp_pipe_reg, is used to enable/disable the pipeline registers of the FPU.

```
stall_div_sqrt =  stall_div | stall_sqrt;
w_fp_pipe_reg  = ~stall_div_sqrt;                   // fp pipe_reg write enable
```

Up to now, we have discussed all cases of pipeline stalls. We summarize the stall signals as follows:

1. stall_lw: caused by integer data dependency with lw;
2. stall_lwc1: caused by float data dependency with lwc1;
3. stall_swc1: caused by float data dependency with swc1;

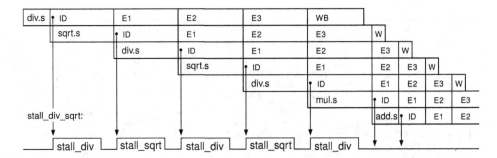

Figure 10.20 Pipeline stalls caused by floating-point division and square root

4. `stall_fp`: caused by float data dependency between float instructions; and
5. `stall_div_sqrt`: caused by Newton–Raphson iterations.

Every signal disables the writings to the PC and the IF/ID pipeline registers. The last signal also disables the pipeline registers of the FPU. Then, we can get the expression for the PC and IF/ID pipeline register write enable, `wpcir`, as below.

```
wpcir        = ~(stall_div_sqrt | stall_others);
stall_others = stall_lw | stall_fp | stall_lwc1 | stall_swc1;
```

As discussed in Chapter 8, in order to prevent an instruction from executing twice or more, we must cancel the instruction during the stall. This can be done by disabling all the four write signals:

```
wreg = (i_add | i_sub | i_and | i_or  | i_xor | i_sll |
        i_srl | i_sra | i_addi | i_andi | i_ori | i_xori |
        i_lw  | i_lui | i_jal) & wpcir;                    // int rf
wmem = (i_sw  | i_swc1) & wpcir;                           // memory
wf   = (i_fadd | i_fsub | i_fmul | i_fdiv | i_fsqrt) & wpcir; // fp rf x
wfpr =  i_lwc1 & wpcir;                                    // fp rf y
```

10.4 Pipelined CPU/FPU Design in Verilog HDL

Based on the descriptions in the previous sections, we design the pipelined CPU that contains an FPU in this section.

10.4.1 Circuit of Pipelined CPU/FPU

Figure 10.21 shows the top block diagram of the CPU. It consists of an IU, an FPU, a two-write-port floating-point register file, and four 2-to-1 multiplexers. The detailed circuits of IU and FPU will be given in the following two sections.

The final float result `wd` calculated by the FPU is written to the float register file via the write port x in the WB stage; `wn` is the destination register number and `ww` is the write enable. The memory data word `wmo` loaded by the `lwc1` instruction is written to the float register file via the write port y in the WB stage; `wrn` is the destination register number and `wwfpr` is the write enable. `qfa` and `qfb` are the two data words read from the float registers `fs` and `ft`, respectively.

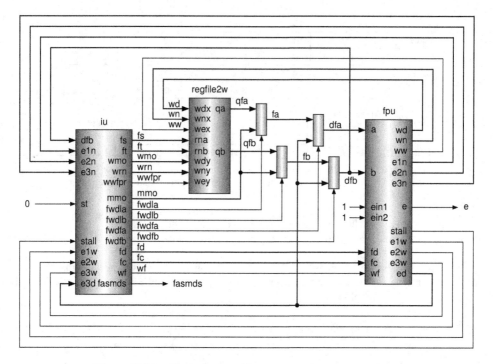

Figure 10.21   Block diagram of CPU with one IU and one FPU

The two multiplexers in the left side are used to bypass mmo, the memory data in the MEM stage. Their selection signals are fwdla and fwdlb, respectively. The two multiplexers in the right side are used to bypass e3d, the float result in the E3 stage. Their selection signals are fwdfa and fwdfb, respectively. The outputs of these two multiplexers, dfa and dfb, are fed to the FPU. dfb is also used by the swc1 instruction to store it to the data memory. The other signals that are connected directly between the IU and the FPU are used for the floating-point operation control or for determining conditions of the pipeline stalls.

There are two output signals that were not used: fasmds and e. The signal fasmds indicates that the instruction in the ID stage is a computational float instruction; e is a pipeline register write enable. There are also three input signals, st, ein1, and ein2, that are connected 0, 1, and 1, respectively. These five signals will be used in the design of other CPUs, which we will describe in later chapters.

10.4.2   Circuit of the Floating-Point Unit

The FPU executes the computational float instructions; its block diagram is shown in Figure 10.22. The FPU consists of four modules for float addition/subtraction, multiplication, division, and square root, which we described in the previous chapter. Other parts include a multiplexer for selecting a float result, the pipeline registers, and some logic gates for generating the control signals.

The input signals are listed as follows: (i) two float inputs a and b; (ii) float operation control fc; (iii) float register file write enable wf, for computational float instructions; (iv) float destination register number fd; and (v) ein1 and ein2 for later use. Both signals are connected to 1 in the current CPU design.

The output signals include (i) float result ed in the E3 stage; (ii) stall signal stall generated by div.s and sqrt.s; (iii) final float result wd in the WB stage; (iv) float register write enable ww in the

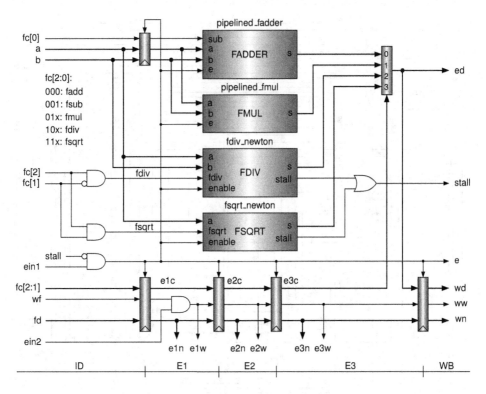

Figure 10.22 Block diagram of the floating-point unit

WB stage; (v) float destination register number wn in the WB stage; and (vi) float register write enables and destination register numbers e1w and e1n, e2w and e2n, and e3w and e3n, in E1, E2, and E3 stages, respectively, which are used for generating multiplexer selection signals and determining the pipeline stall conditions.

10.4.3   Circuit of the Integer Unit

Figure 10.23 shows the detailed circuit of the IU. Compared to the pipelined CPU described in Chapter 8, the new added parts include decoding the float instructions and generating the control signals, such as float operation control, float register write enables, and destination register numbers, multiplexer selection signals for internal forwarding, and signals for pipeline stalls.

10.4.4   Verilog HDL Codes of the Pipelined CPU/FPU

We give the complete Verilog HDL codes of the pipelined CPU with an FPU described above. The following is the code in the top module, which is almost identical to the circuit in Figure 10.21. The modules iu and fpu that it invokes will be listed next; all other modules were provided previously.

```
// pipelined cpu with fpu, instruction memory, and data memory
module fpu_1_iu (clk,memclk,clrn,pc,inst,ealu,malu,walu,wn,wd,ww,stl_lw,
                 stl_fp,stl_lwc1,stl_swc1,stl,cnt_div,cnt_sqrt,
                 e1n,e2n,e3n,e3d,e);
```

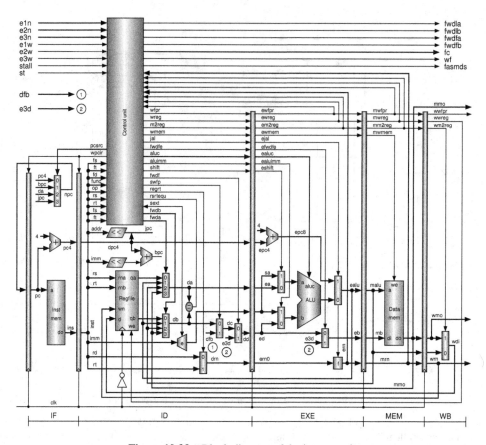

Figure 10.23 Block diagram of the integer unit

```
input         clk, memclk, clrn;                    // clocks and reset
output [31:0] pc, inst, ealu, malu, walu;
output [31:0] e3d, wd;
output  [4:0] e1n, e2n, e3n, wn;
output        ww, stl_lw, stl_fp, stl_lwc1, stl_swc1, stl;
output        e;                      // for multithreading CPU, not used here
output  [4:0] cnt_div, cnt_sqrt;                    // for testing
wire   [31:0] qfa,qfb,fa,fb,dfa,dfb,mmo,wmo;        // for iu
wire    [4:0] fs,ft,fd,wrn;
wire    [2:0] fc;
wire    [1:0] e1c,e2c,e3c;                          // for fpu
wire          fwdla,fwdlb,fwdfa,fwdfb,wf,fasmds,e1w,e2w,e3w,wwfpr;
iu i_u (e1n,e2n,e3n, e1w,e2w,e3w,stl,1'b0,          // st = 0
        dfb,e3d, clk,memclk,clrn,
        fs,ft,wmo,wrn,wwfpr,mmo,fwdla,fwdlb,fwdfa,fwdfb,fd,fc,wf,fasmds,
        pc,inst,ealu,malu,walu,                     // for testing
        stl_lw,stl_fp,stl_lwc1,stl_swc1);           // for testing
regfile2w fpr (fs,ft,wd,wn,ww,wmo,wrn,wwfpr,~clk,clrn,qfa,qfb);
```

```
        mux2x32 fwd_f_load_a (qfa,mmo,fwdla,fa);        // forward lwc1 to fp a
        mux2x32 fwd_f_load_b (qfb,mmo,fwdlb,fb);        // forward lwc1 to fp b
        mux2x32 fwd_f_res_a  (fa,e3d,fwdfa,dfa);        // forward fp res to fp a
        mux2x32 fwd_f_res_b  (fb,e3d,fwdfb,dfb);        // forward fp res to fp b
        fpu fp_unit (dfa,dfb,fc,wf,fd,1'b1,clk,clrn,e3d,wd,wn,ww,
                     stl,eln,elw,e2n,e2w,e3n,e3w,
                     elc,e2c,e3c,cnt_div,cnt_sqrt,e,1'b1);
endmodule
```

Below is the Verilog HDL code of the fpu module that implements the circuit of Figure 10.22. It invokes four modules pipelined_fadder, pipelined_fmul, fdiv_newton, and fsqrt_newton, which were given in the previous chapter.

```
module fpu (a,b,fc,wf,fd,ein1,clk,clrn,ed,wd,wn,ww,st_ds,eln,elw,       // fpu
            e2n,e2w,e3n,e3w,  elc,e2c,e3c,count_div,count_sqrt,e,ein2);
    input          clk, clrn;             // clock and reset
    input   [31:0] a, b;                  // 32-bit fp numbers
    input   [4:0]  fd;                    // fp dest reg number
    input   [2:0]  fc;                    // fp control
    input          wf;                    // write fp regfile
    input          ein1;                  // no_cache_stall
    input          ein2;                  // for canceling E1 inst
    output  [31:0] ed,wd;                 // wd: fp result
    output  [4:0]  count_div,count_sqrt;  // for testing
    output  [4:0]  eln,e2n,e3n,wn;        // reg numbers
    output  [1:0]  elc,e2c,e3c;           // for testing
    output         elw,e2w,e3w,ww;        // write fp regfile
    output         st_ds;                 // stall caused by fdiv or fsqrt
    output         e;                     // ein1 & ~st_ds
    reg     [31:0] wd;
    reg     [31:0] efa,efb;
    reg     [4:0]  eln,e2n,e3n,wn;
    reg     [1:0]  elc,e2c,e3c;
    reg            elw0,e2w,e3w,ww,sub;
    wire    [31:0] s_add,s_mul,s_div,s_sqrt;
    wire    [25:0] reg_x_div,reg_x_sqrt;
    wire           busy_div,stall_div,busy_sqrt,stall_sqrt;
    wire           fdiv  = fc[2] & ~fc[1];
    wire           fsqrt = fc[2] & fc[1];
    assign         elw   = elw0 & ein2;
    assign         e     = ein1 & ~st_ds;
    pipelined_fadder f_add  (efa,efb,sub,2'b0,s_add,clk,clrn,e);
    pipelined_fmul   f_mul  (efa,efb,2'b0,s_mul,clk,clrn,e);
    fdiv_newton      f_div  (a,b,2'b0,fdiv, e,clk,clrn,s_div, busy_div,
                             stall_div, count_div, reg_x_div );
    fsqrt_newton     f_sqrt (a,2'b0,fsqrt,e,clk,clrn,s_sqrt,busy_sqrt,
                             stall_sqrt,count_sqrt,reg_x_sqrt);
    assign st_ds = stall_div | stall_sqrt;
```

```
   mux4x32 fsel (s_add,s_mul,s_div,s_sqrt,e3c,ed);
   always @ (negedge clrn or posedge clk)
     if (!clrn) begin                          // pipeline registers
         sub <= 0;              efa  <= 0;            efb <= 0;
         e1c <= 0;              e1w0 <= 0;            e1n <= 0;
         e2c <= 0;              e2w  <= 0;            e2n <= 0;
         e3c <= 0;              e3w  <= 0;            e3n <= 0;
         wd  <= 0;              ww   <= 0;            wn  <= 0;
     end else if (e) begin
         sub <= fc[0];          efa  <= a;            efb <= b;
         e1c <= fc[2:1];        e1w0 <= wf;           e1n <= fd;
         e2c <= e1c;            e2w  <= e1w;          e2n <= e1n;
         e3c <= e2c;            e3w  <= e2w;          e3n <= e2n;
         wd  <= ed;             ww   <= e3w;          wn  <= e3n;
     end
endmodule
```

The following is the Verilog HDL code of the iu module. It implements the circuit shown in Figure 10.23. The module of the control unit iu_control it invokes is given next.

```
// pipelined cpu with instruction memory, data memory, and interface to fpu
module iu (e1n,e2n,e3n, e1w,e2w,e3w, stall,st,dfb,e3d, clk,memclk,clrn,
           fs,ft,wmo,wrn,wwfpr,mmo,fwdla,fwdlb,fwdfa,fwdfb,fd,fc,wf,fasmds,
           pc,inst,ealu,malu,walu,stall_lw,stall_fp,stall_lwc1,stall_swc1);
   input  [31:0] dfb, e3d;
   input  [4:0] e1n, e2n, e3n;
   input         e1w, e2w, e3w, stall, st;
   input         clk, memclk, clrn;                     // clocks, reset
   output [31:0] pc, inst, ealu, malu, walu;
   output [31:0] mmo, wmo;
   output [4:0] fs, ft, fd, wrn;
   output [2:0] fc;
   output        wwfpr, fwdla, fwdlb, fwdfa, fwdfb, wf, fasmds;
   output        stall_lw, stall_fp, stall_lwc1, stall_swc1;
   wire   [31:0] bpc,jpc,npc,pc4,ins,dpc4,inst,qa,qb,da,db,dimm,dc,dd;
   wire   [31:0] simm,epc8,alua,alub,ealu0,ealu,sa,eb,mmo,wdi;
   wire   [5:0] op,func;
   wire   [4:0] rs,rt,rd,drn,ern;
   wire   [3:0] aluc;
   wire   [1:0] pcsrc,fwda,fwdb;
   wire          wpcir,wreg,m2reg,wmem,aluimm,shift,jal,z;
   reg           ewfpr,ewreg,em2reg,ewmem,ejal,efwdfe,ealuimm,eshift;
   reg           mwfpr,mwreg,mm2reg,mwmem;
   reg           wwfpr,wwreg,wm2reg;
   reg    [31:0] epc4,ea,ed,eimm,malu,mb,wmo,walu;
   reg    [4:0] ern0,mrn,wrn;
   reg    [3:0] ealuc;
   dffe32 program_counter (npc,clk,clrn,wpcir,pc);      // pc
   cla32 pc_plus4 (pc,32'h4,1'b0,pc4);                  // pc+4
```

```
   mux4x32 next_pc (pc4,bpc,da,jpc,pcsrc,npc);              // next pc
   inst_mem i_mem (pc,ins);                                 // inst memory
   dffe32 pc_4_r (pc4,clk,clrn,wpcir,dpc4);                 // pc+4 reg
   dffe32 inst_r (ins,clk,clrn,wpcir,inst);                 // ir
   wire swfp,regrt,sext,fwdf,fwdfe,wfpr;
   assign op   = inst[31:26];
   assign rs   = inst[25:21];
   assign rt   = inst[20:16];
   assign rd   = inst[15:11];
   assign ft   = inst[20:16];
   assign fs   = inst[15:11];
   assign fd   = inst[10:6];
   assign func = inst[5:0];
   assign simm = {{16{sext&inst[15]}},inst[15:0]};
   assign jpc  = {dpc4[31:28],inst[25:0],2'b00};            // jump target
   cla32 br_addr (dpc4,{simm[29:0],2'b00},1'b0,bpc);        // branch target
   regfile rf (rs,rt,wdi,wrn,wwreg,~clk,clrn,qa,qb);        // reg file
   mux4x32 alu_a (qa,ealu,malu,mmo,fwda,da);                // forward A
   mux4x32 alu_b (qb,ealu,malu,mmo,fwdb,db);                // forward B
   mux2x32 store_f (db,dfb,swfp,dc);                        // swc1
   mux2x32 fwd_f_d (dc,e3d,fwdf,dd);                        // forward
   wire    rsrtequ = ~|(da^db);                             // (da == db)
   mux2x5 des_reg_no (rd,rt,regrt,drn);                     // dest reg
   iu_control cu (op,func,rs,rt,fs,ft,rsrtequ,ewfpr,ewreg, // control unit
                  em2reg,ern,mwfpr,mwreg,mm2reg,mrn,e1w,
                  e1n,e2w,e2n,e3w,e3n,stall,st,pcsrc,wpcir,
                  wreg,m2reg,wmem,jal,aluc,aluimm,shift,
                  sext,regrt,fwda,fwdb,swfp,fwdf,fwdfe,wfpr,
                  fwdla,fwdlb,fwdfa,fwdfb,fc,wf,fasmds,
                  stall_lw,stall_fp,stall_lwc1,stall_swc1);
   always @(negedge clrn or posedge clk)                    // ID/EXE regs
     if (!clrn) begin
         ewfpr   <= 0;                  ewreg    <= 0;
         em2reg  <= 0;                  ewmem    <= 0;
         ejal    <= 0;                  ealuimm  <= 0;
         efwdfe  <= 0;                  ealuc    <= 0;
         eshift  <= 0;                  epc4     <= 0;
         ea      <= 0;                  ed       <= 0;
         eimm    <= 0;                  ern0     <= 0;
     end else begin
         ewfpr   <= wfpr;               ewreg    <= wreg;
         em2reg  <= m2reg;              ewmem    <= wmem;
         ejal    <= jal;                ealuimm  <= aluimm;
         efwdfe  <= fwdfe;              ealuc    <= aluc;
         eshift  <= shift;              epc4     <= dpc4;
         ea      <= da;                 ed       <= dd;
         eimm    <= simm;               ern0     <= drn;
     end
```

```
        cla32 ret_addr (epc4,32'h4,1'b0,epc8);              // pc+8
        assign sa = {eimm[5:0],eimm[31:6]};                 // shift amount
        mux2x32 alu_ina (ea,sa,eshift,alua);                // alu input a
        mux2x32 alu_inb (ed,eimm,ealuimm,alub);             // alu input b
        mux2x32 save_pc8 (ealu0,epc8,ejal,ealu);            // pc+8 if jal
        alu al_unit (alua,alub,ealuc,ealu0,z);              // alu
        assign ern = ern0 | {5{ejal}};                      // $31 for jal
        mux2x32 fwd_f_e (ed,e3d,efwdfe,eb);                 // forward
        always @(negedge clrn or posedge clk)               // EXE/MEM regs
          if (!clrn) begin
              mwfpr   <= 0;             mwreg   <= 0;
              mm2reg  <= 0;             mwmem   <= 0;
              malu    <= 0;             mb      <= 0;
              mrn     <= 0;
          end else begin
              mwfpr   <= ewfpr;         mwreg   <= ewreg;
              mm2reg  <= em2reg;        mwmem   <= ewmem;
              malu    <= ealu;          mb      <= eb;
              mrn     <= ern;
          end
        data_mem d_mem (mwmem,malu,mb,memclk,mmo);          // data memory
        always @(negedge clrn or posedge clk)               // MEM/WB regs
          if (!clrn) begin
              wwfpr   <= 0;             wwreg   <= 0;
              wm2reg  <= 0;             wmo     <= 0;
              walu    <= 0;             wrn     <= 0;
          end else begin
              wwfpr   <= mwfpr;         wwreg   <= mwreg;
              wm2reg  <= mm2reg;        wmo     <= mmo;
              walu    <= malu;          wrn     <= mrn;
          end
        mux2x32 wb_sel (walu,wmo,wm2reg,wdi);
endmodule
```

The Verilog HDL code of the IU control unit is listed below. It generates all the control signals, including those for the FPU control.

```
// control unit of pipelined cpu with memories and interface to fpu
module iu_control (op,func,rs,rt,fs,ft,rsrtequ,ewfpr,ewreg,em2reg,ern,mwfpr,
                   mwreg,mm2reg,mrn,e1w,e1n,e2w,e2n,e3w,e3n,stall_div_sqrt,
                   st,pcsrc,wpcir,wreg,m2reg,wmem,jal,aluc,aluimm,shift,
                   sext,regrt,fwda,fwdb,swfp,fwdf,fwdfe,wfpr,fwdla,fwdlb,
                   fwdfa,fwdfb,fc,wf,fasmds,stall_lw,stall_fp,stall_lwc1,
                   stall_swc1);
    input           rsrtequ,ewreg,em2reg,ewfpr, mwreg,mm2reg,mwfpr;
    input           e1w,e2w,e3w,stall_div_sqrt,st;
    input   [5:0]   op,func;
```

```
input   [4:0]  rs,rt,fs,ft,ern,mrn,e1n,e2n,e3n;
output         wpcir,wreg,m2reg,wmem,jal,aluimm,shift,sext,regrt;
output         swfp,fwdf,fwdfe;
output         fwdla,fwdlb,fwdfa,fwdfb;
output         wfpr,wf,fasmds;
output  [3:0]  aluc;
output  [2:0]  fc;
output  [1:0]  pcsrc,fwda,fwdb;
output         stall_lw,stall_fp,stall_lwc1,stall_swc1;
wire           rtype,i_add,i_sub,i_and,i_or,i_xor,i_sll,i_srl,i_sra;
wire           i_jr,i_j,i_jal;
wire           i_addi,i_andi,i_ori,i_xori,i_lw,i_sw,i_beq,i_bne,i_lui;
wire           ftype,i_lwc1,i_swc1,i_fadd,i_fsub,i_fmul,i_fdiv,i_fsqrt;
and(rtype,~op[5],~op[4],~op[3],~op[2],~op[1],~op[0]);        // r format
and(i_add,rtype, func[5],~func[4],~func[3],~func[2],~func[1],~func[0]);
and(i_sub,rtype, func[5],~func[4],~func[3],~func[2], func[1],~func[0]);
and(i_and,rtype, func[5],~func[4],~func[3], func[2],~func[1],~func[0]);
and(i_or, rtype, func[5],~func[4],~func[3], func[2],~func[1], func[0]);
and(i_xor,rtype, func[5],~func[4],~func[3], func[2], func[1],~func[0]);
and(i_sll,rtype,~func[5],~func[4],~func[3],~func[2],~func[1],~func[0]);
and(i_srl,rtype,~func[5],~func[4],~func[3],~func[2], func[1],~func[0]);
and(i_sra,rtype,~func[5],~func[4],~func[3],~func[2], func[1], func[0]);
and(i_jr, rtype,~func[5],~func[4], func[3],~func[2],~func[1],~func[0]);
and(i_addi,~op[5],~op[4], op[3],~op[2],~op[1],~op[0]);       // i format
and(i_andi,~op[5],~op[4], op[3], op[2],~op[1],~op[0]);
and(i_ori, ~op[5],~op[4], op[3], op[2],~op[1], op[0]);
and(i_xori,~op[5],~op[4], op[3], op[2], op[1],~op[0]);
and(i_lw,   op[5],~op[4],~op[3],~op[2], op[1], op[0]);
and(i_sw,   op[5],~op[4], op[3],~op[2], op[1], op[0]);
and(i_beq, ~op[5],~op[4],~op[3], op[2],~op[1],~op[0]);
and(i_bne, ~op[5],~op[4],~op[3], op[2],~op[1], op[0]);
and(i_lui, ~op[5],~op[4], op[3], op[2], op[1], op[0]);
and(i_j,   ~op[5],~op[4],~op[3],~op[2], op[1],~op[0]);       // j format
and(i_jal, ~op[5],~op[4],~op[3],~op[2], op[1], op[0]);
and(ftype,~op[5], op[4],~op[3],~op[2],~op[1], op[0]);        // f format
and(i_lwc1, op[5], op[4],~op[3],~op[2],~op[1], op[0]);
and(i_swc1, op[5], op[4], op[3],~op[2],~op[1], op[0]);
and(i_fadd,ftype,~func[5],~func[4],~func[3],~func[2],~func[1],~func[0]);
and(i_fsub,ftype,~func[5],~func[4],~func[3],~func[2],~func[1], func[0]);
and(i_fmul,ftype,~func[5],~func[4],~func[3],~func[2], func[1],~func[0]);
and(i_fdiv,ftype,~func[5],~func[4],~func[3],~func[2], func[1], func[0]);
and(i_fsqrt,ftype,~func[5],~func[4],~func[3], func[2],~func[1], func[0]);
wire i_rs = i_add  | i_sub | i_and  | i_or  | i_xor  | i_jr  | i_addi |
            i_andi | i_ori | i_xori | i_lw  | i_sw   | i_beq | i_bne  |
            i_lwc1 | i_swc1;
wire i_rt = i_add  | i_sub | i_and  | i_or  | i_xor  | i_sll | i_srl  |
            i_sra  | i_sw  | i_beq  | i_bne;
```

```verilog
assign stall_lw = ewreg & em2reg & (ern != 0) & (i_rs & (ern == rs) |
                                                  i_rt & (ern == rt));
reg    [1:0] fwda, fwdb;
always @ (ewreg or mwreg or ern or mrn or em2reg or mm2reg or rs or
          rt) begin
    fwda = 2'b00;                               // default: no hazards
    if (ewreg & (ern != 0) & (ern == rs) & ~em2reg) begin
        fwda = 2'b01;                           // select exe_alu
    end else begin
        if (mwreg & (mrn != 0) & (mrn == rs) & ~mm2reg) begin
            fwda = 2'b10;                        // select mem_alu
        end else begin
            if (mwreg & (mrn != 0) & (mrn == rs) & mm2reg) begin
                fwda = 2'b11;                    // select mem_lw
            end
        end
    end
    fwdb = 2'b00;                               // default: no hazards
    if (ewreg & (ern != 0) & (ern == rt) & ~em2reg) begin
        fwdb = 2'b01;                           // select exe_alu
    end else begin
        if (mwreg & (mrn != 0) & (mrn == rt) & ~mm2reg) begin
            fwdb = 2'b10;                        // select mem_alu
        end else begin
            if (mwreg & (mrn != 0) & (mrn == rt) & mm2reg) begin
                fwdb = 2'b11;                    // select mem_lw
            end
        end
    end
end
assign wreg  =(i_add  | i_sub  | i_and  | i_or   | i_xor | i_sll  |
               i_srl  | i_sra  | i_addi | i_andi | i_ori | i_xori |
               i_lw   | i_lui  | i_jal) & wpcir;
assign regrt = i_addi | i_andi | i_ori | i_xori | i_lw | i_lui | i_lwc1;
assign jal   = i_jal;
assign m2reg = i_lw;
assign shift = i_sll  | i_srl  | i_sra;
assign aluimm= i_addi | i_andi | i_ori | i_xori | i_lw | i_lui | i_sw |
               i_lwc1 | i_swc1;
assign sext  = i_addi | i_lw   | i_sw | i_beq | i_bne | i_lwc1 | i_swc1;
assign aluc[3] = i_sra;
assign aluc[2] = i_sub | i_or  | i_srl | i_sra | i_ori  | i_lui;
assign aluc[1] = i_xor | i_sll | i_srl | i_sra | i_xori | i_beq |
                 i_bne | i_lui;
assign aluc[0] = i_and | i_or  | i_sll | i_srl | i_sra | i_andi | i_ori;
assign wmem    = (i_sw | i_swc1) & wpcir;
assign pcsrc[1] = i_jr | i_j | i_jal;
```

```
    assign pcsrc[0] = i_beq & rsrtequ | i_bne & ~rsrtequ | i_j | i_jal;
    // fop:  000: fadd  001: fsub  01x: fmul  10x: fdiv  11x: fsqrt
    wire    [2:0] fop;
    assign fop[0]    = i_fsub;                          // fpu control code
    assign fop[1]    = i_fmul | i_fsqrt;
    assign fop[2]    = i_fdiv | i_fsqrt;
    // stall caused by fp data harzards
    wire       i_fs = i_fadd | i_fsub | i_fmul | i_fdiv | i_fsqrt; // use fs
    wire       i_ft = i_fadd | i_fsub | i_fmul | i_fdiv;          // use ft
    assign stall_fp = (e1w & (i_fs & (e1n == fs) | i_ft & (e1n == ft))) |
                      (e2w & (i_fs & (e2n == fs) | i_ft & (e2n == ft)));
    assign fwdfa     = e3w & (e3n == fs);        // forward fpu e3d to fp a
    assign fwdfb     = e3w & (e3n == ft);        // forward fpu e3d to fp b
    assign wfpr      = i_lwc1 & wpcir;           // fp rf y write enable
    assign fwdla     = mwfpr & (mrn == fs);      // forward mmo to fp a
    assign fwdlb     = mwfpr & (mrn == ft);      // forward mmo to fp b
    assign stall_lwc1 = ewfpr & (i_fs & (ern == fs) | i_ft & (ern == ft));
    assign swfp      = i_swc1;                    // select signal
    assign fwdf      = swfp & e3w & (ft == e3n); // forward to id  stage
    assign fwdfe     = swfp & e2w & (ft == e2n); // forward to exe stage
    assign stall_swc1 = swfp & e1w & (ft == e1n); // stall
    wire stall_others = stall_lw | stall_fp | stall_lwc1 | stall_swc1 | st;
    assign wpcir     = ~(stall_div_sqrt | stall_others);
    assign fc        = fop & {3{~stall_others}};
    assign wf        = i_fs & wpcir;             // fp rf x write enable
    assign fasmds    = i_fs;
endmodule
```

10.5   Memory Modules and Pipelined CPU/FPU Test

This section provides the test program and data and gives the simulation waveforms.

10.5.1   Instruction Memory and Data Memory

The instruction memory and data memory can be implemented with general Verilog HDL statements, like what we did in Chapter 8. Here we use the LPM (library of parameterized modules) provided by Altera. The instruction memory is implemented with lpm_rom, see the code below. We will list the contents of the initialization file in the mif format later. It can be converted to the hex format for simulation with ModelSim.

```
module inst_mem (a,inst);
    input   [31:0] a;
    output  [31:0] inst;
    lpm_rom rom (.address(a[7:2]),.q(inst),
                 .inclock(),.outclock(),.memenab());
    defparam rom.lpm_width       = 32,
             rom.lpm_widthad     = 6,
```

```
                rom.lpm_file              = "inst_mem.hex",
                rom.lpm_outdata           = "UNREGISTERED",
                rom.lpm_address_control = "UNREGISTERED";
endmodule
```

The data memory is implemented with lpm_ram_dq. Following is the code. Similarly, we will give the memory initialization file later.

```
module data_mem (we,addr,datain,memclk,dataout);
    input   [31:0] addr,datain;
    input          we,memclk;
    output [31:0] dataout;
    lpm_ram_dq ram (.data(datain),.address(addr[6:2]),.we(we),
                    .inclock(memclk),.outclock(memclk),.q(dataout));
    defparam ram.lpm_width          = 32;
    defparam ram.lpm_widthad        = 5;
    defparam ram.lpm_file           = "data_mem.hex";
    defparam ram.lpm_indata         = "REGISTERED";
    defparam ram.lpm_outdata        = "REGISTERED";
    defparam ram.lpm_address_control = "REGISTERED";
endmodule
```

10.5.2   Test Program of Pipelined CPU/FPU

Below is the test program in the mif format. We have added the floating-point computational instructions and load/store instructions for the FPU. The patterns for checking internal forwarding and pipeline stalls are also prepared.

```
DEPTH = 64;            % Memory depth and width are required        %
WIDTH = 32;            % Enter a decimal number                     %
ADDRESS_RADIX = HEX;  % Address and value radixes are optional      %
DATA_RADIX = HEX;      % Enter BIN, DEC, HEX, or OCT; unless         %
                       % otherwise specified, radixes = HEX         %
CONTENT
BEGIN
[0..3f] : 00000000; % Range--Every address from 0 to 3f = 00000000  %
 0 : 00000820; % (00)        add    $1,   $0, $0  # address of data[0] %
 1 : c4200000; % (04)        lwc1   f0,   0($1)   # load fp data       %
 2 : c4210050; % (08)        lwc1   f1,   80($1)  # load fp data       %
 3 : c4220054; % (0c)        lwc1   f2,   84($1)  # load fp data       %
 4 : c4230058; % (10)        lwc1   f3,   88($1)  # load fp data       %
 5 : c424005c; % (14)        lwc1   f4,   92($1)  # load fp data       %
 6 : 46002100; % (18)        add.s  f4,   f4, f0  # f4: stall 1        %
 7 : 460418c1; % (1c)        sub.s  f3,   f3, f4  # f4: stall 2        %
 8 : 46022082; % (20)        mul.s  f2,   f4, f2  # mul               %
 9 : 46040842; % (24)        mul.s  f1,   f1, f4  # mul               %
 a : e4210070; % (28)        swc1   f1,   112($1) # f1: stall 1        %
```

```
b  : e4220074;  % (2c)            swc1   f2, 116($1)  # store fp data         %
c  : e4230078;  % (30)            swc1   f3, 120($1)  # store fp data         %
d  : e424007c;  % (34)            swc1   f4, 124($1)  # store fp data         %
e  : 20020004;  % (38)            addi   $2,  $0,  4  # counter               %
f  : c4230000;  % (3c) l3:        lwc1   f3,   0($1)  # load fp data          %
10 : c4210050;  % (40)            lwc1   f1,  80($1)  # load fp data          %
11 : 46030840;  % (44)            add.s  f1,  f1, f3  # stall 1               %
12 : 46030841;  % (48)            sub.s  f1,  f1, f3  # stall 2               %
13 : e4210030;  % (4c)            swc1   f1,  48($1)  # stall 1               %
14 : c4050004;  % (50)            lwc1   f5,  04($0)  # load fp data          %
15 : c4060008;  % (54)            lwc1   f6,  08($0)  # load fp data          %
16 : c408000c;  % (58)            lwc1   f8,  12($0)  # load fp data          %
17 : 460629c3;  % (5c)            div.s  f7,  f5, f6  # div                   %
18 : 46004244;  % (60)            sqrt.s f9,  f8       # sqrt                  %
19 : 46004a84;  % (64)            sqrt.s f10, f9       # sqrt                  %
1a : 2042ffff;  % (68)            addi   $2,  $2, -1  # counter - 1           %
1b : 1440fff3;  % (6c)            bne    $2,  $0, 13  # finish?               %
1c : 20210004;  % (70)            addi   $1,  $1,  4  # address+4, delay slot %
1d : 3c010000;  % (74) iu_test:lui $1, 0              # address of data[0]    %
1e : 34240050;  % (78)            ori  $4, $1, 80     # address of data[0]    %
1f : 0c000038;  % (7c) call:      jal  sum            # call function         %
20 : 20050004;  % (80) dslot1: addi $5, $0,  4        # delayed slot(ds)      %
21 : ac820000;  % (84) return: sw  $2, 0($4)          # store result          %
22 : 8c890000;  % (88)            lw   $9, 0($4)       # check sw              %
23 : 01244022;  % (8c)            sub  $8, $9, $4      # sub: $8 <-- $9 - $4   %
24 : 20050003;  % (90)            addi $5, $0,  3      # counter               %
25 : 20a5ffff;  % (94) loop2:   addi $5, $5, -1       # counter - 1           %
26 : 34a8ffff;  % (98)            ori  $8, $5, 0xffff # zero-extend: 0000ffff %
27 : 39085555;  % (9c)            xori $8, $8, 0x5555 # zero-extend: 0000aaaa %
28 : 2009ffff;  % (a0)            addi $9, $0, -1      # sign-extend: ffffffff %
29 : 312affff;  % (a4)            andi $10, $9, 0xffff# zero-extend: 0000ffff %
2a : 01493025;  % (a8)            or   $6, $10, $9    # or:  ffffffff         %
2b : 01494026;  % (ac)            xor  $8, $10, $9    # xor: ffff0000         %
2c : 01463824;  % (b0)            and  $7, $10, $6    # and: 0000ffff         %
2d : 10a00003;  % (b4)            beq  $5, $0, shift  # if $5 = 0, goto shift %
2e : 00000000;  % (b8) dslot2: nop                    # ds                    %
2f : 08000025;  % (bc)            j    loop2          # jump loop2            %
30 : 00000000;  % (c0) dslot3: nop                    # ds                    %
31 : 2005ffff;  % (c4) shift:   addi $5, $0, -1       # $5   = ffffffff       %
32 : 000543c0;  % (c8)            sll  $8, $5, 15     # << 15 = ffff8000      %
33 : 00084400;  % (cc)            sll  $8, $8, 16     # << 16 = 80000000      %
34 : 00084403;  % (d0)            sra  $8, $8, 16     # >>> 16 = ffff8000     %
35 : 000843c2;  % (d4)            srl  $8, $8, 15     # >> 15 = 0001ffff      %
36 : 08000036;  % (d8) finish: j  finish              # dead loop             %
37 : 00000000;  % (dc) dslot4: nop                    # ds                    %
38 : 00004020;  % (e0) sum:     add  $8, $0, $0       # sum                   %
39 : 8c890000;  % (e4) loop:    lw   $9, 0($4)        # load data             %
```

```
3a  :  01094020;  %  (e8)            add  $8, $8, $9      # sum                        %
3b  :  20a5ffff;  %  (ec)            addi $5, $5, -1      # counter - 1                %
3c  :  14a0fffc;  %  (f0)            bne  $5, $0, loop    # finish?                    %
3d  :  20840004;  %  (f4)  dslot5:   addi $4, $4,  4      # address + 4, ds            %
3e  :  03e00008;  %  (f8)            jr   $31             # return                     %
3f  :  00081000;  %  (fc)  dslot6:   sll  $2, $8, 0       # move res. to v0, ds        %
END ;
```

Below is the test data in the mif format for initializing the data memory.

```
DEPTH = 32;              % Memory depth and width are required           %
WIDTH = 32;              % Enter a decimal number                        %
ADDRESS_RADIX = HEX;     % Address and value radixes are optional        %
DATA_RADIX = HEX;        % Enter BIN, DEC, HEX, or OCT; unless            %
                         % otherwise specified, radixes = HEX            %
CONTENT
BEGIN
[0..1f] : 00000000;      % Range--Every address from 0 to 1f = 00000000  %
   0 :     bf800000;     % (00)  1 01111111 00..0 fp -1                   %
   1 :     40800000;     % (04)                                          %
   2 :     40000000;     % (08)                                          %
   3 :     41100000;     % (0c)                                          %
  14 :     40c00000;     % (50)  0 10000001 10..0 data[0]  4.5           %
  15 :     41c00000;     % (54)  0 10000011 10..0 data[1]                %
  16 :     43c00000;     % (58)  0 10000111 10..0 data[2]                %
  17 :     47c00000;     % (5c)  0 10001111 10..0 data[3]                %
END ;
```

10.5.3   Test Waveforms of Pipelined CPU/FPU

Figure 10.24 shows the waveforms when the CPU executes the instructions:

```
5 : c424005c; % (14)        lwc1   f4,   92($1)  # load fp data           %
6 : 46002100; % (18)        add.s  f4,   f4, f0  # f4: stall 1            %
7 : 460418c1; % (1c)        sub.s  f3,   f3, f4  # f4: stall 2            %
```

add.s has a data dependency with lwc1, and sub.s has a data dependency with add.s. We can see that stl_lwc1 is 1 in the ID stage of add.s, and stl_fp is 1 in the ID stage of sub.s. The result of add.s is written in the WB stage with ww = 1. The execution result of add.s is 0x47bfff80, because the two source float operands are 0xbf800000 (in f0) and 0x47c00000 (in f4).

Figure 10.25 shows the waveforms when the CPU executes the following instructions:

```
6 : 46002100; % (18)        add.s  f4,  f4, f0  # f4: stall 1            %
7 : 460418c1; % (1c)        sub.s  f3,  f3, f4  # f4: stall 2            %
8 : 46022082; % (20)        mul.s  f2,  f4, f2  # mul                    %
9 : 46040842; % (24)        mul.s  f1,  f1, f4  # mul                    %
a : e4210070; % (28)        swc1   f1,  112($1) # f1: stall 1            %
```

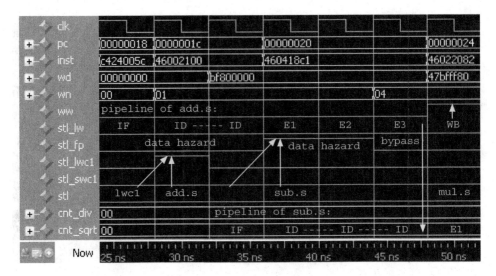

Figure 10.24 Waveform 1 of CPU/FPU (lwc1, add.s, and sub.s)

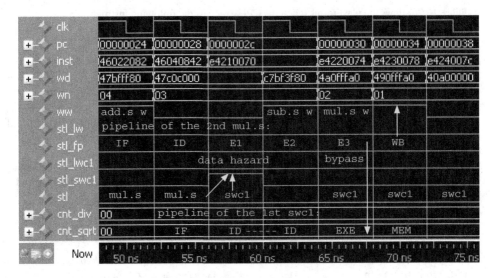

Figure 10.25 Waveform 2 of CPU/FPU (sub.s, mul.s, mul.s, and swc1)

```
b : e4220074; % (2c)    swc1  f2, 116($1)  # store fp data    %
c : e4230078; % (30)    swc1  f3, 120($1)  # store fp data    %
d : e424007c; % (34)    swc1  f4, 124($1)  # store fp data    %
```

Because swc1 has a data dependency with mul.s, the pipeline was stalled by one cycle (stl_swc1 = 1) at the ID stage of swc1. From the signals of ww, wn, and wd, we can see that the four float results were written to the float registers f4, f3, f2, and f1, respectively.

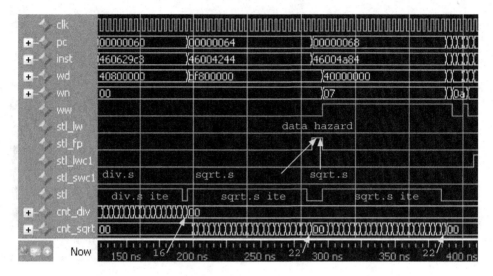

Figure 10.26   Waveform 3 of CPU/FPU (div.s, sqrt.s, and sqrt.s)

Figure 10.26 shows the waveforms when the CPU executes the following instructions:

```
17 : 460629c3;  %  (5c)       div.s   f7,   f5, f6    # div              %
18 : 46004244;  %  (60)       sqrt.s  f9,   f8        # sqrt             %
19 : 46004a84;  %  (64)       sqrt.s  f10,  f9        # sqrt             %
```

These instructions take more cycles for the execution because of the Newton–Raphson iterations. Although we cannot see the values of the counters `cnt_div` and `cnt_sqrt` from the figure, we do know that the values change during every clock cycle (three decimal values of the counters were given by the marks in the bottom of the figure).

For checking the simulation waveforms easily, we list the execution results of some instructions below. These results are correct; we have checked them by executing the programs on an Intel x86 FPU and by simulating the test code on our CPU.

```
0 : 00000820;  %  (00)       add     $1,   $0, $0    # $1   = 00000000           %
1 : c4200000;  %  (04)       lwc1    f0,    0($1)    # f0   = bf800000           %
2 : c4210050;  %  (08)       lwc1    f1,   80($1)    # f1   = 40c00000           %
3 : c4220054;  %  (0c)       lwc1    f2,   84($1)    # f2   = 41c00000           %
4 : c4230058;  %  (10)       lwc1    f3,   88($1)    # f3   = 43c00000           %
5 : c424005c;  %  (14)       lwc1    f4,   92($1)    # f4   = 47c00000           %
6 : 46002100;  %  (18)       add.s   f4,   f4, f0    # f4   = 47bfff80 add.s     %
7 : 460418c1;  %  (1c)       sub.s   f3,   f3, f4    # f3   = c7bf3f80 sub.s     %
8 : 46022082;  %  (20)       mul.s   f2,   f4, f2    # f2   = 4a0fffa0 mul.s     %
9 : 46040842;  %  (24)       mul.s   f1,   f1, f4    # f1   = 490fffa0 mul.s     %
a : e4210070;  %  (28)       swc1    f1,  112($1)    #                           %
b : e4220074;  %  (2c)       swc1    f2,  116($1)    #                           %
c : e4230078;  %  (30)       swc1    f3,  120($1)    #                           %
d : e424007c;  %  (34)       swc1    f4,  124($1)    #                           %
```

```
e  : 20020004; % (38)          addi   $2,  $0,  4   #                            %
f  : c4230000; % (3c) 13:      lwc1   f3,   0($1)   # f3  = bf800000             %
10 : c4210050; % (40)          lwc1   f1,  80($1)   # f1  = 40c00000             %
11 : 46030840; % (44)          add.s  f1,  f1, f3   # f1  = 40a00000 add.s       %
12 : 46030841; % (48)          sub.s  f1,  f1, f3   # f1  = 40c00000 sub.s       %
13 : e4210030; % (4c)          swc1   f1,  48($1)   #                            %
14 : c4050004; % (50)          lwc1   f5,  04($0)   # f5  = 40800000             %
15 : c4060008; % (54)          lwc1   f6,  08($0)   # f6  = 40000000             %
16 : c408000c; % (58)          lwc1   f8,  12($0)   # f8  = 41100000             %
17 : 460629c3; % (5c)          div.s  f7,  f5, f6   # f7  = 40000000 div.s       %
18 : 46004244; % (60)          sqrt.s f9,  f8       # f9  = 40400000 sqrt.s      %
19 : 46004a84; % (64)          sqrt.s f10, f9       # f10 = 3fddb3d7 sqrt.s      %
```

Exercises

10.1 Write a test bench to simulate `regfile2w.v`.

10.2 Design `regfile2w` with schematic capture or write Verilog HDL code in dataflow style.

10.3 Try to add the following instructions to the pipelined CPU described in this chapter. The first two instructions transfer data between the integer register file and the float register file; the last two instructions convert data between integer and float. Their formats are shown in Table 10.2.

```
    mfc1      rt, fs              # rt <-- fs
    mtc1      rt, fs              # fs <-- rt
    cvt.s.w   fd, fs              # fd <-- convert_and_round(fs)
    cvt.w.s   fd, fs              # fd <-- convert_and_round(fs)
```

Table 10.2 Data transfer and data conversion instructions

Inst.	[31:26]	[25:21]	[20:16]	[15:11]	[10:6]	[5:0]	Description
mfc1	010001	00000	rt	fs	00000	000000	Move from CP1
mtc1	010001	00100	rt	fs	00000	000000	Move to CP1
cvt.s.w	010001	10100	00000	fs	fd	100000	Integer to float
cvt.w.s	010001	10000	00000	fs	fd	100000	Float to integer

10.4 Try to add an interrupt/exception mechanism, including the exceptions of the float calculations, to the CPU.

10.5 Try to design the pipelined CPU with a fully pipelined FPU so that such a CPU can also execute `div.s` or `sqrt.s` in every clock cycle. You can do it by using one of the following three methods:
(a) Use a pipelined version of the nonrestoring division and square root algorithms.
(b) Use a huge ROM table so that the precision is good enough through one-time Newton–Raphson algorithms.
(c) Use a large ROM table so that the precision is good enough through two-time Newton–Raphson algorithms and duplicate the iteration circuits.

10.6 Try to design a simple superscalar CPU that can execute an integer instruction and a float instruction in parallel, as shown in Figure 10.27.

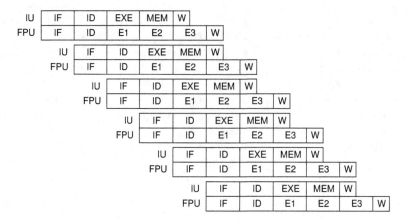

Figure 10.27 Simple superscalar pipelines

10.7 Design a more complex superscalar CPU that can execute a float addition (or subtraction), a float multiplication, a float division, a float square root, and two integer instructions in parallel.

11

Memory Hierarchy and Virtual Memory Management

Memory is a temporary place for storing programs (instructions and data). It is commonly implemented with dynamic random access memory (DRAM). Because DRAM is slower than the CPU (central processing unit), an instruction cache and a data cache are fabricated inside the CPU. Not only the caches but also TLBs (translation lookaside buffers) are fabricated for fast translation from a virtual address to a physical memory address.

This chapter describes the memory structures, cache organizations, virtual memory management, and TLB organizations. The mechanism of the TLB-based MIPS (microprocessor without interlocked pipeline stages) virtual memory management is also introduced.

11.1   Memory

A computer consists of a CPU, the memory, and I/O interfaces. Memory is used to store programs that are being executed by the CPU. There are many types of memory, but we discuss only the following four types of memory in this book.

1. SRAM (static random access memory), which is fast and expensive, is used to design caches and TLBs. Some high-performance computers also use it as the main memory.
2. DRAM, which is large and inexpensive, is mainly used as the computer's main memory.
3. ROM (read-only memory), which is nonvolatile and cheap, is typically used to store the computer's initial start-up program or firmware in embedded systems.
4. CAM (content addressable memory), which is a very special memory, is mainly used to design a fully associative cache or TLB.

Except for ROM, all memories are volatile. It means that when the power supply is off, the contents in the memory will be lost. The contents in such memories are not usable when the power supply is just turned on. Therefore, there must be a ROM in a computer or embedded system.

"Random access" means that any location of the memory can be accessed directly by providing the address of that location. There are some other types of memory that cannot be accessed randomly, the FIFO (first-in first-out) memory, for instance.

Computer Principles and Design in Verilog HDL, First Edition. Yamin Li.
© 2015 Tsinghua University Press. All rights reserved. Published 2015 by John Wiley & Sons Singapore Pte Ltd.
Companion Website: www.wiley.com/go/li/verilog

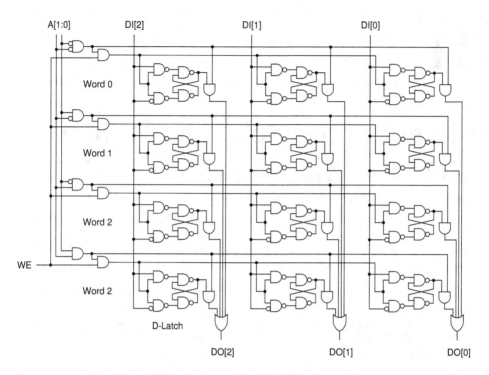

Figure 11.1   Demonstrating SRAM implementation

11.1.1   Static Random Access Memory

In order to understand easily the principle of how the SRAM works, Figure 11.1 illustrates the circuit of a 4 × 3 bit SRAM. There are four words and each word has 3 bits. Each bit is stored in a traditional D-latch. A 2-bit address A[1:0] is decoded to select one word. The content of the selected word is outputted from the data out DO[2:0]. If the write enable WE is an active 1, the data in DI[2:0] will be written to the word selected by A[1:0].

Using a traditional D-latch to hold a bit of information is too expensive. A real SRAM uses a form of two-NOT-gate flip-flop (bistable state) to hold each bit of memory. A memory cell of a SRAM takes 4–6 transistors along with some wiring. Figure 11.2 shows the circuit of a six-transistor cell. Figure 11.2(a) is a simplified drawing for easy understanding. Two NOT gates form a bistable state circuit. The state will be accessed through BL (bit line) if WL (word line), which controls the transistors (switches) of n3 and n4, is active. Figure 11.2(b) shows the detailed circuit where the transistors of p1 and n1 form one NOT gate and p2 and n2 form the other NOT gate. There are six transistors in total.

11.1.2   Dynamic Random Access Memory

Different from SRAM, each memory cell of DRAM is made up of two parts: a transistor and a small capacitor, as shown in Figure 11.3. The capacitor holds the bit of information. The transistor is used to read the capacitor or to change its state. If the capacitor stores electrons, the bit is a 1. If it is empty, the bit is a 0.

The DRAM capacity shown in Figure 11.3 has 64M × 1 bits, which are organized into 8K rows and 8K columns. Accessing these bits needs a 26-bit address, but the DRAM chip provides only a 13-bit address A[12:0]. The 26-bit address is sent to the DRAM twice. The RAS (row access strobe) signal stores the

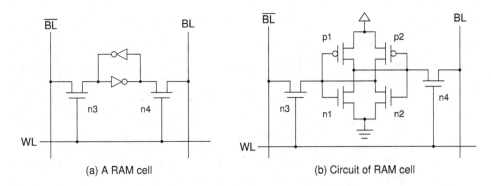

Figure 11.2 Static RAM cell

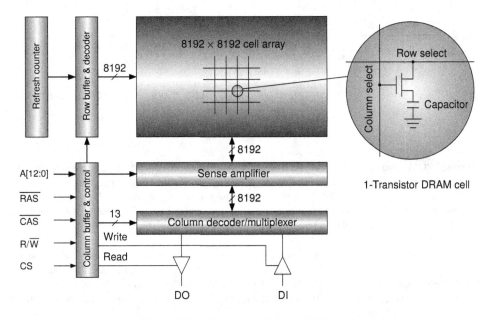

Figure 11.3 Dynamic RAM and DRAM cell

row (high bits of address) and, then, after a short time, the CAS (Column access strobe) stores the column (low bits of address).

The capacitor leaks electrons. Therefore, the memory controller must recharge all the capacitors holding a 1 every few milliseconds before they discharge. This is called a refresh. It takes time and slows down the memory.

11.1.3   Read-Only Memory

The ROM is used to store permanent data, for example, the computer's initial start-up program. It can be also accessed (read) randomly, but, different from RAM (random access memory), ROM is a nonvolatile memory. Once a ROM chip is programmed, its contents are not lost even when the power is turned off.

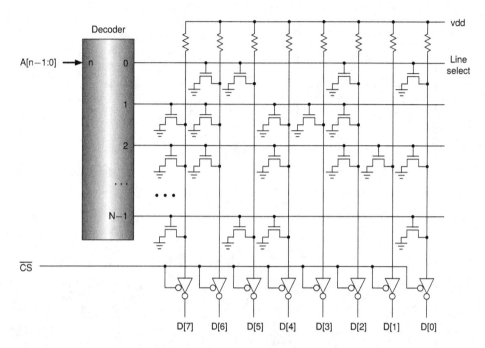

Figure 11.4   Read-only memory

Although ROM is said to be a memory, it is actually a combinational circuit. We can use it to implement some combinational logic, in addition to storing the computer's initial start-up program.

There are many types of ROM, such as MROM (mask read-only memory), PROM (one-time programmable read-only memory), EPROM (erasable programmable read-only memory), EEPROM (electrically erasable programmable read-only memory), and flash memory. Figure 11.4 shows an example of an MROM.

An n-N decoder decodes an n-bit address A[n-1:0] to select one line. There are $N = 2^n$ lines, and each line has 8 bits. Once a line is selected, the transistors in that line will be on, pulling down the bit lines. The 8 bits are outputted from D[7:0] via eight tri-state NOT gates controlled by an active-low $\overline{CS}$ (chip select). For example, if $\overline{CS}$ is 0 and A[n-1:0] is 0, D[7:0] outputs a binary pattern of 01100101. D[7:0] will be high-impedance if $\overline{CS}$ is 1.

11.1.4   Content Addressable Memory

CAM is a very special memory. Figure 11.5 gives the main difference between a CAM and a RAM. Figure 11.5(a) shows a RAM in which a data word stored in a location is returned by supplying that location's address. Figure 11.5(b) shows a CAM which searches the entire memory to see whether a given data word is stored anywhere in it. If the data word is found, the CAM returns the address where the word was found.

Figure 11.6 shows the structure of a CAM. It has four words (rows), and each word has 3 bits (columns). There are 4×3 (or 12) CAM cells. Each bit of the input D[2:0] is converted to a pair of differential signals: SL (search line) and $\overline{SL}$.

At each column (bit), the search line is compared with the bits of all the words stored in the same column of the CAM. If there is any stored bit that does not match the search line, its ML (match line) will be a 0. ML connects all the bits of a word (at a same row) so that it will be a 1 if and only if every bit of

Figure 11.5 Comparison of CAM and RAM

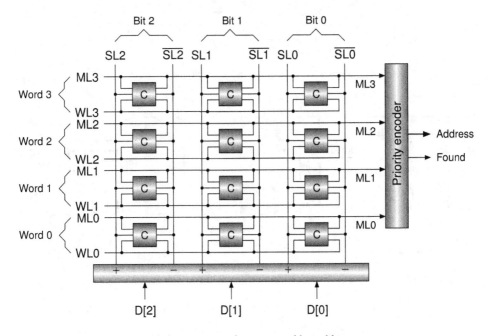

Figure 11.6 Structure of a content-addressable memory

the word is the same as its search line (matched). There is pull-up resistor connected to each match line, which is not shown in the figure. All the match lines can be encoded by a priority encoder or outputted directly.

There are many implementations of the CAM cell. Figure 11.7 shows a possible circuit of the CAM cell. Four transistors, n1, n2, n3, and n4, are added to a standard SRAM cell. If both SL and $\overline{D}$ are 1 (SL and D are not equal), the transistors n1 and n2 will be on, and hence ML outputs a 0. If both $\overline{SL}$ and D are 0 (SL and D are not equal), the transistors n3 and n4 will be on, and hence ML outputs a 0 also. If SL and D are equal (matched), either n1 or n2 is off, and meanwhile, either n3 or n4 is off, and the CAM cell does not pull ML down. WL (word line) is used to update the state of the CAM cell through SL and $\overline{SL}$.

A CAM can be used with a RAM to build a fully associative cache or TLB, as shown in Figure 11.8. The CAM is searched to find if there is a match to the input pattern. If there is a match, the content (data) of the RAM in the corresponding address will be outputted. The hit signal indicates whether data is usable or not. Multiple matches are allowed in the circuit shown in Figure 11.8(a), in which we used a priority encoder and a decoder. Even if there are multiple matches found in the CAM, the priority encoder outputs only one address which has the highest priority.

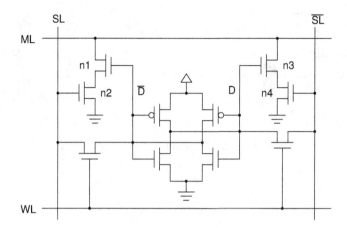

Figure 11.7   Content-addressable memory cell

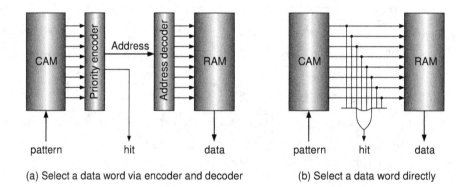

(a) Select a data word via encoder and decoder   (b) Select a data word directly

Figure 11.8   Cache or TLB implemented with a CAM and a RAM

Multiple matches are not allowed in the circuit shown in Figure 11.8(b) where we connect match lines of the CAM directly to the word lines of the RAM. We will discuss the designs of the cache and TLB in detail later.

11.1.5   Asynchronous Memory and Synchronous Memory

Traditional DRAM uses an asynchronous interface, which means that it responds as quickly as possible to changes in control inputs. Referring to Figure 11.9, a synchronous dynamic random access memory (SDRAM) uses a synchronous interface, meaning that the memory access request is registered with a clock signal, allowing it to queue up one request while waiting for another. Adding a register will result in a cycle delay (latency) of the availability of the read data.

Similar to SDRAM, there is also a synchronous static random access memory (SSRAM), whose signals are synchronized by a clock. As mentioned in previous chapters, Altera field-programmable gate arrays (FPGAs) support only synchronous SRAM when using the LPM.

Classic SDRAM operates on single data rate (SDR). A DDR-SDRAM (double data rate synchronous dynamic random access memory) transfers data on both the rising and falling edges of the clock signal.

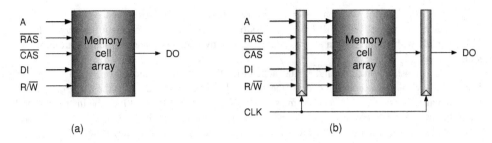

Figure 11.9 (a) Asynchronous DRAM and (b) synchronous DRAM

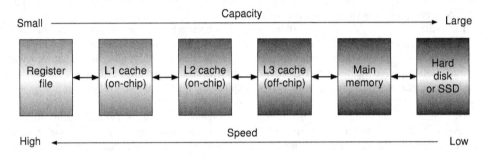

Figure 11.10 Memory hierarchy

Each further generation of DDR, such as DDR2, DDR3, DDR4, and so forth, is twice as fast at data transmission rate as its predecessor.

11.1.6 Memory Hierarchy

Figure 11.10 shows a memory hierarchy used in today's computer systems. It offers very large storage capacity at very low cost and high speed.

The register file can be considered as a memory, and it is the fastest memory. Caches store partial data of memory. Three levels of caches (L1, L2, and L3) are shown in the figure. L1 and L2 are on-chip caches, and L3 is an off-chip cache. Caches are faster than main memory but their capacities are smaller. Main memory stores programs that are currently being executed. Hard disk or solid-state drive (SSD) has the largest capacity but is slowest at speed. It has two functions: one is to store files; the other is, together with main memory, to provide users with a large virtual memory space. Some computer systems may have tapes for only back-up of files.

11.2 Cache Memory

The original meaning of cache is a secure or hidden storage place in which some important things like foods are stored to prevent them from being stolen. A cache in the computer field is a fast storage for storing data that are likely to be used again. It is commonly transparent to software, meaning that operating it is the job of hardware.

The reason of using caches is that there is a big gap in the speed between the CPU and the main memory. Suppose the clock frequency of a pipelined CPU is 1 GHz and therefore the CPU wants to get an instruction from the main memory every nanosecond (ns). Also suppose the main memory is implemented with SDRAM that has an access time of 6 ns. That is, the main memory cannot provide the CPU with an instruction at 1 ns. Data accesses by load and store instructions are in the same situation.

If we use an instruction cache and a data cache inside the CPU and the access cycle time of the caches is also 1 ns, whenever a cache hit occurs (the expected instruction or data is in the cache), it takes 1 ns to get the instruction or data. If a cache miss (not hit) occurs, the CPU must load the instruction or data from the main memory. We use t_c and t_m to denote the cycle time of the cache and the main memory, respectively, and use h to denote the hit ratio; then, the effective memory access time is

$$t = ht_c + (1 - h)(t_c + t_m) = t_c + (1 - h)t_m$$

For example, suppose $h = 98\%$, $t_c = 1$ ns, and $t_m = 6$ ns; then we get $t = 1 + 0.02 \times 6 = 1 + 0.12 = 1.12$ ns. The speedup of using cache is $6/1.12 = 5.36$ times.

The cache is empty in the beginning when a program is starting to be executed. Any missed instruction or data is put in the corresponding cache for later use. The basic unit with which the data are transferred between the cache and the main memory is a cache block or a cache line. A block may be several words in length. Assume that the block size is one word and each instruction in the program is executed only once; the hit ratio of the instruction cache will be 0, and hence the effective access time $t = t_c + t_m$, which is even worse than the case where is no cache.

The benefit of using a cache comes from the program's temporal locality and spatial locality. Temporal locality means that, if a particular memory location is referenced at one point in time, then the same location will be likely referenced again in the near future. Spatial locality means that, if a particular memory location is referenced, the nearby memory locations will be likely referenced also in the near future. It may seem that spatial locality can be exploited with a cache that has a large block, but the memory traffic will increase and some unnecessary data may occupy the cache locations. Using a large cache may improve the hit ratio, but it also increases the cache access time. Suppose that the cache size is the same as the main memory size; then the cache is as slow as the main memory and there is no reason to use a complete copy of the main memory.

The parameters for designing a cache include the total cache size, block size, cache mapping methods, cache block replacement algorithms, cache write strategies in case of a cache hit, and cache write policies in case of a cache miss. We explain some of these parameters below.

11.2.1   Mapping Methods of Caches

Cache mapping methods determine how the cache blocks are accessed with the memory address. There are mainly three mapping methods: direct mapping, set-associative mapping, and fully associative mapping.

11.2.1.1   Direct Mapping Cache

Direct mapping cache has the simplest organization. Let's take an example to explain it. As shown in Figure 11.11, suppose we have a cache of 16 KB in size, the block size is 8 bytes, or two words, the memory address has 32 bits, and the cache output data have also 32 bits.

We divide the address to four parts: Tag, Block, W, and B. B is the byte offset in a word, and therefore it has 2 bits always. W is the word offset in a block. Because there are two words in a block in our example, W has 1 bit. Block is the block index, that is, the address of the cache blocks. Because there are 16 KB/8 B = 2 K = 2^{11} blocks, the index has 11 bits. The total number of bits in these three parts is $11 + 1 + 2 = 14$; it is just the number of bits of the byte address of the cache: $2^{14} = 16$ KB. Tag contains the rest of the bits of the address; in our example it has $32 - 14 = 18$ bits.

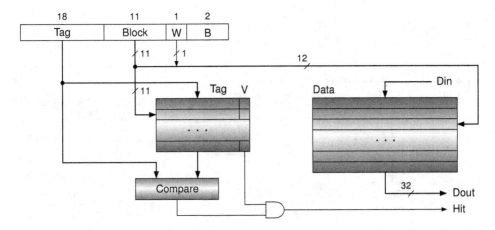

Figure 11.11 Block diagram of direct mapping cache

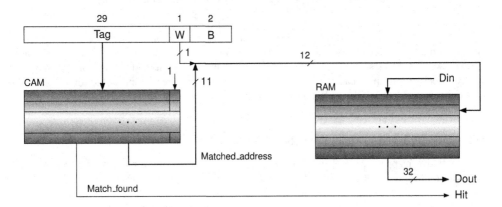

Figure 11.12 Block diagram of a fully associative mapping cache

In addition to the data in a data RAM, a block contains a `Tag` in a tag RAM and a valid bit (`V`) in a valid RAM. The block index is used to access `Tag` and `V`. If the read tag is the same as the `Tag` in the address, and `V` = 1, we say that there is a cache hit; the `Hit` signal will be a 1. The corresponding data word, `Dout`, can be read from the data RAM by using the 12-bit word address which contains the block index and word offset in a block. The `Din` signal is used to update the data RAM on a cache write. The `Tag` of the address and a 1 are written to the tag RAM and valid RAM, respectively, on a cache write. `B` is not used because the output data is a word. The tag is somewhat like the surname of a person, and the block index is like the given name. A person can be identified with a name that consists of a surname and a given name.

11.2.1.2 Fully Associative Mapping Cache

Different from the direct mapping cache, in which a data block can be only put in a particular location pointed by the block index, in a fully associative mapping cache, a data block can be put anywhere in the cache. Referring to Figure 11.12, in a fully associative mapping cache, there is no index bits in the memory address, and all the bits, except for `W` and `B`, belong to the `Tag`. A CAM is searched to see if there is a match to the `Tag` of the memory address and to a 1 for the valid bit.

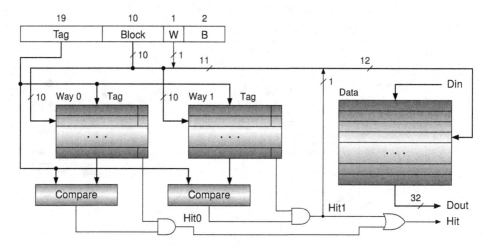

Figure 11.13   Block diagram of set-associative mapping cache

A classic RAM is used to store data blocks; its address is the output of the CAM, attaching the 1-bit W. We can also use the circuit shown in Figure 11.8(b) to implement the fully associative mapping cache. In this case, if the block size is larger than one word, a multiplexer is needed to select one word by W. In a fully associative mapping cache, when the cache is fully filled, a replacement algorithm is needed to determine which block will be replaced for a new incoming block.

11.2.1.3   Set Associative Mapping Cache

Set associative mapping cache has an organization in between a direct mapping cache and a fully associative mapping cache. Figure 11.13 shows the circuit of a two-way set associative mapping cache.

The tag RAM of the direct mapping cache shown in Figure 11.11 is divided to two equal parts, Way 0 and Way 1, which can be accessed in parallel with the block index. Because each part has $2 K/2 = 2^{10}$ blocks, the index has 10 bits, which is one bit less than the index of the direct mapping cache. Correspondingly, the Tag of the address has 19 bits, one bit more than the Tag of the direct mapping cache. If there is any cache hit in either Way 0 (Hit0) or Way 1 (Hit1), we say that a cache hit occurs. The corresponding data word in the data RAM can be accessed with the 10-bit block index, concatenating the 1-bit W and the 1-bit Hit1.

The blocks in both Way 0 and Way 1 addressed by the block index are called a set. That is, we select a set with direct mapping, and select a block inside a set with fully associative mapping. Two-way is the minimum mean; we can increase the number of ways to a number to the power of 2, such as 4-way, 8-way, 16-way, and so on. When the number of ways is equal to the total number of blocks, the set associative cache will become the fully associative cache. If the number of ways is larger than or equal to 4, an encoder is needed to generate the low bits of the address, which is used to access the data RAM. Like the fully associative mapping cache, the set associative mapping cache also needs a replacement algorithm for replacing a block inside a set.

11.2.2   Cache Block Replacement Algorithms

Let's consider a fully associative mapping cache. When we write a new block to the cache, we first check if there is an empty block in the cache. If yes, we write the new block to that block. Otherwise, we must select a block that will be replaced with the new block. Now, the question is which block should be replaced. The set associative mapping cache has also such a problem because all the blocks within a set are fully associatively mapped.

In order to enhance the cache hit ratio, we wish to replace a block that will never be used further or will be used only in distant future. But unfortunately, we can't know what will happen in the future. The purpose of using caches is to shorten the effective memory access time; therefore, the replacement algorithms must be implemented by hardware. We introduce three replacement algorithms next.

11.2.2.1   LRU Replacement Algorithm

Least recently used (LRU) algorithm is a widely used algorithm. Instead of predicting the future, LRU looks back upon the past. Consider a 16-way set associative mapping cache. There are 16 blocks in the set. We prepare a 4-bit counter for each block. The LRU algorithm works as follows:

1. Clear the counters and valid bits of all the blocks when the computer is started up.
2. If a cache miss occurs, replace a block whose counter value is a 0. If there is more than one such block, replace any of them. Set its counter to the largest value 15, and decrease all other counters whose values are not 0. We call such a counter a saturated counter.
3. If there is a hit on a block whose counter value is k, set the counter to 15, and decrease all other counters whose values are larger than k.

If there are eight ways in a set, prepare a 3-bit counter for each block. Similarly, If there are four ways in a set, prepare a 2-bit counter for each block. But for a two-way set associative mapping cache, we need to prepare just a 1-bit counter for a set (not for each block), because if a block is most recently used, the other block must be LRU. Figure 11.14 shows an example of how an LRU counter changes in a four-way set associative mapping cache in which a counter has 2 bits.

11.2.2.2   Random Replacement Algorithm

The LRU algorithm needs to use counters, which will increase the hardware cost. The random replacement algorithm randomly selects a block to be replaced. It does not require keeping any information

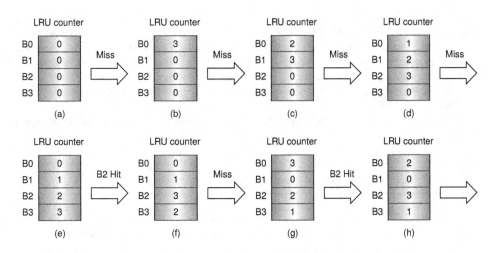

Figure 11.14   LRU counter changes on cache misses and hits

about the access history. The random number can be generated with a hardware counter; and it has the lowest hardware cost (only one counter is required).

11.2.2.3 FIFO Replacement Algorithm

The FIFO replacement algorithm selects a block to be replaced that entered the cache the earliest. This algorithm requires the same hardware cost as the LRU algorithm.

There are several other replacement algorithms that are rarely used in real cache designs. Generally, the LRU is used in two-way or four-way set associative mapping caches. Other caches use the random algorithm. The direct mapping caches do not require the use of any replacement algorithm because the blocks in the cache are directly mapped.

11.2.3   Cache Write Policies

Up to now, we have discussed what happens in a cache during a memory read operation. There are also some policies and strategies on a memory write operation.

11.2.3.1 Cache Hit: Write Through or Copy Back

There are two choices in the case of a memory write and cache hit: write through and copy back (or write back). With the write through policy, both the cache and the memory are updated. The advantage of this policy is that it keeps the consistency of the cache and the memory. This is an important issue in a multiprocessor system where the memory is shared by all CPUs and each CPU has its own cache. The disadvantage is that it increases the memory access traffic.

With the copy back policy, only the cache is updated and, when the updated block is replaced, the contents of the block are written back to the memory. The advantage of this policy is that it reduces the memory access traffic and the disadvantage is the inconsistency between the cache and the memory.

11.2.3.2 Cache Miss: Write Allocate or No Write Allocate

There are also two choices in the case of a memory write and cache miss: write allocate and no write allocate (or write no allocate). With the write allocate strategy, the block is first loaded from the memory and put in the cache, and then the data word is written to the cache. Thus the next write to the same block will result in a cache hit. The reason why we load the block first is that a block contains multiple words but we only update one of them and the tag and valid bit are associated with the whole block. With no write allocate strategy, only the memory is updated. The next write to the same block will cause a cache miss again.

The better combinations of the policies on cache hits and the strategies on cache misses are listed in Table 11.1.

Table 11.1 Combinations of write policies and allocate strategies

Comb.	Write policy on cache hit	Allocate strategy on cache miss
1	Write through	No write allocate (write no allocate)
2	Copy back (write back)	Write allocate

11.2.4 Data Cache Design in Verilog HDL

In this section, we design a simple data cache and give its Verilog HDL (hardware description language) code. The data cache is located in between the CPU and the memory, as shown in Figure 11.15.

We prefix the names of the signals that connect the CPU and cache with a character "p" (processor) and the names of the signals connecting cache and the memory with "m" (memory). Ignoring the prefixes "p" and "m," a is a 32-bit memory address; dout is a 32-bit data out; din is a 32-bit data in; strobe is a 1-bit strobe indicating a memory access; rw indicates that the memory access is a read (rw = 0) or a write (rw = 1); and ready indicates whether ready or not. Although we use the name p_ready, its meaning is that the data for the CPU is ready. uncached = 1 if I/O is accessed. It prohibits updating the cache.

We use the direct mapping method to design the cache. The cache size is 256 bytes, and the block size is one word or 4 bytes. The memory address has 32 bits. Therefore, there are 64 blocks, and the block index has 6 bits and the tag has 24 bits. We use the write-through policy and the write allocate strategy.

The detailed circuit of the data cache is shown in Figure 11.16. There are three cache RAMs: valid RAM, tag RAM, and data RAM. Valid RAM needs to be cleared, whereas the others need not. Two

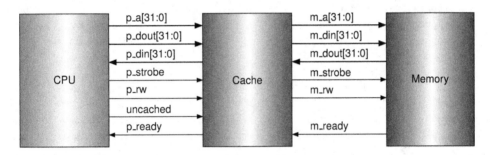

Figure 11.15 Cache in between the CPU and the main memory

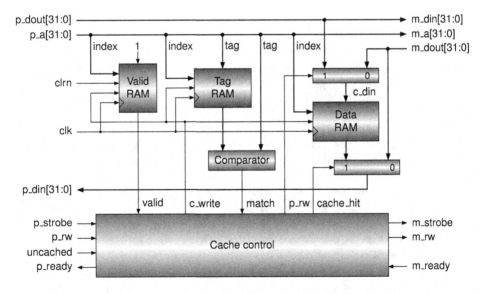

Figure 11.16 Block diagram of the write-through cache

multiplexers are used in the cache data input port and CPU data input port. The data written to cache (c_din) comes from either the CPU (for store instruction) or memory (for cache miss). The data sent to the CPU (p_din) comes from either cache (for cache hit) or memory (for cache miss). Below is the Verilog HDL code of the data cache.

```verilog
module d_cache (  // direct mapping, 2^6 blocks, 1 word/block, write-through
    input    [31:0] p_a,              // cpu address
    input    [31:0] p_dout,           // cpu data out  to mem
    output   [31:0] p_din,            // cpu data in from mem
    input           p_strobe,         // cpu strobe
    input           p_rw,             // cpu read/write command
    input           uncached,         // uncached
    output          p_ready,          // ready (to cpu)
    input           clk, clrn,        // clock and reset
    output   [31:0] m_a,              // mem address
    input    [31:0] m_dout,           // mem data out  to cpu
    output   [31:0] m_din,            // mem data in from cpu
    output          m_strobe,         // mem strobe
    output          m_rw,             // mem read/write
    input           m_ready           // mem ready
                 );
    reg             d_valid [0:63];   // 1-bit valid
    reg      [23:0] d_tags  [0:63];   // 24-bit tag
    reg      [31:0] d_data  [0:63];   // 32-bit data
    wire     [23:0] tag = p_a[31:8];  // address tag
    wire     [31:0] c_din;            // data to cache
    wire      [5:0] index = p_a[7:2]; // block index
    wire            c_write;          // cache write
    integer         i;
    always @ (posedge clk or negedge clrn)
        if (!clrn) begin
            for (i=0; i<64; i=i+1)
                d_valid[i] <= 0;      // clear valid
        end else if (c_write)
            d_valid[index] <= 1;      // write valid
    always @ (posedge clk)
        if (c_write) begin
            d_tags[index] <= tag;     // write address tag
            d_data[index] <= c_din;   // write data
        end
    wire            valid = d_valid[index];   // read cache valid
    wire     [23:0] tagout = d_tags[index];   // read cache tag
    wire     [31:0] c_dout = d_data[index];   // read cache data
    wire cache_hit  = p_strobe & valid & (tagout == tag);    // cache hit
    wire cache_miss = p_strobe & (!valid | (tagout != tag)); // cache miss
    assign m_din    = p_dout;         // mem <- cpu data
    assign m_a      = p_a;            // mem <- cpu address
    assign m_rw     = p_rw;           // write through
```

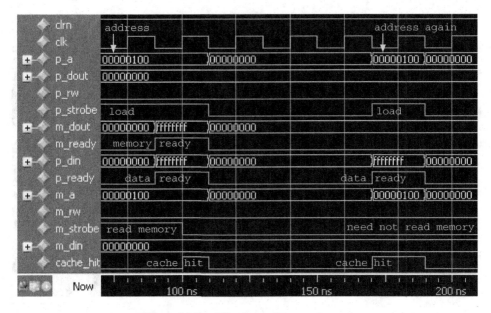

Figure 11.17   Waveform of data cache read

```
assign m_strobe = p_rw | cache_miss;          // also read on miss
assign p_ready  = ~p_rw & cache_hit |         // read and hit or
                  (cache_miss | p_rw) & m_ready;  // write and mem ready
assign c_write  = ~uncached & (p_rw | cache_miss & m_ready);  // write
assign c_din    = p_rw?      p_dout : m_dout;  // data from cpu or mem
assign p_din    = cache_hit? c_dout : m_dout;  // data from cache or mem
endmodule
```

Figure 11.17 shows the simulation waveforms of the cache reads. The CPU loads a data word from the memory location 0x00000100. This results in a cache miss. The missed data word is loaded from the memory and written to the cache. When the CPU loads the data word from the same memory location again, a cache hit occurs.

Figure 11.18 shows the simulation waveforms of the cache write. The CPU writes a data word 0x5555aaaa to the memory location 0x00000104. Because we used the write-through policy, both cache and memory were updated. When the CPU loads a data word from the same memory location, a cache hit occurs and the CPU got 0x5555aaaa with one clock cycle.

11.3   Virtual Memory Management and TLB Design

Virtual memory, as the name suggests, is not a real memory. The virtual memory technique is widely used in modern computers. It allows processes[1] to use more memory than the real memory. When a process accesses a virtual memory, a memory management unit (MMU) translates the virtual address into a physical address. In order to speed up this translation, a special cache, called a TLB, is fabricated in modern CPU chips.

---

[1] Simply, a process is a program in execution. It is an essential concept in operating systems.

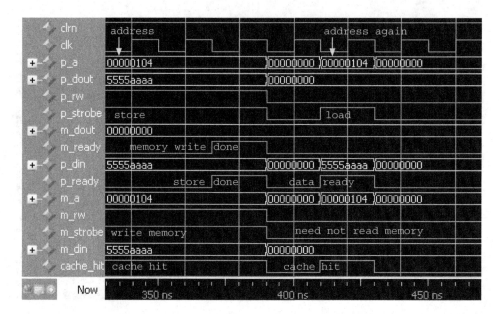

Figure 11.18 Waveform of data cache write

This section introduces the virtual memory management and gives the Verilog HDL code for a design example of an eight-entry TLB.

11.3.1 Relationship of Main Memory and Virtual Memory

The main memory, commonly implemented with DRAM chips, is a real storage in which programs (instructions and data) can be stored for execution by the CPU. Main memory is also called a physical memory or real memory. Modern operating systems support multitasking. That is, multiple processes can reside in the main memory simultaneously. It is obvious that these processes reside in different areas of the main memory. It is also the case that two processes run the same program at the same time.

However, a process uses a virtual address to access its virtual memory. Referring to Table 11.2, there are two processes, Process A and Process B, that have the same virtual address space, staring from 0. The instruction at location 0x0104 is an `lw $1, 0x0a00($0)` instruction in both the processes. It loads the data word in the virtual address 0x0a00. The data words of the two processes are different. Figure 11.19 shows a possible layout of the main memory for the two processes.

Processes A and B are mapped to the main memory starting at physical addresses 0x2000 and 0x6000, respectively. When the CPU executes process B, the virtual address 0x0104 in the program counter is translated by the ITLB (instruction translation lookaside buffer) to the real memory address 0x6104 where `lw $1, 0x0a00($0)` is stored. Once the load instruction is fetched, the arithmetic logic unit (ALU) calculates the virtual address of the data word. The calculated virtual address 0x0a00 is then translated to the real address 0x6a00 by the DTLB (data translation lookaside buffer). Finally, the data word in 0x6a00 main memory location, 0x55555555, is loaded and written to a data register.

If the CPU switches to execute Process A, the real addresses of the instructions and data translated by MMU are different from the real addresses of Process B, although the virtual address of Process A is the same as that of Process B. From this example, we see that the process number is required for the address translations.

Table 11.2 Two processes that have the same virtual addresses

Process A		Process B	
Virtual address	Instruction or data	Virtual address	Instruction or data
0x0000:	· · ·	0x0000:	· · ·
· · ·	· · ·	· · ·	· · ·
0x0104:	lw $1, 0x0a00($0)	0x0104:	lw $1, 0x0a00($0)
· · ·	· · ·	· · ·	· · ·
0x0a00:	0xcccccccc	0x0a00:	0x55555555
0x0a04:	0x00000000	0x0a04:	0x00000000

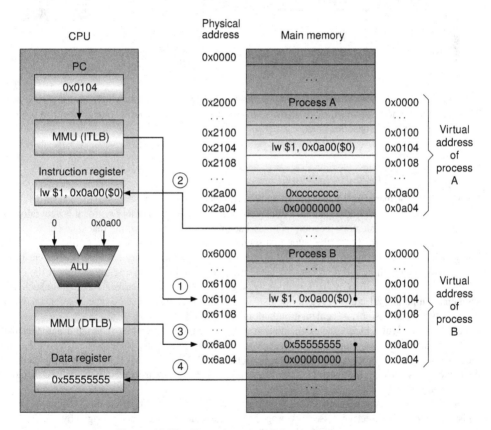

Figure 11.19 Two processes residing in main memory

The virtual memory space of a process can be even larger than the real memory capacity. Referring to Figure 11.20, there are two processes and each has a 4 GB virtual memory space, but only a portion of a process is residing in the main memory and the rest is in the hard disk. The MMU is responsible for translating the virtual address with the process number to the real address.

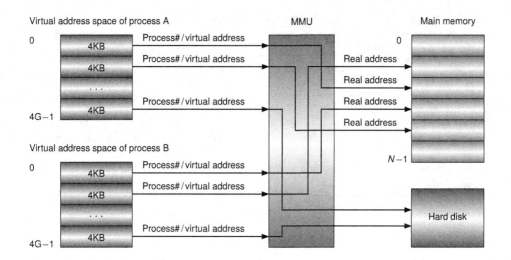

Figure 11.20   Concept of virtual memory management

In Figure 11.20, the virtual memory space is divided into several portions which have the same size. There are two methods for an MMU to do the address translation: segmentation management and paging management, which we describe next.

11.3.2   Segmentation Management

With segmentation management, the main memory is divided into segments based on the natural divisions of a program such as individual routines or data tables, so the size of a memory segment is generally not fixed. A reference to a memory location includes a segment and an offset within that segment.

As shown as in Figure 11.21, the value in a segment base register indicates the starting address of a segment in the main memory that is added to the offset to form a memory address. A segment has a length associated with it, and a process is allowed to make a reference to a segment only if the offset is within the range specified by the length of the segment. Otherwise, a hardware exception such as a segmentation fault is raised. A segment has also a set of permissions, and if the type of reference is not allowed by the permissions, a segmentation fault is again raised.

Referring to Figure 11.22, the segmentation management may generate external fragments that are too small to be used. Figure 11.22(a) shows that there are initially three processes in the main memory—Process A, Process B, and Process C. In Figure 11.22(b), Process B has been completed and its memory was released. In Figure 11.22(c), a new process, Process D, is located to the unused memory, and an external fragmentation appears.

Generally, the segment register is visible to the programmer; for example, in Intel x86 the segment registers can be used directly in assembly programs. A segment can be further managed by paging. Intel 80386 and later versions are examples in which the calculated address is not a real address; it needs to be translated to real address by the MMU of paging management.

11.3.3   Paging Management

With paging management, both the virtual memory and real memory are divided into pages of fixed size. A page is the smallest unit with which the MMU maps a virtual page to a physical page. Therefore, a

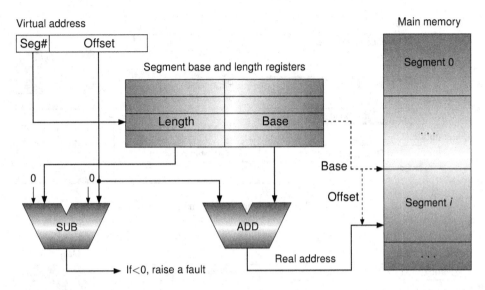

Figure 11.21   Memory management by segmentation

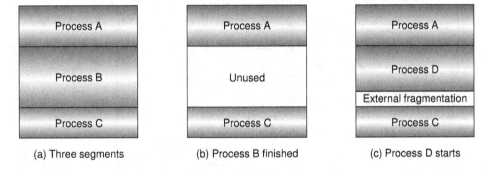

(a) Three segments   (b) Process B finished   (c) Process D starts

Figure 11.22   External fragmentation in segmentation management

virtual address consists of two parts: a virtual page number (VPN) and offset within that page. Similarly, the real memory address includes a real page number (RPN) and an offset which is the same as that in a virtual address. The MMU needs only to translate VPN to RPN.

Referring to Figure 11.23, this translation is done with a page table, which stores mainly RPNs. The VPN is used as the address of the page table. The concatenation of the RPN and offset forms a real address.

The offset in Figure 11.23 has 12 bits, meaning that the page size is 4 KB and the page table has 2^{20} = 1M entries. Suppose an entry has 32 bits (RPN and a set of permissions); then a page table takes a large, contiguous memory of 4 MB. To solve this problem, we can use two-level page tables, as shown in Figure 11.24.

The first-level page table is called a page directory; it stores the physical base address of the second-level page table (or simply page table). If we allocate 10 bits to address the page directory and 10 bits to address the page table, then the page directory has 1K entries and 4 KB (2^{10} × 4), a page table has 4 KB (2^{10} × 4), and there are 1K such page tables. With two-level page tables, it becomes possible

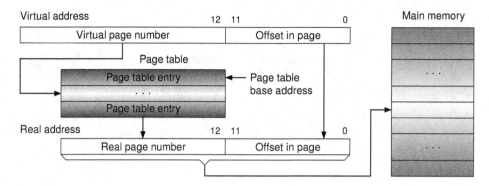

Figure 11.23   Memory management by paging

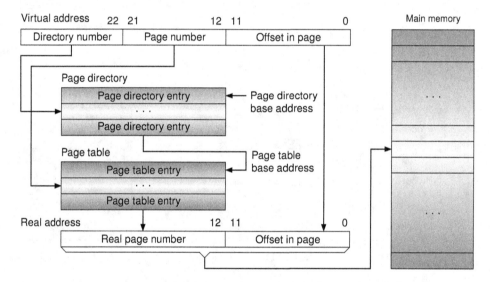

Figure 11.24   Two-level paging management

to allow portions of the page tables to be kept in the main memory and there is no necessity to store the page tables in contiguous memory.

Paging management and segmentation management can be used together. Figure 11.25 illustrates one example of paging and segmentation management in which the translated address by segmentation management is translated further by paging management.

11.3.4   Translation Lookaside Buffer Design

The virtual address translations described above takes a long time. For example, in the two-level paging management scheme, loading a data word needs access to the main memory three times: (i) access the page directory to get the RPN of a page table; (ii) access the page table to get the RPN of the data word; and (iii) access main memory to get the data word. In order to speed up the translation, almost all the modern CPUs fabricate TLBs in the CPU chips. The organization of a TLB is very similar to that of a

Virtual address

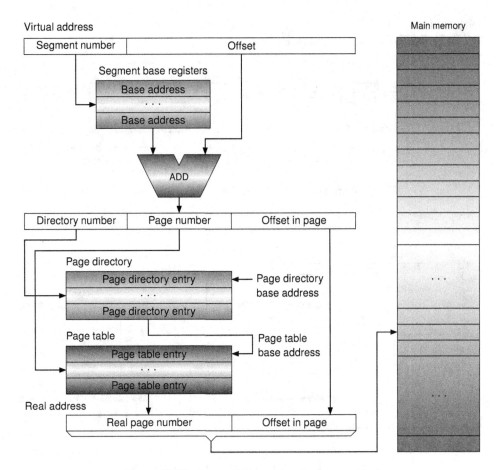

Figure 11.25 Segmentation and paging management

data cache. The data cache stores data blocks in cache RAM, while the TLB stores RPNs. Figure 11.26 shows a block diagram of a TLB that uses the fully associative mapping method. A CAM is used to search for the VPN of the virtual address.

Figure 11.27 shows the detailed circuit of an eight-entry fully associative mapping TLB. The CAM stores VPNs and valid bits. It is searched with the VPN (`pattern`) of the virtual address and a 1 (valid bit). If there is match, the CAM outputs `vpn_index` (matched address) and `vpn_found` as the TLB hit signal (`tlb_hit`). A CAM line, indexed by `addr`, can be updated with `pattern`. The `addr` comes from either `index` or a random number generator. There are two write enable signals, `tlbwr` (write TLB entry at random) and `tlbwi` (write TLB entry at index).

The RAM stores page table entries (PTEs), copied from the page table. A PTE may contain the following information: `P`, `D`, `A`, `AC`, and `RPN`. `P` (present) indicates whether the corresponding page is in the memory. If not, a page fault exception will occur. `D` (dirty) indicates whether a page is modified. If yes, this page must be written back to the hard disk when it is replaced. `A` (accessed) indicates whether a page has been accessed (read or write) recently. This bit is used, together with `D`, to implement page replacement algorithms. `AC` (access control) bits specify some access permissions, such as writable, executable, and cacheable. The RPN is the base address of the page in memory. `vpn_index` is the read address of the RAM, and `addr` is the write address. For the write operation that happens on TLB miss, `pte_in`, which comes from page table, will be written to the RAM.

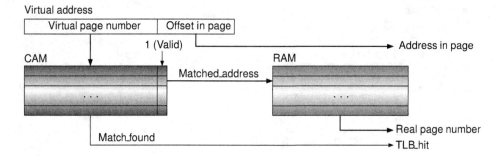

Figure 11.26   Block diagram of TLB using CAM

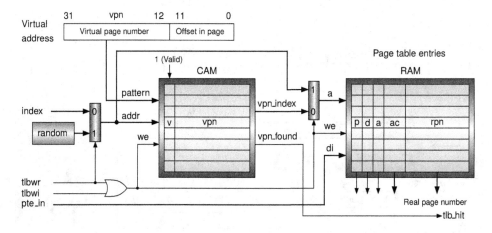

Figure 11.27   Implementation of an eight-entry TLB

Figure 11.28 shows the schematic diagram of the TLB. It invokes two memory modules: an 8-word × 24-bit RAM module `ram8x24`, and an 8-word × 21-bit CAM module `cam8x21`, which will be given next.

Below is the Verilog HDL code of the `ram8x24` module. We use the standard Verilog HDL statements to implement this RAM module.

```
module ram8x24 (address,data,clk,we,q);     // ram for tlb
    input    [2:0] address;                 // address
    input    [23:0] data;                   // data in
    input          clk;                     // clock
    input          we;                      // write enable
    output [23:0] q;                        // data out
    reg    [23:0] ram [0:7];                // ram cells: 8 words * 24 bits
    always @(posedge clk) begin
        if (we) ram[address] <= data;       // write ram
    end
    assign q = ram[address];                // read ram
```

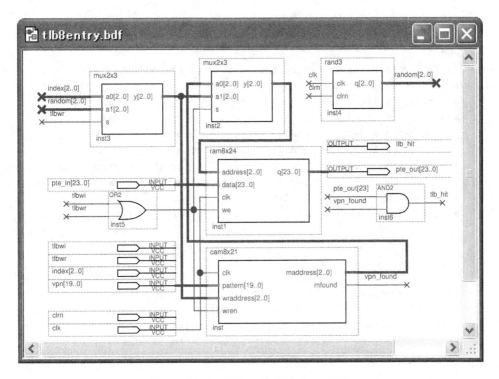

Figure 11.28   Schematic diagram of TLB

```
    integer i;
    initial begin
        for (i = 0; i < 8; i = i + 1)
            ram[i] = 24'h0;                    // initialization
    end
endmodule
```

Implementing a CAM module is little bit complex. If you have an FPGA chip that supports the CAM, you can use it directly. Otherwise, you have to design it yourself. The transistor-level circuit of the CAM was given in the first section of this chapter. Here we implement the CAM in the behavioral-style Verilog HDL, as listed below. The key part of the code is the comparison between the input pattern and the content of each TLB line. The eight match lines are encoded to a 3-bit line number, which is the matched index.

```
module cam8x21 (clk,pattern,wraddress,wren,maddress,mfound);          // cam
    input         clk;             // clock
    input         wren;            // cam write enable
    input [19:0]  pattern;         // vpn to be compared to all 8 lines
    input  [2:0]  wraddress;       // write address
    output [2:0]  maddress;        // matched address
    output        mfound;          // a match was found
    reg    [20:0] ram [0:7];       // ram 8-line * 21-bit: valid (1), vpn (20)
    // write cam, update a line with pattern, valid bit <- 1
```

```
    always @ (posedge clk) begin
        if (wren) ram[wraddress] <= {1'b1,pattern};     // valid, pattern
    end
    // fully associative search, should be implemented with CAM cells
    wire    [7:0] match_line;                            // match line
    assign match_line[7] = (ram[7] == {1'b1,pattern}); // valid, pattern
    assign match_line[6] = (ram[6] == {1'b1,pattern}); // valid, pattern
    assign match_line[5] = (ram[5] == {1'b1,pattern}); // valid, pattern
    assign match_line[4] = (ram[4] == {1'b1,pattern}); // valid, pattern
    assign match_line[3] = (ram[3] == {1'b1,pattern}); // valid, pattern
    assign match_line[2] = (ram[2] == {1'b1,pattern}); // valid, pattern
    assign match_line[1] = (ram[1] == {1'b1,pattern}); // valid, pattern
    assign match_line[0] = (ram[0] == {1'b1,pattern}); // valid, pattern
    assign mfound        = |match_line;                 // a match was found
    // encoder for matched address, no multiple-match is allowed
    assign maddress[2] = match_line[7] | match_line[6] |
                         match_line[5] | match_line[4];
    assign maddress[1] = match_line[7] | match_line[6] |
                         match_line[3] | match_line[2];
    assign maddress[0] = match_line[7] | match_line[5] |
                         match_line[3] | match_line[1];
    // initialize cam, mainly clear valid bit of each line
    integer i;
    initial begin
        for (i = 0; i < 8; i = i + 1)
            ram[i] = 0;
    end
endmodule
```

Figure 11.29 shows the waveform when the TLB is updated with tlbi (write TLB entry at index). The VPNs of 0x80000, 0x80001, 0x80002, and 0x80003 and the corresponding RPNs of 0xf0000 (the low 20 bits of pte_in), 0xf0001, 0xf0002, and 0xf0003 are written to TLB lines of 0, 1, 2, and 3, respectively, on the rising edge of the clock. Once a TLB line is updated, a TLB hit occurs and the corresponding RPN is outputted.

Figure 11.30 shows the waveform when the TLB is updated with tlbr (write TLB entry at random). The random index number is generated by a counter. First, the TLB entry 0 was updated with VPN = 0x80008 and the corresponding RPN = 0xf0008, and therefore there is a TLB hit on this line. Then the TLB entry 2 was updated with VPN = 0x80009 and the corresponding RPN = 0xf0009, and therefore there is also a TLB hit on this line.

11.3.5 Accessing Cache and TLB in Parallel

Suppose we use a *k*-way set associative mapping cache and the tag in the cache is the most significant bits of the real address that is translated by a TLB, as shown as in Figure 11.31. What we want is that the cache and TLB be accessible simultaneously, that is, the block index used to access the cache block should be in the offset of the virtual address that need not to be translated by the TLB. Otherwise, we have to first get the real address through accessing the TLB and then use it to access the cache, resulting in sequentially accessing the TLB and cache. This will slow down the speed of data accesses.

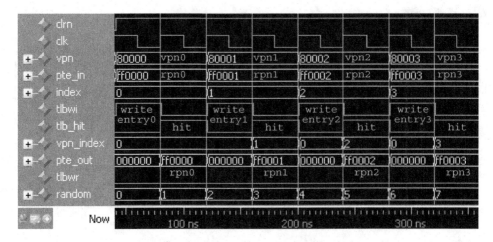

Figure 11.29 Waveform 1 of an eight-entry TLB (tlbwi)

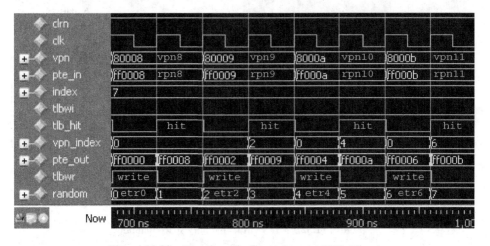

Figure 11.30 Waveform 2 of an eight-entry TLB (tlbwr)

Therefore, the condition under which the cache and TLB can be accessed in parallel is that the cache size should not be larger than $2^m k$ bytes where k is the number of ways in a set and 2^m is the page size in bytes in the paging management scheme. For example, for $m = 12$ (the page size is 4 KB) and $k = 8$ (eight-way set associative cache), the cache size should not exceed 32 KB.

The discussions above are based on a real address cache in which both the cache tag and block index use the real address. We can also design a virtual address cache in which both the cache tag and block index use the virtual address. Furthermore, we can combine the cache and TLB together in the virtual address cache. That is, the PTEs are treated as data.

11.4 Mechanism of TLB-Based MIPS Memory Management

The MIPS CPU uses a TLB-based address translation mechanism. It provides special instructions and control registers for the maintenance of the TLB. This section describes the MIPS virtual address mapping, TLB organization, CP0 registers, and instructions related to the TLB.

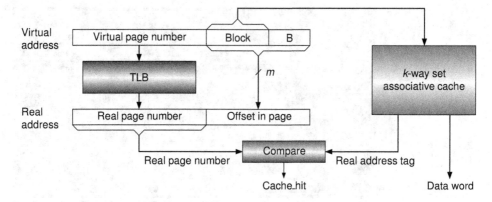

Figure 11.31   Conditions for cache and TLB accesses in parallel

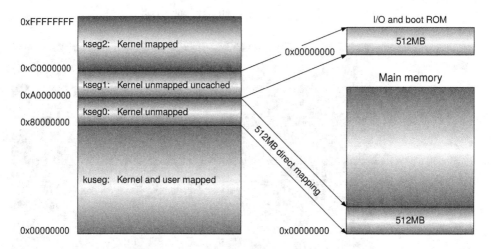

Figure 11.32   MIPS virtual memory space mapping

11.4.1   MIPS Virtual Memory Space

A MIPS CPU runs at one of two privilege modes: kernel mode and user mode.[2] With the CPU in user mode, a program has access only to the general-purpose register file, floating-point register file, and up to 2 GB virtual memory. When the CPU runs in kernel mode, the program has access to 4 GB virtual memory and all register files including CP0 registers. Privileged instructions can be executed only when the CPU is in kernel mode.

Figure 11.32 shows the MIPS virtual memory space mapping. The 4 GB virtual memory space is divided into four segments: kuseg (kernel and user segment), kseg0 (kernel segment 0), kseg1 (kernel segment 1), and kseg2 (kernel segment 2).

1. The kuseg segment (0x00000000 to 0x7fffffff) has 2 GB. This is the only segment that is accessible in the user mode. Kernel mode has also access to it. The virtual address in this segment will be translated to real address by the TLB.

---

[2] A third mode, supervisor mode, was defined but has never been used in operating systems.

2. The kseg0 segment (0x80000000 to 0x9fffffff) has 512 MB. It is directly mapped to the low 512 MB of the main memory (0x00000000 to 0x1fffffff). The operating system code will usually be placed in this segment. That is, even if the TLB was not initialized yet, the operating system still has real memory for use.

3. The kseg1 segment (0xa0000000 to 0xbfffffff) has 512 MB. It is also directly mapped to 0x00000000 to 0x1fffffff, but this area is most often used for I/O addresses and the boot ROM. Therefore, access to this area of memory does not use the cache (cache should be disabled automatically).

4. The kseg2 segment (0xc0000000 to 0xffffffff) has 1 GB. This area is accessible only in kernel mode and is translated to real address by the TLB.

Although the name "segment" is used for identifying the virtual memory areas, the MIPS MMU uses the paging management scheme.

11.4.2   Organization of MIPS TLB

In MIPS architecture, virtual addresses are extended with an 8-bit address space identifier (ASID) to distinguish between processes. The ASID is stored in each TLB entry so that the CPU can move from one process to another (called context switching) without having to invalidate TLB entries.

The MIPS TLB has a fully associative mapping structure that is used to translate virtual addresses. Figure 11.33 shows a MIPS TLB entry. Each entry contains two parts: a virtual part and a real part. The virtual part includes an ASID, a VPN (VPN2), a global bit (G), and a Mask field. Real part contains a pair of real entries, entry 0 and entry 1, each of which contains a physical page frame number (PFN), a cache coherence field (C), a dirty (D) bit, and a valid (V) bit.

VPN2 is actually VPN/2 because each entry maps two physical pages. The least significant bit of the VPN is used to select one real entry from the pair of entries. If we name this bit as S, then if S = 1, select real entry 1; otherwise select real entry 0. The global bit (G) is used for physical pages that are globally shared by all processes. Thus, if the global bit is set, the ASID is ignored. The Mask field specifies the page size. The smallest page size is 4 KB, and the largest page size is 256 MB. The possible page sizes are shown in Figure 11.34.

The virtual address bits [31:13] are stored in the VPN2 field of the TLB entry shown in Figure 11.33; the Mask field is used to mask the low bits of the VPN2. For example, if the page size is 4 KB, the mask pattern is 0000000000000000000; if the page size is 16 KB, the mask pattern is 000000000000000011; and if the page size is 256 MB, the mask pattern is 0001111111111111111.

In the real part of a TLB entry, the valid bit (V) determines the final success of the translation. If the valid bit is off, the entry is not valid and a TLB miss exception is raised (referring to Table 6.1). The dirty bit (D) is marked on whenever the page is written to. If the dirty bit is off and the reference is a store, a TLB modified exception is raised (also referring to Table 6.1).

Figure 11.35 shows the block diagram of the address translation based on MIPS TLB. TLB uses the fully associative mapping scheme, all the ASIDs and VPN2s in the TLB are compared with the process identifier and the VPN2 of the virtual address, respectively. The current process identifier (ASID) is stored in the MIPS EntryHi register, which we will describe later.

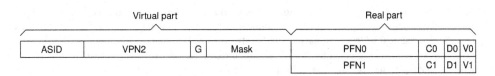

Figure 11.33   Contents of MIPS TLB entry

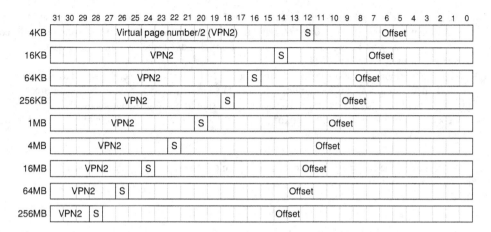

Figure 11.34   Page sizes and format of virtual address

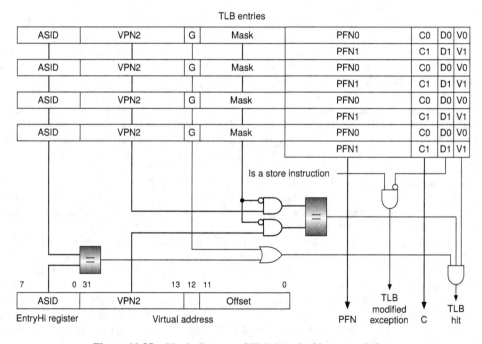

Figure 11.35   Block diagram of TLB-based address translation

If there is an address match with a valid entry and no modified exception, the final physical address is obtained by concatenating the appropriate number of bits of PFN and the page offset, as shown as in Figure 11.36. The real address can have up to 36 bits (up to 64 TB main memory is supported).

The corresponding cache coherency bits C are outputted at the same time and determine how a page is cached by the hardware, as shown in Table 11.3.

The undefined C can be defined by the MIPS CPU makers. Table 11.4 shows the definitions of C in IDT79RC32355 CPU, produced by IDT.

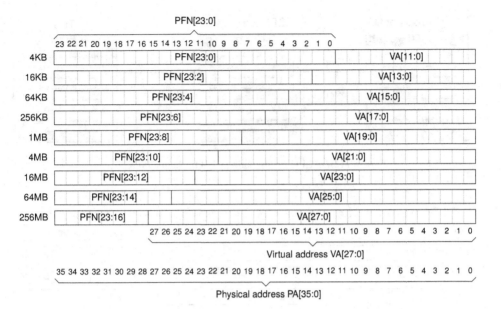

Figure 11.36   Page sizes and format of physical address

Table 11.3   Cache coherency attributes

Cache coherency (C)	Attribute
0–1	Undefined, available for implement dependent use
2	Memory page is uncacheable (I/O space for instance)
3	Memory page is cacheable (normal memory)
4–7	Undefined, available for implement dependent use

Table 11.4   IDT79RC32355 cache coherency attributes

Cache coherency (C)	Attribute
0	Cacheable, write through, no write allocate
1	Cacheable, write through, write allocate
2	Uncacheable
3	Cacheable, write/copy back
4–7	Reserved

11.4.3   TLB Manipulation by Software

The MIPS TLB is software-managed; there are instructions to manage the TLB's contents. MIPS provides four such instructions: (i) tlbp, which probes the TLB to see if a particular translation is in there and puts the search result on a CP0 register; (ii) tlbr, which reads the contents of a TLB entry into CP0 registers; (iii) tlbwi, which writes a specific TLB entry with the contents of CP0 registers; and (iv)

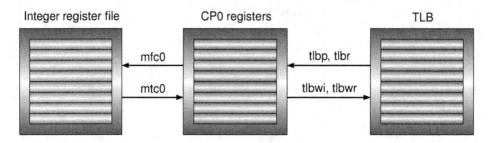

Figure 11.37   CPU manipulating TLB through CP0 registers

	31 30 29 28 27 26 25 24 23 22 21 20 19 18 17 16 15 14 13 12 11 10 9 8 7 6 5 4 3 2 1 0

EntryHi — 31..13: VPN2; 12 11 10 9 8: 0 0 0 0 0; 7..0: ASID

EntryLo0 — 31 30: 0 0; 29..6: PFN; 5..3: C; 2: D; 1: V; 0: G

EntryLo1 — 31 30: 0 0; 29..6: PFN; 5..3: C; 2: D; 1: V; 0: G

PageMask — 31 30 29: 0 0 0; 28..13: Mask; 12..0: 0 0 0 0 0 0 0 0 0 0 0 0 0

Index — 31: P; 30..n: 0 0 0 0 0 0 0 0 0 0 0 0 0 0 0 0 0 0 0 0 0 0 0 0 0 0 0; n−1..0: Index

Random — 31..n: 0 0 0 0 0 0 0 0 0 0 0 0 0 0 0 0 0 0 0 0 0 0 0 0 0 0 0 0; n−1..0: Random

Wired — 31..n: 0 0 0 0 0 0 0 0 0 0 0 0 0 0 0 0 0 0 0 0 0 0 0 0 0 0 0 0; n−1..0: Wired

Figure 11.38   MIPS TLB manipulation registers of CP0

tlbwr, which writes a random TLB entry with the contents of CP0 registers. Then we can use the mfc0 and mtc0 instructions to transfer data between a CP0 register and an IU register in the general-purpose register file, as shown in Figure 11.37.

Figure 11.38 shows the CP0 registers that are used to manipulate the TLB. The fields in EntryHi, EntryLo0, EntryLo1, and PageMask registers correspond exactly to the fields of the TLB entry. However, there are two G bits: G0 in EntryLo0 and G1 in EntryLo1, but there is only one G bit in a TLB entry. MIPS defines G = G0 & G1 on a TLB write and G0 = G1 = G on a TLB read. In other words, G will be set if and only if both G0 and G1 are set. If G0 and G1 disagree, G will not be set, and a read after a write will result in both G0 and G1, becoming an agreed 0.

The PageMask register holds a mask pattern that defines the page size for each TLB entry, as shown in Table 11.5. It may seem that there is room for defining more page sizes, such as 8 KB, 32 KB, and so on, but such settings are treated as "undefined" by MIPS definitions. The S in the table is a bit of the virtual address that is used to select one real entry from a pair of real entries. An implementation may not use the PageMask register. In such case, the page size is the fixed 4 KB.

The Index register is used to return the index of a matched TLB entry for tlbp instruction and to provide a pointer to a TLB entry for tlbr and tlbwi instructions. If there are N entries in a TLB, the width of the index field should be $\lceil \log_2 N \rceil$. For example, 5 bits are required for a TLB with 20 entries.

The Random register holds a random number generated by hardware that is used to index the TLB for the tlbwr instruction. The width of the random field is calculated in the same manner as that described for the Index register above.

The Wired register is used to specify the boundary between the wired (reserved) and random entries in the TLB, as shown in Figure 11.39. The wired entries cannot be replaced with a tlbwr instruction.

Table 11.5 Mask values in PageMask register

Page size	28	27	26	25	24	23	22	21	20	19	18	17	16	15	14	13	S
4KB	0	0	0	0	0	0	0	0	0	0	0	0	0	0	0	0	VA[12]
16KB	0	0	0	0	0	0	0	0	0	0	0	0	0	0	1	1	VA[14]
64KB	0	0	0	0	0	0	0	0	0	0	0	0	1	1	1	1	VA[16]
256KB	0	0	0	0	0	0	0	0	0	0	1	1	1	1	1	1	VA[18]
1MB	0	0	0	0	0	0	0	0	1	1	1	1	1	1	1	1	VA[20]
4MB	0	0	0	0	0	0	1	1	1	1	1	1	1	1	1	1	VA[22]
16MB	0	0	0	0	1	1	1	1	1	1	1	1	1	1	1	1	VA[24]
64MB	0	0	1	1	1	1	1	1	1	1	1	1	1	1	1	1	VA[26]
256MB	1	1	1	1	1	1	1	1	1	1	1	1	1	1	1	1	VA[28]

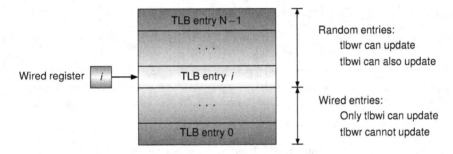

Figure 11.39 The meaning of wired register

```
      31 30 29 28 27 26 25 24 23 22 21 20 19 18 17 16 15 14 13 12 11 10 9  8  7  6  5  4  3  2  1  0
tlbp  | 0  1  0  0  0  0 | 1 | 0  0  0  0  0  0  0  0  0  0  0  0  0  0  0  0  0  0  0 | 0  0  1  0  0  0 |
tlbr  | 0  1  0  0  0  0 | 1 | 0  0  0  0  0  0  0  0  0  0  0  0  0  0  0  0  0  0  0 | 0  0  0  0  0  1 |
tlbwi | 0  1  0  0  0  0 | 1 | 0  0  0  0  0  0  0  0  0  0  0  0  0  0  0  0  0  0  0 | 0  0  0  0  1  0 |
tlbwr | 0  1  0  0  0  0 | 1 | 0  0  0  0  0  0  0  0  0  0  0  0  0  0  0  0  0  0  0 | 0  0  0  1  1  0 |
```

Figure 11.40 Formats of TLB manipulating instructions

We have described the mfc0 and mtc0 instructions in Chapter 6, and shown their formats in Figure 6.10. The formats of tlbp, tlbr, tlbwi, and tlbwr instructions are shown in Figure 11.40.

The tlbp instruction probes the TLB to see whether there is an entry that matches EntryHi. If so, the index of the entry that matches is written to the index field of the Index register, and the P (probe failure) bit is cleared. Otherwise, P is set.

The tlbr instruction reads the contents of a TLB entry into EntryHi, EntryLo0, EntryLo1, and Page-Mask registers. The entry is pointed to by the index in the Index register. As mentioned above, the single G bit in the TLB entry will be written to both EntryLo0 and EntryLo1.

The tlbwi instruction writes the contents of the EntryHi, EntryLo0, EntryLo1, and PageMask registers into a TLB entry pointed to by Index in the Index register. As mentioned above, the logical AND of two G bits from EntryLo0 and EntryLo1 will become the G bit of the TLB entry.

The tlbwr instruction is similar to tlbwi. The differences are that the TLB entry is pointed to by the Random in the Random register and it cannot write to the wired entries.

Exercises

11.1 The technique of interleaved memory can speed up the memory access. In an interleaved memory, the data in consecutive addresses reside in different banks. Suppose there are $N = 2^n$ memory banks, and the memory location i is in the ($i \% N$)th bank. Figure 11.41 shows a possible organization of a four-way interleaved memory, and Figure 11.42 shows the read timing of the memory. Design and simulate the control circuit of the four-way interleaved memory.

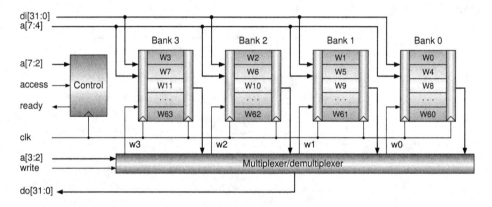

Figure 11.41   Four-way interleaved memory

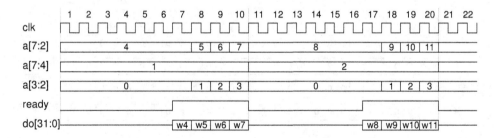

Figure 11.42   Read timing of four-way interleaved memory

11.2 Explain why the cache, MMU, and TLB are required in modern computers.

11.3 Explain why the effective memory access time of using a cache is $t = t_c + (1 - h)t_m$, but not $t = ht_c + (1 - h)t_m$, where t_c and t_m are the cycle times of cache and main memory, respectively, and h is the hit ratio.

11.4 Explain the reasons why the combination given in Table 11.1 is better.

11.5 Design a two-way set associative mapping cache with the LRU algorithm in Verilog HDL.

11.6 Design a fully associative mapping TLB with the CAM components provided by FPGA chips.

11.7   Write a MIPS assembly program for initializing a TLB.

11.8   Can you design a virtual address cache combined with TLB?

11.9   Simulate cache performance by trace-driven simulation under different configurations by changing the cache size, block size, mapping methods, block replacement algorithms, and write policies. And draw the performance figures with gnuplot—a tool for representing data graphically.

12

Design of Pipelined CPU with Caches and TLBs in Verilog HDL

This chapter describes the design of a complete pipelined CPU (central processing unit) which contains an integer unit (IU), a floating-point unit (FPU), an instruction cache (ICache), a data cache (DCache), an instruction translation lookaside buffer (ITLB), and a data translation lookaside buffer (DTLB).

The caches are transparent to programmers and controlled by hardware. If a cache miss occurs, the hardware takes the responsibility for loading instructions or data from main memory. But the TLBs (translation lookaside buffers) are controlled by software through the TLB manipulation instructions. If a TLB miss occurs, an exception is generated. The CPU acknowledges the exception by transferring control to an exception handler in which a tlbwi or tlbwr instruction replaces a TLB entry with a page table entry which is maintained by the operating system kernel.

We use one memory module for storing both instructions and data, including the page table; therefore an arbiter is needed for arbitrating the simultaneous memory access requests caused by the misses of ICache and DCache that may happen at the same time. We give the complete Verilog HDL (hardware description language) code of the CPU and show the simulation waveforms in this chapter.

12.1 Overall Structure of Caches and TLBs

We introduced the circuit designs of the cache and TLB in Chapter 11. This section describes how to use them with the IU, FPU, and main memory together to provide programs with an efficient memory hierarchy. Figure 12.1 shows the overall structure of caches and TLBs that will be implemented in our pipelined CPU. As mentioned above, the cache is controlled by hardware, and the TLB is manipulated by software. The cache hit (or miss) signals are used to control the pipeline stalls for waiting for the main memory's response; and the TLB hit (or miss) signals are used as the exception signals to let the CPU update the TLB entry with the contents of CP0 registers in the interrupt/exception handler.

The purpose of the TLB is to speed up the translation from virtual address to physical address. Because in the pipelined CPU instruction fetch and data access may happen in the same clock cycle, we use two dedicated TLBs (ITLB and DTLB) for translating instruction virtual address and data virtual address simultaneously. For simplicity, the cache size is set to the fixed 4 KB (12-bit offset within a page), and hence it need not use the CP0 PageMask register. The fully associative mapping method is used, and there are eight entries in each of the TLBs.

The purpose of the cache is to speed up the memory access. For the same reason as above, we use two dedicated caches (ICache and DCache) for buffering instructions and data. The caches use the simplest direct mapping method, and the cache tag uses the upper bits of the translated physical address. The

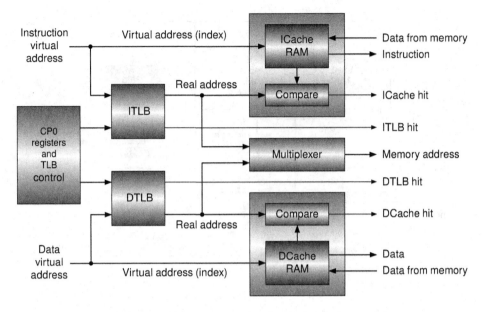

Figure 12.1   Dedicated TLBs and caches for instructions and data

cache size is only 256 bytes in our pipelined CPU so that the cache block index can use the bits of the untranslated virtual address (within the page offset).[1]

12.2   Design of Circuits Related to Caches

This section describes the instruction cache design, the interface circuit between the main memory and caches, and the pipeline halt circuit for cache misses.

12.2.1   Instruction Cache and Its Verilog HDL Codes

Figure 12.2 shows the block diagram of the instruction cache. It is similar to but simpler than the data cache given in Figure 11.28, because the CPU does not write data to the instruction cache. For simplicity, we use a block size of 4 bytes. The cache size is 256 bytes or 64 words.

There are three cache RAMs: valid RAM, tag RAM, and data RAM. Valid RAM needs to be cleared, and the others need not. A multiplexer is used in the cache data output port: the instruction sent to the CPU (p_din) comes from either cache (for cache hit) or memory (for cache miss). p_a is the physical address, but the cache block index is within the page offset so that it comes from virtual address directly. The p_ready signal informs the CPU if an instruction is ready. Below is the Verilog HDL code of the instruction cache.

```
module i_cache (                    // direct mapping, 2^6 blocks, 1 word/block
    input   [31:0] p_a,                          // cpu address
    output  [31:0] p_din,                        // cpu data from mem
    input          p_strobe,                     // cpu strobe
```

---

[1] The cache size can be up to 4 KB for a direct mapping cache which uses the virtual address as index.

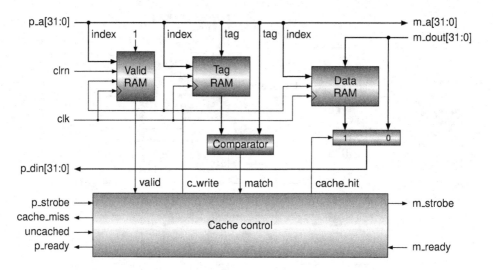

Figure 12.2   Block diagram of instruction cache

```
input           uncached,                   // uncached
output          p_ready,                    // ready (to cpu)
output          cache_miss,                 // cache miss
input           clk, clrn,                  // clock and reset
output [31:0]   m_a,                        // mem address
input  [31:0]   m_dout,                     // mem data out to cpu
output          m_strobe,                   // mem strobe
input           m_ready                     // mem ready
           );
reg             d_valid [0:63];             // 1-bit valid
reg    [23:0]   d_tags  [0:63];             // 24-bit tag
reg    [31:0]   d_data  [0:63];             // 32-bit data
wire    [5:0]   index = p_a[7:2];           // block index
wire   [23:0]   tag = p_a[31:8];            // address tag
wire            c_write;                    // cache write
wire   [31:0]   c_din;                      // data to cache
integer i;
always @ (posedge clk or negedge clrn)
    if (!clrn) begin
        for (i=0; i<64; i=i+1)
            d_valid[i] <= 0;                // cleat valid
    end else if (c_write)
        d_valid[index] <= 1;                // write valid
always @ (posedge clk)
    if (c_write) begin
        d_tags[index] <= tag;               // write address tag
        d_data[index] <= c_din;             // write data
    end
```

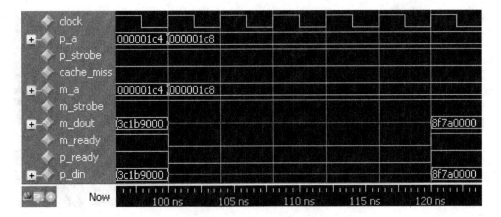

Figure 12.3 Waveform of instruction cache miss

```
    wire          valid = d_valid[index];          // read cache valid
    wire    [23:0] tagout = d_tags[index];          // read cache tag
    wire    [31:0] c_dout = d_data[index];          // read cache data
    wire    cache_hit  = p_strobe & valid & (tagout == tag);  // cache hit
    assign cache_miss = p_strobe & (!valid | (tagout != tag)); // cache miss
    assign m_a        = p_a;                         // mem <-- cpu address
    assign m_strobe = cache_miss ;                   // read on miss
    assign p_ready  = cache_hit | cache_miss & m_ready;     // data ready
    assign c_write  = cache_miss & ~uncached & m_ready;    // write cache
    assign c_din    = m_dout;                        // data from mem
    assign p_din    = cache_hit? c_dout : m_dout;   // data from cache or mem
endmodule
```

Figure 12.3 shows the simulation waveform when a cache miss occurs. The memory address 0x000001c8 is issued at 100 ns. The memory access takes 6 clock cycles, and then the instruction 0x8f7a0000 is ready for CPU use (p_ready = 1). This waveform is picked up from the waveforms of the final CPU's simulation.

Figure 12.4 shows the simulation waveform when the instruction cache hits. We can see that an instruction is available on every clock cycle (p_ready = 1). This waveform is also picked up from the waveforms of the final CPU's simulation.

12.2.2 Memory Interface to Instruction and Data Caches

The two caches, instruction cache and data cache, enable the CPU to fetch instruction and access memory data at the same time. But there is only one main memory module for storing both instructions and data. Therefore, a mechanism is needed to arbitrate the memory access requests, issued by the instruction cache and data cache at the same time, as shown in Figure 12.5.

The arbitrator consists of a 2-to-1 multiplexer (mux) and a 1-to-2 demultiplexer (demux). The selection signal of the multiplexer is sel_i (select instruction), which is the output signal c_miss of the instruction cache. This means that the instruction cache has a higher priority to access the main memory than the data cache.

As shown in Figure 12.6, a demultiplexer is the opposite of a multiplexer. A multiplexer selects one of multiple inputs. A demultiplexer directs a single input to one of multiple outputs. The demultiplexer in

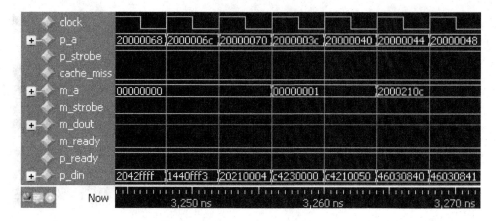

Figure 12.4   Waveform of instruction cache hit

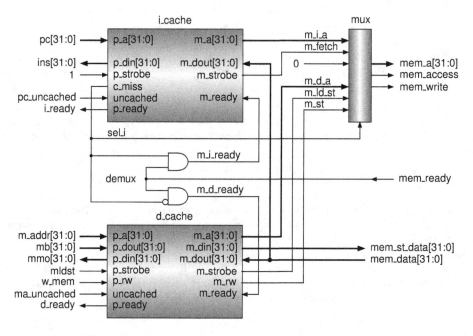

Figure 12.5   Arbitrating competing requests for memory access

Figure 12.5 consists of two AND gates, it sets either `m_i_ready` for instruction cache or `m_d_ready` for data cache whenever `mem_ready` (main memory ready) is active. This portion of the Verilog HDL code is listed below. A more general demultiplexer design method can be found in Chapter 2.

```
// mux, i_cache has higher priority than d_cache
wire        sel_i = i_cache_miss;                // fetch inst first
assign      mem_a = sel_i ? m_i_a   : m_d_a;     // mem address
assign mem_access = sel_i ? m_fetch : m_ld_st;   // mem access
```

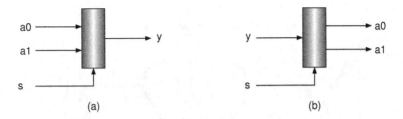

Figure 12.6 (a) Multiplexer and (b) demultiplexer

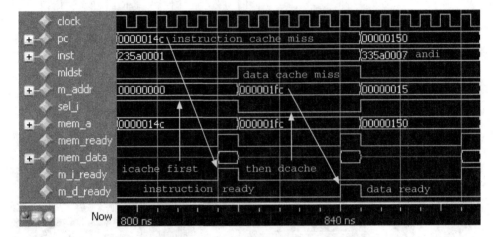

Figure 12.7 Waveform of arbitrating requests for memory access

```
assign mem_write  = sel_i ? 1'b0    : m_st;        // mem write enable
// demux the main mem ready
assign m_i_ready  = mem_ready & sel_i;             // ready for inst
assign m_d_ready  = mem_ready & ~sel_i;            // ready for mem data
```

Figure 12.7 shows the arbitration waveforms of the competing memory access requests issued by the instruction cache and data cache at the same time when both instruction cache and data cache are miss. The sel_i signal arbitrates the requests. The instruction cache's request (pc = 0x0000014c) is serviced first (m_i_ready = 1). Although the data cache's request (m_addr = 0x000001fc and mldst = 1) has a lower priority, it is serviced (m_d_ready = 1) before servicing the next instruction cache's request (pc = 0x00000150). This waveform is also picked up from the waveforms of the final CPU's simulation.

12.2.3 Pipeline Halt Circuit for Cache Misses

The caches are controlled by hardware. When a cache miss occurs, we stop the pipeline to wait for loading instruction or data from main memory. The signal expression of the pipeline stall caused by cache misses is as follows:

```
cache_stall = ~i_ready & ~itlb_exce |     // inst not ready & itlb hit
         mldst & ~d_ready & ~dtlb_exce;   // data not ready & dtlb hit
```

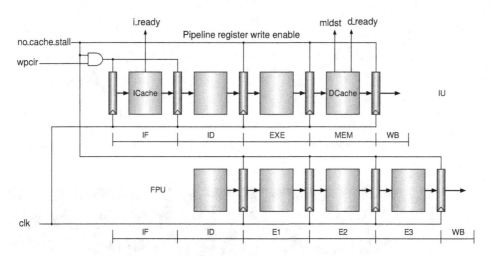

Figure 12.8   Pipeline stall caused by cache misses

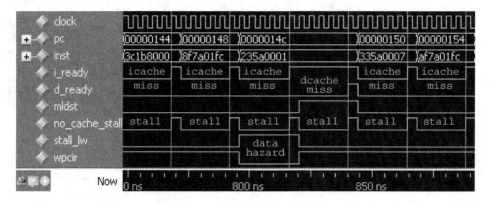

Figure 12.9   Waveform of pipeline stall caused by cache misses

where i_ready and d_ready indicate that an instruction and a data word is ready for use, respectively; mldst indicates that the instruction in MEM stage is a load or store. A stall can happen only if there are no TLB misses. Then the signal no_cache_stall (inverse of cache_stall) is used to stop the pipeline. We introduced the wpcir signal in Chapter 10, which enables writing the program counter (PC) and IF/ID pipeline registers. Different from wpcir, no_cache_stall enables writing all the pipeline registers, as shown in Figure 12.8.

Figure 12.9 shows the waveforms of the pipeline halt caused by an instruction cache miss first, and then a data cache miss (there are no TLB miss exceptions).

12.3   Design of Circuits Related to TLB

We introduced the TLB and gave a schematic circuit of an eight-entry, fully associative TLB (tlb_8_entry) in Chapter 11. This section describes how to use it to translate the virtual instruction address and virtual data address.

12.3.1   Instruction TLB (ITLB) and Data TLB (DTLB)

Both ITLB and DTLB use `tlb_8_entry`. Below is its Verilog HDL code that implements the circuit shown in Figure 11.28. In our TLB, each entry contains only one real page number; therefore, we use VPN, not VPN2, as the virtual tag of the TLB.

```
module tlb_8_entry (pte_in,tlbwi,tlbwr,index,vpn,        // 8-entry fully tlb
                    clk,clrn,pte_out,tlb_hit);
    input   [23:0] pte_in;           // page table entry from cp0 entrylo reg
    input          tlbwi;            // write tlb by index
    input          tlbwr;            // write tlb by random
    input   [2:0] index;             // tlb entry index from cp0 index reg
    input   [19:0] vpn;              // virtual page #
    input          clk, clrn;        // clock and reset
    output  [23:0] pte_out;          // physical page frame # and attributes
    output         tlb_hit;          // tlb hit
    wire    [2:0] random;            // random #
    wire    [2:0] w_idx;             // random # or index
    wire    [2:0] ram_idx;           // ram address
    wire    [2:0] vpn_index;         // matched address
    wire          tlbw = tlbwi | tlbwr;
    rand3    rdm (clk,clrn,random);                       // random # generator
    mux2x3   w_address (index,random,tlbwr,w_idx);        // write address
    mux2x3   ram_address (vpn_index,w_idx,tlbw,ram_idx);  // ram address
    ram8x24  rpn (ram_idx,pte_in,clk,tlbw,pte_out);       // ram
    cam8x21  valid_tag (clk,vpn,w_idx,tlbw,vpn_index,tlb_hit);        // cam
endmodule
```

Referring to Figure 12.10, the ITLB translates a virtual PC (v_pc) to a real PC. Because we use 4 KB as the page size, the virtual page number is v_pc[31:12]. The translated physical page number is contained in `ipte_out` (instruction page table entry). Similarly, the DTLB translates the virtual data memory address (malu) to a real memory address. The virtual page number is malu[31:12]. The translated physical page number is contained in `dpte_out` (data page table entry).

Different from the caches, which are controlled by hardware, the TLBs are managed by software. When a TLB miss occurs, the CPU enters the exception handler to update a TLB entry with a page table entry. When the CPU executes the `tlbwi` or `tlbwr` instruction, the content of CP0 EntryHi register will be written to the virtual part of a TLB entry. Hence we have

```
ipattern = (itlbwi | itlbwr) ? enthi[19:0] : v_pc[31:12];
dpattern = (dtlbwi | dtlbwr) ? enthi[19:0] : malu[31:12];
```

where `ipattern` and `dpattern` are the VPN inputs of ITLB and DTLB, respectively. The real part of the TLB entry is replaced with the content of the EntryLo register.

The content in CP0 Index register is used by the `tlbwi` instruction to select a TLB entry which will be updated. The content in CP0 Random register is used by the `tlbwr` instruction to select a TLB entry. We implement the CP0 Random register inside the TLB and let it be inaccessible to the CPU.

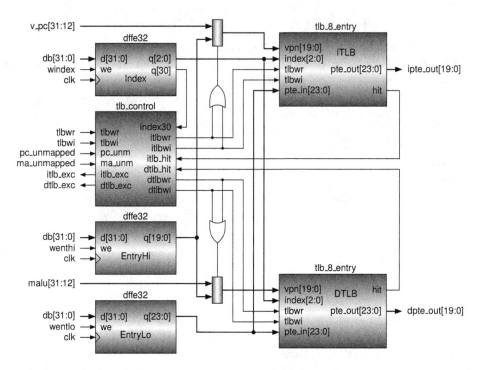

Figure 12.10   Signals of instruction TLB and data TLB

Table 12.1   Address mapping of the MIPS architecture

Segment	Virtual address	Physical address	TLBed	Cached	I/O
kuseg	00000000 ... 0000 01111111 ... 1111	00100000 ... 0000 –	Yes	Yes	No
kseg0	10000000 ... 0000 10011111 ... 1111	00000000 ... 0000 00011111 ... 1111	No	Yes	No
kseg1	10100000 ... 0000 10111111 ... 1111	00000000 ... 0000 00011111 ... 1111	No	No	Yes
kseg2	11000000 ... 0000 11111111 ... 1111	00100000 ... 0000 –	Yes	Yes	No

12.3.2   Generating Exception Signals during TLB Misses

A TLB miss exception happens only when the TLB is used. Recall that the MIPS virtual space from 0x80000000 to 0xbfffffff is unmapped, meaning that translating a virtual address in this space need not use a TLB. Table 12.1 lists the address mapping of the MIPS architecture.

Based on Table 12.1, we have the following expressions represented in Verilog HDL, where itlb_exc and dtlb_exc are the exception signals caused by the ITLB and DTLB misses, respectively; v_pc is the instruction virtual address; malu is the data virtual address in the MEM stage; and mldst indicates if the instruction in the MEM stage is a load or a store.

```
// mapped or unmapped
    pc_unmapped = v_pc[31]  &  ~v_pc[30];     // 10xx...xx; v_pc: va of inst
    ma_unmapped = malu[31]  &  ~malu[30];     // 10xx...xx; malu: va of data
// cached or uncached
    pc_uncached = pc_unmapped & v_pc[29];     // 101x...xx; v_pc: va of inst
    ma_uncached = ma_unmapped & malu[29];     // 101x...xx; malu: va of data
// real addresses
    pc      = pc_unmapped ? {3'b0,v_pc[28:0]} : {ipte_out[19:0],v_pc[11:0]};
    m_addr = ma_unmapped ? {3'b0,malu[28:0]} : {dpte_out[19:0],malu[11:0]};
// exceptions
    itlb_exc  = ~itlb_hit & ~pc_unmapped;
    dtlb_exc  = ~dtlb_hit & ~ma_unmapped & mldst;
    itlb_exce =  itlb_exc & sta[4];           // & exc enable
    dtlb_exce =  dtlb_exc & sta[5];           // & exc enable
// io: i/o or boot rom
    io = pc_uncached | ma_uncached;           // 101x...xx
```

The virtual addresses of v_pc and malu are translated into physical addresses of pc and m_addr, respectively. ipte_out and dpte_out are the real part (real page frame number) of the ITLB and DTLB, respectively.

12.3.3 Registers Related to TLB Miss Exceptions

Figure 12.11 shows the formats of the CP0 registers that are used to maintain the TLB entries. Note that the register formats are not the same as that defined by MIPS. We used a 1-bit D field in the Index register to indicate updating the DTLB (D = 1) or ITLB (D = 0) with the tlbwi or tlbwr instruction. Also, we used an 8-bit ExcEna to enable or disable interrupt and exceptions.

The EntryHi register holds the process ID and virtual page number that will be written to the virtual part of a TLB entry. EntryLo holds the physical page entry that will be written to the real part of a TLB entry. The physical page entry comes from the page table whose address is given by the Context register. The Context register contains PTEBase and BadVPN. PTEBase is the physical base address of the page table. Writing to the Context register with a mtc0 instruction will only update the PTEBase field and have no effect upon BadVPN. BadVPN is updated with VPN by hardware when a TLB miss occurs, as shown in the code below.

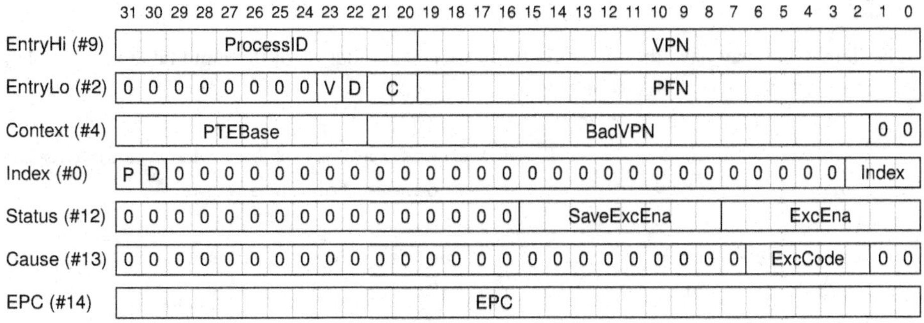

Figure 12.11 TLB-related CP0 registers

```
if        (wcontx)    contx[31:22] <= db[31:22];              // PTEBase
if        (itlb_exce) contx[21:0]  <= {v_pc[31:12],2'b00};    // BadVPN
else if  (dtlb_exce) contx[21:0]  <= {malu[31:12],2'b00};    // BadVPN
```

The Index register contains an index of a TLB entry, a 1-bit D field which indicates DTLB or ITLB, and a 1-bit P field which is not used in our design. Therefore we can use the following program to update a TLB entry pointed by the index field of the Index register. Whether to update DTLB or ITLB is determined by the D field in the Index register.

```
mfc0  $27, c0_context        # address of pte
lw    $26, 0x0($27)          # pte
mtc0  $26, c0_entry_lo       # move to cp0 entry_lo
sll   $26, $27, 10           # get bad vpn
srl   $26, $26, 12           # for cp0 entry_hi
mtc0  $26, c0_entry_hi       # move to entry_hi
tlbwi                        # update tlb
```

The Status register contains the interrupt/exceptions enable mask (ExcEna) and saved enable mask (SaveExcEna). There are six exceptions/interrupt in our CPU: (i) interrupt; (ii) Syscall; (ii) unimplemented instruction; (iv) overflow; (v) ITLB miss; and (vi) DTLB miss. A 1 in the ExcEna field will enable the corresponding exception or interrupt. SaveExcEna is used for saving ExcEna.

The exception program counter (EPC) contains the return address from the exception/interrupt handler. The Cause register holds the encoding of the exceptions and interrupt (ExcCode). It is generated by hardware when an exception or interrupt occurs. It can be used to let the control transfer to a corresponding place to handle the exception or interrupt through a jump table, as illustrated in the program below.

```
mfc0  $26, c0_cause          # read cp0 Cause reg
andi  $26, $26, 0x1c         # get ExcCode, 3 bits here
lui   $27, 0x8000            #
or    $27, $27, $26          #
lw    $27, j_table($27)      # get address from jump table
nop                          #
jr    $27                    # jump to that address
nop                          #
```

The jump table may have the following format: the start address of the jump table is j_table, and one address takes one word in the memory.

```
j_table:                     # address table for exception and interrupt
.word     int_entry          # 0. address for interrupt
.word     sys_entry          # 1. address for Syscall
.word     uni_entry          # 2. address for Unimplemented instruction
.word     ovf_entry          # 3. address for Overflow
.word     itlb_entry         # 4. address for itlb miss
.word     dtlb_entry         # 5. address for dtlb miss
```

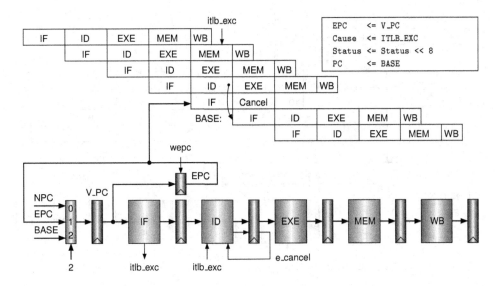

Figure 12.12 ITLB miss in ordinary case

12.3.4 Circuit Dealing with TLB Miss Exceptions

The ITLB is used in the instruction fetch (IF) stage for translating a virtual address v_pc to a real address pc. When an ITLB miss exception (itlb_exc) occurs, the control should be transferred to the exception handler. This can be done by writing the entry address of the exception handler to the PC. It is very important to note that this entry address should not be translated by the ITLB further. Therefore, we arrange the exception handler in the kseg0 segment (0x80000000–0x9fffffff), which is directly mapped to area 0x00000000–0x1fffffff of the main memory by setting the three leftmost bits to 0.

Referring to Figure 12.12, v_pc at which the ITLB miss exception occurs should be saved to the CP0 EPC register in order to enable it to return from the exception handler and fetch instruction from that address again.

The CPU's control unit in the ID stage detects the ITLB miss exception itlb_exc. It has the responsibility to do the following five operations at the same time if the ITLB miss exception is enabled:

1. Save the address v_pc at which the ITLB miss occurs to EPC;
2. Write a 4 (ExcCode of ITLB miss) to the ExcCode field of the Cause register;
3. Shift the content of the Status register to the left by 8 bits;
4. Generate a cancel signal; and
5. Write the entry address of the exception handler (BASE) to the v_pc register.

However, if the IF stage is located in a delayed slot, that is, the instruction in the ID stage is a jump or a branch, the address of the jump or branch instruction should be saved into EPC, because the instruction needs to be executed whenever returning from the exception handler. Referring to Figure 12.13, v_pc points to the instruction located in the delay slot. The address of the jump or branch instruction is in the PCD pipeline register. At the end of the ID stage, the control is transferred to the exception handler, and the content in the PCD is saved to EPC.

Now, consider the cases of DTLB miss. DTLB is used in the memory access (MEM) stage for translating a virtual address malu to a real address m_addr for a memory access instruction. When a DTLB miss exception (dtlb_exc) occurs, the control should be transferred to the exception handler.

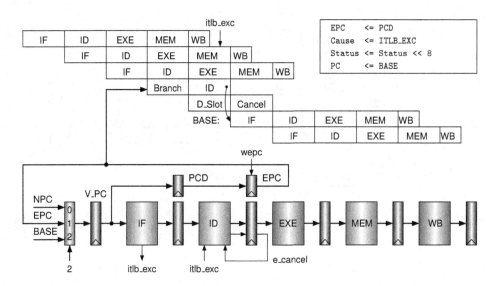

Figure 12.13   ITLB miss in delay slot

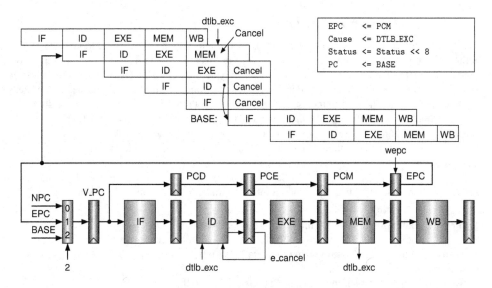

Figure 12.14   DTLB miss in the ordinary case

Referring to Figure 12.14, PCM in the MEM stage at which the DTLB miss occurs should be saved to the EPC register in order to be able to return from the exception handler and fetch instruction from that address again.

Because the DTLB miss occurs in the MEM stage, the circuit for dealing with the exception is more complex than that of ITLB miss which occurs in the IF stage. The complex things include not only adding multiple pipeline registers for saving the PC but also canceling many more instructions that are already in the pipeline.

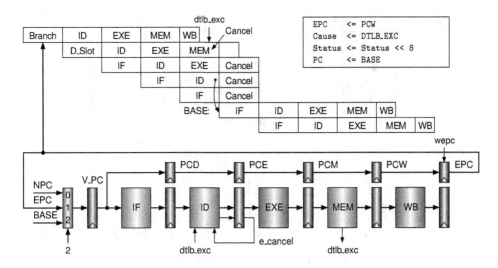

Figure 12.15 DTLB miss in delay slot

Table 12.2 Return address for EPC

itlb_exce	dtlb_exce	isbr	wisbr	EPC	sepc[1:0]
1	x	0	x	V_PC	0 0
1	x	1	x	PCD	0 1
0	1	x	0	PCM	1 0
0	1	x	1	PCW	1 1

We will describe soon how to cancel instructions. Before that, let's see the case where the instruction that caused a DTLB miss exception is in a delay slot. Referring to Figure 12.15, if the DTLB miss exception occurs in the delay slot, the address of the jump or branch instruction should be saved into the EPC. There are four instructions that need to be canceled.

We summarize the return address that should be written to the EPC in these four TLB miss exceptions in Table 12.2. We let itlb_exc to have a higher priority to be serviced than dtlb_exc. itlb_exce and dtlb_exce are the enabled miss exceptions of ITLB and DTLB, respectively; isbr indicates whether the instruction in the ID stage is a jump or a branch; and wisbr is the signal in the WB stage that was passed through the pipeline registers from isbr in the ID stage. We use a multiplexer to select a suitable return address for EPC input; sepc is the selection signal of the multiplexer. From Table 12.2, we get the expressions of the selection signal sepc as given below.

```
sepc[1] = ~itlb_exce & dtlb_exce;
sepc[0] =  itlb_exce & isbr | ~itlb_exce & dtlb_exce & wisbr;
```

There is also a multiplexer for the input of V_PC register. Input 2 of the multiplexer is the entry address of the exception handler; input 1 is the EPC used for returning from the exception handler; and input 0 is the next PC of the ordinary case. The expressions of the selection signal selpc of the multiplexer is given below.

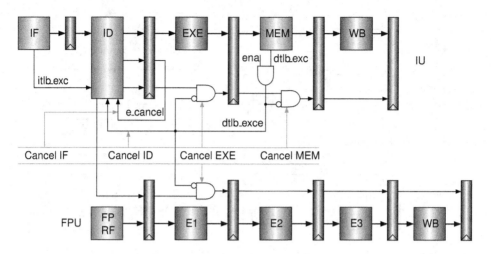

Figure 12.16   Cancel instructions due to DTLB miss

```
// selpc:  0 0 : npc
//         0 1 : epc
//         1 0 : exc_base
//         1 1 : x
selpc[1] = exce;                                    // go to handler
selpc[0] = i_eret;                                  // epc return
```

Now we describe how to cancel instructions when a TLB miss exception occurs. Canceling an instruction can be done by prohibiting the write signals that cause the change of the CPU's state. As discussed above, four instructions must be canceled when a DTLB miss occurs. Figure 12.16 shows the circuit that cancels the four instructions in the IF, ID, EXE, and MEM stages.

The instruction in the ID stage can be canceled inside the control unit. The instruction in the IF stage can be canceled after it enters the ID stage by the e_cancel signal. But the instructions in the EXE and MEM stages must be canceled outside the control unit. We just show an AND gate in each stage in the IU and FPU. The write signals in the MEM stage include mwreg (write IU register file), mwmem (write memory), and mwfpr (write FPU register file for lwc1 instruction). The write signals in the EXE stage include ewreg, ewmem, and ewfpr. The write signals in the ID stage include wreg, wmem, wfpr, wf (write FPU register file for float computational instructions), and the write signals issued by the mtc0 instruction.

The write signal in FPU's E2 stage is e2w. Because the instruction in the MEM stage is a memory reference instruction, there is no need to cancel e2w. The write signal in the E1 stage is e1w; it should be canceled because of the DTLB miss.

12.4   Design of CPU with Caches and TLBs

This section describes the design of the pipelined CPU with FPU, ITLB, ICache, DTLB, and DCache in Verilog HDL and gives the simulation waveforms.

12.4.1   Structure of CPU with Caches and TLBs

According to the discussions in previous sections, we give the simplified overall block diagram of the pipelined CPU in Figure 12.17.

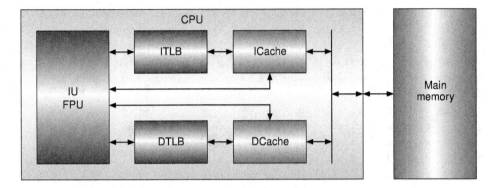

Figure 12.17   Block diagram of CPU with dedicated caches and TLBs

The CPU consists of an IU, an FPU, an instruction cache, a data cache, an ITLB, and a DTLB. There is an interface to the main memory. The caches are controlled by hardware; the TLBs are controlled by both hardware (generating exception signals) and software (manipulating TLB entries).

Different from the pipelined CPU with FPU described in Chapter 10, the initial value of the PC is 0x80000000, which is the starting address of the kseg0. It will be mapped to the real address 0x00000000 without using ITLB.

12.4.2   Verilog HDL Codes of CPU with Caches and TLBs

Below is the top Verilog HDL code, which invokes the CPU and memory modules. We also provide the I/O space for the kseg1 segment, but that was not tested. Some signals are not necessarily outputted, but we put them there for testing their waveforms.

```
module cpu_cache_tlb_memory (clk,memclk,clrn,v_pc,pc,inst,ealu,malu,wdi,wn,
    wd,ww,stall_lw,stall_fp,stall_lwc1,stall_swc1,stall,m_a,m_d_r,m_d_w,
    m_access,m_write,m_ready);                 // cpu + cache + tlb + memory
    input          clk, memclk, clrn;          // clocks and reset
    output [31:0] v_pc;                         // virtual pc
    output [31:0] pc;                           // real pc
    output [31:0] inst;                         // instruction
    output [31:0] ealu;                         // alu output in exe stage
    output [31:0] malu;                         // alu output in mem stage
    output [31:0] wdi;                          // data to iu reg file
    output [31:0] wd;                           // data to fpu reg file
    output   [4:0] wn;                          // fpu reg file's write number
    output        ww;                           // fpu reg file's write enable
    output        stall_lw;                     // stall by lw
    output        stall_fp;                     // stall by fp data dependency
    output        stall_lwc1;                   // stall by lwc1
    output        stall_swc1;                   // stall by swc1
    output        stall;                        // stall by div and sqrt
    output [31:0] m_a;                          // main memory address
    output [31:0] m_d_r;                        // main memory data read
```

```
    output [31:0] m_d_w;                      // main memory data write
    output        m_access;                   // main memory access
    output        m_write;                    // main memory write enable
    output        m_ready;                    // main memory ready
    wire          io;                         // inst rom or data i/o
    // cpu
    cpu_cache_tlb cpucachetlb (
        clk,memclk,clrn,v_pc,pc,inst,ealu,malu,wdi,wn,wd,
        ww,stall_lw,stall_fp,stall_lwc1,stall_swc1,stall,
        m_a,m_d_r,m_d_w,m_access,m_write,m_ready,io);
    // i/o, ignored
    wire [31:0]   io_d_r  = 0;                 // ignored io here
    wire          io_ready = 1;                // ignored io here
    wire [31:0]   mem_d_r;
    wire          mem_ready;
    wire          mem_access = m_access & ~io;
    wire           io_access = m_access & io;
    wire          mem_write  = m_write  & ~io;
    wire           io_write  = m_write  & io;
    assign        m_d_r    = (io) ? io_d_r   : mem_d_r;
    assign        m_ready  = (io) ? io_ready : mem_ready;
    // main memory
    physical_memory mem (m_a,mem_d_r,m_d_w,mem_access,m_write,mem_ready,
                        clk,memclk,clrn);
endmodule
```

The following is the CPU module that invokes the IU, FPU, and two-write-port floating-point register file modules. It also contains the multiplexers for the purpose of internal forwarding.

```
module cpu_cache_tlb (clk,memclk,clrn,v_pc,pc,inst,ealu,malu,wdi,wn,wd,
                      ww,stall_lw,stall_fp,stall_lwc1,stall_swc1,stall,
                      mem_a,mem_data,mem_st_data,mem_access,mem_write,
                      mem_ready,io);                // cpu + cache + tlb
    input         clk, memclk, clrn;                // clocks and reset
    input  [31:0] mem_data;                         // main memory data read
    input         mem_ready;                         // main memory ready
    output [31:0] v_pc;                             // virtual pc
    output [31:0] pc;                               // real pc
    output [31:0] inst;                             // instruction
    output [31:0] ealu;                            // alu output in exe stage
    output [31:0] malu;                            // alu output in mem stage
    output [31:0] wdi;                             // data to iu reg file
    output [31:0] wd;                              // data to fpu reg file
    output [31:0] mem_a;                           // main memory address
    output [31:0] mem_st_data;                     // main memory data write
    output  [4:0] wn;                              // fpu reg file's write number
    output        ww;                              // fpu reg file's write enable
```

```
    output          stall_lw;                   // stall by lw
    output          stall_fp;                   // stall by fp data dependency
    output          stall_lwc1;                 // stall by lwc1
    output          stall_swc1;                 // stall by swc1
    output          stall;                      // stall by div and sqrt
    output          mem_access;                 // main memory access
    output          mem_write;                  // main memory write enable
    output          io;                         // inst rom or data i/o
    wire    [31:0]  e3d;                        // fpu data in e3 stage
    wire    [31:0]  qfa,qfb,fa,fb,dfa,dfb;      // for iu
    wire    [31:0]  mmo,wmo;
    wire     [4:0]  count_div,count_sqrt;
    wire     [4:0]  e1n,e2n,e3n,wrn,fs,ft,fd;
    wire     [2:0]  fc;
    wire     [1:0]  e1c,e2c,e3c;                // for fpu
    wire            fwdla,fwdlb,fwdfa,fwdfb;
    wire            wf,fasmds,e1w,e2w,e3w,wwfpr;
    wire            no_cache_stall,dtlb;
    wire            e;                          // for multithreading CPU
    iu_cache_tlb i_u (e1n,e2n,e3n, e1w,e2w,e3w, stall,1'b0,dfb,e3d,clk,
                    memclk,clrn,no_cache_stall,dtlb,fs,ft,wmo,wrn,wwfpr,mmo,
                    fwdla,fwdlb,fwdfa,fwdfb,fd,fc,wf,fasmds,v_pc,pc,inst,ealu,
                    malu,wdi,stall_lw,stall_fp,stall_lwc1,stall_swc1,mem_a,
                    mem_data,mem_st_data,mem_access,mem_write,mem_ready,io);
    regfile2w fpr (fs,ft,wd,wn,ww,wmo,wrn,wwfpr,~clk,clrn,qfa,qfb);
    mux2x32 fwd_f_load_a (qfa,mmo,fwdla,fa);    // forward lwc1 to fp a
    mux2x32 fwd_f_load_b (qfb,mmo,fwdlb,fb);    // forward lwc1 to fp b
    mux2x32 fwd_f_res_a  (fa,e3d,fwdfa,dfa);    // forward fp res to fp a
    mux2x32 fwd_f_res_b  (fb,e3d,fwdfb,dfb);    // forward fp res to fp b
    fpu  fp_unit (dfa,dfb,fc,wf,fd,no_cache_stall,clk,clrn,e3d,wd,wn,ww,
                    stall,e1n,e1w,e2n,e2w,e3n,e3w,e1c,e2c,e3c,count_div,
                    count_sqrt,e,dtlb);
endmodule
```

The IU module is listed below; it consists of a virtual PC, an ITLB, an ICache, a general-purpose register file, a control unit, an ALU, a DTLB, a DCache, CP0 registers, pipeline registers, the registers for saving v_pc, and multiplexers. This module also arbitrates the competing memory access requests issued by ICache and DCache at the same time.

```
module iu_cache_tlb (e1n,e2n,e3n, e1w,e2w,e3w, stall,st,dfb,e3d,clk,
                    memclk,clrn,no_cache_stall,dtlb,fs,ft,wmo,wrn,wwfpr,
                    mmo,fwdla,fwdlb,fwdfa,fwdfb,fd,fc,wf,fasmds,v_pc,pc,
                    inst,ealu,malu,wdi,stall_lw,stall_fp,stall_lwc1,
                    stall_swc1,mem_a,mem_data,mem_st_data,mem_access,
                    mem_write,mem_ready,io);   // iu + cache + tlb
    input         clk, memclk, clrn;           // clocks and reset
    input  [31:0] mem_data;                    // main mem data read
    input  [31:0] dfb;                         // fpu b in id stage
```

```
input   [31:0] e3d;                        // fpu data in e3 stage
input    [4:0] e1n, e2n, e3n;              // fp reg numbers
input          e1w, e2w, e3w;              // fp reg write enables
input          stall;                      // stall by div and sqrt
input          st;                         // not used
input          mem_ready;                  // main mem ready
output  [31:0] v_pc;                       // virtual pc
output  [31:0] pc;                         // real pc
output  [31:0] inst;                       // inst
output  [31:0] ealu;                       // alu output in exe stage
output  [31:0] malu;                       // alu output in mem stage
output  [31:0] wdi;                        // data to iu reg file
output  [31:0] mmo;                        // main mem out in mem stage
output  [31:0] wmo;                        // main mem out in wb stage
output  [31:0] mem_a;                      // main mem address
output  [31:0] mem_st_data;                // main mem data write
output   [4:0] fs, ft, fd;                 // fpu fs, ft, fd
output.  [4:0] wrn;                        // regfile2w write # for lwc1
output   [2:0] fc;                         // fpu op
output         mem_access;                 // main mem access
output         mem_write;                  // main mem write enable
output         wwfpr;                      // regfile2w we for lwc1
output         fwdla;                      // forward lwc1 data to a
output         fwdlb;                       // forward lwc1 data to b
output         fwdfa;                      // forward fp data to b
output         fwdfb;                      // forward fp data to b
output         wf;                         // write fp regfile, id stage
output         fasmds;                     // is fp calc, id stage
output         stall_lw;                   // stall by lw
output         stall_fp;                   // stall by fp data depend.
output         stall_lwc1;                 // stall by lwc1
output         stall_swc1;                 // stall by swc1
output         no_cache_stall;             // no cache stall
output         dtlb;                       // dtlb hit
output         io;                         // inst rom or data i/o
parameter      exc_base = 32'h80000008;    // int/exc handler entry
wire    [31:0] bpc,jpc,npc,pc4,ins,dpc4,inst,qa,qb,da,db,dc,dd;
wire    [31:0] simm,epc8,alua,alub,ealu0,ealu1,ealu,sa,eb;
wire     [5:0] op,func;
wire     [4:0] rs,rt,rd,fs,ft,fd,drn,ern;
wire     [3:0] aluc;
wire     [1:0] pcsrc,fwda,fwdb;
wire           wpcir,wreg,m2reg,wmem,aluimm,shift,jal;
wire     [4:0] e1n,e2n,e3n;
reg            ewfpr,ewreg,em2reg,ewmem,ejal,efwdfe,ealuimm,eshift;
reg            mwfpr,mwreg,mm2reg,mwmem;
reg            wwfpr,wwreg,wm2reg;
```

```
reg     [31:0] epc4,ea,ed,eimm,malu,mb,wmo,walu;
reg      [4:0] ern0,mrn,wrn;
reg      [3:0] ealuc;
wire    [23:0] ipte_out;
wire           itlb_hit;
wire    [31:0] pcd;
wire    [31:0] index;                         // cp0 reg 0: index
wire    [31:0] entlo;                         // cp0 reg 2: entry lo
reg     [31:0] contx;                         // cp0 reg 4: context
wire    [31:0] enthi;                         // cp0 reg 9: entry hi
wire           windex,wentlo,wcontx,wenthi;   // write enables
wire           rc0,wc0;                       // read,write c0 res
wire     [1:0] c0rn;                          // c0 reg # for mux
wire           swfp,regrt,sext,fwdf,fwdfe,wfpr;
wire           i_ready,i_cache_miss;
wire           tlbwi,tlbwr;
wire           wepc,wcau,wsta,isbr,cancel,exce,ldst;
wire     [1:0] sepc,selpc;
wire    [31:0] sta,cau,epc,sta_in,cau_in,epc_in;
wire    [31:0] stalr,epcin,cause,c0reg,next_pc;
reg     [31:0] pce;
reg      [1:0] ec0rn;
reg            erc0,ecancel,eisbr,eldst;
reg     [31:0] pcm;
reg            misbr,mldst;
wire    [23:0] dpte_out;
wire           dtlb_hit;
reg     [31:0] pcw;
reg            wisbr;
wire           m_fetch,m_ld_st,m_st;
wire    [31:0] m_i_a,m_d_a;
wire           itlb_exc, dtlb_exc;
wire           itlb_exce,dtlb_exce;
wire           m_i_ready;
wire           m_d_ready;
// IF stage
vpc v_p_c (next_pc,clk,clrn,wpcir&no_cache_stall,v_pc);    // vpc
cla32 pc_plus4 (v_pc,32'h4,1'b0,pc4);                      // vpc+4
mux4x32 nextpc (pc4,bpc,da,jpc,pcsrc,npc);                 // next pc
wire           itlbwi = tlbwi & ~index[30];                // itlb write
wire           itlbwr = tlbwr & ~index[30];
wire           dtlbwi = tlbwi & index[30];                 // dtlb write
wire           dtlbwr = tlbwr & index[30];
wire    [19:0] ipattern = (itlbwi | itlbwr)? enthi[19:0] : v_pc[31:12];
wire           pc_unmapped = v_pc[31] & ~v_pc[30];         // 10xx...xx
wire           pc_uncached = pc_unmapped & v_pc[29];       // 101x...xx
assign pc = pc_unmapped?{3'b0,v_pc[28:0]} : {ipte_out[19:0],v_pc[11:0]};
```

```
tlb_8_entry itlb (entlo[23:0],itlbwi,itlbwr,index[2:0],ipattern, // itlb
                  clk,clrn,ipte_out,itlb_hit);
assign          itlb_exc = ~itlb_hit & ~pc_unmapped;
i_cache icache (pc,ins,1'b1,pc_uncached,i_ready,i_cache_miss,  // icache
                clk,clrn,m_i_a,mem_data,m_fetch,m_i_ready);
// IF/ID pipeline registers
dffe32 pc_4_r (pc4, clk,clrn,wpcir&no_cache_stall,dpc4);        // pc4
dffe32 inst_r (ins, clk,clrn,wpcir&no_cache_stall,inst);        // ir
dffe32 pcd_r  (v_pc,clk,clrn,wpcir&no_cache_stall,pcd);         // pcd
// ID stage
assign op   = inst[31:26];
assign rs   = inst[25:21];
assign rt   = inst[20:16];
assign rd   = inst[15:11];
assign ft   = inst[20:16];
assign fs   = inst[15:11];
assign fd   = inst[10:6];
assign func = inst[5:0];
assign simm = {{16{sext&inst[15]}},inst[15:0]};
assign jpc  = {dpc4[31:28],inst[25:0],2'b00};          // jump target
cla32 br_addr (dpc4,{simm[29:0],2'b00},1'b0,bpc);      // branch target
regfile rf (rs,rt,wdi,wrn,wwreg,~clk,clrn,qa,qb);      // reg file
mux4x32 alu_a(qa,ealu,malu,mmo,fwda,da);              // forward A
mux4x32 alu_b(qb,ealu,malu,mmo,fwdb,db);              // forward B
mux2x32 store_f (db,dfb,swfp,dc);                     // swc1
mux2x32 fwd_f_d (dc,e3d,fwdf,dd);                     // forward fp result
wire rsrtequ = ~|(da^db);                             // (da == db)
mux2x5 des_reg_no (rd,rt,regrt,drn);                  // destination reg
iu_cache_tlb_cu cu (op,func,rs,rt,rd,fs,ft,rsrtequ,   // control unit
             ewfpr,ewreg,em2reg,ern,mwfpr,mwreg,mm2reg,mrn,e1w,e1n,e2w,
             e2n,e3w,e3n,stall,st,pcsrc,wpcir,wreg,m2reg,wmem,jal,aluc,
             sta,aluimm,shift,sext,regrt,fwda,fwdb,swfp,fwdf,fwdfe,
             wfpr,fwdla,fwdlb,fwdfa,fwdfb,fc,wf,fasmds,stall_lw,
             stall_fp,stall_lwc1,stall_swc1,windex,wentlo,wcontx,
             wenthi,rc0,wc0,tlbwi,tlbwr,c0rn,wepc,wcau,wsta,isbr,sepc,
             cancel,cause,exce,selpc,ldst,wisbr,ecancel,itlb_exc,
             dtlb_exc,itlb_exce,dtlb_exce);
assign dtlb = ~dtlb_exce;                                       // dtlb hit
dffe32 c0_Index (db,clk,clrn,windex&no_cache_stall,index);  // index
dffe32 c0_Entlo (db,clk,clrn,wentlo&no_cache_stall,entlo);  // entlo
dffe32 c0_Enthi (db,clk,clrn,wenthi&no_cache_stall,enthi);  // enthi
always @(negedge clrn or posedge clk)                          // contx
   if (!clrn) begin
      contx <= 0;
   end else begin
      if        (wcontx)   contx[31:22] <= db[31:22];           // PTEBase
      if        (itlb_exce) contx[21:0] <= {v_pc[31:12],2'b00}; // BadVPN
```

```
          else if (dtlb_exce) contx[21:0] <= {malu[31:12],2'b00};
     end
   dffe32  c0_status (sta_in,clk,clrn,wsta&no_cache_stall,sta);  // sta
   dffe32  c0_cause  (cau_in,clk,clrn,wcau&no_cache_stall,cau);  // cau
   dffe32  c0_epc    (epc_in,clk,clrn,wepc&no_cache_stall,epc);  // epc
   mux2x32 sta_mx (stalr,db,wc0,sta_in);                // mux for status reg
   mux2x32 cau_mx (cause,db,wc0,cau_in);                // mux for cause reg
   mux2x32 epc_mx (epcin,db,wc0,epc_in);                // mux for epc reg
   mux2x32 sta_lr ({8'h0,sta[31:8]},{sta[23:0],8'h0},exce,stalr);
   mux4x32 epc_04 (v_pc,pcd,pcm,pcw,sepc,epcin);        // epc source
   mux4x32 irq_pc (npc,epc,exc_base,32'h0,selpc,next_pc); // for pc
   mux4x32 fromc0 (contx,sta,cau,epc,ec0rn,c0reg);      // for mfc0
   // ID/EXE pipeline registers
   always @(negedge clrn or posedge clk) begin
       if (!clrn) begin
             ewfpr    <= 0;             ewreg    <= 0;
             eldst    <= 0;             ewmem    <= 0;
             ejal     <= 0;             ealuimm <= 0;
             efwdfe   <= 0;             ealuc    <= 0;
             eshift   <= 0;             epc4     <= 0;
             ea       <= 0;             ed       <= 0;
             eimm     <= 0;             ern0     <= 0;
             erc0     <= 0;             ec0rn    <= 0;
             ecancel <= 0;             eisbr    <= 0;
             pce      <= 0;             em2reg   <= 0;
       end else if (no_cache_stall) begin
             ewfpr    <= wfpr;          ewreg    <= wreg;
             ewmem    <= wmem;          eldst    <= ldst;
             ejal     <= jal;           ealuimm <= aluimm;
             efwdfe   <= fwdfe;         ealuc    <= aluc;
             eshift   <= shift;         epc4     <= dpc4;
             ea       <= da;            ed       <= dd;
             eimm     <= simm;          ern0     <= drn;
             erc0     <= rc0;           ec0rn    <= c0rn;
             ecancel <= cancel;        eisbr    <= isbr;
             pce      <= pcd;           em2reg   <= m2reg;
       end
   end
   // EXE stage
   cla32 ret_addr (epc4,32'h4,1'b0,epc8);                   // pc+8
   assign  sa = {eimm[5:0],eimm[31:6]};                     // shift amount
   mux2x32 alu_ina (ea,sa,eshift,alua);                     // alu input A
   mux2x32 alu_inb (eb,eimm,ealuimm,alub);                  // alu input B
   mux2x32 save_pc8 (ealu0,epc8,ejal,ealu1);                // pc+8 if jal
   mux2x32 read_cr0 (ealu1,c0reg,erc0,ealu);                // read c0 regs
   wire z;
   alu al_unit (alua,alub,ealuc,ealu0,z);                   // alu
```

```verilog
assign        ern = ern0 | {5{ejal}};                // $31 for jal
mux2x32 fwd_f_e (ed,e3d,efwdfe,eb);                  // forward fp result
// EXE/MEM pipeline registers
always @(negedge clrn or posedge clk) begin
    if (!clrn) begin
        mwfpr   <= 0;           mwreg  <= 0;
        mldst   <= 0;           mwmem  <= 0;
        malu    <= 0;           mb     <= 0;
        mrn     <= 0;           misbr  <= 0;
        pcm     <= 0;           mm2reg <= 0;
    end else if (no_cache_stall) begin
        mwfpr   <= ewfpr & ~dtlb_exce;               // cancel exe
        mwreg   <= ewreg & ~dtlb_exce;               // cancel exe
        mwmem   <= ewmem & ~dtlb_exce;               // cancel exe
        mldst   <= eldst & ~dtlb_exce;               // cancel exe
        malu    <= ealu;        mb     <= eb;
        mrn     <= ern;         misbr  <= eisbr;
        pcm     <= pce;         mm2reg <= em2reg;
    end
end
// MEM stage
wire   [19:0] dpattern = (dtlbwi | dtlbwr) ? enthi[19:0] : malu[31:12];
wire          ma_unmapped = malu[31] & ~malu[30]; // 10xx...xx
wire          ma_uncached = ma_unmapped & malu[29]; // 101x...xx
wire   [31:0] m_addr = ma_unmapped? {3'b0,malu[28:0]} :
                                    {dpte_out[19:0],malu[11:0]};
tlb_8_entry d_tlb (entlo[23:0],dtlbwi,dtlbwr,index[2:0],         // dtlb
            dpattern,clk,clrn,dpte_out,dtlb_hit);
assign        dtlb_exc = ~dtlb_hit & ~ma_unmapped & mldst;
wire          d_ready;
wire          w_mem = mwmem & ~dtlb_exce;         // cancel mem (sw/swc1)
d_cache dcache (m_addr,mb,mmo,mldst,w_mem,ma_uncached,d_ready, // dcache
            clk,clrn,m_d_a,mem_data,mem_st_data,m_ld_st,
            m_st,m_d_ready);
assign        io = pc_uncached | ma_uncached;     // 101x...xx
// MEM/WB pipeline registers
always @(negedge clrn or posedge clk) begin
    if (!clrn) begin
        wwfpr   <= 0;           wwreg  <= 0;
        wm2reg  <= 0;           wmo    <= 0;
        walu    <= 0;           wrn    <= 0;
        pcw     <= 0;           wisbr  <= 0;
    end else if (no_cache_stall) begin
        wwfpr   <= mwfpr & ~dtlb_exce;               // cancel mem
        wwreg   <= mwreg & ~dtlb_exce;               // cancel mem
        wm2reg  <= mm2reg;      wmo    <= mmo;
```

```
            walu    <= malu;        wrn   <= mrn;
            pcw     <= pcm;         wisbr <= misbr;
        end
    end
    // WB stage
    mux2x32 wb_sel (walu,wmo,wm2reg,wdi);
    // mux, i_cache has higher priority than d_cache
    wire         sel_i = i_cache_miss;                // fetch inst first
    assign       mem_a = sel_i ? m_i_a   : m_d_a;     // mem address
    assign mem_access = sel_i ? m_fetch : m_ld_st;    // mem access
    assign mem_write  = sel_i ? 1'b0    : m_st;       // mem write enable
    // demux the main mem ready
    assign m_i_ready  = mem_ready & sel_i;            // ready for inst
    assign m_d_ready  = mem_ready & ~sel_i;           // ready for mem data
    assign no_cache_stall = ~(~i_ready & ~itlb_exce | // no itlb miss exce
                    mldst & ~d_ready & ~dtlb_exce);   // no dtlb miss exce
endmodule
```

The following code implements the VPC (virtual PC). It is set to 0x80000000 on a reset, pointing to the starting address of the kseg0 segment.

```
module vpc (d,clk,clrn,e,q);                    // virtual program counter
    input      [31:0] d;                        // input d
    input             e;                        // e: enable
    input             clk, clrn;                // clock and reset
    output reg [31:0] q;                        // output q
    always @(negedge clrn or posedge clk)
        if (!clrn)                              // if reset
            q <= 32'h8000_0000;                 // kseg0 starting address
        else if (e) q <= d;                     // save d if enabled
endmodule
```

The control unit module is listed below. It decodes instructions and generates all the control signals for the CPU, including the signals for dealing with the TLB miss exceptions and for pipeline stalls caused by data dependencies and cache misses.

```
module iu_cache_tlb_cu (op,func,rs,rt,rd,fs,ft,rsrtequ,ewfpr,ewreg,em2reg,
                ern,mwfpr,mwreg,mm2reg,mrn,e1w,e1n,e2w,e2n,e3w,e3n,
                stall_div_sqrt,st,pcsrc,wpcir,wreg,m2reg,wmem,jal,aluc,
                sta,aluimm,shift,sext,regrt,fwda,fwdb,swfp,fwdf,fwdfe,
                wfpr,fwdla,fwdlb,fwdfa,fwdfb,fc,wf,fasmds,stall_lw,
                stall_fp,stall_lwc1,stall_swc1,windex,wentlo,wcontx,
                wenthi,rc0,wc0,tlbwi,tlbwr,c0rn,wepc,wcau,wsta,isbr,sepc,
                cancel,cause,exce,selpc,ldst,wisbr,ecancel,itlb_exc,
                dtlb_exc,itlb_exce,dtlb_exce);          // control unit
    input        rsrtequ, ewreg,em2reg,ewfpr, mwreg,mm2reg,mwfpr;
    input        e1w,e2w,e3w,stall_div_sqrt,st;
    input  [5:0] op,func;
```

```
input    [4:0]  rs,rt,rd,fs,ft,ern,mrn,e1n,e2n,e3n;
input    [31:0] sta;  // IM[7:0] : x,x,dtlb_exc,itlb_exc,ov,unimpl,sys,int
output          wpcir,wreg,m2reg,wmem,jal,aluimm,shift,sext,regrt;
output          swfp,fwdf,fwdfe;
output          fwdla,fwdlb,fwdfa,fwdfb;
output          wfpr,wf,fasmds;
output   [1:0]  pcsrc,fwda,fwdb;
output   [3:0]  aluc;
output   [2:0]  fc;                // fp op
output          stall_lw,stall_fp,stall_lwc1,stall_swc1; // stalls
output          rc0;               // read  c0 regs, mfc0
output          wc0;               // write c0 regs, mtc0
output          tlbwi;             // write tlb by index
output          tlbwr;             // write tlb by random
output   [1:0]  c0rn,sepc,selpc;   // mux select signals
output          windex;            // cp0 index   write enable
output          wentlo;            // cp0 entrylo write enable
output          wcontx;            // cp0 context write enable
output          wenthi;            // cp0 entryhi write enable
output          wepc;              // cp0 epc     write enable
output          wcau;              // cp0 cause   write enable
output          wsta;              // cp0 status  write enable
output          isbr;              // inst in id stage is a jump or branch
output          cancel;            // cancel in id stage
output          exce;              // itlb_exce | dtlb_exce, masked
output          ldst;              // inst in id stage is a load or store
output [31:0]   cause;             // cp0 cause reg
input           wisbr;             // inst in wb stage is a jump or branch
input           ecancel;           // cancel in exe stage
input           itlb_exc;          // itlb miss exception
input           dtlb_exc;          // dtlb miss exception
output          itlb_exce;         // itlb miss exception & its enable
output          dtlb_exce;         // dtlb miss exception & its enable
wire            stall_others;
wire            rtype,i_add,i_sub,i_and,i_or,i_xor,i_sll,i_srl,i_sra,i_jr;
wire            i_addi,i_andi,i_ori,i_xori,i_beq,i_bne,i_lw,i_sw,i_j;
wire            i_jal,i_eret,i_mtc0,i_mfc0;
wire            ftype,i_lwc1,i_swc1,i_fadd,i_fsub,i_fmul,i_fdiv,i_fsqrt;
assign          itlb_exce = itlb_exc & sta[4];            // & exc enable
assign          dtlb_exce = dtlb_exc & sta[5];            // & exc enable
wire            no_dtlb_exce = ~dtlb_exce;
assign ldst = (i_lw | i_sw | i_lwc1 | i_swc1) & ~ecancel & no_dtlb_exce;
assign isbr = i_beq | i_bne | i_j | i_jal;
// itlb_exce dtlb_exce isbr wisbr EPC   sepc[1:0]
// 1         x         0    x     V_PC  0 0
// 1         x         1    x     PCD   0 1
// 0         1         x    0     PCM   1 0
```

```
// 0          1         x    1       PCW    1 1
assign sepc[1]= ~itlb_exce & dtlb_exce;
assign sepc[0]=  itlb_exce & isbr | ~itlb_exce & dtlb_exce & wisbr;
assign exce    =  itlb_exce | dtlb_exce;
assign cancel =  exce;
// selpc:  0 0 : npc
//         0 1 : epc
//         1 0 : exc_base
//         1 1 : x
assign selpc[1] = exce;                                // go to handler
assign selpc[0] = i_eret;                              // epc return
// op      rs    rt    rd            func
// 010000 00100 xxxxx xxxxx 00000 000000 mtc0 rt, rd; c0[rd] <- gpr[rt]
// 010000 00000 xxxxx xxxxx 00000 000000 mfc0 rt, rd; gpr[rt] <- c0[rd]
// 010000 10000 00000 00000 00000 000010 tlbwi
// 010000 10000 00000 00000 00000 000110 tlbwr
// 010000 10000 00000 00000 00000 011000 eret
assign i_mtc0 = (op==6'h10) & (rs==5'h04) & (func==6'h00)& no_dtlb_exce;
assign i_mfc0 = (op==6'h10) & (rs==5'h00) & (func==6'h00);
assign i_eret = (op==6'h10) & (rs==5'h10) & (func==6'h18);
assign tlbwi  = (op==6'h10) & (rs==5'h10) & (func==6'h02);
assign tlbwr  = (op==6'h10) & (rs==5'h10) & (func==6'h06);
assign windex = i_mtc0 & (rd==5'h00);                  // write index
assign wentlo = i_mtc0 & (rd==5'h02);                  // write entry_lo
assign wcontx = i_mtc0 & (rd==5'h04);                  // write context
assign wenthi = i_mtc0 & (rd==5'h09);                  // write entry_hi
assign wsta   = i_mtc0 & (rd==5'h0c) | exce | i_eret;  // write status
assign wcau   = i_mtc0 & (rd==5'h0d) | exce;           // write cause
assign wepc   = i_mtc0 & (rd==5'h0e) | exce;           // write epc
//wire rcontx = i_mfc0 & (rd==5'h04);                  // read  context
wire rstatus  = i_mfc0 & (rd==5'h0c);                  // read  status
wire rcause   = i_mfc0 & (rd==5'h0d);                  // read  cause
wire repc     = i_mfc0 & (rd==5'h0e);                  // read  epc
assign c0rn[1]= rcause  | repc;            // c0rn:  00    01    10    11
assign c0rn[0]= rstatus | repc;            //        contx sta   cau   epc
assign rc0    = i_mfc0;                    // read  c0 regs
assign wc0    = i_mtc0;                    // write c0 regs
wire [2:0] exccode;                        // test itlb_exc and dtlb_exc
//     000  interrupt                      // not  test here
//     001  syscall                        // not  test here
//     010  unimpl. inst.                  // not  test here
//     011  overflow                       // not  test here
//     100  itlb_exc                       // test here
//     101  dtlb_exc                       // test here
assign exccode[2] = itlb_exc | dtlb_exc;
assign exccode[1] = 0;
assign exccode[0] = dtlb_exc;
```

```verilog
assign cause      = {27'h0,exccode,2'b00};
and(rtype,~op[5],~op[4],~op[3],~op[2],~op[1],~op[0]);           // r format
and(i_add,rtype, func[5],~func[4],~func[3],~func[2],~func[1],~func[0]);
and(i_sub,rtype, func[5],~func[4],~func[3],~func[2], func[1],~func[0]);
and(i_and,rtype, func[5],~func[4],~func[3], func[2],~func[1],~func[0]);
and(i_or, rtype, func[5],~func[4],~func[3], func[2],~func[1], func[0]);
and(i_xor,rtype, func[5],~func[4],~func[3], func[2], func[1],~func[0]);
and(i_sll,rtype,~func[5],~func[4],~func[3],~func[2],~func[1],~func[0]);
and(i_srl,rtype,~func[5],~func[4],~func[3],~func[2], func[1],~func[0]);
and(i_sra,rtype,~func[5],~func[4],~func[3],~func[2], func[1], func[0]);
and(i_jr, rtype,~func[5],~func[4], func[3],~func[2],~func[1],~func[0]);
and(i_addi,~op[5],~op[4], op[3],~op[2],~op[1],~op[0]);         // i format
and(i_andi,~op[5],~op[4], op[3], op[2],~op[1],~op[0]);
and(i_ori, ~op[5],~op[4], op[3], op[2],~op[1], op[0]);
and(i_xori,~op[5],~op[4], op[3], op[2], op[1],~op[0]);
and(i_lw,   op[5],~op[4],~op[3],~op[2], op[1], op[0]);
and(i_sw,   op[5],~op[4], op[3],~op[2], op[1], op[0]);
and(i_beq, ~op[5],~op[4],~op[3], op[2],~op[1],~op[0]);
and(i_bne, ~op[5],~op[4],~op[3], op[2],~op[1], op[0]);
and(i_lui, ~op[5],~op[4], op[3], op[2], op[1], op[0]);
and(i_j,   ~op[5],~op[4],~op[3],~op[2], op[1],~op[0]);         // j format
and(i_jal, ~op[5],~op[4],~op[3],~op[2], op[1], op[0]);
and(ftype, ~op[5], op[4],~op[3],~op[2],~op[1], op[0]);         // f format
and(i_lwc1, op[5], op[4],~op[3],~op[2],~op[1], op[0]);
and(i_swc1, op[5], op[4], op[3],~op[2],~op[1], op[0]);
and(i_fadd,ftype,~func[5],~func[4],~func[3],~func[2],~func[1],~func[0]);
and(i_fsub,ftype,~func[5],~func[4],~func[3],~func[2],~func[1], func[0]);
and(i_fmul,ftype,~func[5],~func[4],~func[3],~func[2], func[1],~func[0]);
and(i_fdiv,ftype,~func[5],~func[4],~func[3],~func[2], func[1], func[0]);
and(i_fsqrt,ftype,~func[5],~func[4],~func[3], func[2],~func[1],~func[0]);
wire i_rs = i_add  | i_sub  | i_and  | i_or  | i_xor  | i_jr  | i_addi |
            i_andi | i_ori  | i_xori | i_lw  | i_sw   | i_beq | i_bne  |
            i_lwc1 | i_swc1;
wire i_rt = i_add  | i_sub  | i_and  | i_or  | i_xor  | i_sll | i_srl  |
            i_sra  | i_sw   | i_beq  | i_bne | i_mtc0;
assign stall_lw = ewreg & em2reg & (ern != 0) & (i_rs & (ern == rs) |
                                                  i_rt & (ern == rt));
reg [1:0] fwda, fwdb;
always @ (ewreg or mwreg or ern or mrn or em2reg or mm2reg or rs or
          rt) begin
    fwda = 2'b00;                                 // default: no hazards
    if (ewreg & (ern != 0) & (ern == rs) & ~em2reg) begin
        fwda = 2'b01;                             // select exe_alu
    end else begin
        if (mwreg & (mrn != 0) & (mrn == rs) & ~mm2reg) begin
            fwda = 2'b10;                         // select mem_alu
        end else begin
```

```
               if (mwreg & (mrn != 0) & (mrn == rs) & mm2reg) begin
                   fwda = 2'b11;                          // select mem_lw
               end
           end
       end
       fwdb = 2'b00;                                      // default: no hazards
       if (ewreg & (ern != 0) & (ern == rt) & ~em2reg) begin
           fwdb = 2'b01;                                  // select exe_alu
       end else begin
           if (mwreg & (mrn != 0) & (mrn == rt) & ~mm2reg) begin
               fwdb = 2'b10;                              // select mem_alu
           end else begin
               if (mwreg & (mrn != 0) & (mrn == rt) & mm2reg) begin
                   fwdb = 2'b11;                          // select mem_lw
               end
           end
       end
   end
   assign wreg   =(i_add  | i_sub  | i_and  | i_or  | i_xor | i_sll  |
                   i_srl  | i_sra  | i_addi | i_andi | i_ori | i_xori |
                   i_lw   | i_lui  | i_jal  | i_mfc0) &
                   wpcir  & ~ecancel & no_dtlb_exce;
   assign regrt  = i_addi | i_andi | i_ori | i_xori | i_lw | i_lui |
                   i_lwc1 | i_mfc0;
   assign jal    = i_jal;
   assign m2reg  = i_lw;
   assign shift  = i_sll | i_srl  | i_sra;
   assign aluimm = i_addi | i_andi | i_ori | i_xori | i_lw | i_lui |
                   i_sw | i_lwc1 | i_swc1;
   assign sext   = i_addi | i_lw | i_sw | i_beq | i_bne | i_lwc1 | i_swc1;
   assign aluc[3] = i_sra;
   assign aluc[2] = i_sub | i_or  | i_srl | i_sra | i_ori  | i_lui;
   assign aluc[1] = i_xor | i_sll | i_srl | i_sra | i_xori | i_beq  |
                    i_bne | i_lui;
   assign aluc[0] = i_and | i_or | i_sll | i_srl | i_sra | i_andi | i_ori;
   assign wmem    = (i_sw | i_swc1) & wpcir & ~ecancel & no_dtlb_exce;
   assign pcsrc[1] = i_jr | i_j | i_jal;
   assign pcsrc[0] = i_beq & rsrtequ | i_bne & ~rsrtequ | i_j | i_jal;
   // fop:   000: fadd  001: fsub  01x: fmul  10x: fdiv  11x: fsqrt
   wire [2:0]  fop;
   assign fop[0] = i_fsub;                               // fpu operation control code
   assign fop[1] = i_fmul | i_fsqrt;
   assign fop[2] = i_fdiv | i_fsqrt;
   // stall caused by fp data harzards
   wire i_fs    = i_fadd | i_fsub | i_fmul | i_fdiv | i_fsqrt; // use fs
   wire i_ft    = i_fadd | i_fsub | i_fmul | i_fdiv;          // use ft
   assign stall_fp = (e1w & (i_fs & (e1n == fs) | i_ft & (e1n == ft))) |
                     (e2w & (i_fs & (e2n == fs) | i_ft & (e2n == ft)));
```

```
    assign fwdfa = e3w & (e3n == fs);                // forward fpu e3d to fp a
    assign fwdfb = e3w & (e3n == ft);                // forward fpu e3d to fp b
    assign wfpr  = i_lwc1 & wpcir & ~ecancel & no_dtlb_exce; // write fp reg
    assign fwdla = mwfpr & (mrn == fs);              // forward mmo to fp a
    assign fwdlb = mwfpr & (mrn == ft);              // forward mmo to fp b
    assign stall_lwc1 = ewfpr & (i_fs & (ern == fs) | i_ft & (ern == ft));
    assign swfp  = i_swc1;                           // select signal
    assign fwdf  = swfp & e3w & (ft == e3n);         // forward to id  stage
    assign fwdfe = swfp & e2w & (ft == e2n);         // forward to exe stage
    assign stall_swc1 = swfp & e1w & (ft == e1n); // stall
    assign wpcir = ~(stall_div_sqrt | stall_others);
    assign stall_others = stall_lw | stall_fp | stall_lwc1 | stall_swc1 |st;
    assign fc    = fop & {3{~stall_others}};         // fp operation control
    assign wf    = i_fs & wpcir & ~ecancel & no_dtlb_exce;  // write fp reg
    assign fasmds = i_fs;
endmodule
```

In order to test TLBs, we enlarged the usage of the memory space. Four physical memory modules were prepared: they are

1. 0x00000000–0x000001ff: initialization and interrupt/exception handler;
2. 0x10000000–0x100001ff: page table;
3. 0x20000000–0x200001ff: test program; and
4. 0x20002000–0x200021ff: test data.

Except for the first memory module, the rest of the modules are accessed through DTLB or ITLB. Below is the Verilog HDL code for the memory modules, including the emulation of the memory timing control. We used Altera LPMs to implement the memory modules.

```
module physical_memory (a,dout,din,strobe,rw,ready,clk,memclk,clrn);
    input         clk, memclk, clrn;               // clocks and reset
    input  [31:0] a;                               // memory address
    output [31:0] dout;                            // data out
    input  [31:0] din;                             // data in
    input         strobe;                          // strobe
    input         rw;                              // read/write
    output        ready;                           // memory ready
    wire   [31:0] mem_data_out0;
    wire   [31:0] mem_data_out1;
    wire   [31:0] mem_data_out2;
    wire   [31:0] mem_data_out3;
    // for memory ready
    reg    [2:0]  wait_counter;
    reg           ready;
    always @ (negedge clrn or posedge clk) begin
        if (!clrn) begin
            wait_counter <= 3'b0;
        end else begin
```

```
                    if (strobe) begin
                        if (wait_counter == 3'h5) begin          // 6 clock cycles
                            ready <= 1;                           // ready
                            wait_counter <= 3'b0;
                        end else begin
                            ready <= 0;
                            wait_counter <= wait_counter + 3'b1;
                        end
                    end else begin
                        ready <= 0;
                        wait_counter <= 3'b0;
                    end
                end
            end
end
// 31 30 29 28 ... 15 14 13 12 ...  3  2  1  0
//  0  0  0  0      0  0  0  0       0  0  0  0   (0) 0x0000_0000
//  0  0  0  1      0  0  0  0       0  0  0  0   (1) 0x1000_0000
//  0  0  1  0      0  0  0  0       0  0  0  0   (2) 0x2000_0000
//  0  0  1  0      0  0  1  0       0  0  0  0   (3) 0x2000_2000
wire    [31:0] m_out32 = a[13] ? mem_data_out3 : mem_data_out2;
wire    [31:0] m_out10 = a[28] ? mem_data_out1 : mem_data_out0;
wire    [31:0] mem_out = a[29] ? m_out32       : m_out10;
assign         dout    = ready ? mem_out       : 32'hzzzz_zzzz;
// (0) 0x0000_0000- (virtual address 0x8000_0000-)
wire           write_enable0 = ~a[29] & ~a[28] & rw;
lpm_ram_dq ram0 (.data(din),.address(a[8:2]),.we(write_enable0),
                 .inclock(memclk),.outclock(memclk&strobe),
                 .q(mem_data_out0));
defparam ram0.lpm_width          = 32;
defparam ram0.lpm_widthad        =  7;
defparam ram0.lpm_file           = "cpu_cache_tlb_0.hex";
defparam ram0.lpm_indata         = "REGISTERED";
defparam ram0.lpm_outdata        = "REGISTERED";
defparam ram0.lpm_address_control = "REGISTERED";
// (1) 0x1000_0000- (virtual address 0x9000_0000-)
wire           write_enable1 = ~a[29] & a[28] & rw;
lpm_ram_dq ram1 (.data(din),.address(a[8:2]),.we(write_enable1),
                 .inclock(memclk),.outclock(memclk&strobe),
                 .q(mem_data_out1));
defparam ram1.lpm_width          = 32;
defparam ram1.lpm_widthad        =  7;
defparam ram1.lpm_file           = "cpu_cache_tlb_1.hex";
defparam ram1.lpm_indata         = "REGISTERED";
defparam ram1.lpm_outdata        = "REGISTERED";
defparam ram1.lpm_address_control = "REGISTERED";
// (2) 0x2000_0000- (mapped va 0x0000_0000-)
wire           write_enable2 = a[29] & ~a[13] & rw;
```

```
    lpm_ram_dq ram2 (.data(din),.address(a[8:2]),.we(write_enable2),
                    .inclock(memclk),.outclock(memclk&strobe),
                    .q(mem_data_out2));
    defparam ram2.lpm_width         = 32;
    defparam ram2.lpm_widthad       = 7;
    defparam ram2.lpm_file          = "cpu_cache_tlb_2.hex";
    defparam ram2.lpm_indata        = "REGISTERED";
    defparam ram2.lpm_outdata       = "REGISTERED";
    defparam ram2.lpm_address_control = "REGISTERED";
    // (3) 0x2000_2000- (mapped va 0x0000_1000-)
    wire            write_enable3 = a[29] & a[13] & rw;
    lpm_ram_dq ram3 (.data(din),.address(a[8:2]),.we(write_enable3),
                    .inclock(memclk),.outclock(memclk&strobe),
                    .q(mem_data_out3));
    defparam ram3.lpm_width         = 32;
    defparam ram3.lpm_widthad       = 7;
    defparam ram3.lpm_file          = "cpu_cache_tlb_3.hex";
    defparam ram3.lpm_indata        = "REGISTERED";
    defparam ram3.lpm_outdata       = "REGISTERED";
    defparam ram3.lpm_address_control = "REGISTERED";
endmodule
```

12.5   Simulation Waveforms of CPU with Caches and TLBs

The first memory module (physical address 0x00000000–0x000001ff, corresponding to the virtual address 0x80000000–0x800001ff) is initialized with the following program that initializes the ITLB and provides the exception handler. The handler uses a jump table to jump to itlb_entry or dtlb_entry based on ExcCode. The jump table is provided here.

```
DEPTH = 128;         % Memory depth and width are required        %
WIDTH = 32;          % Enter a decimal number                     %
ADDRESS_RADIX = HEX; % Address and value radixes are optional     %
DATA_RADIX = HEX;    % Enter BIN, DEC, HEX, or OCT; unless         %
CONTENT              % otherwise specified, radixes = HEX          %
BEGIN
    % physical address = 0x0000_0000                              %
    % reset entry,  va = 0x8000_0000                              %
 0: 08000070; %(80000000) j    initialize_itlb  # jump to init itlb  %
 1: 00000000; %(80000004) nop                                     %
    % exc_base:                             # exception handler entry %
 2: 401a6800; %(80000008) mfc0 $26, c0_cause    # read cp0 cause reg  %
 3: 335a001c; %(8000000c) andi $26, $26, 0x1c   # get exccode, 3 bits %
 4: 3c1b8000; %(80000010) lui  $27, 0x8000      #                     %
 5: 037ad825; %(80000014) or   $27, $27, $26    #                     %
 6: 8f7b0040; %(80000018) lw   $27, j_table($27)# get addr from j table %
 7: 00000000; %(8000001c) nop                   #                     %
 8: 03600008; %(80000020) jr   $27              # jump to that address %
```

```
 9: 00000000; %(80000024) nop                    #                       %
[a..f]: 0;
     % j_table:          # address table for exception and interrupt     %
10: 80000000; %(80000040) int_entry # 0. address for interrupt           %
11: 80000000; %(80000044) sys_entry # 1. address for syscall             %
12: 80000000; %(80000048) uni_entry # 2. address for unimpl. inst.       %
13: 80000000; %(8000004c) ovf_entry # 3. address for overflow            %
14: 800000c0; %(80000050) itlb_entry # 4. address for itlb miss          %
15: 80000140; %(80000054) dtlb_entry # 5. address for dtlb miss          %
16: 80000000; %(80000058)                                                %
17: 80000000; %(8000005c)                                                %
[18..2f]: 0;
     % itlb_entry:                                                       %
30: 3c1b8000; %(800000c0) lui  $27, 0x8000       # 0x800001f8: counter   %
31: 8f7a01f8; %(800000c4) lw   $26, 0x1f8($27)   # load itlb index counter %
32: 235a0001; %(800000c8) addi $26, $26, 1       # index + 1             %
33: 335a0007; %(800000cc) andi $26, $26, 7       # 3-bit index           %
34: af7a01fc; %(800000d0) sw   $26, 0x1fc($27)   # store index           %
35: 3c1b0000; %(800000d4) lui  $27, 0x0000       # itlb tag              %
36: 037ad025; %(800000d8) or   $26, $27, $26     # itlb tag and index    %
37: 409a0000; %(800000dc) mtc0 $26, c0_index     # move to cp0 index     %
38: 401b2000; %(800000e0) mfc0 $27, c0_context   # address of pte        %
39: 8f7a0000; %(800000e4) lw   $26, 0x0($27)     # pte                   %
3a: 409a1000; %(800000e8) mtc0 $26, c0_entry_lo # move to cp0 entry_lo   %
3b: 001bd280; %(800000ec) sll  $26, $27, 10      # get bad vpn           %
3c: 001ad302; %(800000f0) srl  $26, $26, 12      # for cp0 entry_hi      %
3d: 409a4800; %(800000f4) mtc0 $26, c0_entry_hi # move to entry_hi       %
3e: 42000002; %(800000f8) tlbwi                  # update itlb           %
3f: 42000018; %(800000fc) eret                   # return from exception %
40: 00000000; %(80000100) nop                    #                       %
[41..4f]: 0;
     % dtlb_entry:                                                       %
50: 3c1b8000; %(80000140) lui  $27, 0x8000       # 0x800001fc: counter   %
51: 8f7a01fc; %(80000144) lw   $26, 0x1fc($27)   # load dtlb index counter %
52: 235a0001; %(80000148) addi $26, $26, 1       # index + 1             %
53: 335a0007; %(8000014c) andi $26, $26, 7       # 3-bit index           %
54: af7a01fc; %(80000150) sw   $26, 0x1fc($27)   # store index           %
55: 3c1b4000; %(80000154) lui  $27, 0x4000       # dtlb tag              %
56: 037ad025; %(80000158) or   $26, $27, $26     # dtlb tag and index    %
57: 409a0000; %(8000015c) mtc0 $26, c0_index     # move to cp0 index     %
58: 401b2000; %(80000160) mfc0 $27, c0_context   # address of pte        %
59: 8f7a0000; %(80000164) lw   $26, 0x0($27)     # pte                   %
5a: 409a1000; %(80000168) mtc0 $26, c0_entry_lo # move to cp0 entry_lo   %
5b: 001bd280; %(8000016c) sll  $26, $27, 10      # get bad vpn           %
5c: 001ad302; %(80000170) srl  $26, $26, 12      # for cp0 entry_hi      %
5d: 409a4800; %(80000174) mtc0 $26, c0_entry_hi # move to entry_hi       %
5e: 42000002; %(80000178) tlbwi                  # update dtlb           %
```

```
5f: 42000018; %(8000017c)  eret                  # return from exception      %
60: 00000000; %(80000180)  nop                   #                            %
[61..6f]: 0;
    % initialize_itlb:                                                        %
70: 40800000; %(800001c0)  mtc0 $0,   c0_index   # c0_index <-- 0 (itlb[0])   %
71: 3c1b9000; %(800001c4)  lui  $27,  0x9000     # page table base            %
72: 8f7a0000; %(800001c8)  lw   $26,  0x0($27)   # 1st entry of page table    %
73: 409a1000; %(800001cc)  mtc0 $26,  c0_entry_lo # c0_entrylo <-- v,d,c,pfn  %
74: 3c1a0000; %(800001d0)  lui  $26,  0x0        # va (=0) for c0_entry_hi    %
75: 409a4800; %(800001d4)  mtc0 $26,  c0_entry_hi # c0_entry_hi <-- vpn (0)   %
76: 42000002; %(800001d8)  tlbwi                 # write itlb for user prog   %
77: 409b2000; %(800001dc)  mtc0 $27,  c0_context # c0_context <-- ptebase     %
78: 341a003f; %(800001e0)  ori  $26,  $0, 0x3f   # enable exceptions          %
79: 409a6000; %(800001e4)  mtc0 $26,  c0_status  # c0_status <-- 0..0111111    %
7a: 3c010000; %(800001e8)  lui  $1,   0x0        # va = 0x0000_0000           %
7b: 00200008; %(800001ec)  jr   $1               # jump to user program       %
7c: 00000000; %(800001f0)  nop                   #                            %
7d: 00000000; %(800001f4)  nop                   #                            %
7e: 00000000; %(800001f8)  .data 0               # itlb index counter         %
7f: 00000000; %(800001fc)  .data 0               # dtlb index counter         %
END ;
```

The second memory module (physical address 0x10000000–0x100001ff, mapped area by DTLB) is initialized with the following data elements which are the contents of the page table. The valid bit is also stored in the page table entry. When a TLB miss occurs, the exception handler reads a word from this page table to fill the real part of a TLB entry. The page table is manipulated by the operating system.

```
DEPTH = 128;           % Memory depth and width are required                 %
WIDTH = 32;            % Enter a decimal number                              %
ADDRESS_RADIX = HEX;   % Address and value radixes are optional              %
DATA_RADIX = HEX;      % Enter BIN, DEC, HEX, or OCT; unless                 %
CONTENT                % otherwise specified, radixes = HEX                  %
BEGIN
    % physical address = 0x1000_0000 %
    % page table,     va = 0x9000_0000 %
 0: 00820000; %(90000000) va: 00000000 --> pa: 20000000; 1 of 8: valid bit %
 1: 00820002; %(90000004) va: 00001000 --> pa: 20002000; 1 of 8: valid bit %
 2: 00820001; %(90000008) va: 00002000 --> pa: 20001000; 1 of 8: valid bit %
 3: 008200f0; %(9000000c) va: 00003000 --> pa: 200f0000; 1 of 8: valid bit %
[4..7f]: 00000000;
END ;
```

The third memory module (physical address 0x20000000–0x200001ff, mapped area by ITLB) is initialized with the following program which is used to test the CPU including FPU, especially to verify the TLBs and caches.

```
DEPTH = 128;           % Memory depth and width are required                 %
WIDTH = 32;            % Enter a decimal number                              %
ADDRESS_RADIX = HEX;   % Address and value radixes are optional              %
```

```
DATA_RADIX = HEX;      % Enter BIN, DEC, HEX, or OCT; unless          %
CONTENT                % otherwise specified, radixes = HEX           %
BEGIN
    % physical address = 0x2000_0000                                  %
    % test program  va = 0x0000_0000                                  %
 0: 20011100; %(00000000)       addi  $1, $0,0x1100 # address of data[0]  %
 1: c4200000; %(00000004)       lwc1  f0,  0x0($1)  # load fp data         %
 2: c4210050; %(00000008)       lwc1  f1, 0x50($1)  # load fp data         %
 3: c4220054; %(0000000c) ·     lwc1  f2, 0x54($1)  # load fp data         %
 4: c4230058; %(00000010)       lwc1  f3, 0x58($1)  # load fp data         %
 5: c424005c; %(00000014)       lwc1  f4, 0x5c($1)  # load fp data         %
 6: 46002100; %(00000018)       add.s f4,  f4,  f0  # f4: stall 1          %
 7: 460418c1; %(0000001c)       sub.s f3,  f3,  f4  # f4: stall 2          %
 8: 46022082; %(00000020)       mul.s f2,  f4,  f2  # mul                  %
 9: 46040842; %(00000024)       mul.s f1,  f1,  f4  # mul                  %
 a: e4210070; %(00000028)       swc1  f1, 0x70($1)  # f1: stall 1          %
 b: e4220074; %(0000002c)       swc1  f2, 0x74($1)  # store fp data        %
 c: e4230078; %(00000030)       swc1  f3, 0x78($1)  # store fp data        %
 d: e424007c; %(00000034)       swc1  f4, 0x7c($1)  # store fp data        %
 e: 20020004; %(00000038)       addi  $2,  $0,   4  # counter              %
 f: c4230000; %(0000003c)13:    lwc1  f3, 0x0($1)   # load fp data         %
10: c4210050; %(00000040)       lwc1  f1, 0x50($1)  # load fp data         %
11: 46030840; %(00000044)       add.s f1,  f1,  f3  # stall 1              %
12: 46030841; %(00000048)       sub.s f1,  f1,  f3  # stall 2              %
13: e4210030; %(0000004c)       swc1  f1,  0x30($1) # stall 1              %
14: c4051104; %(00000050)       lwc1  f5,0x1104($0) # load fp data         %
15: c4061108; %(00000054)       lwc1  f6,0x1108($0) # load fp data         %
16: c408110c; %(00000058)       lwc1  f8,0x110c($0) # load fp data         %
17: 460629c3; %(0000005c)       div.s f7,  f5,  f6  # div                  %
18: 46004244; %(00000060)       sqrt.s f9,  f8      # sqrt                 %
19: 46004a84; %(00000064)       sqrt.s f10, f9      # sqrt                 %
1a: 2042ffff; %(00000068)       addi  $2,  $2,  -1  # counter - 1          %
1b: 1440fff3; %(0000006c)       bne   $2,  $0, 13   # finish?              %
1c: 20210004; %(00000070)       addi  $1,  $1,   4  # address+4, delay slot %
1d: 3c010000; %(00000074)iu_test: lui $1, 0         # address of data[0]   %
1e: 34241150; %(00000078)       ori $4, $1, 0x1150  # address of data[0]   %
1f: 0c000038; %(0000007c)call: jal  sum             # call function        %
20: 20050004; %(00000080)dslot1: addi $5,$0,4       # delayed slot(ds)     %
21: ac820000; %(00000084)return: sw $2, 0($4)       # store result         %
22: 8c890000; %(00000088)       lw  $9, 0($4)       # check sw             %
23: 01244022; %(0000008c)       sub $8, $9, $4      # sub: $8 <-- $9 - $4  %
24: 20050003; %(00000090)       addi $5, $0,  3     # counter              %
25: 20a5ffff; %(00000094)loop2: addi $5,$5,-1       # counter - 1          %
26: 34a8ffff; %(00000098)       ori $8, $5, 0xffff  # zero-extend: 0000ffff %
27: 39085555; %(0000009c)       xori $8, $8, 0x5555 # zero-extend: 0000aaaa %
28: 2009ffff; %(000000a0)       addi $9, $0, -1     # sign-extend: ffffffff %
29: 312affff; %(000000a4)       andi $10, $9,0xffff # zero-extend: 0000ffff %
```

```
2a: 01493025; %(000000a8)      or   $6, $10, $9    # or:  ffffffff        %
2b: 01494026; %(000000ac)      xor  $8, $10, $9    # xor: ffff0000        %
2c: 01463824; %(000000b0)      and  $7, $10, $6    # and: 0000ffff        %
2d: 10a00003; %(000000b4)      beq  $5, $0, shift  # if $5 = 0, goto shift %
2e: 00000000; %(000000b8)dslot2: nop              # ds                    %
2f: 08000025; %(000000bc)      j    loop2          # jump loop2            %
30: 00000000; %(000000c0)dslot3: nop              # ds                    %
31: 2005ffff; %(000000c4)shift: addi $5,$0,-1     # $5    = ffffffff       %
32: 000543c0; %(000000c8)      sll  $8, $5, 15     # <<  15 = ffff8000     %
33: 00084400; %(000000cc)      sll  $8, $8, 16     # <<  16 = 80000000     %
34: 00084403; %(000000d0)      sra  $8, $8, 16     # >>> 16 = ffff8000     %
35: 00084c32; %(000000d4)      srl  $8, $8, 15     # >>  15 = 0001ffff     %
36: 08000036; %(000000d8)finish: j finish         # dead loop             %
37: 00000000; %(000000dc)dslot4: nop              # ds                    %
38: 00004020; %(000000e0)sum: add  $8, $0, $0     # sum                   %
39: 8c890000; %(000000e4)loop: lw $9, 0($4)       # load data             %
3a: 01094020; %(000000e8)      add  $8, $8, $9     # sum                   %
3b: 20a5ffff; %(000000ec)      addi $5, $5, -1     # counter - 1           %
3c: 14a0fffc; %(000000f0)      bne  $5, $0, loop   # finish?               %
3d: 20840004; %(000000f4)dslot5: addi $4, $4, 4   # address + 4, ds       %
3e: 03e00008; %(000000f8)      jr   $31            # return                %
3f: 00081000; %(000000fc)dslot6: sll $2, $8, 0    # move res. to v0, ds   %
[40..7f]: 0;
END ;
```

The last memory module (physical address 0x20002000–0x200021ff, mapped area by DTLB) contains the test data used by the test program.

```
DEPTH = 128;          % Memory depth and width are required           %
WIDTH = 32;           % Enter a decimal number                        %
ADDRESS_RADIX = HEX;  % Address and value radixes are optional        %
DATA_RADIX = HEX;     % Enter BIN, DEC, HEX, or OCT; unless           %
CONTENT               % otherwise specified, radixes = HEX            %
BEGIN
    % physical address = 0x2000_2000                                  %
    % test data     va = 0x0000_2000                                  %
[0..3f]:   0; %(00002000..200020fc) 0                                 %
40: bf800000; %(00002100) 1 01111111 00..0 fp -1                      %
41: 40800000; %(00002104)                                             %
42: 40000000; %(00002108)                                             %
43: 41100000; %(0000210c)                                             %
[44..53]:  0; %(00002110..2000214c) 0                                 %
54: 40c00000; %(00002150) 0 10000001 10..0 data[0] 4.5               %
55: 41c00000; %(00002154) 0 10000011 10..0 data[1]                   %
56: 43c00000; %(00002158) 0 10000111 10..0 data[2]                   %
57: 47c00000; %(0000215c) 0 10001111 10..0 data[3]                   %
[58..7f]:  0; %(00002160..200021fc) 0                                 %
END ;
```

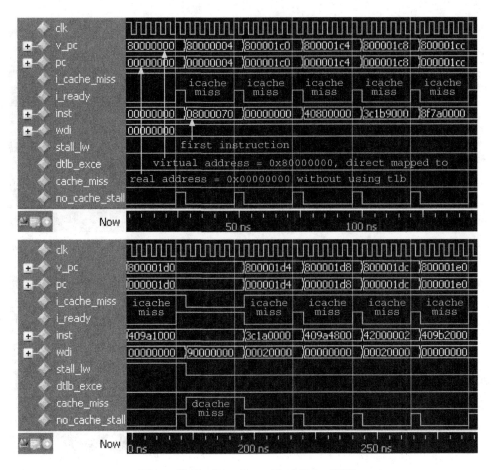

Figure 12.18   Waveform of initializing ITLB

Figure 12.18 shows the simulation waveforms when the CPU executes the initialization program. Because the first instruction is a jump, the instruction in the delay slot is also executed before jumping to the target address. We can see that the virtual PC is translated to physical PC by directed mapping (it does not use ITLB). We can also see that instruction cache is miss (i_cache_miss = 1) and main memory is accessed. The i_ready signal indicates that the instruction is ready. There is also a data cache miss (cache_miss = 1), caused by a load instruction. The instructions executed are listed below for easy checking.

```
80000000 : 08000070; %  j    initialize_itlb  # jump to init itlb      %
80000004 : 00000000; %  nop                                            %
800001c0 : 40800000; %  mtc0 $0,   c0_index    # c0_index <-- 0 (itlb[0]) %
800001c4 : 3c1b9000; %  lui  $27, 0x9000       # page table base        %
800001c8 : 8f7a0000; %  lw   $26, 0x0($27)     # 1st entry of page table %
800001cc : 409a1000; %  mtc0 $26, c0_entry_lo  # c0_entrylo <-- v,d,c,pfn %
800001d0 : 3c1a0000; %  lui  $26, 0x0          # va (=0) for c0_entry_hi %
800001d4 : 409a4800; %  mtc0 $26, c0_entry_hi  # c0_entry_hi <-- vpn (0) %
```

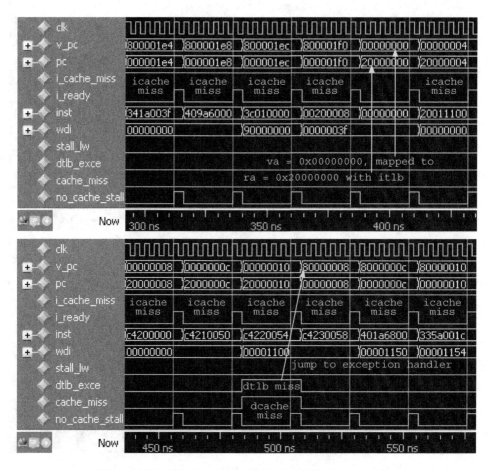

Figure 12.19   Waveform of dealing with DTLB miss exception

```
800001d8 : 42000002; %  tlbwi                     # write itlb for user prog   %
800001dc : 409b2000; %  mtc0 $27, c0_context      # c0_context <-- ptebase     %
800001e0 : 341a003f; %  ori  $26, $0, 0x3f        # enable exceptions          %
```

Figure 12.19 shows the simulation waveforms when the control is transferred to the test program starting at the virtual address of 0x00000000. This address is translated to 0x20000000 by the ITLB. The instruction cache is still a miss. The lwc1 f0, 0x0($1) instruction (0xc4200000) at v_pc = 0x00000004 causes a DTLB miss exception in the MEM stage (dtlb_exce = 1). This instruction and its follow-up instructions in the pipeline are canceled, and the control is transferred to the exception handler at the virtual address 0x80000008. The instructions shown in the figure are listed below.

```
800001e4 : 409a6000; %  mtc0 $26, c0_status       # c0_status <-- 0..0111111   %
800001e8 : 3c010000; %  lui  $1,  0x0             # va = 0x0000_0000           %
800001ec : 00200008; %  jr   $1                    # jump to user program       %
800001f0 : 00000000; %  nop                        #                            %
```

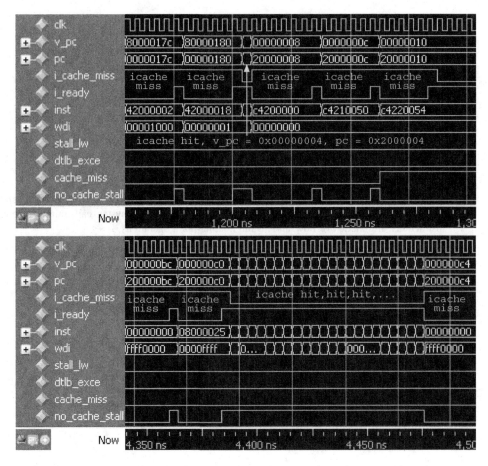

Figure 12.20   Waveforms of returning from exception and cache hits

```
00000000 : 20011100; %  addi  $1,  $0,0x1100    # address of data[0]      %
00000004 : c4200000; %  lwc1  f0,  0x0($1)      # load fp data            %
00000008 : c4210050; %  lwc1  f1,  0x50($1)     # load fp data            %
0000000c : c4220054; %  lwc1  f2,  0x54($1)     # load fp data            %
00000010 : c4230058; %  lwc1  f3,  0x58($1)     # load fp data            %
80000008 : 401a6800; %  mfc0  $26, c0_cause     # read cp0 cause reg      %
8000000c : 335a001c; %  andi  $26, $26, 0x1c    # get exccode, 3 bits     %
```

Figure 12.20 shows the simulation waveforms when returning from the exception. The load instruction at v_pc = 0x00000004 is re-executed and the instruction cache has a hit (the instruction fetch takes one clock cycle). The upper part of the figure shows the waveform when the following instructions are executed:

```
8000017c : 42000018; %  eret                    # return from exception   %
80000180 : 00000000; %  nop                     #                         %
00000004 : c4200000; %  lwc1  f0,  0x0($1)      # load fp data            %
```

```
00000008 : c4210050; %  lwc1  f1, 0x50($1)    # load fp data          %
0000000c : c4220054; %  lwc1  f2, 0x54($1)    # load fp data          %
00000010 : c4230058; %  lwc1  f3, 0x58($1)    # load fp data          %
```

The lower part shows the waveforms for the instructions in `loop2` of the test program that are hit on the instruction cache.

Exercises

12.1 Design a computer with a four-way interleaved memory.

12.2 The CPU described in this chapter can deal with only ITLB and DTLB miss exceptions. Redesign the CPU so that it can also deal with the interrupt and the exceptions of system call, unimplemented instructions, arithmetic overflow, and invalid floating-point operations.

12.3 Redesign the CPU in which the ITLB and DTLB use both EntryLo0 and EntryLo1.

12.4 The TLB has a random register, therefore it can implement random replacement algorithm easily. Try to implement the LRU algorithm in software.

12.5 How does an operating system manage the page table?

13

Multithreading CPU and Multicore CPU Design in Verilog HDL

The CPUs (central processing units) we have designed up to now can execute only one instruction stream at any time; we call them single-thread CPUs. Of course, a single-thread CPU can also execute multiple threads by means of context switching, which is an essential feature of a time-sharing, multitask operating system.

A CPU that can execute multiple threads simultaneously is called a multithreading CPU. That is, multiple instructions coming from multiple threads are executed in parallel. In order to do that, we must provide multiple functional units, such as arithmetic logic units (ALUs) and floating-point units (FPUs), in multithreading CPUs. The main feature of multithreading CPUs is that the multiple threads share all the functional units. The instruction cache and data cache are also shared by all threads. This feature makes the design of multithreading CPUs a little bit complex, but, as a result, the functional units and caches can be better utilized.

Multicore CPUs are different from the multithreading CPUs. In a multicore CPU, there are multiple cores. A core is virtually a simple CPU. The multicore has an academic name: chip multiprocessor; it implements a small-scale multiprocessor on a chip. Multicore CPUs are suitable for executing multiple tasks with a task running on a core. But, a task that runs on one core cannot use the functional units of other cores. This results in poor utilization of the functional units if there are not enough tasks to be run at the same time. Each core has its own L1 instruction and data caches. The L2 on-chip caches may be dedicated or shared.

Referring to Figure 13.1, a core in a multicore CPU can be a multithreading CPU. For example, if a core in a quad-core CPU can execute four threads simultaneously, the multicore CPU can execute 16 threads at the same time. Such a CPU is called a multithreading multicore or a multicore multithreading CPU.

This chapter discusses the design issues of multithreading and multicore CPUs, and gives two simple design examples: one for a multithreading CPU, and the other for a multicore CPU. The Verilog HDL (hardware description language) codes and the simulation waveforms of the two CPUs are also provided.

13.1 Overview of Multithreading CPUs

This section describes the basic concept of the multiple threads and the general structure of multithreading CPUs.

Computer Principles and Design in Verilog HDL, First Edition. Yamin Li.
© 2015 Tsinghua University Press. All rights reserved. Published 2015 by John Wiley & Sons Singapore Pte Ltd.
Companion Website: www.wiley.com/go/li/verilog

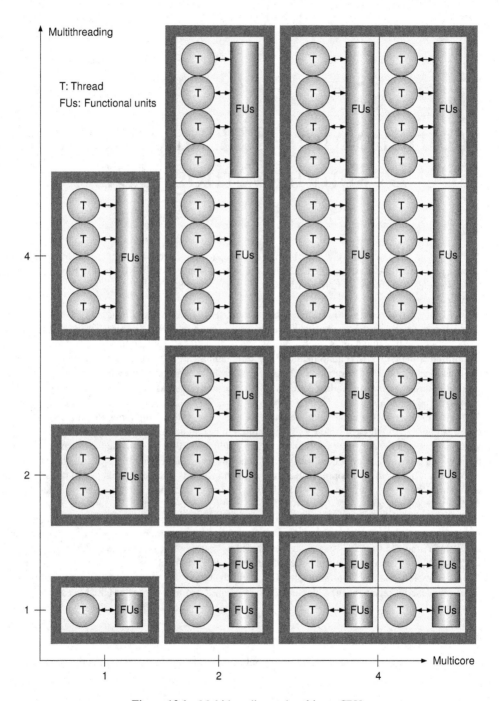

Figure 13.1   Multithreading and multicore CPUs

13.1.1 Basic Concept of Multiple Threads

A thread is a sequential execution stream of instructions, and is contained inside a process. The inputs at the entry of a thread may be the outputs of other threads, but the thread is independent of other threads during the execution. The output results may be fed to other threads as the inputs. Thus, multiple threads can be executed in parallel whenever their inputs are available. Figure 13.2 shows the concept of multithreading in which thread D and thread E can run simultaneously.

The single-thread CPU executes multiple threads sequentially by means of context switching. It is usually implemented by periodic interrupt. The interrupt handler saves the states of the interrupted thread, restores the states of another thread that will be executed, and transfers control to that thread, as shown as in Figure 13.3(a). The idea of context switching is that the interrupt handler must save and restore the states of threads so that the execution can be resumed from the same point. The states of a thread include all registers that the thread may be using. At any time, there is only one thread that is being executed.

In contrast, the multithreading CPU can execute multiple threads in parallel. As shown in Figure 13.3(b), there are two threads that are executed simultaneously in the time periods t_1, t_3, and t_5. The time periods of t_2 and t_4 are for the main memory access due to the cache misses. The multithreading CPU also needs context switching because there may be more threads or processes than what the CPU can execute simultaneously.

13.1.2 Basic Structure of Multithreading CPUs

The main function of multithreading CPUs is to simultaneously execute instructions that come from multiple threads on the shared functional units. Figure 13.4 shows a possible organization of a multithreading CPU.

There are multiple program counters (PCs) in the multithreading CPU, and multiple instructions can be fetched from instruction caches. The instruction caches can be implemented with multiple banks, and each bank provides instructions for a thread. Then the fetched instructions are scheduled and dispatched to the functional units for execution. The functional units may contain integer units (IUs), FPUs, integer multiplication units (MULs), integer division units (DIVs), load/store units (LSUs), and

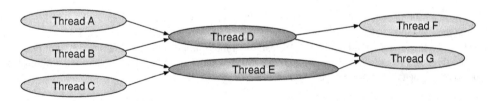

Figure 13.2 Basic concept of multithreading

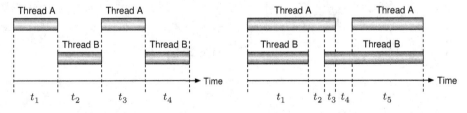

(a) Single-thresd CPU executes threads sequentially   (b) Multithreading CPU executes threads in parallel

Figure 13.3 Sequential multithreading and parallel multithreading

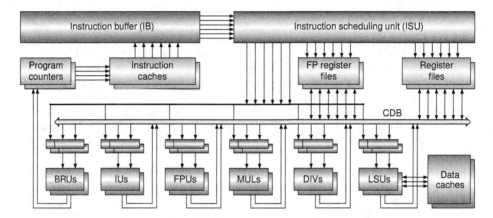

Figure 13.4   Organization of a multithreading CPU

branch units (BRUs). Although a common data bus (CDB) is drawn in the figure, it can be implemented with separate buses or a crossbar interconnect.

13.2   Multithreading CPU Design

This section describes the design of a very simple two-way multithreading CPU. The simplified block diagram of the CPU is shown in Figure 13.5.

The CPU can execute two threads in parallel. Three functional units are shared by the two threads. Because there are two ALUs, there will be no competition for the IUs. The key point of the design is on the competition for an FPU.

13.2.1   Thread Selection Method

There is only one FPU in our multithreading CPU. When two threads attempt to use the FPU at the same time, we must select one thread. The selection can be done based on a counter. The counter value indicates which thread will be selected. We can use a multiple-bits counter to assign different priorities to the threads. For example, if we use a 3-bit counter which has eight different values, the ratio of the priorities of the two threads can be assigned as 5 to 3.

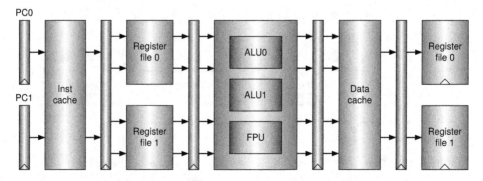

Figure 13.5   Block diagram of a two-thread CPU

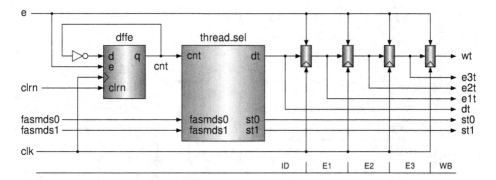

Figure 13.6 Schematic circuit of thread selection

Table 13.1 Method of thread selection

Input			Output		
cnt	fasmds0	fasmds1	dt	st0	st1
0	1	0	0	0	0
0	1	1	0	0	1
0	0	1	1	0	0
1	0	1	1	0	0
1	1	1	1	1	0
1	1	0	0	0	0
x	0	0	0	0	0

Figure 13.6 shows the schematic circuit of the thread selection (selthread module) for using an FPU. We let the two threads have the same priority; therefore we used a 1-bit counter which is implemented with a D flip-flop (dffe) and a NOT gate. The enable signal e is used to control the pipeline. The thread_sel module determines which thread will be selected based on the counter value cnt and two requests, fasmds0 and fasmds1. When a thread dt is selected, the pipeline of the other thread must be stalled if it is requesting the use of the FPU. The selected thread number dt in the ID stage will pass through the FPU pipeline stages. Table 13.1 is the truth table of the module.

The signals st0 and st1 indicate whether thread 0 and thread 1 should be stalled, respectively. According to the table, we can write the expressions of the output signals in Verilog HDL format as shown below.

```
dt  = ~fasmds0 & fasmds1 | cnt & fasmds1;        // selected thread
st0 =  cnt & fasmds0 & fasmds1;                  // stall thread 0
st1 = ~cnt & fasmds0 & fasmds1;                  // stall thread 1
```

13.2.2   Circuit of the Multithreading CPU

Our simple two-way multithreading CPU shares an FPU. Figure 13.7 shows the detailed circuit of the CPU. There are two IUs and two floating-point register files for the use of two threads. There is an FPU that is shared by two threads.

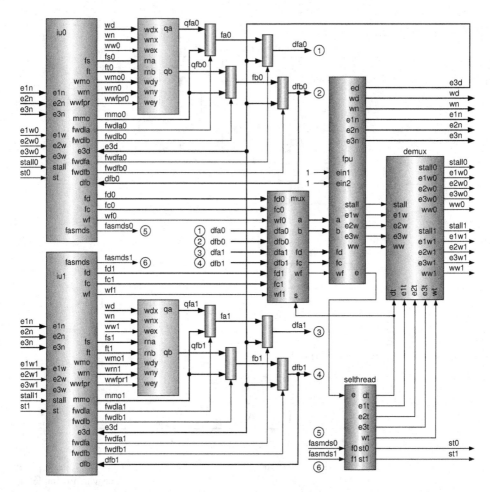

Figure 13.7 Block diagram of multithreading CPU with two IUs and one FPU

To simplify the design, we used an IU for one thread. The detailed circuit of the IU was shown in Figure 10.23, and the circuit of the FPU was shown in Figure 10.22, in Chapter 10. As described in the previous section, the selthread module selects a thread whenever there is a competition for the FPU. The multiplexer mux in the center of the figure selects the input data for the FPU. The demultiplexer demux apportions the outputs of the FPU to each of the two IUs. Only the write enable and stall signals need to be apportioned. The signals of stall, e1w, e2w, e3w, and ww are in stages of ID, E1, E2, E3, and WB, respectively.

13.2.3 Verilog HDL Codes of the Multithreading CPU

Below is the Verilog HDL code of the simple two-way multithreading CPU. It invokes the modules of iu and fpu, which were already given in Chapter 10. A real multithreading CPU is more complex than this.

```
// two-thread cpu with fpu, instruction memory, and data memory
module fpu_2_iu (clrn,memclk,clk,pc0,inst0,ealu0,malu0,walu0,ww0,stl_lw0,
```

```
                 stl_lwc10,stl_swc10,stl_fp0,stall0,st0,pc1,inst1,ealu1,
                 malu1,walu1,ww1,stl_lw1,stl_lwc11,stl_swc11,stl_fp1,stall1,
                 st1,wn,wd,cnt_div,cnt_sqrt,e1n,e2n,e3n,e3d,e);
input            clk, memclk, clrn;                 // clocks and reset
output [31:0] pc0, inst0, ealu0, malu0, walu0;
output [31:0] pc1, inst1, ealu1, malu1, walu1;
output [31:0] e3d, wd;
output  [4:0] e1n, e2n, e3n, wn;
output        ww0, stl_lw0, stl_lwc10, stl_swc10, stl_fp0, stall0, st0;
output        ww1, stl_lw1, stl_lwc11, stl_swc11, stl_fp1, stall1, st1;
output        e;                      // for multithreading CPU, not used here
output  [4:0] cnt_div,cnt_sqrt;
wire   [31:0] qfa0,qfb0,fa0,fb0,dfa0,dfb0,mmo0,wmo0;
wire   [31:0] qfa1,qfb1,fa1,fb1,dfa1,dfb1,mmo1,wmo1;
wire    [4:0] fs0,ft0,fd0,wrn0;
wire    [4:0] fs1,ft1,fd1,wrn1;
wire    [2:0] fc0;
wire    [2:0] fc1;
wire          fwdla0,fwdlb0,fwdfa0,fwdfb0,wf0,fasmds0;
wire          fwdla1,fwdlb1,fwdfa1,fwdfb1,wf1,fasmds1;
wire   [31:0] dfa,dfb;                               // fp inputs a and b
wire    [4:0] fd;                                    // fp dest reg #
wire    [2:0] fc;                                    // fp operation code
wire    [1:0] e1c,e2c,e3c;
wire          wf;                                    // fp regfile we
wire          stall0,e1w0,e2w0,e3w0,ww0;
wire          stall1,e1w1,e2w1,e3w1,ww1;
wire          dt;
reg           cnt,e1t,e2t,e3t,wt;
// thread 0
iu th0 (e1n,e2n,e3n, e1w0,e2w0,e3w0, stall0,st0,
        dfb0,e3d, clk,memclk,clrn,
        fs0,ft0,wmo0,wrn0,wwfpr0,mmo0,fwdla0,fwdlb0,fwdfa0,fwdfb0,
        fd0,fc0,wf0,fasmds0,pc0,inst0,ealu0,malu0,walu0,
        stl_lw0,stl_fp0,stl_lwc10,stl_swc10);
regfile2w fpr0 (fs0,ft0,wd,wn,ww0,wmo0,wrn0,wwfpr0,
                ~clk,clrn,qfa0,qfb0);
mux2x32 fwd_f_load_a0 (qfa0,mmo0,fwdla0,fa0);  // forward lwc1 to fp a
mux2x32 fwd_f_load_b0 (qfb0,mmo0,fwdlb0,fb0);  // forward lwc1 to fp b
mux2x32 fwd_f_res_a0  (fa0,e3d,fwdfa0,dfa0);   // forward fp res to fp a
mux2x32 fwd_f_res_b0  (fb0,e3d,fwdfb0,dfb0);   // forward fp res to fp b
// thread 1
iu th1 (e1n,e2n,e3n, e1w1,e2w1,e3w1, stall1,st1,
        dfb1,e3d, clk,memclk,clrn,
        fs1,ft1,wmo1,wrn1,wwfpr1,mmo1,fwdla1,fwdlb1,fwdfa1,fwdfb1,
        fd1,fc1,wf1,fasmds1,pc1,inst1,ealu1,malu1,walu1,
        stl_lw1,stl_fp1,stl_lwc11,stl_swc11);
```

```
       regfile2w fpr1 (fs1,ft1,wd,wn,ww1,wmo1,wrn1,wwfpr1,
                       ~clk,clrn,qfa1,qfb1);
       mux2x32 fwd_f_load_a1 (qfa1,mmo1,fwdla1,fa1);   // forward lwc1 to fp a
       mux2x32 fwd_f_load_b1 (qfb1,mmo1,fwdlb1,fb1);   // forward lwc1 to fp b
       mux2x32 fwd_f_res_a1  (fa1,e3d,fwdfa1,dfa1);    // forward fp res to fp a
       mux2x32 fwd_f_res_b1  (fb1,e3d,fwdfb1,dfb1);    // forward fp res to fp b
       // fpu
       wire   [31:0] s_sqrt;                           // fp output
       wire   [25:0] sqrt_x;                           // x_i
       fpu fp_unit (dfa,dfb,fc,wf,fd,1'b1,clk,clrn,
                    e3d,wd,wn,ww,stall,e1n,e1w,e2n,
                    e2w,e3n,e3w,e1c,e2c,e3c,cnt_div,
                    cnt_sqrt,e,1'b1);                   // no dtlb, hit = 1
       // mux: fpu selects thread 0 or thread 1
       assign dfa = dt? dfa1 : dfa0;
       assign dfb = dt? dfb1 : dfb0;
       assign fd  = dt? fd1  : fd0;
       assign wf  = dt? wf1  : wf0;
       assign fc  = dt? fc1  : fc0;
       // demux: for thread 0;         for thread 1
       assign stall0 = stall & ~dt;  assign stall1 = stall & dt;  // ID stage
       assign   e1w0 =   e1w & ~e1t;  assign   e1w1 =   e1w & e1t; // E1 stage
       assign   e2w0 =   e2w & ~e2t;  assign   e2w1 =   e2w & e2t; // E2 stage
       assign   e3w0 =   e3w & ~e3t;  assign   e3w1 =   e3w & e3t; // E3 stage
       assign    ww0 =    ww & ~wt;   assign    ww1 =    ww & wt;  // WB stage
       // thread selection
       assign   dt  = ~fasmds0 & fasmds1 | cnt & fasmds1;    // selected thread
       assign   st0 =  cnt & fasmds0 & fasmds1;              // stall thread 0
       assign   st1 = ~cnt & fasmds0 & fasmds1;              // stall thread 1
       // count for thread selection
       always @(negedge clrn or posedge clk) begin
           if (!clrn) begin
               cnt <= 0;
           end else if (e) begin
               cnt <= ~cnt;
           end
       end
       // pipelined thread info
       always @(negedge clrn or posedge clk) begin
           if (!clrn) begin
               e1t <= 0;        e2t <= 0;        e3t <= 0;        wt  <= 0;
           end else if (e) begin
               e1t <= dt;       e2t <= e1t;      e3t <= e2t;      wt  <= e3t;
           end
       end
endmodule
```

13.2.4 Simulation Waveforms of Multithreading CPU

The two threads execute the same test program given in Chapter 10. The integer instructions of the two threads can be executed simultaneously. Competition occurs when the two threads execute floating-point instructions. Figure 13.8 shows the waveforms when the two threads execute the following instructions:

```
5 : c424005c; % (14)     lwc1   f4,   92($1)   # load fp data         %
6 : 46002100; % (18)     add.s  f4,   f4, f0   # f4: stall 1          %
7 : 460418c1; % (1c)     sub.s  f3,   f3, f4   # f4: stall 2          %
```

The waveforms in the figure have three parts. The upper part shows the waveforms of thread 0; the middle part shows the waveforms of thread 1; and the lower part shows the waveforms of the FPU. The waveform of the 1-bit counter is shown at the bottom of the figure. There is data dependency between add.s and lwc1, and the pipelines stall for one clock cycle (stl_lwc10 = 1 and stl_lwc11 = 1). The counter has a value 0 in the next cycle, and therefore add.s of thread 0 is executed first and the pipeline of the thread 1 stalls (st1 = 1). After that, add.s of thread 1 is executed and the pipeline of the thread 0 stalls (st0 = 1). Because sub.s has a data dependency with add.s, the pipelines stall for two cycles.

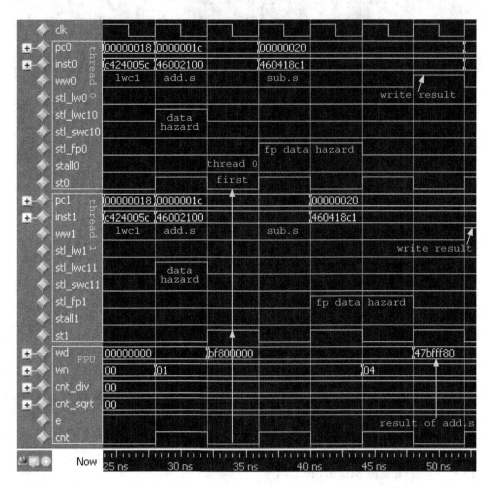

Figure 13.8 Waveform of the multithreading CPU (lwc1, sub.s, and add.s)

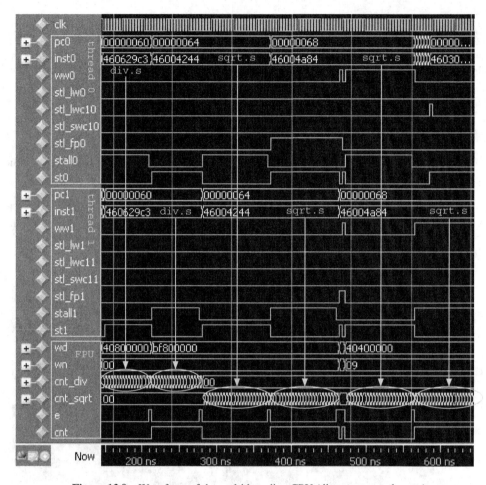

Figure 13.9   Waveform of the multithreading CPU (div.s, sqrt.s, and sqrt.s)

The FPU result (wd = 0x47bfff80) will be written to the FP register file of the thread 0. In the next cycle, the same result is written to the FP register file of thread 1.

Figure 13.9 shows the waveforms when the two threads execute the following instructions that have long execution time due to the Newton–Raphson iterations.

```
17 : 460629c3; % (5c)    div.s  f7,  f5, f6   # div       %
18 : 46004244; % (60)    sqrt.s f9,  f8        # sqrt      %
19 : 46004a84; % (64)    sqrt.s f10, f9        # sqrt      %
```

The instruction div.s of thread 0 is executed first (st1 = 1), and then div.s of thread 1 is executed (st0 = 1). cnt_div shows the counter values when the CPU executes these two instructions. The two sqrt.s instructions are executed in a similar manner.

13.3   Overview of Multicore CPUs

This section describes the basic concept of the multiple cores and gives the general organizations of multicore CPUs.

13.3.1   Basic Concept of Multicore CPUs

Advances in IC fabrication technology increase both the integration density and the clock speed. But, as the clock frequency increases, the clock skew problem becomes more serious. Clock skew means that the clock signal arrives at different components at different times, mainly due to the signal propagation delay on wires.

We know that the speed of light in vacuum is 299,792,458 meters per second (mps). The propagation speed of an electronic signal is $299,792,458 \times 0.64 = 191,867,173.12$ mps. Although the use of a clock tree network can alleviate the clock skew problem, the signal propagation delay will limit the CPU's performance. Suppose that the clock frequency in a CPU is 4 GHz; then in a clock cycle, an electronic signal can propagate $191,867,173.12 \times 100/4,000,000,000 \approx 4.8$ cm.

Since 2003, when the Intel Pentium 4 Gallatin clock frequency reached 3.2–3.46 GHz, the clock frequency has not improved much up to now. In order to achieve higher performance, CPU makers pay their attention from time to space. One of the results such effort is the multicore CPU, also known as a chip multiprocessor, that implements a small-scale multiprocessor on a chip die, as shown in Figure 13.10.

In a multicore CPU, there are several identical cores, and each core can be considered as a traditional CPU. Instead of increasing the clock frequency, multicore CPUs improve performance by means of executing multiple, independent tasks on multiple cores simultaneously.

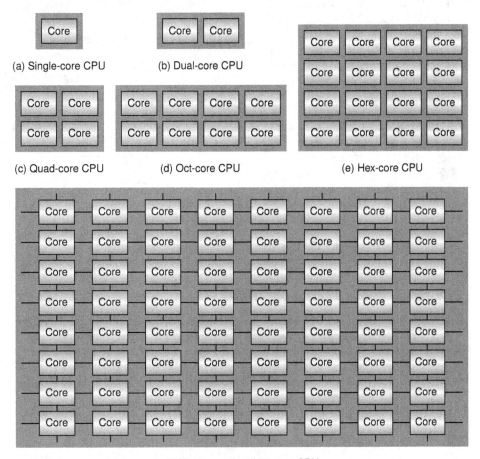

(f) Mesh-connected 64-core CPU

Figure 13.10   Various multicore CPUs

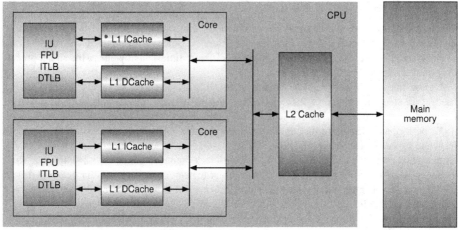

(a) Multicore CPU with a shared L2 cache

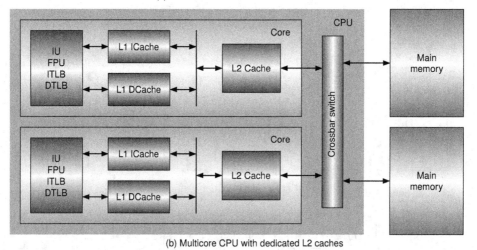

(b) Multicore CPU with dedicated L2 caches

Figure 13.11   Cache architecture in multicore CPUs

13.3.2   Basic Structure of Multicore CPUs

All the cores in a multicore CPU share the main memory and have their own level 1 (L1) instruction and data caches. But the L2 cache(s) may or may not be shared by the cores, as shown in Figure 13.11. Figure 13.11(a) shows a dual-core CPU in which the two cores share an L2 cache. Figure 13.11(b) shows a dual-core CPU in which each core has a dedicated L2 cache. By designing a crossbar switch for the memory controller, two main memory modules can be accessed simultaneously if their memory addresses point to different memory modules.

13.4   Multicore CPU Design

This section describes the design of a simple dual-core CPU and gives its Verilog HDL code and simulation waveforms.

13.4.1 Overall Structure of the Multicore CPU

Figure 13.12 shows a simplified block diagram of the dual-core CPU. Each core in the CPU contains translation lookaside buffers (TLBs) and caches; there is only one memory access port.

Core 1 and Core 2 in the figure are the same as the CPU designed in Chapter 12. The new added component is the circuit of competing for access to the single main memory module. We use a counter to arbitrate the memory access requests of Core 1 and Core 2. Table 13.2 is the truth table for the signal select1. If select1 is a 1, the request of Core 1 is selected; otherwise, the request of Core 2 is selected. If the counter (cnt) is a 0, the request of Core 1 has a higher priority than that of Core 2; otherwise, the request of Core 2 has a higher priority.

The instructions mem_access1 and mem_access2 are the memory access requests of the Core 1 and Core 2, respectively. cnt is a 1-bit counter. From the table, we can have the following expression of the select1 in Verilog HDL format:

```
select1 = ~cnt & mem_access1 | cnt & ~mem_access2;
```

This signal is used not only for selecting the request from cores but also for admeasuring the memory ready signal to the cores, as shown in Figure 13.13.

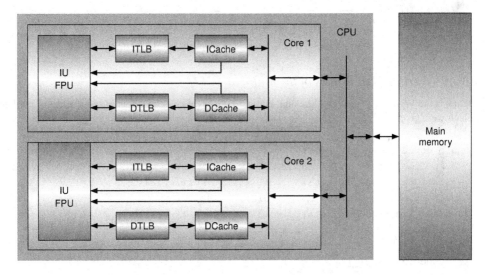

Figure 13.12   Block diagram of a dual-core CPU

Table 13.2   Arbitration of memory accesses

cnt	mem_access1	mem_access2	select1
x	0	0	x
0	0	1	0
0	1	0	1
0	1	1	1
1	0	1	0
1	1	0	1
1	1	1	0

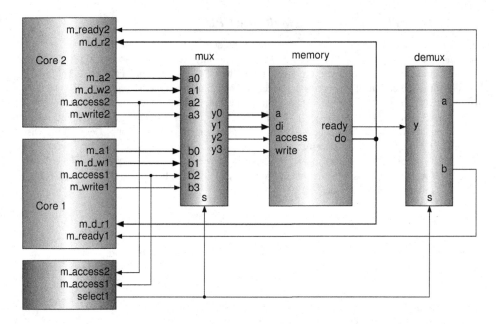

Figure 13.13   Block diagram of memory access arbitration

13.4.2   *Verilog HDL Codes of the Multicore CPU*

Below is the Verilog HDL code of the dual-core CPU. It invokes a core module and a memory module, which were given in Chapter 12.

```
module core2_cache_tlb_memory    // dual-core cpu with cache and tlb + memory
    (v_pc1,pc1,inst1,ealu1,malu1,wdi1,wn1,wd1,ww1,
    stall_lw1,stall_fp1,stall_lwc11,stall_swc11,stall1,
    v_pc2,pc2,inst2,ealu2,malu2,wdi2,wn2,wd2,ww2,
    stall_lw2,stall_fp2,stall_lwc12,stall_swc12,stall2,
    clk,memclk,clrn,m_a,m_d_r,m_d_w,m_access,m_write,m_ready);
    input         clk, memclk, clrn;        // clocks and reset
    output [31:0] v_pc1, v_pc2;             // virtual pc
    output [31:0] pc1, pc2;                 // real pc
    output [31:0] inst1, inst2;             // instruction
    output [31:0] ealu1, ealu2;             // exe stage result
    output [31:0] malu1, malu2;             // mem stage result
    output [31:0] wdi1, wdi2;               // wb  stage result
    output [31:0] wd1, wd2;                 // fp result to be written back
    output  [4:0] wn1, wn2;                 // fp dest register number
    output        ww1, ww2;                 // fp register file write enable
    output        stall_lw1, stall_lw2;     // stall by lw
    output        stall_fp1, stall_fp2;     // stall by fp data hazard
    output        stall_lwc11, stall_lwc12; // stall by lwc1
    output        stall_swc11, stall_swc12; // stall by swc1
    output        stall1, stall2;           // pipeline stall
    output [31:0] m_a;                      // memory address
    output [31:0] m_d_r;                    // memory data read
```

```
    output [31:0] m_d_w;                    // memory data write
    output        m_access;                 // memory access
    output        m_write;                  // memory write enable
    output        m_ready;                  // memory ready
    wire          io1,io2;                  // i/o, not used
    wire   [31:0] m_a1,m_a2;
    wire   [31:0] m_d_w1,m_d_w2;
    wire          m_access1,m_access2;
    wire          m_write1,m_write2;
    reg           cnt;                      // counter
    // core1
    cpu_cache_tlb core1
      (clk,memclk,clrn,v_pc1,pc1,inst1,ealu1,malu1,wdi1,wn1,wd1,ww1,
       stall_lw1,stall_fp1,stall_lwc11,stall_swc11,stall1,
       m_a1,m_d_r,m_d_w1,m_access1,m_write1,m_ready1,io1);
    // core2
    cpu_cache_tlb core2
      (clk,memclk,clrn,v_pc2,pc2,inst2,ealu2,malu2,wdi2,wn2,wd2,ww2,
       stall_lw2,stall_fp2,stall_lwc12,stall_swc12,stall2,
       m_a2,m_d_r,m_d_w2,m_access2,m_write2,m_ready2,io2);
    // counter for arbitration on memory access collision
    always @(negedge clrn or posedge clk) begin
        if (!clrn) begin
            cnt <= 0;
        end else if (m_ready) begin
            cnt <= ~cnt;
        end
    end
    wire select1 = ~cnt & m_access1 | cnt & ~m_access2;
    // mux
    assign m_a      = select1 ? m_a1      : m_a2;
    assign m_d_w    = select1 ? m_d_w1    : m_d_w2;
    assign m_access = select1 ? m_access1 : m_access2;
    assign m_write  = select1 ? m_write1  : m_write2;
    // demux
    assign m_ready1 =  select1 & m_ready;
    assign m_ready2 = ~select1 & m_ready;
    // main memory
    physical_memory mem (m_a,m_d_r,m_d_w,m_access,m_write,m_ready,clk,
                         memclk,clrn);
endmodule
```

13.4.3 Simulation Waveforms of the Multicore CPU

The two cores execute the same test program given in Chapter 12. Figures 13.14 and 13.15 show the simulation waveforms.

Figure 13.14 shows the waveforms when the two cores execute the following instructions. Signal select1 = 1 means that the memory request of Core 1 was selected; otherwise, the request of Core 2 is selected.

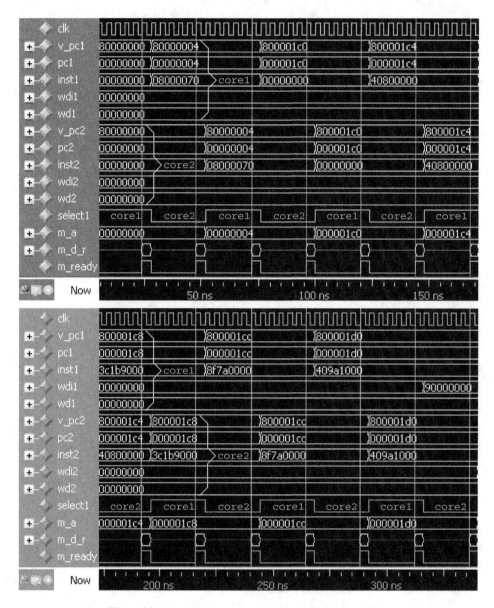

Figure 13.14   Waveform of a dual-core CPU (caches miss)

```
80000000 : 08000070; %  j    initialize_itlb  # jump to init itlb          %
80000004 : 00000000; %  nop                                                 %
800001c0 : 40800000; %  mtc0 $0,  c0_index     # c0_index <-- 0 (itlb[0])   %
800001c4 : 3c1b9000; %  lui  $27, 0x9000       # page table base            %
800001c8 : 8f7a0000; %  lw   $26, 0x0($27)     # 1st entry of page table    %
800001cc : 409a1000; %  mtc0 $26, c0_entry_lo  # c0_entrylo <-- v,d,c,pfn   %
800001d0 : 3c1a0000; %  lui  $26, 0x0          # va (=0) for c0_entry_hi    %
```

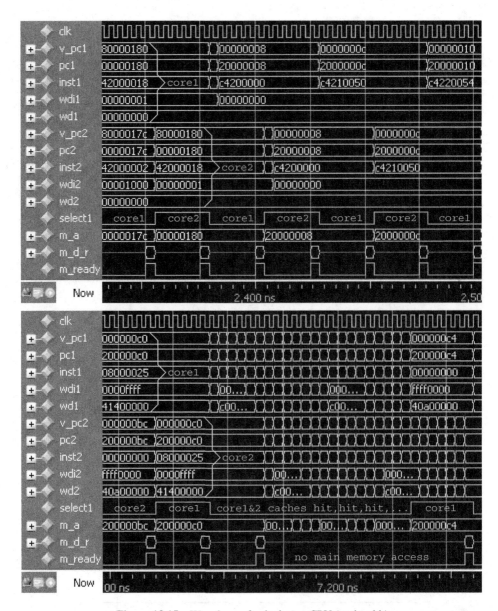

Figure 13.15   Waveform of a dual-core CPU (caches hit)

The upper part of Figure 13.15 shows the waveforms when the two cores execute the following instructions. The lower part shows the waveforms when the two cores have the cache hits (executing instructions in loop2 of the test program).

```
80000180 : 00000000; %  nop                         #                              %
00000004 : c4200000; %  lwc1  f0,  0x0($1)          # load fp data                 %
00000008 : c4210050; %  lwc1  f1,  0x50($1)         # load fp data                 %
0000000c : c4220054; %  lwc1  f2,  0x54($1)         # load fp data                 %
```

Table 13.3 Cache coherence problem in multicore CPUs

Time	Action	Core 1 cache	Core 2 cache	Memory x
t0				0x55555555
t1	Core 1 reads x	0x55555555		0x55555555
t2	Core 2 writes x	0x55555555	0x55555555	0x55555555
t3	Core 1 writes x	0xaaaaaaaa	0x55555555	0xaaaaaaaa

One of challenges in the design of multicore CPUs is in solving the cache coherence problem. Referring to Table 13.3, at time t0, the content of the memory location x is 0x55555555 and there are no copies in the data caches of the two cores. Core 1 loads data from x in t1; Core 2 loads data from x in t2; Up to now, there is no problem; the contents in the caches and memory are consistent. But in t3, Core 1 writes its result 0xaaaaaaaa to x, and the memory is also updated with the new result if a write-through policy is used. Then, the content of x in the cache of Core 2 is not the most up-to-date version, causing a problem of cache incoherence.

Generally, a bus-snooping cache coherence protocol is used on bus-based systems, and a directory-based cache coherence protocol is used on an arbitrary interconnection network, like a mesh or a torus, that is scalable to many cores.

Exercises

13.1 Design a thread selector for a two-way multithreading CPU by using a 3-bit counter to let the ratio of the two thread's priorities to be 5 to 3.

13.2 Design a two-way multithreading CPU with caches and TLBs.

13.3 Design a dual-core CPU with a shared L2 cache.

13.4 Design a two-way multithreading dual-core CPU with one memory port.

13.5 Add a crossbar switch to the CPU above so that two memory modules can be accessed simultaneously.

13.6 To solve the cache coherence problem, the state of a cache block could be assigned to one of the states shown in Table 13.4.

Table 13.4 Cache block states for cache coherence

2-State	3-State	4-State	5-State
I (invalid)	I (invalid)	I (invalid)	I (invalid)
V (valid)	S (shared)	S (shared clean)	S (shared clean)
	M (modified)	M (exclusive modified)	M (exclusive modified)
		E (exclusive clean)	E (exclusive clean)
			O (shared modified)

Investigate these protocols and give their possible state transition diagrams.

13.7 Investigate bus-snooping cache coherence protocols and directory-based cache coherence protocols.

14

Input/Output Interface Controller Design in Verilog HDL

A computer system consists of a computer, software, and input/output (I/O) devices. A computer consists of a CPU (central processing unit), memory, and I/O interfaces. An I/O interface is a controller circuit required to transfer data between an I/O device and the CPU.

This chapter introduces the basic technologies related to I/O interface design, describes the methods of data error detection and correction, and gives some I/O interface design examples, including the UART (universal asynchronous receiver transmitter), PS/2 (personal system/2) keyboard and mouse, VGA (video graphics array) controller, I2C (inter integrated circuit) serial bus, and PCI (peripheral component interconnect) parallel bus.

14.1 Overview of Input/Output Interface Controllers

The speeds of I/O devices vary and are significantly different from those of CPU. Some I/O devices use serial signals that are not suitable to be connected directly to the CPU. Therefore, I/O interfaces are needed to solve these differences between the CPU and I/O devices. This section introduces some basic technologies related to the I/O interface design.

14.1.1 Input/Output Address Space and Instructions

A CPU accesses I/O devices through I/O interfaces. Some CPUs have dedicated instructions and I/O space for performing I/O operations, as shown in Figure 14.1(a). The x86 CPUs are examples that have in and out I/O instructions. Other CPUs use the same set of instructions and memory space to access I/O and memory, as shown in Figure 14.1(b). We call this mode a memory mapped I/O. Most RISC (reduced instruction set computer) CPUs use this mode. For example, the MIPS CPUs use load and store instructions and the virtual address space of 0xa0000000–0xbfffffff to access the I/O devices.

14.1.2 Input/Output Polling and Interrupt

Generally, there are data registers, control registers, and state registers in an I/O interface. The basic I/O operations are to read/write data from/to I/O devices. Before performing such operations, the CPU must know whether the I/O devices are ready or not. There are two common methods for detecting when I/O devices are ready, namely software polling and hardware interrupt, as shown in Figure 14.2.

Computer Principles and Design in Verilog HDL, First Edition. Yamin Li.
© 2015 Tsinghua University Press. All rights reserved. Published 2015 by John Wiley & Sons Singapore Pte Ltd.
Companion Website: www.wiley.com/go/li/verilog

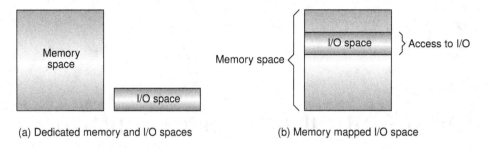

(a) Dedicated memory and I/O spaces   (b) Memory mapped I/O space

Figure 14.1   Relationship between memory space and I/O space

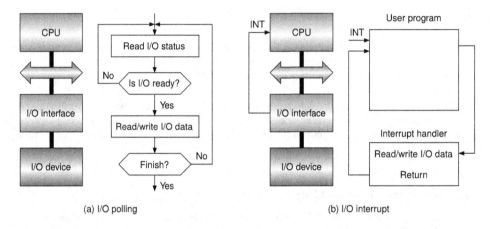

(a) I/O polling   (b) I/O interrupt

Figure 14.2   I/O polling and I/O interrupt

In software polling, the software is responsible for checking the status of I/O devices and initiating transactions when the device is ready. The disadvantage of software polling is the wastage of CPU time on the not-ready waiting loop.

As described in Chapters 6 and 8, in hardware interrupt the CPU does its own work. When an I/O device becomes ready, it informs the CPU by an interrupt signal (via I/O interface). Whenever receiving the interrupt request, the CPU transfers control to an interrupt handler (also called an interrupt service routine) to deal with the I/O transactions with the device. After finishing the transactions, the control is transferred back to the interrupted program.

14.1.3   Direct Memory Access (DMA)

Transferring data between I/O devices and memory can be performed by executing load/store or in/out instructions by the CPU. If there is huge amount data to be transferred, this software method takes a long time. In most modern CPUs, a direct memory access controller (DMAC) is fabricated that can generate addresses and initiate memory read and write cycles, as shown in Figure 14.3.

A DMAC contains several registers, including an address register which specifies the start address of memory; a byte count register, which specifies the number of bytes to be transferred; and control registers, which specify the I/O port and the direction of the transfer (reading from the I/O device or writing to the I/O device). The CPU can write data to these registers. Once the DMAC is initialized, it sends a signal to

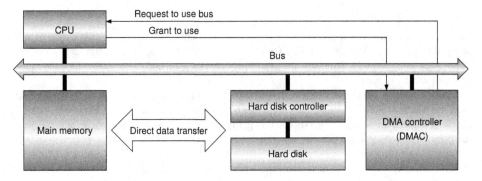

Figure 14.3 Direct memory access

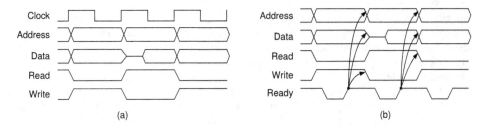

	(a)			(b)

Figure 14.4 (a) Synchronous and (b) asynchronous buses

the CPU to request to use the memory bus. The CPU acknowledges the request by sending a grant signal to DMAC and releasing the bus (floating the output signals). Then the DMAC takes over the control of the bus for the data transfer. Once the specified amount of data is transferred, the DMAC releases the bus back to the CPU.

14.1.4 Synchronization Schemes of Buses

Referring to Figure 14.4, there are two synchronization schemes of buses, namely synchronous scheme and asynchronous scheme.

In synchronous buses, the bus master (usually the CPU) initiates an action based on a clock signal and does not care about the bus slave (memory or an I/O device, for example). The clock speed should be determined so that any action can be completed within a clock cycle. In asynchronous buses, there is no clock signal. Instead, the bus master waits for a ready signal becoming active, issued by the bus slave. Once the bus slave becomes ready, the bus master deactivates the signals of the current action and can initiate a next action. Asynchronous buses are flexible for I/O devices whose speeds vary.

14.2 Error Detection and Correction

Data bit errors may occur during data transmission from the source to a receiver due to some reasons, the communication channel noise for instance. This section introduces two error detection techniques (parity check and cyclic redundancy check (CRC)) and an error correction technique (extended Hamming code).

14.2.1   Parity Check

Parity check appends an extra bit, a 0 or a 1, to the original small-size data bits, usually a byte, to indicates the parity (odd or even) of the data bits. The odd parity checking appends a bit so that the total number of bits with value 1 is odd. Similarly, the even parity checking appends a bit so that the total number of bits with value 1 is even.

As an example, if the original data byte is 01000111, there are four 1's. When odd parity checking is used, the parity bit is a 1, making the total number of bits with value 1 odd (five). However, if even parity checking is used, the parity bit is a 0. The odd parity bit p_o and the even parity bit p_e can be calculated easily by the following logic expressions where d_i is the ith bit in a byte for $0 \le i \le 7$.

$$p_o = \overline{d_7 \oplus d_6 \oplus d_5 \oplus d_4 \oplus d_3 \oplus d_2 \oplus d_1 \oplus d_0}$$

$$p_e = d_7 \oplus d_6 \oplus d_5 \oplus d_4 \oplus d_3 \oplus d_2 \oplus d_1 \oplus d_0$$

A bit error means that the value of a bit was inverted unexpectedly. Parity check technique can detect odd-bit errors; it cannot detect even-bit errors. For example, if the two bits d_i and d_j, $i \ne j$, in a byte are error bits, the parity check cannot find these errors because $\overline{d_i} \oplus \overline{d_j} = d_i \oplus d_j$. Therefore, the parity check technique works on the basis of the fact that the probability of 1-bit error is much larger than that of others. Note that the error may occur also on the parity bit itself.

Parity check technique can detect 1-bit error but cannot know which bit is the error bit. Therefore, it is impossible to correct the error bit by using parity checking. The RAID (redundant arrays of inexpensive/independent disks) uses the parity check technique to recover the data of one faulty disk from other nonfaulty disks. This recovery is possible because we know which disk is faulty. The failure of a disk is different from that when a data bit is inverted unexpectedly.

14.2.2   Error Correction Code (Extended Hamming Code)

In some cases, we wish to know not only whether a 1-bit error is occurring but also which bit is the error bit. Thus, the error can be corrected by inverting that bit. The code that can do such correction is called an error correcting code (ECC).

The Hamming code, proposed by Richard Hamming, is one of the ECCs. A Hamming code has m parity bits. The maximum number of data bits is k, with $k = 2^m - m - 1$. The length of the encoded word is $n = k + m$. Each of parity bits is placed in the bit position 2^t with $0 \le t \le m - 1$. As an example, Figure 14.5 shows the case of $m = 4, k = 11$, and $n = 15$. p_i is a parity check bit with $i = 2^t$ and $0 \le t \le 3$. d_j is data bit with $1 \le j \le 15$ and $j \ne i$.

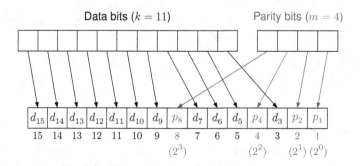

Figure 14.5   Data and parity check bits of Hamming code

d_{15}	d_{14}	d_{13}	d_{12}	d_{11}	d_{10}	d_9	p_8	d_7	d_6	d_5	p_4	d_3	p_2	p_1	
2^3	1	1	1	1	1	1	1	1	0	0	0	0	0	0	0	p_8
2^2	1	1	1	1	0	0	0	0	1	1	1	1	0	0	0	p_4
2^1	1	1	0	0	1	1	0	0	1	1	0	0	1	1	0	p_2
2^0	1	0	1	0	1	0	1	0	1	0	1	0	1	0	1	p_1

$$(2^3) \qquad\qquad (2^2) \quad (2^1)\,(2^0)$$

Figure 14.6 Calculating parity check bits

The parity bit p_i is generated with $p_i = \oplus\{d_j\} \mid j\ \&\ i = i$. That is

$$p_8 = d_{15} \oplus d_{14} \oplus d_{13} \oplus d_{12} \oplus d_{11} \oplus d_{10} \oplus d_9$$
$$p_4 = d_{15} \oplus d_{14} \oplus d_{13} \oplus d_{12} \oplus d_7 \oplus d_6 \oplus d_5$$
$$p_2 = d_{15} \oplus d_{14} \oplus d_{11} \oplus d_{10} \oplus d_7 \oplus d_6 \oplus d_3$$
$$p_1 = d_{15} \oplus d_{13} \oplus d_{11} \oplus d_9 \oplus d_7 \oplus d_5 \oplus d_3$$

For easy understanding, we illustrate the calculations above in Figure 14.6. The bit numbers are given in the binary format (four bits in vertical direction and the most significant bit (MSB) in the top position). p_8, whose bit number is 1000 (the MSB is a 1), is calculated from d_j with j's binary number contains a 1 in the MSB position. Other parity bits, p_4, p_2, and p_1, are calculated in a similar manner.

We get the checking code $c = c_8 c_4 c_2 c_1$ from the following calculations:

$$c_8 = d_{15} \oplus d_{14} \oplus d_{13} \oplus d_{12} \oplus d_{11} \oplus d_{10} \oplus d_9 \oplus p_8$$
$$c_4 = d_{15} \oplus d_{14} \oplus d_{13} \oplus d_{12} \oplus d_7 \oplus d_6 \oplus d_5 \oplus p_4$$
$$c_2 = d_{15} \oplus d_{14} \oplus d_{11} \oplus d_{10} \oplus d_7 \oplus d_6 \oplus d_3 \oplus p_2$$
$$c_1 = d_{15} \oplus d_{13} \oplus d_{11} \oplus d_9 \oplus d_7 \oplus d_5 \oplus d_3 \oplus p_1$$

If there is no error, $c = 0000$, because $p \oplus p = 0$. If there is a 1-bit error, c points to the error bit. For example, if d_{13} is an error, $c = 1101$, because d_{13} appears in the calculations of p_8, p_4, and p_1. If we invert d_{13}, then the error is corrected.

What will happen if two bits are in error? Let's see an example. Suppose that the bits d_9 and d_7 are error. Because d_9 appears in the calculations of p_8 and p_1, and d_7 appears in the calculations of p_4, p_2, and p_1, we get $c = 1110$, that is, $1001 \oplus 0111 = 1110$, pointing out that d_{14} is an error bit. Instead of correcting the error bits d_9 and d_7, if we invert d_{14}, it will make things worse.

The solution to this problem is to use an extended Hamming code. Referring to Figure 14.7, it appends one more parity bit p_0, which is calculated as follows:

$$P_0 = d_{15} \oplus d_{14} \oplus d_{13} \oplus d_{12} \oplus d_{11} \oplus d_{10} \oplus d_9 \oplus p_8 \oplus d_7 \oplus d_6 \oplus d_5 \oplus p_4 \oplus d_3 \oplus p_2 \oplus p_1$$

On checking, calculate

$$c_0 = d_{15} \oplus d_{14} \oplus d_{13} \oplus d_{12} \oplus d_{11} \oplus d_{10} \oplus d_9 \oplus p_8 \oplus d_7 \oplus d_6 \oplus d_5 \oplus p_4 \oplus d_3 \oplus p_2 \oplus p_1 \oplus p_0$$

Thus, if there is a 2-bit error, $c \neq 0$ but $c_0 = 0$. In such a case, we just report that there are errors but we do not correct them. If there is a 1-bit error, $c \neq 0$ and $c_0 \neq 0$, the error bit can be corrected by inverting the bit c. Please consider the cases in which there are more bit errors. The extended Hamming code sometimes are called a SECDED (single error correction and double error detection), but this name does not reflect the whole meaning of the extended Hamming code.

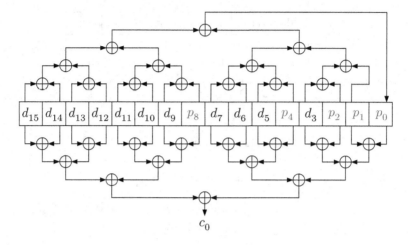

Figure 14.7   Data and parity check bits of extended Hamming code

14.2.3   Cyclic Redundancy Check

Parity checking is applied to small-size data, a byte for instance, to detect the odd-bit errors. A CRC is also an error detecting code applied to large-size data. Commonly, the large-size data is written (sent) or read (received) serially, and the CRC code is calculated during the serial data transmission. Two examples of using CRCs are the CRC checksum for CD/DVD and the CRC code for the link-level transmissions on networks.

The calculation of the CRC code is based on a modulo-2 carry-less polynomial division. Assume that we have an m-bit data information whose absolute is M (dividend). Also assume that we have a d-bit divisor whose absolute is D. The dividend M is shifted to the left by $d - 1$ bits. By dividing the shifted data by D, we get $(d - 1)$-bit remainder R. That is

$$M \times 2^{d-1} = Q \times D + R$$

where Q is the quotient that is discarded. The subtraction in modulo-2 carry-less division can be done with XOR (exclusive OR). The sender concatenates M and R, that is, $N = M \times 2^{d-1} + R$, and sends N to a receiver. The receiver divides N by D. If there are no errors, the remainder should be 0, because

$$M \times 2^{d-1} + R = Q \times D + R + R = Q \times D + 0$$

Note that $R + R = 0$. The divisor D is reassembled from the coefficients of a polynomial generator G. For example, if we define $G = x^3 + x + 1$, then $D = 1011$. An example in Table 14.1 shows $M = 110001011111, D = 1011$, and $R = 110$.

In the division operation, the remainder (the dividend initially) is checked from the MSB. If the d-bit MSBs of the remainder is larger than or equal to the d-bit divisor D, then we subtract the divisor from the MSBs of the remainder and get a new remainder. The subtraction is done with the XOR operation. At the final stage, the MSBs of the remainders become zero, and we get the $(d - 1)$-bit CRC code R.

Because $x \oplus 0 = x$ and $x \oplus 1 = \bar{x}$, we can perform the XOR operation on every bit of the remainder starting from the MSB: if the MSB is a 0, then $x \oplus 0 = x$ (i.e., unchanged); otherwise, $x \oplus 1 = \bar{x}$ (i.e., subtracted). Therefore, we have the circuit shown in Figure 14.8 for calculating the CRC code when $G = x^3 + x + 1$ ($D = 1011$).

Three DFFs (D flip-flops) contain the CRC code and are cleared initially. The dividend M enters the circuit sequentially from the MSB. There are three 1's in divisor D, but only two XOR gates are needed

Table 14.1 CRC calculation example

	Sender			Receiver	
	MSBs of remainder	R		MSBs of remainder	R
M	1 1 0 0 0 1 0 1 1 1 1 1	0 0 0	M	1 1 0 0 0 1 0 1 1 1 1 1	1 1 0
⊕	1 0 1 1		⊕	1 0 1 1	
=	0 1 1 1 0 1 0 1 1 1 1 1	0 0 0	=	0 1 1 1 0 1 0 1 1 1 1 1	1 1 0
⊕	0 1 0 1 1		⊕	0 1 0 1 1	
=	0 0 1 0 1 1 0 1 1 1 1 1	0 0 0	=	0 0 1 0 1 1 0 1 1 1 1 1	1 1 0
⊕	0 0 1 0 1 1		⊕	0 0 1 0 1 1	
=	0 0 0 0 0 0 0 1 1 1 1 1	0 0 0	=	0 0 0 0 0 0 0 1 1 1 1 1	1 1 0
⊕	0 0 0 0 0 0 0 1 0 1 1		⊕	0 0 0 0 0 0 0 1 0 1 1	
=	0 0 0 0 0 0 0 0 1 0 0 1	0 0 0	=	0 0 0 0 0 0 0 0 1 0 0 1	1 1 0
⊕	0 0 0 0 0 0 0 0 1 0 1 1		⊕	0 0 0 0 0 0 0 0 1 0 1 1	
=	0 0 0 0 0 0 0 0 0 0 1 0	0 0 0	=	0 0 0 0 0 0 0 0 0 0 1 0	1 1 0
⊕	0 0 0 0 0 0 0 0 0 0 1 0	1 1	⊕	0 0 0 0 0 0 0 0 0 0 1 0	1 1
=	0 0 0 0 0 0 0 0 0 0 0 0	1 1 0	=	0 0 0 0 0 0 0 0 0 0 0 0	0 0 0

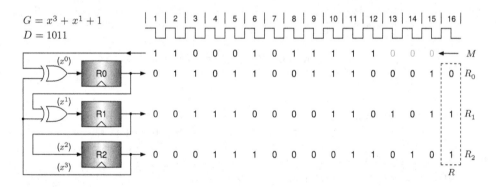

Figure 14.8 Example of a polynomial division circuit

because the calculation on x^3 is either $1 \oplus 1 = 0$ or $0 \oplus 0 = 0$. In both cases, the result value of the bit is always 0, which makes the MSBs of the remainder to become 0.

As an example, in the figure we let $M = 110001011111$. The CRC code $R = 110$ is available at the clock cycle 16 after extra 3-bit 0's are entered in the circuit. As shown in Table 14.2, what was done in the first three clock cycles (clock cycles 1–3) is just to shift the 1-bit dividend M_0 ($\oplus$ 0) to the left by 3 bits. In clock cycle 4, we get the CRC code $R = 011$ with $D = 1011$ and $M \times 2^3 = 1000$.

Therefore, we can use the CRC calculation circuit shown in Figure 14.9 in which M_0 ($\oplus$ 0) is used immediately for generating the CRC code R. At the same dividend $M = 110001011111$, we get the CRC code $R = 110$ at clock cycle 13, $d - 1 = 3$ cycles earlier than the circuit of Figure 14.8.

Below is the Verilog HDL (hardware description language) code that implements the circuit shown in Figure 14.9.

Table 14.2 Polynomial division operations in first three cycles

Clock cycle	Remainder (CRC codes)			Dividend	Extra zeroes		
	R_2	R_1	R_0	M_0			
0	0	0	0	1	0	0	0
1	0	0	1	0	0	0	
2	0	1	0	0	0		
3	1	0	0	0			
4	0	1	1				

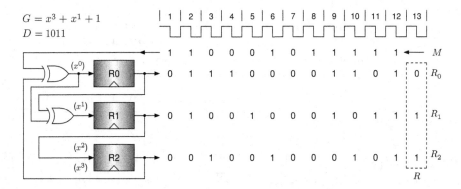

Figure 14.9 CRC calculation circuit

```
module crc3 (clk,clrn,m,crc);              // G = x^3 + x + 1 (D = 1011)
    input           clk, clrn;             // clock and reset
    input           m;                     // message or dividend
    output reg [2:0] crc;                  // 3-bit crc code
    wire            m_xor_crc2 = m ^ crc[2];
    always @ (posedge clk or negedge clrn) begin
        if (!clrn) begin
            crc <= 0;
        end else begin
            crc[0] <=           m_xor_crc2;     // 1
            crc[1] <= crc[0] ^ m_xor_crc2;     // x
            crc[2] <= crc[1];
        end
    end
endmodule
```

The simulation waveform is given in Figure 14.10 with $M = 110001011111$. We can see that the CRC code $R = 6$ is available immediately after the last bit of M entered the circuit.

Some common CRC generator polynomials are listed in Table 14.3.

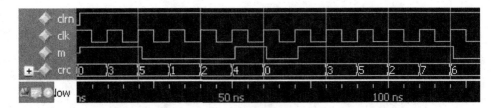

Figure 14.10   Waveform of the CRC calculation circuit

Table 14.3   Common CRC generator polynomials

CRC standard	CRC generator polynomial
CRC-1	$x + 1$ (parity check)
CRC-8-CCITT	$x^8 + x^2 + x + 1$
CRC-12	$x^{12} + x^{11} + x^3 + x^2 + x + 1$
CRC-16-CCITT	$x^{16} + x^{12} + x^5 + 1$
CRC-16-IBM	$x^{16} + x^{15} + x^2 + 1$
CRC-16-DNP	$x^{16} + x^{13} + x^{12} + x^{11} + x^{10} + x^8 + x^6 + x^5 + x^2 + 1$
CRC-32-IEEE	$x^{32} + x^{26} + x^{23} + x^{22} + x^{16} + x^{12} + x^{11} + x^{10} + x^8 + x^7 + x^5 + x^4 + x^2 + x + 1$
CRC-64-ISO	$x^{64} + x^4 + x^3 + x + 1$
CRC-64-ECMA	$x^{64} + x^{62} + x^{57} + x^{55} + x^{54} + x^{53} + x^{52} + x^{47} + x^{46} + x^{45} + x^{40} + x^{39} + x^{38} +$ $x^{37} + x^{35} + x^{33} + x^{32} + x^{31} + x^{29} + x^{27} + x^{24} + x^{23} + x^{22} + x^{21} + x^{19} + x^{17} +$ $x^{13} + x^{12} + x^{10} + x^9 + x^7 + x^4 + x + 1$

Below is the Verilog HDL code that implements the circuit of CRC-32-IEEE.

```
module crc32 (clk,clrn,m,crc);  // G = x^{32} + x^{26} + x^{23} + x^{22} +
    // x^{16} + x^{12} + x^{11} + x^{10} + x^{8} + x^{7} + x^{5} + x^{4} +
    // x^{2} + x + 1
    // D = 1_0000_0100_1100_0001_0001_1101_1011_0111 = 0x104c11db7
    input           clk, clrn;                  // clock and reset
    input           m;                          // message or dividend
    output reg [31:0] crc;                      // 32-bit crc code
    wire            m_xor_crc31 = m ^ crc[31];
    always @ (posedge clk or negedge clrn) begin
        if (!clrn) begin
            crc <= 0;
        end else begin
            crc[00] <=             m_xor_crc31;         // x^{00}
            crc[01] <= crc[00]  ^  m_xor_crc31;         // x^{01}
            crc[02] <= crc[01]  ^  m_xor_crc31;         // x^{02}
            crc[03] <= crc[02];
            crc[04] <= crc[03]  ^  m_xor_crc31;         // x^{04}
            crc[05] <= crc[04]  ^  m_xor_crc31;         // x^{05}
            crc[06] <= crc[05];
```

```
                  crc[07] <= crc[06] ^ m_xor_crc31;        // x^{07}
                  crc[08] <= crc[07] ^ m_xor_crc31;        // x^{08}
                  crc[09] <= crc[08];
                  crc[10] <= crc[09] ^ m_xor_crc31;        // x^{10}
                  crc[11] <= crc[10] ^ m_xor_crc31;        // x^{11}
                  crc[12] <= crc[11] ^ m_xor_crc31;        // x^{12}
                  crc[13] <= crc[12];
                  crc[14] <= crc[13];
                  crc[15] <= crc[14];
                  crc[16] <= crc[15] ^ m_xor_crc31;        // x^{16}
                  crc[17] <= crc[16];
                  crc[18] <= crc[17];
                  crc[19] <= crc[18];
                  crc[20] <= crc[19];
                  crc[21] <= crc[20];
                  crc[22] <= crc[21] ^ m_xor_crc31;        // x^{22}
                  crc[23] <= crc[22] ^ m_xor_crc31;        // x^{23}
                  crc[24] <= crc[23];
                  crc[25] <= crc[24];
                  crc[26] <= crc[25] ^ m_xor_crc31;        // x^{26}
                  crc[27] <= crc[26];
                  crc[28] <= crc[27];
                  crc[29] <= crc[28];
                  crc[30] <= crc[29];
                  crc[31] <= crc[30];
              end
          end
      endmodule
```

The simulation waveform of the CRC-32-IEEE code is given in Figure 14.11 with $M = D =$ 0x104c11db7. Because $M = D$, the remainder of M/D would be 0. From the figure, we can see that the CRC code $R = 0$ is available immediately after the last bit of M entered the circuit. CRC-32-IEEE is used in Ethernet, MPEG-2, PNG, rar, zip, POSIX checksum, and more.

14.3   Universal Asynchronous Receiver Transmitter

A UART can send and receive SDA asynchronously. The data frame format of UARTs is shown in Figure 14.12. A data frame contains a start bit, 5–8 data bits, a parity bit (optional), and 1–2 stop bits. The start bit is logic low, and the stop bits are logic high. The LSB (least significant bit) of the data bits is transmitted first.

Figure 14.13 shows the signal connection of the two UARTs that communicate each other. The signals txd and rxd are used to send data and receive data, respectively. gnd is a ground signal.

The UART translates parallel data to SDA for sending; it also translates SDA to parallel data for receiving. Inside a UART, there are several registers that are used to store status and data for sending and receiving. The CPU can read and write these registers through I/O ports.

In order to transmit data correctly between two UARTs, both sides must make an agreement on the data frame format and the transmission speed. The data frame concerns the number of data bits, the number of stop bits, and the parity bit (use or not use, even checking or odd checking if used). The transmission

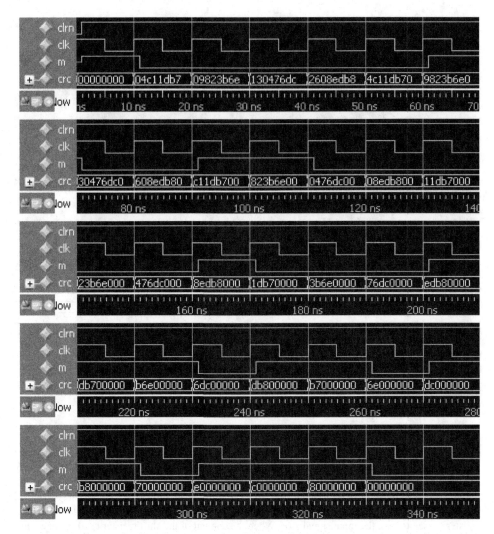

Figure 14.11 Waveform of CRC-32-IEEE calculation circuit

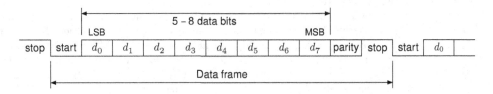

Figure 14.12 Data frame format of UART

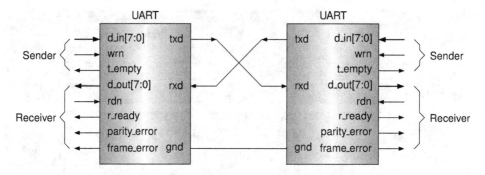

Figure 14.13   Connection between two UARTs

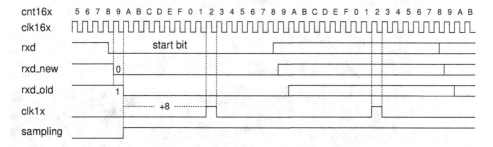

Figure 14.14   Timing of the UART receiver

speed is specified with the baud rate, which defines how many times a signal changes (or can change) per second. The common baud rates are 110, 150, 300, 600, 1200, 2400, 4800, 9600, 19200, 38400, 57600, 115200, and 230400.

We first describe the receiver design of the UART. Figure 14.14 shows the timing chart for detecting the start bit. We use a clock signal `clk16x`, whose frequency is 16 times the baud rate. A 2-bit shift register is used to save the input signal `rxd`. When the content of the shift register is 10 (`rxd_old = 1` and `rxd_new = 0`), that is, a falling edge is detected, we let `sampling = 1`, to start to sample `rxd`.

A 4-bit counter `cnt16x` is used to generate a sampling pulse `clk1x`. During `sampling = 1`, we let the rising edge of `clk1x` to appear at about the center place of one `rxd` bit, and the width of the `clk1x` pulse to be one `clk16x` clock cycle.

The `sampling` signal keeps active until a whole frame is received. As shown in Figure 14.15, in order to know when a frame is received, we use a 4-bit counter, namely `no_bits_rcvd`, to count the number of `rxd` bits received that contains from the `start` bit to the `stop` bit, that is, totally 11 bits. The counter is also used as a bit index to a receive buffer `r_buffer` for storing an `rxd` bit. If the value of the counter reaches 11, we deactivate the `sampling` signal, save the 8-bit data to a register `r_data`, and issue a signal `r_ready` to inform the CPU to get data from the `r_data` register.

The design of the UART transmitter is described below. CPU can send a data byte to UART when a `t_data` register of UART is empty, indicated by a `t_empty` signal. Figure 14.16 shows the timing when UART sends the `start` bit and the LSB of the data byte 0xe1. The data byte in `t_data` is sent to a register `t_buffer`, and the `sending` signal becomes active for sending each bit of the data byte to `txd`. Then the CPU can send the next data byte, 0x55 in the figure, to the `t_data` register. Further data bytes cannot be sent by CPU because the `t_data` register is not empty.

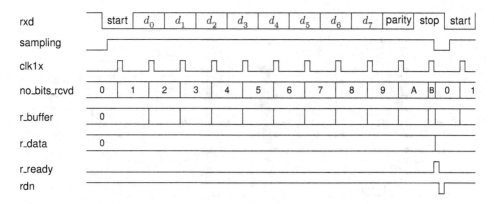

Figure 14.15   Sampling clock of UART receiver

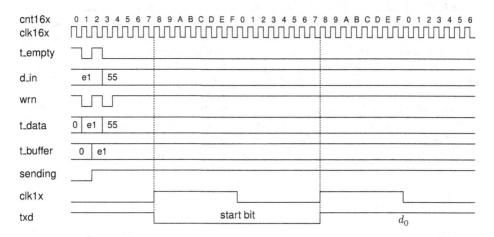

Figure 14.16   Timing of UART sender

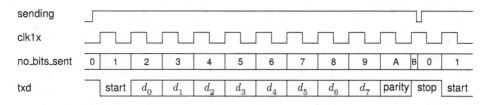

Figure 14.17   Sending clock of UART sender

clk1x is the clock signal for sending a bit to txd. The rising edge of clk1x appears in the time point when the value of the cnt16x counter becomes 8. Referring to Figure 14.17, we also prepare a counter, namely no_bits_sent, to control the sending of a frame to txd. CPU sends only 8-bit data byte to UART; the bits of start, parity, and stop are generated by hardware.

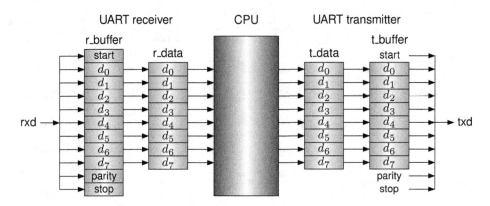

Figure 14.18 Double buffers for receiver and transmitter in a UART

The UART operates on a full-duplex mode, meaning that the transmitter and receiver can operate in parallel. We use double buffers for the receiver and transmitter, as shown as in Figure 14.18. Thus, even if a received data byte is not yet read by the CPU, the receiver can also receive the next frame. Similarly, even when the current data byte is being sent, the CPU can send the next data byte to the UART.

According to the discussion above, we have the following Verilog HDL code for implementing the full-duplex UART. It is the top module and invokes the receiver module and the transmitter module.

```
module uart (clk16x,clrn,        // universal asynchronous receiver transmitter
            rdn,d_out,r_ready,rxd,parity_error,frame_error,wrn,d_in,
            t_empty,txd,cnt16x,no_bits_rcvd,r_buffer,r_clk1x,sampling,
            r_data,no_bits_sent,t_buffer,t_clk1x,sending,t_data);
    input           clk16x, clrn;                    // baud rate * 16 clock
    // signals for receiver
    input           rdn;                             // cpu read, active low
    input           rxd;                             // uart rxd
    output  [7:0] d_out;                             // data byte to cpu
    output          r_ready;                         // receiver is ready
    output          parity_error;                    // parity check error
    output          frame_error;                     // data frame error
    // signals for transmitter
    input   [7:0] d_in;                              // data byte from cpu
    input           wrn;                             // cpu write, active low
    output          txd;                             // uart txd
    output          t_empty;                         // transmitter empty
    // for test (internal signals)
    output [10:0] r_buffer;                          // 11-bit frame
    output  [7:0] r_data;                            // received data bits
    output  [7:0] t_buffer;                          // register for sending
    output  [7:0] t_data;                            // register d_in
    output  [3:0] cnt16x;                            // x16 clock counter
    output  [3:0] no_bits_rcvd;                      // # of bits received
    output  [3:0] no_bits_sent;                      // number of bits sent
    output          sampling;                        // sampling an rxd bit
```

```
    output          r_clk1x;                    // clock for sampling rxd
    output          sending;                    // sending a txd bit
    output          t_clk1x;                    // clock for sending txd
    reg [3:0] cnt16x;                           // x16 clock counter
    // a 4-bit counter
    always @ (posedge clk16x or negedge clrn) begin
        if (!clrn) begin                        // on reset
            cnt16x <= 4'd0;                      // clear counter
        end else begin
            cnt16x <= cnt16x + 4'd1;             // counter++
        end
    end
    // receiver
    uart_rx recver (clk16x, clrn, rdn, d_out, r_ready, rxd, parity_error,
                    frame_error, cnt16x, r_data, no_bits_rcvd, r_buffer,
                    r_clk1x, sampling);
    // transmitter
    uart_tx sender (clk16x, clrn, wrn, d_in,  t_empty, txd, cnt16x,
                    no_bits_sent, t_buffer, t_clk1x, sending, t_data);
endmodule
```

Below is the Verilog HDL code that implements the UART receiver.

```
module uart_rx (clk16x, clrn, rdn, d_out, r_ready, rxd, parity_error,
                frame_error, cnt16x, r_data, no_bits_rcvd, r_buffer,
                clk1x, sampling);                // uart receiver
    input         [3:0] cnt16x;                  // x16 clock counter
    input               clk16x, clrn;            // baud rate * 16 clock
    input               rdn;                     // cpu read, active low
    input               rxd;                     // uart rxd
    output        [7:0] d_out;                   // data byte to cpu
    output reg          r_ready;                 // receiver is ready
    output reg          parity_error;            // parity check error
    output reg          frame_error;             // data frame error
    output reg [10:0] r_buffer;                   // 11-bit frame
    output reg    [7:0] r_data;                    // received data bits
    output reg    [3:0] no_bits_rcvd;              // # of bits received
    output reg          clk1x;                   // clock for sampling rxd
    output reg          sampling;                // sampling an rxd bit
    reg           [3:0] sampling_place;          // center of an rxd bit
    reg                 rxd_new;                 // registered rxd
    reg                 rxd_old;                 // registered rxd_new
    // latch 2 rxd bits for detecting a falling edge
    always @ (posedge clk16x or negedge clrn) begin
        if (!clrn) begin
            rxd_old <= 1;                        // stop bits
            rxd_new <= 1;                        // stop bits
```

```
      end else begin
          rxd_old <= rxd_new;                  // shift registers
          rxd_new <= rxd;                      // shift registers
      end
end
// detect start bit and generate sampling signal
always @ (posedge clk16x or negedge clrn) begin
    if (!clrn) begin
        sampling <= 0;                         // stop sampling
    end else begin
        if (rxd_old && !rxd_new) begin         // had a negative edge
            if (!sampling)                     // if not sampling yet
               sampling_place <= cnt16x + 4'd8; // +8: center place
            sampling <= 1;                     // sampling please
        end else begin
            if (no_bits_rcvd == 4'd11)         // got one frame
               sampling <= 0;                  // stop sampling
        end
    end
end
// sampling clock: clk1x
always @ (posedge clk16x or negedge clrn) begin
    if (!clrn) begin
        clk1x <= 0;
    end else begin
        if (sampling) begin                    // if sampling
            if (cnt16x == sampling_place)      // at the center place
               clk1x <= 1;                     // generate a pos edge
            if (cnt16x == sampling_place + 4'd1)
               clk1x <= 0;                     // one x16 cycle pulse
        end else clk1x <= 0;                   // stop clk1x
    end
end
// number of bits received
always @ (posedge clk1x or negedge sampling) begin
    if (!sampling) begin
        no_bits_rcvd <= 4'd0;                  // clear counter
    end else begin
        no_bits_rcvd <= no_bits_rcvd + 4'd1;   // counter++
        r_buffer[no_bits_rcvd] <= rxd;         // save rxd to r_buffer
    end
end
// one frame, rdn clears r_ready
always @ (posedge clk16x or negedge clrn or negedge rdn) begin
    if (!clrn) begin                           // on a reset
        r_ready <= 0;                          // clear ready
        parity_error <= 0;                     // clear parity error
```

```verilog
            frame_error  <= 0;                     // clear frame error
            r_buffer     <= 0;                     // clear r_buffer
            r_data       <= 0;                     // clear r_data
        end else begin
            if (!rdn) begin                        // on a read
                r_ready       <= 0;                // clear ready
                parity_error  <= 0;                // clear parity error
                frame_error   <= 0;                // clear frame error
            end else begin
                if (no_bits_rcvd == 4'd11) begin   // got a frame
                    r_data <= r_buffer[8:1];       // extract data byte
                    r_ready <= 1;                  // receiver ready
                    if (^r_buffer[9:1]) begin      // parity check (even)
                        parity_error <= 1;         // parity error
                    end
                    if (!r_buffer[10]) begin       // if no stop bit
                        frame_error <= 1;          // frame error
                    end
                end
            end
        end
    end
    assign d_out = !rdn ? r_data : 8'hz;           // data byte output
endmodule
```

Below is the Verilog HDL code that implements the UART transmitter.

```verilog
module uart_tx (clk16x,clrn,wrn,d_in,t_empty,txd,cnt16x,no_bits_sent,
                t_buffer,clk1x,sending,t_data);   // uart transmitter
    input       [7:0]  d_in;                       // data byte from cpu
    input       [3:0]  cnt16x;                      // x16 clock counter
    input              clk16x,clrn;                 // baud rate * 16 clock
    input              wrn;                          // cpu write, active low
    output reg         txd;                          // uart txd
    output reg         t_empty;                      // transmitter empty
    output reg  [3:0]  no_bits_sent;                 // number of bits sent
    output reg  [7:0]  t_data;                        // reg d_in
    output reg  [7:0]  t_buffer;                      // reg for sending
    output             clk1x;                         // clock for sending txd
    output reg         sending;                       // sending a txd bit
    reg                load_t_buffer;                 // load t_buffer
    // load data to t_data, then to t_buffer, and generate sending signal
    always @ (posedge clk16x or negedge clrn or negedge wrn) begin
        if (!clrn) begin                             // on a reset
            sending       <= 0;                      // clear sending
            t_empty       <= 1;                      // transmitter is ready
```

```
            load_t_buffer <= 0;                  // clear load_t_buffer
            t_data        <= 0;                  // clear t_data
            t_buffer      <= 0;                  // clear t_buffer
        end else begin
            if (!wrn) begin                      // cpu write
                t_data  <= d_in;                 // load t_data
                t_empty <= 0;                    // transmitter is busy
                load_t_buffer <= 1;              // ready to load t_buffer
            end else begin
                if (!sending) begin              // not sending
                    if (load_t_buffer) begin     // d2t ready
                        sending <= 1;            // sending please
                        t_buffer <= t_data;      // load t_buffer
                        t_empty <= 1;            // transmitter is ready
                        load_t_buffer <= 0;      // clear load_t_buffer
                    end
                end else begin                   // sending
                    if (no_bits_sent == 4'd11)   // sent a frame
                        sending <= 0;            // clear sending
                end
            end
        end
    end
    // send a frame: [start, d0, d1, ..., d7, parity, stop]
    assign clk1x = cnt16x[3];                    // clock for sending txd
    always @ (posedge clk1x or negedge sending) begin
        if (!sending) begin                      // if not sending
            no_bits_sent <= 4'd0;                // clear counter
            txd <= 1;                            // stop bits
        end else begin                           // sending
            case (no_bits_sent)                  // sending serially
                0: txd <= 0;                     // sending start bit
                1: txd <= t_buffer[0];           // sending data bit 0
                2: txd <= t_buffer[1];           // sending data bit 1
                3: txd <= t_buffer[2];           // sending data bit 2
                4: txd <= t_buffer[3];           // sending data bit 3
                5: txd <= t_buffer[4];           // sending data bit 4
                6: txd <= t_buffer[5];           // sending data bit 5
                7: txd <= t_buffer[6];           // sending data bit 6
                8: txd <= t_buffer[7];           // sending data bit 7
                9: txd <= ^t_buffer;             // sending parity (even)
                default: txd <= 1;               // sending stop bit(s)
            endcase
            no_bits_sent <= no_bits_sent + 4'd1; // counter++
        end
    end
endmodule
```

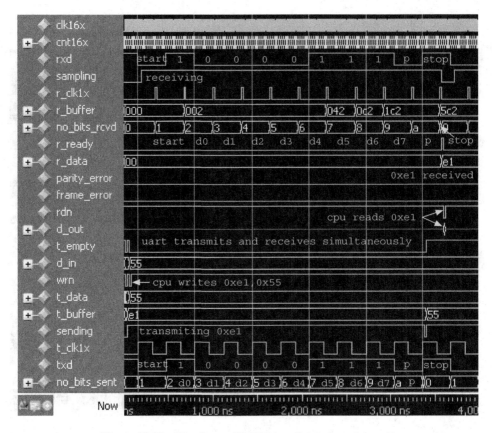

Figure 14.19   Waveform of UART (sending and receiving e1)

Figure 14.19 shows the waveform when the UART sends and receives the data byte 0xe1. Figure 14.20 shows the waveform when the UART sends and receives the data byte 0x55. In the simulation, we connect txd to rxd so that the UART receives what is sent (operates in the full-duplex mode).

14.4   PS/2 Keyboard/Mouse Interface Design

The PS/2 keyboard/mouse protocol was developed by IBM and used first in IBM PS/2. This section describes this protocol briefly, and gives a design example of the PS/2 keyboard interface.

14.4.1   PS/2 Keyboard Interface

When a key on the keyboard is pressed, the keyboard's controller sends out a packet of information, known as a scan code. A scan code consists of a make code and a break code. A make code is sent when a key is pressed or held down. A break code is sent when a key is released. For example, when the key [A] is pressed and released immediately, the scan code is 1c f0 1c in hexadecimal. The make code is 1c and the break code is f0 1c (preceding the make code with a byte f0). Table 14.4 gives three other examples of the scan code.

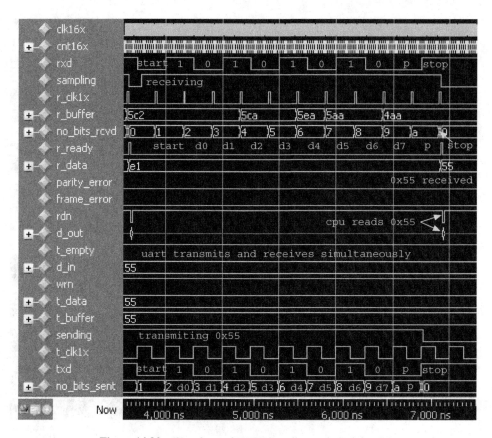

Figure 14.20   Waveform of UART (sending and receiving 55)

Table 14.4   Keyboard make code and break code examples

Key type	Action	Scan code
Standard	Press [A], release [A]	1c f0 1c
Extended	Press [Delete], release [Delete]	e0 71 e0 f0 71
Combination	Press [L Shift] [A], release [A] [L Shift]	12 1c f0 1c f0 12
Combination	Press [R Ctrl] [A], release [A] [R Ctrl]	e0 14 1c f0 1c e0 f0 14

Most make codes are a single byte in length, with the exception of some of the extended keys (e.g., [Right Ctrl], [Right Alt], and Arrow keys). The extended keys are recognizable by checking the e0 prefix to their make codes. If a key is held down without being released, the make code for that key will be sent continuously. For example, when the key [A] is pressed, held down for a while, and then released, the scan code is 1c 1c ⋯ 1c 1c f0 1c. In the next section, we will show a hardware and software design that displays the scan codes on a VGA in response to pressing, holding down, and releasing any key on a keyboard.

All data are transmitted one byte at a time, and each byte is sent in a frame. As shown in Figure 14.21, there are two serial signals, namely ps2_clk and ps2_data, in the PS/2 protocol. A PS/2 interface

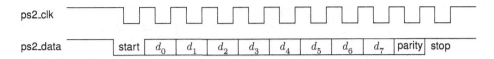

Figure 14.21   Data frame format of a PS/2 keyboard

controller can read `ps2_data` when `ps2_clk` is low. A data frame consists of 11 bits: 1 start bit, 8 data bits, the LSB first; 1 parity bit (odd parity); and 1 stop bit.

The `ps2_clk` frequency is in the range 10–16.7 kHz. This means that `ps2_clk` must be high for 30–50 μs and low for 30–50 μs. The time from the rising edge of `ps2_clk` to a `ps2_data` transition must be at least 5 μs. The time from a `ps2_data` transition to the falling edge of `ps2_clk` must be at least 5 μs but no greater than 25 μs.

`ps2_clk` cannot be used as a clock signal directly to register the `ps2_data` to DFFs because of the rather ugly waveform of `ps2_clk`. In our interface design, we use a system clock, 50 MHz for instance, to detect the falling edge of `ps2_clk`. If it occurs, we use the system clock to register the `ps2_data` to a frame buffer. Once a frame is received, the data byte is placed in a queue (first-in first-out, FIFO). The interface is implemented with the following Verilog HDL code.

```
module ps2_keyboard (clk,clrn,ps2_clk,ps2_data,rdn,data,ready,overflow);
    input         clk, clrn;                    // 50 MHz
    input         ps2_clk;                      // ps2 clock
    input         ps2_data;                     // ps2 data
    input         rdn;                          // read, active low
    output [7:0]  data;                         // 8-bit code
    output        ready;                        // code ready
    output reg    overflow;                     // fifo overflow
    reg    [9:0]  buffer;                       // ps2_data bits
    reg    [7:0]  fifo[7:0];                    // circular fifo
    reg    [3:0]  count;                        // count ps2_data bits
    reg    [2:0]  w_ptr,r_ptr;                  // fifo w/r pointers
    reg    [1:0]  ps2_clk_sync;                 // for detecting falling edge
    always @ (posedge clk)
        ps2_clk_sync <= {ps2_clk_sync[0],ps2_clk};
    wire sampling = ps2_clk_sync[1] &
                    ~ps2_clk_sync[0];           // had a falling edge
    always @ (posedge clk) begin
        if (!clrn) begin                        // on reset
            count    <= 0;                      // clear count
            w_ptr    <= 0;                      // clear w_ptr
            r_ptr    <= 0;                      // clear r_ptr
            overflow <= 0;                      // clear overflow
        end else if (sampling) begin            // if sampling
            if (count == 4'd10) begin           // if got one frame
                if ((buffer[0] == 0) && (ps2_data) && (^buffer[9:1])) begin
                    if ((w_ptr + 3'b1) != r_ptr) begin
                        fifo[w_ptr] <= buffer[8:1];
                        w_ptr <= w_ptr + 3'b1; // w_ptr++
```

```
                    end else begin
                        overflow <= 1;             // overflow
                    end
                end
                count <= 0;                         // for next frame
            end else begin                          // else
                buffer[count] <= ps2_data;          // store ps2_data
                count <= count + 4'b1;              // count++
            end
        end
        if (!rdn && ready) begin                    // on cpu read
            r_ptr <= r_ptr + 3'b1;                  // r_ptr++
            overflow <= 0;                          // clear overflow
        end
    end
    assign ready = (w_ptr != r_ptr);                // fifo is not empty
    assign data  = fifo[r_ptr];                     // code byte
endmodule
```

Figure 14.22 shows the simulation waveform when the interface receives a byte of 0x4b, the make code of key [L]. It should be noted that, in a real case, the ps2_clk frequency is much lower than that of clk.

Also note that the signals of ps2_clk and ps2_data should be bidirectional, meaning that the CPU can send commands to the keyboard.

14.4.2 PS/2 Mouse Interface

PS/2 mouse uses the same data frame format as that of the keyboard. The data contents of the three-button PS/2 mouse are shown in Figure 14.23.

The horizontal and vertical movement values of the mouse are represented by the 9-bit 2's complement integers of $X_8 \cdots X_0$ and $Y_8 \cdots Y_0$, respectively. The range of values that can be expressed is -256 to $+255$. If this range is exceeded, the appropriate overflow bit, X Ov or Y Ov, is set. $Z_7 \cdots Z_0$ is a 2's complement value that represents the scrolling wheel's movement since the last data report. The valid values of $Z_7 \cdots Z_0$ are in the range -8 to $+7$, meaning that the number is actually represented only by

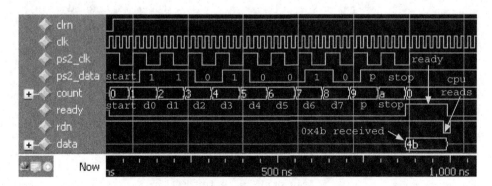

Figure 14.22   Waveform of a PS/2 keyboard

Byte 1	Y Ov	X Ov	Y_8	X_8	1	M-button	R-button	L-button
Byte 2	X_7	X_6	X_5	X_4	X_3	X_2	X_1	X_0
Byte 3	Y_7	Y_6	Y_5	Y_4	Y_3	Y_2	Y_1	Y_0
Byte 4	Z_7	Z_6	Z_5	Z_4	Z_3	Z_2	Z_1	Z_0

Figure 14.23 Data frame contents of a PS/2 mouse

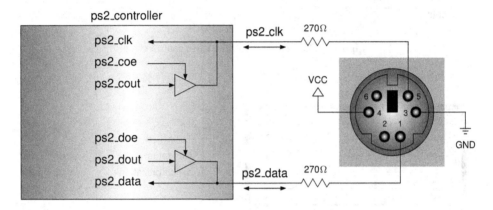

Figure 14.24 PS/2 keyboard/mouse connect

the least significant four bits; the upper four bits act only as sign extension. The M-Button, R-Button, or L-Button is set to 1 if the middle, right, or left button is pressed; reset to 0 if the button is not pressed.

By default, the PS/2 mouse works in the stream mode. In this mode, it produces a stream of packets indicating mouse movements and button presses. The CPU can send a command (a byte) to change the mouse's working mode. The command 0xf0 lets the mouse to work in the remote mode, in which the mouse only sends a packet when the host requests one, using the 0xeb command. The command 0xee lets the mouse to work in the echo mode, in which everything the host sends is echoed back, until either a reset (0xff) or clear echo mode (0xec) is received.

The CPU can send commands to mouse, and vice versa. Therefore, the signals of ps2_clk and ps2_data should be bidirectional. Figure 14.24 shows the bidirectional signal control inside a PS/2 interface controller.

The important component is a tristate gate, which can be implemented with bufif0 or bufif1, as described in Chapter 2. We can also implement it with a general Verilog HDL statement as shown below.

```
inout   ps2_clk;                               // bi-directional ps2_clk
inout   ps2_data;                              // bi-directional ps2_data
assign ps2_clk  = ps2_coe ? ps2_cout : 1'bz;   // z: high-impedance
assign ps2_data = ps2_doe ? ps2_dout : 1'bz;   // z: high-impedance
```

Figure 14.25 shows the timing chart when the CPU sends a command to the mouse. In the waveforms of ps2_clk and ps2_data, the dotted line parts are sent by the CPU and the solid-line parts are sent by the mouse.

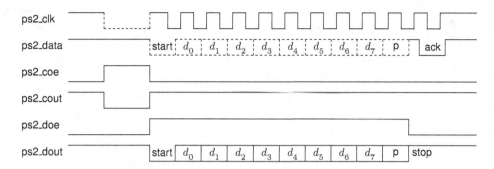

Figure 14.25   Timing of PS/2 mouse control signals

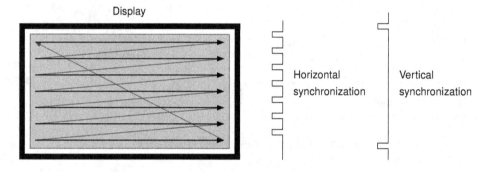

Figure 14.26   VGA display

14.5   Video Graphics Array (VGA) Interface Design

VGA is a popular display standard developed by IBM. This section describes the design of a VGA interface and gives a hardware and software design that displays scan codes on VGA in response to pressing, holding down, and releasing any key on a keyboard. A simple CPU is used for executing the software.

14.5.1   VGA Interface Design in Verilog HDL

The resolution of the standard VGA is 640×480 pixels, that is, there are 640 pixels in a line and 480 lines. Figure 14.26 shows the VGA display (cathode ray tube (CRT) or liquid crystal display (LCD)) with synchronization signals.

Figure 14.27 shows the timing chart of the synchronization signals. hs and vs are the horizontal and vertical synchronization signals, respectively. rgb is the color pixels. The upper part shows the timing of a line. The unit of the numbers is a pixel. We can see that, although in a line 640 pixels must be provided, one line takes $96 + 48 + 640 + 16 = 800$ pixel time. The lower part shows the timing of a frame. The unit of the numbers is line. One frame takes $2 + 33 + 480 + 10 = 525$ line time.

The refresh rate defined by the VGA standard is 60 Hz, meaning that in 1 s, 60 frames must be displayed. Thus, we can calculate the frequency of the clock that is used to send pixels: $800 \times 525 \times 60 = 25.2$ MHz.

Figure 14.28 shows the signals of a VGA interface controller. It generates a 19-bit pixel address and gets a 24-bit pixel color. The pixel address consists of a 9-bit row address and a 10-bit column address.

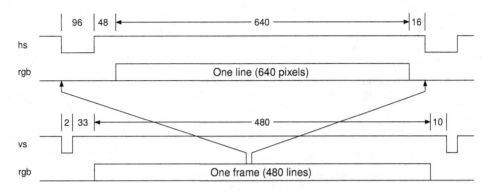

Figure 14.27 HS/VS timing of a standard VGA

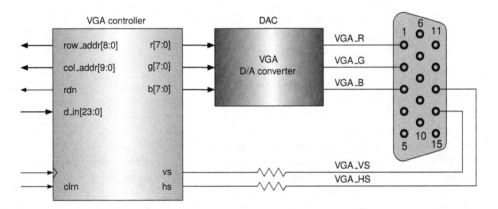

Figure 14.28 Signals of VGA interface controller

The 24-bit pixel color consists 8-bit red, 8-bit green, and 8-bit blue colors. The 24-bit digital colors must be converted to analog signals, which are connected to a VGA 15-pin connector. The digital-to-analog conversion can be done with a DAC (digital-to-analog converter), as shown in Figure 14.28. The DAC can be implemented simply with an array of resistors.

Based on the discussions above, we can develop the Verilog HDL code to implement the VGA interface controller, which is listed below.

```
module vgac (vga_clk,clrn,d_in,row_addr,col_addr,rdn,r,g,b,hs,vs);   // vgac
    input      [23:0] d_in;        // rrrrrrrr_gggggggg_bbbbbbbb, pixel
    input             vga_clk;     // 25MHz
    input             clrn;
    output reg [8:0] row_addr;     // pixel ram row address, 480 (512) lines
    output reg [9:0] col_addr;     // pixel ram col address, 640 (1024) pixels
    output reg [7:0] r,g,b;        // red, green, blue colors, 8-bit for each
    output reg        rdn;         // read pixel RAM (active low)
    output reg        hs,vs;       // horizontal and vertical synchronization
    // h_count: vga horizontal counter (0-799 pixels)
```

```verilog
    reg [9:0] h_count;
    always @ (posedge vga_clk or negedge clrn) begin
        if (!clrn) begin
            h_count <= 10'h0;
        end else if (h_count == 10'd799) begin
            h_count <= 10'h0;
        end else begin
            h_count <= h_count + 10'h1;
        end
    end
    // v_count: vga vertical counter (0-524 lines)
    reg [9:0] v_count;
    always @ (posedge vga_clk or negedge clrn) begin
        if (!clrn) begin
            v_count <= 10'h0;
        end else if (h_count == 10'd799) begin
            if (v_count == 10'd524) begin
                v_count <= 10'h0;
            end else begin
                v_count <= v_count + 10'h1;
            end
        end
    end
    // signals, will be latched for outputs
    wire [9:0] row   = v_count - 10'd35;       // pixel ram row address
    wire [9:0] col   = h_count - 10'd143;      // pixel ram col address
    wire       h_sync = (h_count > 10'd95);    //  96 -> 799
    wire       v_sync = (v_count > 10'd1);     //   2 -> 524
    wire       read   = (h_count > 10'd142) && // 143 -> 782 =
                        (h_count < 10'd783) && //                640 pixels
                        (v_count > 10'd34) &&  //  35 -> 514 =
                        (v_count < 10'd515);   //                480 lines
    // vga signals
    always @ (posedge vga_clk) begin
        row_addr <= row[8:0];                  // pixel ram row address
        col_addr <= col;                       // pixel ram col address
        rdn      <= ~read;                     // read pixel (active low)
        hs       <= h_sync;                    // horizontal synch
        vs       <= v_sync;                    // vertical  synch
        r        <= rdn ? 8'h0 : d_in[23:16];  // 8-bit red
        g        <= rdn ? 8'h0 : d_in[15:08];  // 8-bit green
        b        <= rdn ? 8'h0 : d_in[07:00];  // 8-bit blue
    end
endmodule
```

We used two counters, h_count and v_count, to count the pixels within a line and the lines in a frame, respectively. The pixel address and the synchronization signals are generated based on these two counters. The 24-bit pixel color is latched and outputted through r, g, and b signals.

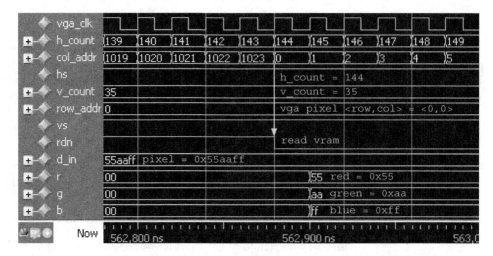

Figure 14.29   Waveform 1 of the VGA interface controller (first pixel)

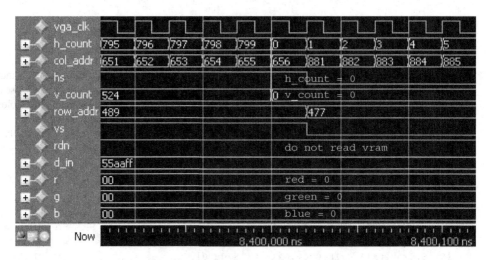

Figure 14.30   Waveform 2 of the VGA interface controller (one frame)

Figure 14.29 shows the simulation waveform of the code when the controller reads the first pixel. During v_count = 35 and h_count = 144, both the row address (row_addr) and column address (col_addr) are 0, and the read signal (rdn) is active. The pixel color is sent out during the next clock cycle.

Figure 14.30 shows the simulation waveform of the code when the controller finished refreshing one frame and starts refreshing the next frame. We can see that the two counters change their values to 0. Both the synchronization signals output 0. The controller does not read pixel (rdn is inactive).

In the simulation, we let the pixel color to be 0x55aaff. This color can be read from a VRAM (video random access memory). We call this a graphics mode. The graphics mode needs a large amount of VRAM for storing the screen image. If we just want to display ASCII characters on a VGA, we can adopt a text mode. This mode requires only a small amount of memory for storing the character's ASCII

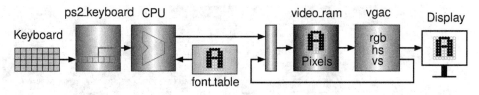

(a) Graphics mode (store pixels to RAM)

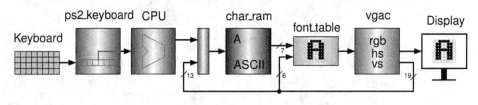

(b) Text mode (store ASCII codes to RAM)

Figure 14.31   Graphics mode and text mode of VGA

codes. Meanwhile, it also needs a font table for storing the shapes of the ASCII characters. The structures of the graphics mode and text mode are shown in Figure 14.31.

The graphics mode requires a video memory of $640 \times 480 \times 24 = 7,372,800$ bits, or 900 KB, where 24 is the bit number of a pixel color. The design of the text mode is a little bit more complex than the that of the graphics mode because the font table is accessed on every VGA controller's read cycle. Suppose a character is displayed with an 8×8 dot matrix. If we want to display each character with a unique color in the text mode, the character RAM (random access memory) must have $640/8 \times 480/8 \times (24 + 7) =$ $148,800$ bits, or about 18.2 KB, where 7 is the bit number of a character ASCII code. The font table has $8 \times 8 \times 128 = 8,192$ bits, that is, totally 156,992 bits, or about 19.2 KB. If the characters are displayed with monocolor, then it needs only $80 \times 60 \times 7 + 8,192 = 41,792$ bits, or about 5.1 KB.

The following C code (font8.c) defines a font table (dot matrices) for 96 ASCII characters. There are 128 characters, but the first 32 characters are not printable.

```
const unsigned char Font[][8] = {                   /*  char  ascii  */
    {0x00,0x00,0x00,0x00,0x00,0x00,0x00,0x00},      /* <SPACE>  20   */
    {0x18,0x18,0x18,0x18,0x00,0x18,0x18,0x00},      /*    !     21   */
    {0x6c,0x6c,0x48,0x00,0x00,0x00,0x00,0x00},      /*    "     22   */
    {0x6c,0x6c,0xfe,0x6c,0xfe,0x6c,0x6c,0x00},      /*    #     23   */
    {0x18,0x7e,0xd8,0x7e,0x1b,0x7e,0x18,0x00},      /*    $     24   */
    {0x62,0x66,0x0c,0x18,0x30,0x66,0x46,0x00},      /*    %     25   */
    {0x38,0x6c,0x68,0x76,0xdc,0xcc,0x76,0x00},      /*    &     26   */
    {0x18,0x18,0x30,0x00,0x00,0x00,0x00,0x00},      /*    '     27   */
    {0x0c,0x18,0x30,0x30,0x30,0x18,0x0c,0x00},      /*    (     28   */
    {0x30,0x18,0x0c,0x0c,0x0c,0x18,0x30,0x00},      /*    )     29   */
    {0x00,0x6c,0x38,0xfe,0x38,0x6c,0x00,0x00},      /*    *     2a   */
    {0x00,0x18,0x18,0x7e,0x18,0x18,0x00,0x00},      /*    +     2b   */
    {0x00,0x00,0x00,0x00,0x00,0x18,0x18,0x10},      /*    ,     2c   */
    {0x00,0x00,0x00,0x7e,0x00,0x00,0x00,0x00},      /*    -     2d   */
    {0x00,0x00,0x00,0x00,0x00,0x18,0x18,0x00},      /*    .     2e   */
```

```
{0x02,0x06,0x0c,0x18,0x30,0x60,0x40,0x00},      /*    /      2f    */
{0x3c,0x66,0x6e,0x76,0x66,0x66,0x3c,0x00},      /*    0      30    */
{0x18,0x18,0x38,0x18,0x18,0x18,0x3c,0x00},      /*    1      31    */
{0x7c,0x06,0x06,0x3c,0x60,0x60,0x7c,0x00},      /*    2      32    */
{0x7c,0x06,0x06,0x3c,0x06,0x06,0x7c,0x00},      /*    3      33    */
{0x66,0x66,0x66,0x7e,0x06,0x06,0x06,0x00},      /*    4      34    */
{0x7e,0x60,0x60,0x7c,0x06,0x06,0x7c,0x00},      /*    5      35    */
{0x3c,0x60,0x60,0x7c,0x66,0x66,0x3c,0x00},      /*    6      36    */
{0x7e,0x06,0x0c,0x18,0x18,0x18,0x18,0x00},      /*    7      37    */
{0x3c,0x66,0x66,0x3c,0x66,0x66,0x3c,0x00},      /*    8      38    */
{0x3c,0x66,0x66,0x3e,0x06,0x06,0x3c,0x00},      /*    9      39    */
{0x00,0x18,0x18,0x00,0x18,0x18,0x00,0x00},      /*    :      3a    */
{0x00,0x00,0x18,0x18,0x00,0x18,0x18,0x10},      /*    ;      3b    */
{0x0c,0x18,0x30,0x60,0x30,0x18,0x0c,0x00},      /*    <      3c    */
{0x00,0x00,0x7e,0x00,0x7e,0x00,0x00,0x00},      /*    =      3d    */
{0x30,0x18,0x0c,0x06,0x0c,0x18,0x30,0x00},      /*    >      3e    */
{0x3c,0x66,0x06,0x1c,0x18,0x00,0x18,0x00},      /*    ?      3f    */
{0x3c,0x66,0x6e,0x6a,0x6e,0x60,0x3e,0x00},      /*    @      40    */
{0x3c,0x66,0x66,0x7e,0x66,0x66,0x66,0x00},      /*    A      41    */
{0x7c,0x66,0x66,0x7c,0x66,0x66,0x7c,0x00},      /*    B      42    */
{0x3c,0x66,0x60,0x60,0x60,0x66,0x3c,0x00},      /*    C      43    */
{0x7c,0x66,0x66,0x66,0x66,0x66,0x7c,0x00},      /*    D      44    */
{0x7e,0x60,0x60,0x7c,0x60,0x60,0x7e,0x00},      /*    E      45    */
{0x7e,0x60,0x60,0x7c,0x60,0x60,0x60,0x00},      /*    F      46    */
{0x3c,0x66,0x60,0x6e,0x66,0x66,0x3c,0x00},      /*    G      47    */
{0x66,0x66,0x66,0x7e,0x66,0x66,0x66,0x00},      /*    H      48    */
{0x3c,0x18,0x18,0x18,0x18,0x18,0x3c,0x00},      /*    I      49    */
{0x3e,0x0c,0x0c,0x0c,0x0c,0x6c,0x38,0x00},      /*    J      4a    */
{0x66,0x6c,0x78,0x70,0x78,0x6c,0x66,0x00},      /*    K      4b    */
{0x60,0x60,0x60,0x60,0x60,0x60,0x7e,0x00},      /*    L      4c    */
{0xc6,0xee,0xfe,0xd6,0xc6,0xc6,0xc6,0x00},      /*    M      4d    */
{0x66,0x66,0x76,0x7e,0x6e,0x66,0x66,0x00},      /*    N      4e    */
{0x3c,0x66,0x66,0x66,0x66,0x66,0x3c,0x00},      /*    O      4f    */
{0x7c,0x66,0x66,0x7c,0x60,0x60,0x60,0x00},      /*    P      50    */
{0x3c,0x66,0x66,0x66,0x6e,0x66,0x3e,0x00},      /*    Q      51    */
{0x7c,0x66,0x66,0x7c,0x66,0x66,0x66,0x00},      /*    R      52    */
{0x3e,0x60,0x60,0x3c,0x06,0x06,0x7c,0x00},      /*    S      53    */
{0x7e,0x18,0x18,0x18,0x18,0x18,0x18,0x00},      /*    T      54    */
{0x66,0x66,0x66,0x66,0x66,0x66,0x3c,0x00},      /*    U      55    */
{0x66,0x66,0x66,0x66,0x3c,0x3c,0x18,0x00},      /*    V      56    */
{0xc6,0xc6,0xd6,0xd6,0xfe,0xee,0x44,0x00},      /*    W      57    */
{0x66,0x66,0x3c,0x18,0x3c,0x66,0x66,0x00},      /*    X      58    */
{0x66,0x66,0x66,0x3c,0x18,0x18,0x18,0x00},      /*    Y      59    */
{0x7e,0x06,0x0c,0x18,0x30,0x60,0x7e,0x00},      /*    Z      5a    */
{0x3c,0x30,0x30,0x30,0x30,0x30,0x3c,0x00},      /*    [      5b    */
{0x40,0x60,0x30,0x18,0x0c,0x06,0x02,0x00},      /*    \      5c    */
{0x3c,0x0c,0x0c,0x0c,0x0c,0x0c,0x3c,0x00},      /*    ]      5d    */
```

```
{0x10,0x38,0x6c,0x00,0x00,0x00,0x00,0x00},          /*      ^       5e    */
{0x00,0x00,0x00,0x00,0x00,0x00,0x00,0xff},          /*      _       5f    */
{0x18,0x18,0x0c,0x00,0x00,0x00,0x00,0x00},          /*      `       60    */
{0x00,0x00,0x3c,0x06,0x3e,0x66,0x3a,0x00},          /*      a       61    */
{0x60,0x60,0x7c,0x66,0x66,0x66,0x7c,0x00},          /*      b       62    */
{0x00,0x00,0x3c,0x66,0x60,0x66,0x3c,0x00},          /*      c       63    */
{0x06,0x06,0x3e,0x66,0x66,0x66,0x3e,0x00},          /*      d       64    */
{0x00,0x00,0x3c,0x66,0x7c,0x60,0x3c,0x00},          /*      e       65    */
{0x0e,0x18,0x18,0x3e,0x18,0x18,0x18,0x00},          /*      f       66    */
{0x00,0x00,0x3e,0x66,0x66,0x3e,0x06,0x3c},          /*      g       67    */
{0x60,0x60,0x7c,0x66,0x66,0x66,0x66,0x00},          /*      h       68    */
{0x18,0x00,0x18,0x18,0x18,0x18,0x18,0x00},          /*      i       69    */
{0x18,0x00,0x18,0x18,0x18,0x18,0x18,0x70},          /*      j       6a    */
{0x60,0x60,0x66,0x6c,0x78,0x6c,0x66,0x00},          /*      k       6b    */
{0x30,0x30,0x30,0x30,0x30,0x30,0x1c,0x00},          /*      l       6c    */
{0x00,0x00,0xcc,0xfe,0xd6,0xc6,0xc6,0x00},          /*      m       6d    */
{0x00,0x00,0x7c,0x66,0x66,0x66,0x66,0x00},          /*      n       6e    */
{0x00,0x00,0x3c,0x66,0x66,0x66,0x3c,0x00},          /*      o       6f    */
{0x00,0x00,0x7c,0x66,0x66,0x7c,0x60,0x60},          /*      p       70    */
{0x00,0x00,0x3e,0x66,0x66,0x3e,0x06,0x06},          /*      q       71    */
{0x00,0x00,0x36,0x38,0x30,0x30,0x30,0x00},          /*      r       72    */
{0x00,0x00,0x3e,0x60,0x3c,0x06,0x7c,0x00},          /*      s       73    */
{0x18,0x18,0x3c,0x18,0x18,0x18,0x0c,0x00},          /*      t       74    */
{0x00,0x00,0x66,0x66,0x66,0x66,0x3c,0x00},          /*      u       75    */
{0x00,0x00,0x66,0x66,0x66,0x3c,0x18,0x00},          /*      v       76    */
{0x00,0x00,0xc6,0xd6,0xd6,0x7c,0x28,0x00},          /*      w       77    */
{0x00,0x00,0x66,0x3c,0x18,0x3c,0x66,0x00},          /*      x       78    */
{0x00,0x00,0x66,0x66,0x66,0x3e,0x06,0x7c},          /*      y       79    */
{0x00,0x00,0x7e,0x0c,0x18,0x30,0x7e,0x00},          /*      z       7a    */
{0x1c,0x30,0x30,0x60,0x30,0x30,0x1c,0x00},          /*      {       7b    */
{0x18,0x18,0x18,0x18,0x18,0x18,0x18,0x00},          /*      |       7c    */
{0x38,0x0c,0x0c,0x06,0x0c,0x0c,0x38,0x00},          /*      }       7d    */
{0x00,0x32,0x4c,0x00,0x00,0x00,0x00,0x00},          /*      ~       7e    */
{0xff,0xff,0xff,0xff,0xff,0xff,0xff,0xff}           /*   <DEL>      7f    */
};
```

The following C code (font_show.c) displays the font shapes. If a dot in dot matrices is a 1, the program prints out a character "O", and a blank space otherwise.

```
#include <stdio.h>                              // display ascii font shape
extern unsigned char Font[][8];                 // 128 - 32 = 96 characters
main() {
    int chars_per_line = 16;                    // fit to book text width
    int char_no;
    unsigned char char_row_bitmap;
    int row, col;
    int i;
```

```
for (char_no = 0; char_no < 96; char_no += chars_per_line) {
    for (row = 0; row < 8; row++) {
        for (i = 0; i < chars_per_line; i++) {
            if ((char_no + i) < 96) {
                char_row_bitmap = Font[char_no + i][row];
                for (col = 7; col >= 0; col--) {
                    if (((char_row_bitmap >> col) & 1) == 1) {
                        printf ("O");         // a dot
                    } else {
                        printf (" ");         // blank
                    }
                }
            }
        }
        printf ("\n");                        // next row
    }
}
```

The execution result of the program is shown in Figure 14.32. In order for it to look better, we reduced the vertical space between lines.

Instead of reading a pixel directly from the VRAM in the graphics mode, in text mode, the VGA interface controller reads a dot from the font table, based on the ASCII code stored in the character RAM as well as the row and column pixel addresses. Figure 14.33 shows the method for getting a dot.

The VGA interface controller outputs a 9-bit pixel row address (row_addr) and a 10-bit pixel column address (col_addr). The three LSBs of row_addr is the font row address (font_row) of the font table, and the rest of the bits form a character row address (char_row) of the character RAM. Similarly, the three LSBs of col_addr is the font column address (font_col) of the font table, and the rest of the bits form a character column address (char_col) of the character RAM. The character base address of the font table is ascii, which is read from the character RAM with an address formed with char_row and char_col.

Figure 14.32 ASCII font example

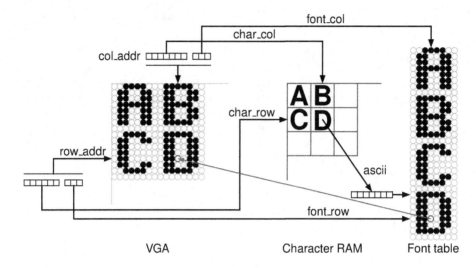

Figure 14.33   Character address and font address

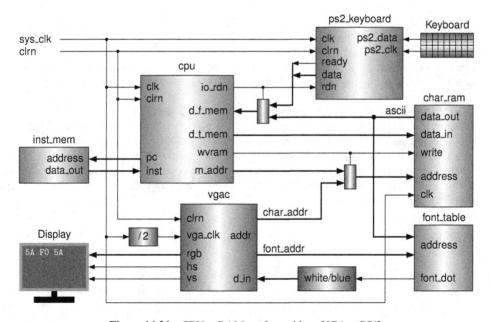

Figure 14.34   CPU + RAMs + font table + VGA + PS/2

14.5.2   Displaying Scan Codes of Keyboard Keys on VGA

This section gives a design example of a computer system that consists of a CPU, an instruction memory for storing program, a character memory for storing character ASCII codes, a font table, a PS/2 keyboard controller, and a VGA interface controller. A PS/2 keyboard and a VGA display are connected to their controllers, respectively. Figure 14.34 shows the organization of the system. This system can be implemented on the Altera DE2-115 FPGA (field programmable gate array) board.

The program reads data bytes (scan code) from the PS/2 controller and converts a data byte into two hexadecimal characters and appends a space. The hexadecimal character is represented with its ASCII code so that it is printable. The program stores these three bytes into the character RAM. This procedure is repeated until all the bytes in a scan code are dealt with. The VGA controller can access the character RAM and then get a dot from the font table for displaying.

The memory space ranging from 0xc0000000 to 0xdfffffff is assigned to VRAM, but we use this space as the character RAM. The memory space range from 0xa0000000 to 0xbfffffff is assigned to I/O. In the system, there is only one I/O read port for reading the keyboard; this port uses all the I/O space.

We implement the system with the I/O polling mechanism. The CPU reads keyboard data in which there is a bit indicating whether a data byte is ready or not. The assembly program is listed below.

```
.text                               # code segment
main:
      lui    $3,    0xc000          # vram space: c0000000 - dfffffff
      lui    $4,    0xa000          # i/o  space: a0000000 - bfffffff
read_kbd:
      lw     $5,    0($4)           # read kbd: {0,ready,byte}
      andi   $6,    $5,    0x100    # check if ready
      beq    $6,    $0,    read_kbd # if no key pressed, wait
      andi   $6,    $5,    0xff     # ready, get data
      srl    $5,    $6,    4        # first digit
      addi   $7,    $5,    -10
      srl    $7,    $7,    31
      beq    $7,    $0,    abcdef1
      addi   $5,    $5,    0x30     # to ascii [0-9]
      j      print1
abcdef1:
      addi   $5,    $5,    0x37     # to ascii [a-f]
print1:
      jal    display               # display char
      andi   $5,    $6,    0xf      # second digit
      addi   $7,    $5,    -10
      srl    $7,    $7,    31
      beq    $7,    $0,    abcdef2
      addi   $5,    $5,    0x30     # to ascii [0-9]
      j      print2
abcdef2:
      addi   $5,    $5,    0x37     # to ascii [a-f]
print2:
      jal    display               # display char
      addi   $5,    $0,    0x20     # [Space]
print3:
      jal    display               # display char
      j      read_kbd              # check next
display:
      sw     $5,    0($3)           # to display
      addi   $3,    $3,    4
      jr     $ra
.end
```

This assembly program will be translated into binary format and stored in the instruction memory. The following Verilog HDL code implements the system shown in Figure 14.34. It invokes the VGA controller module, font table module, PS/2 keyboard module, CPU module, and instruction memory module. The character RAM is implemented with the general Verilog HDL `reg` statement inside this top module.

```verilog
module display_scan_codes (sys_clk,clrn,ps2_clk,ps2_data,r,g,b,hs,vs,
                           vga_clk,blankn,syncn);  // display key scan codes
    input         sys_clk, clrn;       // sys_clk: 50MHz
    input         ps2_clk, ps2_data;  // kbd clk and data
    output [7:0]  r, g, b;             // vga red,green,blue colors
    output        hs, vs;              // vga h and v synchronization
    output        vga_clk;             // for ADV7123 VGA DAC
    output        blankn;              // for ADV7123 VGA DAC
    output        syncn;               // for ADV7123 VGA DAC
    wire          font_dot;            // font dot
    wire   [31:0] inst,pc,d_t_mem,cpu_mem_a,d_f_mem;
    wire          write,read,io_rdn,io_wrn,wvram,rvram,ready,overflow;
    wire   [7:0]  key_data;            // kbd code byte
    // vga_clk
    reg           vga_clk = 1;
    always @(posedge sys_clk) begin
        vga_clk <= ~vga_clk;           // vga_clk: 25MHz
    end
    // vgac
    assign blankn = 1;
    assign syncn  = 0;
    wire   [8:0]  row_addr;            // pixel ram row addr, 480 (512) lines
    wire   [9:0]  col_addr;            // pixel ram col addr, 640 (1024) pixels
    wire          vga_rdn;             // in vgac, rd_a = {row[8:0],col[9:0]}
    wire   [23:0] vga_pixel = font_dot? 24'hffffff : 24'h0000ff; //white/blue
    vgac vga_24 (vga_clk,clrn,vga_pixel,row_addr,col_addr,vga_rdn,
                 r,g,b,hs,vs);
    wire   [5:0]  char_row = row_addr[8:3];               // char row
    wire   [2:0]  font_row = row_addr[2:0];               // font row
    wire   [6:0]  char_col = col_addr[9:3];               // char col
    wire   [2:0]  font_col = col_addr[2:0];               // font col
    // char ram, 640/8 = 80 = 64 + 16; 480/8 = 60;
    wire   [12:0] vga_cram_addr = (char_row<<6) + (char_row<<4) + char_col;
    wire   [12:0] char_ram_addr = wvram? cpu_mem_a[14:2] : vga_cram_addr;
    reg    [6:0]  char_ram [0:4799];                      // 80 * 60 = 4800
    wire   [6:0]  ascii = char_ram[char_ram_addr];
    always @(posedge sys_clk) begin
        if (wvram) begin
            char_ram[char_ram_addr] <= d_t_mem[6:0];
        end
    end
    // font_table 128 x 8 x 8 x 1
```

```
   wire [12:0] ft_a = {ascii,font_row,font_col};           // ascii,row,col
   font_table ft (ft_a,font_dot);
   // ps2_keyboard
   ps2_keyboard kbd (sys_clk,clrn,ps2_clk,ps2_data,io_rdn,key_data,ready,
                     overflow);
   // cpu
   single_cycle_cpu_io cpu (sys_clk,clrn,pc,inst,cpu_mem_a,d_f_mem,
                            d_t_mem,write,io_rdn,io_wrn,rvram,wvram);
   assign d_f_mem = io_rdn? {25'h0,ascii} : {23'h0,ready,key_data};
   // instruction memory
   scinstmem_make_code_break_code imem (pc,inst);
endmodule
```

We have already given the Verilog HDL codes of the VGA controller module and the PS/2 keyboard module. Like the character RAM, the font table is also implemented with the general Verilog HDL statement. Because the font table is a ROM, it is not necessary to use reg; instead, the wire statement is used. The font table module is too long to show here, and actually it is generated with a C program. Thus we just list the C program here.

```
#include <stdio.h>  // gen_font_table_v.c, generate font table (verilog HDL)
extern unsigned char Font[][8];
main() {
    unsigned char char_row_bitmap;
    int addr = 0;
    int char_no, i, j;
    printf ("module font_table (a,d);\n");
    printf ("    input   [12:0] a; // 8*8*128=2^3*2^3*2^7\n");
    printf ("    output         d; // font dot\n");
    printf ("    wire           rom [0:8191];\n");
    printf ("    assign         d = rom[a];\n\n");
    for (char_no = 0; char_no < 128; char_no++) {
        for (i = 0; i < 8; i++) {                           // 8 rows per char
            if (char_no >= 0x20) {                          // <space>
                char_row_bitmap = Font[char_no - 0x20][i];
            } else {
                char_row_bitmap = 0;
            }
            for (j = 7; j >= 0; j--) {                      // 8 pixels per row
                printf ("    assign rom[13'h%04x] = ", addr++);
                if (((char_row_bitmap >> j) & 1) == 1) {
                    printf ("1;\n");                         // a dot
                } else {
                    printf ("0;\n");                         // blank
                }
            }
        }
    }
    printf ("endmodule\n");
}
```

The following Verilog HDL code implements the CPU. It is a simple single-cycle CPU written in behavioral style. The signal wvram is the character RAM write enable, and io_rdn is the keyboard read signal, active-low.

```verilog
module single_cycle_cpu_io (clk,clrn,pc,inst,m_addr,d_f_mem,d_t_mem,write,
                            io_rdn,io_wrn,rvram,wvram);  // cpu kbd i/o
    input   clk, clrn;                          // clock and reset
    input   [31:0] inst;                        // instruction
    input   [31:0] d_f_mem;                     // load data
    output  [31:0] pc;                          // program counter
    output  [31:0] m_addr;                      // mem or i/o addr
    output  [31:0] d_t_mem;                     // store data
    output         write;                       // data memory write
    output         wvram;                       // vram write
    output         rvram;                       // vram read
    output         io_wrn;                      // i/o write
    output         io_rdn;                      // i/o read
    // control signals
    reg            wreg;                         // write regfile
    reg            wmem,rmem;                    // write/read memory
    reg     [31:0] alu_out;                      // alu output
    reg     [4:0] dest_rn;                       // dest reg number
    reg     [31:0] next_pc;                      // next pc
    wire    [31:0] pc_plus_4 = pc + 4;           // pc + 4
    // instruction format
    wire    [05:00] opcode = inst[31:26];
    wire    [04:00] rs     = inst[25:21];
    wire    [04:00] rt     = inst[20:16];
    wire    [04:00] rd     = inst[15:11];
    wire    [04:00] sa     = inst[10:06];
    wire    [05:00] func   = inst[05:00];
    wire    [15:00] imm    = inst[15:00];
    wire    [25:00] addr   = inst[25:00];
    wire           sign    = inst[15];
    wire    [31:00] offset = {{14{sign}},imm,2'b00};
    wire    [31:00] j_addr = {pc_plus_4[31:28],addr,2'b00};
    // instruction decode
    wire i_add  = (opcode == 6'h00) & (func == 6'h20);  // add
    wire i_sub  = (opcode == 6'h00) & (func == 6'h22);  // sub
    wire i_and  = (opcode == 6'h00) & (func == 6'h24);  // and
    wire i_or   = (opcode == 6'h00) & (func == 6'h25);  // or
    wire i_xor  = (opcode == 6'h00) & (func == 6'h26);  // xor
    wire i_sll  = (opcode == 6'h00) & (func == 6'h00);  // sll
    wire i_srl  = (opcode == 6'h00) & (func == 6'h02);  // srl
    wire i_sra  = (opcode == 6'h00) & (func == 6'h03);  // sra
    wire i_jr   = (opcode == 6'h00) & (func == 6'h08);  // jr
    wire i_addi = (opcode == 6'h08);                    // addi
```

```verilog
wire i_andi = (opcode == 6'h0c);                    // andi
wire i_ori  = (opcode == 6'h0d);                    // ori
wire i_xori = (opcode == 6'h0e);                    // xori
wire i_lw   = (opcode == 6'h23);                    // lw
wire i_sw   = (opcode == 6'h2b);                    // sw
wire i_beq  = (opcode == 6'h04);                    // beq
wire i_bne  = (opcode == 6'h05);                    // bne
wire i_lui  = (opcode == 6'h0f);                    // lui
wire i_j    = (opcode == 6'h02);                    // j
wire i_jal  = (opcode == 6'h03);                    // jal
// pc
reg [31:0] pc;
always @ (posedge clk or negedge clrn) begin
    if (!clrn) pc <= 0;
    else       pc <= next_pc;
end
// data written to register file
wire   [31:0] data_2_rf = i_lw ? d_f_mem : alu_out;
// register file
reg    [31:0] regfile [1:31];                       // $1 - $31
wire   [31:0] a = (rs==0) ? 0 : regfile[rs];        // read port
wire   [31:0] b = (rt==0) ? 0 : regfile[rt];        // read port
always @ (posedge clk) begin
    if (wreg && (dest_rn != 0)) begin
        regfile[dest_rn] <= data_2_rf;              // write port
    end
end
wire   io_space = alu_out[31] &                     // i/o space:
                  ~alu_out[30] &                    // a0000000-bfffffff
                  alu_out[29];
wire   vr_space = alu_out[31] &                     // vram space:
                  alu_out[30] &                     // c0000000-dfffffff
                  ~alu_out[29];
// output signals
assign write    =   wmem & ~io_space & ~vr_space;   // data memory write
assign d_t_mem  =   b;                              // data to store
assign m_addr   =   alu_out;                        // memory address
assign io_rdn   = ~(rmem & io_space);               // i/o read
assign io_wrn   = ~(wmem & io_space);               // i/o write
assign wvram    =   wmem & vr_space;                // video ram write
assign rvram    =   rmem & vr_space;                // video ram read
// control signals, will be combinational circuit
always @(*) begin
    alu_out = 0;                                    // alu output
    dest_rn = rd;                                   // dest reg number
    wreg    = 0;                                    // write regfile
    wmem    = 0;                                    // write memory (sw)
```

```
    rmem    = 0;                                         // read  memory (lw)
    next_pc = pc_plus_4;
    case (1'b1)
        i_add: begin                                     // add
            alu_out = a + b;
            wreg    = 1; end
        i_sub: begin                                     // sub
            alu_out = a - b;
            wreg    = 1; end
        i_and: begin                                     // and
            alu_out = a & b;
            wreg    = 1; end
        i_or: begin                                      // or
            alu_out = a | b;
            wreg    = 1; end
        i_xor: begin                                     // xor
            alu_out = a ^ b;
            wreg    = 1; end
        i_sll: begin                                     // sll
            alu_out = b << sa;
            wreg    = 1; end
        i_srl: begin                                     // srl
            alu_out = b >> sa;
            wreg    = 1; end
        i_sra: begin                                     // sra
            alu_out = $signed(b) >>> sa;
            wreg    = 1; end
        i_jr: begin                                      // jr
            next_pc = a; end
        i_addi: begin                                    // addi
            alu_out = a + {{16{sign}},imm};
            dest_rn = rt;
            wreg    = 1; end
        i_andi: begin                                    // andi
            alu_out = a & {16'h0,imm};
            dest_rn = rt;
            wreg    = 1; end
        i_ori: begin                                     // ori
            alu_out = a | {16'h0,imm};
            dest_rn = rt;
            wreg    = 1; end
        i_xori: begin                                    // xori
            alu_out = a ^ {16'h0,imm};
            dest_rn = rt;
            wreg    = 1; end
        i_lw: begin                                      // lw
            alu_out = a + {{16{sign}},imm};
```

```
                    dest_rn = rt;
                    rmem    = 1;
                    wreg    = 1; end
                i_sw: begin                                    // sw
                    alu_out = a + {{16{sign}},imm};
                    wmem    = 1; end
                i_beq: begin                                   // beq
                    if (a == b)
                       next_pc = pc_plus_4 + offset; end
                i_bne: begin                                   // bne
                    if (a != b)
                       next_pc = pc_plus_4 + offset; end
                i_lui: begin                                   // lui
                    alu_out = {imm,16'h0};
                    dest_rn = rt;
                    wreg    = 1; end
                i_j: begin                                     // j
                    next_pc = j_addr; end
                i_jal: begin                                   // jal
                    alu_out = pc_plus_4;
                    wreg    = 1;
                    dest_rn = 5'd31;
                    next_pc = j_addr; end
                default: ;
            endcase
        end
endmodule
```

The following Verilog HDL code implements the instruction memory. Like the font table module, it is also a ROM. The memory is initialized with the binary code of the assembly program given at the beginning of this section.

```
module scinstmem_make_code_break_code (a,inst);     // instruction memory, rom
    input   [31:0] a;                               // address
    output  [31:0] inst;                            // instruction
    wire    [31:0] rom [0:63];                       // rom cells: 64 words * 32 bits
    assign inst = rom[a[7:2]];                       // use word address to read rom
    // rom[word_addr] = instruction
    assign rom[6'h00] = 32'b00111100000000011110000000000000;
    assign rom[6'h01] = 32'b00111100000000100101000000000000;
    assign rom[6'h02] = 32'b10001100100001010000000000000000;
    assign rom[6'h03] = 32'b00110000010100110000000000100000;
    assign rom[6'h04] = 32'b00010000011000001111111111111101;
    assign rom[6'h05] = 32'b00110000010100110000000000111111;
    assign rom[6'h06] = 32'b00000000000001100010010000000010;
    assign rom[6'h07] = 32'b00010000010100111111111111110110;
    assign rom[6'h08] = 32'b00000000000001110011111111000010;
    assign rom[6'h09] = 32'b00010000011100000000000000000010;
```

```
assign rom[6'h0a] = 32'b00100000010100101000000000110000;
assign rom[6'h0b] = 32'b00001000000000000000000000001101;
assign rom[6'h0c] = 32'b00100000010100101000000000110111;
assign rom[6'h0d] = 32'b00001100000000000000000000011001;
assign rom[6'h0e] = 32'b00110000110001010000000000001111;
assign rom[6'h0f] = 32'b00100000010100111111111111110110;
assign rom[6'h10] = 32'b00000000000001110011111111000010;
assign rom[6'h11] = 32'b00010000111000000000000000000010;
assign rom[6'h12] = 32'b00100000010100101000000000110000;
assign rom[6'h13] = 32'b00001000000000000000000000010101;
assign rom[6'h14] = 32'b00100000010100101000000000110111;
assign rom[6'h15] = 32'b00001100000000000000000000011001;
assign rom[6'h16] = 32'b00100000000001010000000000100000;
assign rom[6'h17] = 32'b00001100000000000000000000011001;
assign rom[6'h18] = 32'b00001000000000000000000000000010;
assign rom[6'h19] = 32'b10101100011001010000000000000000;
assign rom[6'h1a] = 32'b00100000011000110000000000000100;
assign rom[6'h1b] = 32'b00000011111000000000000000001000;
assign rom[6'h1c] = 32'b00000000000000000000000000000000;
assign rom[6'h1d] = 32'b00000000000000000000000000000000;
assign rom[6'h1e] = 32'b00000000000000000000000000000000;
assign rom[6'h1f] = 32'b00000000000000000000000000000000;
endmodule
```

The system was implemented on the Altera DE2-115 FPGA board. Except the signals for the VGA DAC chip, we did not use any special libraries or even external memory chips; therefore, this system is easy to be ported to other FPGA boards. Below we show some key scan codes that come from the VGA output of the system. The " ... " means that a key is held down for a while and the make code is repeated. The make code of the [Pause] key is not repeated.

```
[Enter]...        5a ... 5a f0 5a
[Space]...        29 ... 29 f0 29
[Esc]...          76 ... 76 f0 76
[L Shift]...      12 ... 12 f0 12
[R Shift]...      59 ... 59 f0 59
[L Ctrl]...       14 ... 14 f0 14
[R Ctrl]...       e0 14 ... e0 14 e0 f0 14
[L Arrow]...      e0 6b ... e0 6b e0 f0 6b
[L Alt]...        11 ... 11 f0 11
[R Alt]...        e0 11 ... e0 11 e0 f0 11
[R Arrow]...      e0 74 ... e0 74 e0 f0 74
[U Arrow]...      e0 75 ... e0 75 e0 f0 75
[D Arrow]...      e0 72 ... e0 72 e0 f0 72
[L Windows]...    e0 1f ... e0 1f e0 f0 1f
[R Windows]...    e0 27 ... e0 27 e0 f0 27
[Delete]...       e0 71 ... e0 71 e0 f0 71
[Pause]...        e1 14 77 e1 f0 14 f0 77  // no break code, no repeat
```

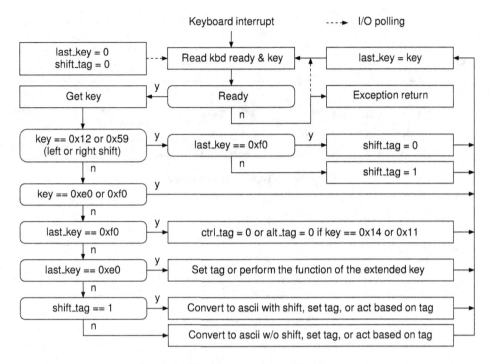

Figure 14.35   Flowchart for dealing with keys

The key scan codes can be dealt with by a program whose flow chart is shown in Figure 14.35. It deals with both the regular and extended keys.

14.6   Input/Output Buses

In a computer system, there are multiple I/O devices and their corresponding I/O interfaces. The I/O interfaces are usually connected together to a bus that consists of the address bus, data bus, read/write control signals, interrupt request and acknowledge, and signals for bus arbitration. In parallel buses, the address bus and data bus can be separated or combined together, called a multiplexed address data bus, in order to decrease the number of bus signals. Serial buses use only a small number of signals: for example, the I2C bus has only two signals (pins). This section introduces the I2C serial bus and the PCI parallel bus.

14.6.1   I2C Serial Bus and I2C EEPROM Access

The I2C or I^2C is a bidirectional two-wire serial bus that provides the communication link between integrated circuits (ICs). The two wires are serial clock (SCL) and serial data (SDA), and they are pulled up by resistors, as shown in Figure 14.36. The I2C bus provides two roles for ICs: master and slave. A master IC generates the clock and initiates communication with slaves. A slave IC receives the clock and responds when addressed by the master.

A master can be a transmitter that sends data to a slave (the slave is a receiver). A master can be also a receiver that receives data from a slave (the slave is a transmitter). I2C supports multiple masters, but in the following descriptions we assume that there is only one master.

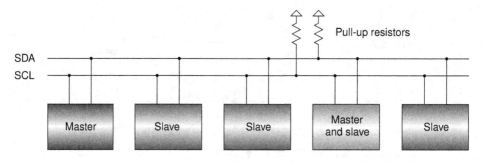

Figure 14.36   I2C serial bus

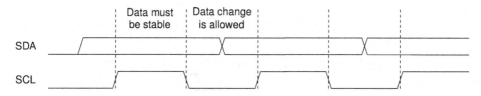

Figure 14.37   I2C bus signals

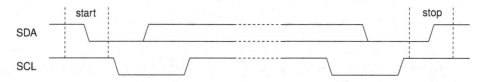

Figure 14.38   Start bit and stop bit of I2C bus

The I2C bus speeds are the 100 kbps (kilobits per second) standard mode, 400 kbps fast mode, 1 Mbps fast mode plus, and 3.4 Mbps high-speed mode. Figure 14.37 shows the relationship between SDA and SCL. The data on the SDA line must be stable during the high period of the clock. The data can change only when the clock signal on the SCL line is low.

A transition (high to low or low to high) on the SDA line is allowed to appear while the clock signal on the SCL line is high. This transition is not explained as the data but as a special condition. A high to low transition on the SDA line while SCL is high is defined as start condition (S), and a low to high transition on the SDA line while SCL is high is defined as stop condition (P), as shown in Figure 14.38. These conditions are always generated by the master.

Once the start condition is issued by the master, we say that the I2C bus is busy. The bus will become free after the stop condition. The unit of data transfer is an 8-bit byte. Referring to Figure 14.39, the data byte is transmitted by the transmitter with the MSB first. Then, the transmitter releases the SDA line (float high), and the receiver sends an acknowledge (A) by bringing the SDA line low during the ninth clock pulse of the SCL. A not-acknowledge ($\overline{A}$), a high on SDA line, may also be sent for the following reasons:

1. The receiver does not exist at all (no low drive on pulled up SDA line);
2. The receiver is busy for performing some real-time functions;
3. The receiver does not understand the meaning of the received data;
4. The master (receiver) tells the slave (transmitter) to stop the transmission.

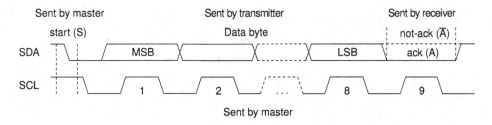

Figure 14.39   ACK of I2C bus

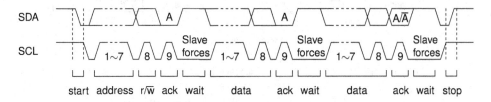

Figure 14.40   Data transfer on I2C bus

All the clock pulses, including the 9th pulse for transmitting or receiving the acknowledgment, are generated by the master.

Data transfers follow the format shown in Figure 14.40. The first byte after the start condition contains the 7-bit slave address and the data direction ($r/\overline{w}$)—a 0 indicates a transmission (write) and a 1 indicates a request for data (read). Some address patterns are used to extend the slave address to 10 bits.

If the slave needs more time to store a received byte or prepare a byte to be transmitted, it can hold the SCL line low after acknowledgment of a byte to force the master into a wait state until the slave is ready for the next byte transfer. This is the reason why the SCL line is bidirectional. The not-acknowledge $\overline{A}$ lets the transmitter to wait logically, while forcing the SCL line low by the slave lets the master wait physically. A data transfer is always terminated by a stop condition generated by the master.

There are multiple slaves on the I2C bus; each slave has its own address. The master must send the address of the slave to which the master wants to talk on I2C bus. Figure 14.41(a) shows the master-transmitter addressing a slave-receiver with a 7-bit address. The direction bit is a 0, indicating a transmission. Two data bytes are transferred by the master to the slave. The acknowledge or not-acknowledge bits are sent by the receiver.

Figure 14.41(b) shows the master reading a slave immediately after the first byte. The data bytes are sent by the slave. The master has the responsibility of sending an acknowledge or not-acknowledge bit to the slave after receiving a data byte.

Figure 14.41(c) shows a combined transfer case in which the master writes a data byte to the slave first and then reads a data byte from the slave. The slave address and the direction bit 0 are sent to the slave. The 8-bit byte, marked with "Reg number/mem address", will be explained by the receiver as a register number or a memory address. For example, if the master wants to read data from an I2C EEPROM, (electrically erasable programmable read-only memory) it must send the memory address first. This address will be saved to a register inside the EEPROM. Then the master changes the direction from write to read. This can be done by sending another start condition, namely the repeated start condition (Sr). Following the Sr, the same slave address and the direction bit 1 are sent to the slave. After acknowledging this byte, the slave sends the data byte to the master. The master terminates the transfer by sending a not-acknowledge followed by a stop condition. The I2C can write or read multiple bytes in a transfer procedure in which only the last byte is acknowledged with a not-acknowledge.

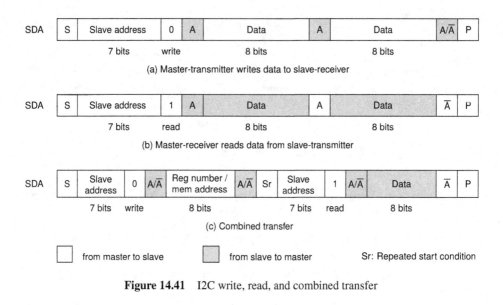

Figure 14.41   I2C write, read, and combined transfer

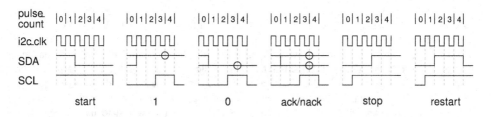

Figure 14.42   A method for generating I2C signals

We have briefly described the I2C serial bus as above. For a complete description, read the I2C specification and user manual. In the following, we give a design example of the I2C master controller and show the case of using it to write and read an I2C EEPROM.

We let the speed of the I2C in our design to be 400 kbps. An I2C clock, namely i2c_clk, is used that is 5 times faster (2 MHz). Figure 14.42 shows the patterns of the I2C signals and conditions. The data value (1, 0, and acknowledge or not-acknowledge) of the SDA in the circulated place will be sampled. The repeated start condition (restart in the figure) consists of a stop condition and immediately a start condition.

Figure 14.43(a) shows the signals of the I2C master controller. In addition to the SDA and SCL, there are signals that are connected to the CPU. d_in[7:0] and d_out[7:0] are 8-bit data lines; addr[1:0] are 2-bit address lines; wrn and rdn are active-low write and read lines, respectively; csn is an active-low chip select line; and clk is a clock line.

The I2C master controller was implemented as a finite state machine. Figure 14.43(b) shows the state transition diagram. Initially, the controller stays at the IDLE state, which does nothing. In the START and STOP states, a start condition and a stop condition are sent out, respectively. In the TX and RX states, the controller sends and receives a data byte, respectively. At the end of the START, TX, or RX state, if the SCL line is forced low by the slave, the controller will enter the WAIT state and stay at the WAIT state until the slave releases the SCL line.

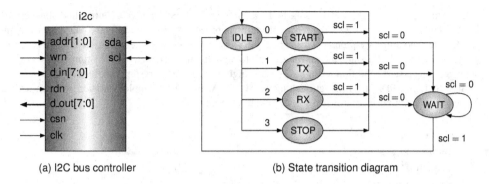

(a) I2C bus controller   (b) State transition diagram

Figure 14.43   I2C controller and its state transition diagram

Table 14.5   Address of I2C bus master controller

addr[1:0]	rdn = 0	wrn = 0
0	Read data	Go to START state
1	Read status	Go to TX state, latch d_in[7:0]
2		Go to RX state, latch d_in[0] (ack/nack)
3		Go to STOP state

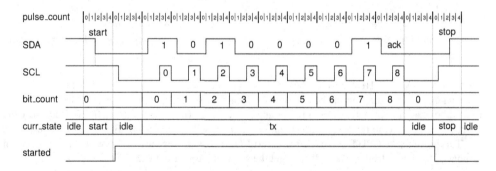

Figure 14.44   I2C bus communication waveform

The transition from the IDLE state to a next state is controlled by program through an I/O write operation, as shown in Table 14.5. A data byte is sent to the controller when entering the TX state, and an acknowledge or not-acknowledge bit is sent to the controller when entering to RX state. The program can also read I2C data or status through an I/O read operation.

Figure 14.44 shows the timing chart when the controller transmits a data byte 0xa1 to a slave. We use curr_state to represent the current state of the controller. The state can be changed only when the value of pulse_count is 4. Initially, the controller stays at the IDLE state. After the START state, the controller enters the IDLE state again, and then enters the TX state in which the 8-bit data byte is sent out and an acknowledge is received. Next, the controller enters the IDLE, STOP, and IDLE states in sequence.

In the first and last IDLE states, both the SCL and SDA are high (SCL and SDA output high impedance). It means that the I2C bus is free. But during the transfer procedure, both the SCL and SDA are low in the IDLE state. Therefore, we prepare a `started` (busy) signal to indicate whether a transfer is in progress. At the end of the START state, the `started` is set to high. It will be cleared in the beginning of the STOP state. Sending a start condition when `started` is high will generate a restart condition.

Both SCL and SDA are bidirectional signals. In the Verilog HDL implementation, we can use the `inout` statement to declare this two signals, as shown below.

```
inout   i2c_scl;                            // i2c clock
inout   i2c_sda;                            // i2c data
assign i2c_scl = scl_h ? 1'bz : 1'b0;       // z or 0
assign i2c_sda = sda_h ? 1'bz : 1'b0;       // z or 0
```

Two control signals, namely `scl_h` and `sda_h`, are used to control the outputs of `i2c_scl` and `i2c_sda`, respectively. When the control signal is high, the corresponding signal outputs a float high, or a 0 otherwise.

Below is a method of writing a test bench to assign a value to an `inout` signal. In the test bench, the `inout` signal itself is declared as the `wire` type. We prepare another signal of the `reg` type and `assign` it to the `inout` signal. Then we can change the value of the `reg` signal for the simulation: see the example below.

```
wire    i2c_scl, i2c_sda;                   // i2c clock and data
reg     scl_in,  sda_in;                    // i2c clock and data inputs
assign i2c_scl = scl_in;                    // i2c clock
assign i2c_sda = sda_in;                    // i2c data
initial begin
    #32500 sda_in = 0;                      // an acknowledgement
    #2500  sda_in = 1'hz;                   // end of the acknowledgement
end
```

Below is the Verilog HDL code of the I2C bus controller. We prepare a `ready` signal that tells CPU when an I/O write instruction can be executed to change the state from IDLE to a next state. The `ready` is a 1 if both the current state (`curr_state`) and the next state (`next_state`) are IDLE. Once the CPU executes an I/O write in the IDLE state, the `next_state` will not be the IDLE, so the `ready` becomes a 0. This prevents the CPU from executing a next I/O write while the current I/O write has not yet been responded by the controller. The CPU can get the value of the `ready` by the I/O status read operation.

```
`define IDLE  0                                 // i2c master controller
`define START 1                                 // define states
`define TX    2
`define RX    3
`define STOP  4
`define WAIT  5
module i2c (clk,clrn,csn,addr,wrn,d_in,rdn,d_out,i2c_sda,i2c_scl);
    input         clk, clrn;                    // system clock, 50MHz
    input         csn;                          // chip select
    input  [1:0]  addr;                         // address
    input         wrn;                          // write, active low
    input  [7:0]  d_in;                         // data to be sent
```

```
input          rdn;                          // read, active low
output  [7:0] d_out;                         // received data or status
inout          i2c_sda;                      // i2c sda, input/output
inout          i2c_scl;                      // i2c scl, input/output
reg     [2:0] prev_state = `IDLE;            // previous state
reg     [2:0] curr_state = `IDLE;            // current state
reg     [2:0] next_state = `IDLE;            // next state
reg     [2:0] pulse_count = 0;               // counting i2c_clk cycles
reg     [3:0] bit_count   = 0;               // counting number of bits
reg     [7:0] in_buf;                        // data to be sent
reg          tx_ack;                         // ack sent be master
reg     [7:0] out_buf;                       // data received
reg          rx_ack;                         // ack received
reg          txd;                            // bit to be sent
reg          scl_h;                          // high impedance
reg          sda_h;                          // high impedance
reg          started;                        // = 0 if stop issued
// i2c clock
reg     [4:0] clk_count = 0;                 // clk / 25 = 2MHz
reg          i2c_clk   = 1;                  // 2MHz
always @(posedge clk or negedge clrn) begin
    if (!clrn) begin
        clk_count <= 0;
    end else begin
        if  (clk_count == 24) clk_count <= 0;
        else                  clk_count <= clk_count + 5'd1;
        if  (clk_count <= 12) i2c_clk   <= 1;
        else                  i2c_clk   <= 0;
    end
end
// pulse_count and bit_count
always @(posedge i2c_clk) begin
    prev_state <= curr_state;
    if (pulse_count == 4)
        pulse_count <= 0;                           // pulse_count
    else if ((curr_state != `WAIT) || i2c_scl)
        pulse_count <= pulse_count + 3'd1;
    if ((bit_count == 8) && (pulse_count == 4))
        bit_count <= 0;                             // bit_count
    else if ((curr_state == `TX) || (curr_state == `RX)) begin
        if (pulse_count == 4) bit_count <= bit_count + 4'd1;
    end else bit_count <= 0;
end
// next state and data inputs
always @(posedge clk or negedge clrn) begin
    if (!clrn) begin
        started <= 0;
```

```verilog
        end else begin
            case (curr_state)
                `IDLE:  begin
                    if ((!csn) && (!wrn)) begin
                        case (addr)
                            2'd0: begin next_state <= `START; end
                            2'd1: begin next_state <= `TX;
                                        in_buf <= d_in;          end
                            2'd2: begin next_state <= `RX;
                                        tx_ack <= d_in[0];       end
                            2'd3: begin next_state <= `STOP;  end
                        endcase
                    end
                    if (prev_state == `START) started <= 1;
                end
                `START: begin
                    if (prev_state == `START) begin
                        case (pulse_count)
                            3'd3: if (i2c_scl == 0) next_state <= `WAIT;
                            3'd4: if (i2c_scl == 0) next_state <= `WAIT;
                                  else              next_state <= `IDLE;
                            default: ;
                        endcase
                    end
                end
                `TX: begin
                    if ((bit_count == 8) && (prev_state == `TX)) begin
                        case (pulse_count)
                            3'd3: if (i2c_scl == 0) next_state <= `WAIT;
                            3'd4: if (i2c_scl == 0) next_state <= `WAIT;
                                  else              next_state <= `IDLE;
                            default: ;
                        endcase
                    end
                end
                `RX: begin
                    if ((bit_count == 8) && (prev_state == `RX)) begin
                        case (pulse_count)
                            3'd3: if (i2c_scl == 0) next_state <= `WAIT;
                            3'd4: if (i2c_scl == 0) next_state <= `WAIT;
                                  else              next_state <= `IDLE;
                            default: ;
                        endcase
                    end
                end
                `STOP: begin
                    if ((pulse_count == 4) && (prev_state == `STOP))
```

```
                        next_state <= 'IDLE;
                    started <= 0;
                end
                'WAIT: begin
                    if (i2c_scl != 0) next_state <= 'IDLE;
                end
            endcase
        end
    end
    // current state
    always @(posedge i2c_clk) begin
        if (pulse_count == 4) curr_state <= next_state;
    end
    // transfer data via i2c bus
    assign i2c_scl = scl_h ? 1'bz : 1'b0;
    assign i2c_sda = sda_h ? 1'bz : 1'b0;
    always @(posedge i2c_clk) begin
        case (curr_state)
            'IDLE: begin
                if (started) begin                  // started
                    sda_h <= 0; scl_h <= 0;
                end else begin                      // stopped
                    sda_h <= 1; scl_h <= 1;
                end end
            'START: begin                           // send start bit
                if (started) begin                  // send re-start bit
                    case (pulse_count)
                        3'd0: begin scl_h <= 1; sda_h <= 0; end
                        3'd1: begin scl_h <= 1; sda_h <= 1; end
                        3'd2: begin scl_h <= 1; sda_h <= 1; end
                        3'd3: begin scl_h <= 1; sda_h <= 1; end
                        3'd4: begin scl_h <= 1; sda_h <= 0; end
                    endcase
                end else begin                      // send start bit
                    case (pulse_count)
                        3'd0: begin scl_h <= 1; sda_h <= 1; end
                        3'd1: begin scl_h <= 1; sda_h <= 0; end
                        3'd2: begin scl_h <= 1; sda_h <= 0; end
                        3'd3: begin scl_h <= 1; sda_h <= 0; end
                        3'd4: begin scl_h <= 1; sda_h <= 0; end
                    endcase
                end end
            'TX: begin
                if (bit_count == 8) begin           // receive ack/nack
                    case (pulse_count)
                        3'd0: begin scl_h <= 0; sda_h <= 1; end
                        3'd1: begin scl_h <= 0; sda_h <= 1; end
```

```
            3'd2: begin scl_h <= 1; sda_h <= 1; end
            3'd3: begin scl_h <= 1; sda_h <= 1;
                        rx_ack <= i2c_sda;          end
            3'd4: begin scl_h <= 0; sda_h <= 1; end
        endcase
    end else begin                      // send data bit
        rx_ack <= 1;                    // no ack
        case (pulse_count)
            3'd0: begin scl_h <= 0;
                        sda_h <= in_buf[7-bit_count]; end
            3'd1: begin scl_h <= 0;
                        sda_h <= in_buf[7-bit_count]; end
            3'd2: begin scl_h <= 1;
                        sda_h <= in_buf[7-bit_count]; end
            3'd3: begin scl_h <= 1;
                        sda_h <= in_buf[7-bit_count]; end
            3'd4: begin scl_h <= 0;
                        sda_h <= in_buf[7-bit_count]; end
        endcase
    end end
`RX: begin
    if (bit_count == 8) begin           // send ack/nack
        case (pulse_count)
            3'd0: begin scl_h <= 0; sda_h <= tx_ack; end
            3'd1: begin scl_h <= 0; sda_h <= tx_ack; end
            3'd2: begin scl_h <= 1; sda_h <= tx_ack; end
            3'd3: begin scl_h <= 1; sda_h <= tx_ack; end
            3'd4: begin scl_h <= 0; sda_h <= tx_ack; end
        endcase
    end else begin                          // receive data bit
        case (pulse_count)
            3'd0: begin scl_h <= 0; sda_h <= 1;      end
            3'd1: begin scl_h <= 0; sda_h <= 1;      end
            3'd2: begin scl_h <= 1; sda_h <= 1;      end
            3'd3: begin scl_h <= 1; sda_h <= 1;
                        out_buf[7-bit_count] <= i2c_sda; end
            3'd4: begin scl_h <= 0; sda_h <= 1;      end
        endcase
    end end
`STOP: begin                            // send stop bit
    case (pulse_count)
        3'd0: begin scl_h <= 1; sda_h <= 0; end
        3'd1: begin scl_h <= 1; sda_h <= 0; end
        3'd2: begin scl_h <= 1; sda_h <= 1; end
        3'd3: begin scl_h <= 1; sda_h <= 1; end
        3'd4: begin scl_h <= 1; sda_h <= 1; end
    endcase end
```

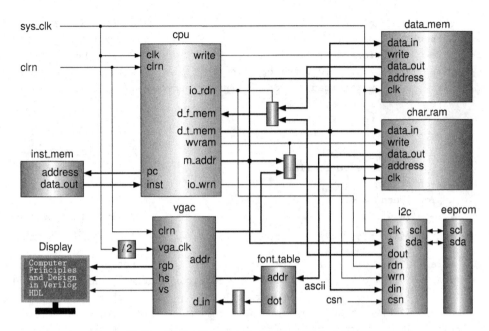

Figure 14.45   CPU + RAMs + font table + VGA + I2C EEPROM

```
        'WAIT:           begin scl_h <= 1; sda_h <= 1; end
    endcase
end
// read from host
wire        ready  = (curr_state == 'IDLE) && (next_state == 'IDLE);
wire [7:0] status = {2'd0,ready,1'b0,rx_ack,curr_state};
assign      d_out  = (addr == 0) ? out_buf : status;
endmodule
```

The simulation waveforms will be given later after we describe a computer system in which the I2C master controller and an I2C EEPROM are included. Figure 14.45 shows a schematic of the system that was implemented on the DE2-115 FPGA board.

The system consists of a CPU, instruction and data memories, a character RAM that stores ASCII codes of characters for display, a font table that stores ASCII fonts, a VGA controller, an I2C master controller, and an I2C EEPROM which is 24LC32A produced by Microchip Technology Inc. What this system does is to execute a program in the instruction memory to read a string from the data memory, write it into the EEPROM, read it back, and display it on the VGA. Below is the Verilog HDL code that implements the system shown in Figure 14.45.

```
module display_i2c_eeprom (sys_clk,clrn,i2c_scl,i2c_sda,r,g,b,hs,vs,vga_clk,
                    blankn,syncn); // write & display chars in eeprom
    input        sys_clk, clrn;              // sys_clk: 50MHz
    inout        i2c_scl, i2c_sda;           // i2c clk and data
    output [7:0] r, g, b;                    // vga r,g,b colors
    output       hs, vs;                     // vga h and v synch
    output       vga_clk;                    // for ADV7123 VGA DAC
```

```
output          blankn;                              // for ADV7123 VGA DAC
output          syncn;                               // for ADV7123 VGA DAC
wire            font_dot;                            // font dot
wire    [31:0]  inst,pc,d_t_mem,cpu_mem_a,d_f_mem;
wire            write,read,io_rdn,io_wrn,wvram,rvram;
wire            ready,overflow;
reg             vga_clk = 1;                         // vga_clk: 25MHz
always @(posedge sys_clk)
    vga_clk <= ~vga_clk;                             // vga_clk: 25MHz
// vga interface controller
assign blankn = 1;
assign syncn  = 0;
wire    [8:0]   row_addr;                            // pixel ram row addr
wire    [9:0]   col_addr;                            // pixel ram col addr
wire            vga_rdn;
wire    [23:0]  vga_pixel = font_dot? 24'hffffff : 24'h0000ff; //white/blue
vgac vga (vga_clk,clrn,vga_pixel,row_addr,col_addr,vga_rdn,r,g,b,hs,vs);
wire    [5:0]   char_row = row_addr[8:3];            // char row
wire    [2:0]   font_row = row_addr[2:0];            // font row
wire    [6:0]   char_col = col_addr[9:3];            // char col
wire    [2:0]   font_col = col_addr[2:0];            // font col
// character ram, 640/8 = 80 = 64 + 16; 480/8 = 60;
wire    [12:0]  vga_cram_addr = {1'b0,char_row,6'h0} +
                               {3'b0,char_row,4'h0} + {6'h0,char_col};
wire    [12:0]  char_ram_addr = wvram ? cpu_mem_a[14:2] : vga_cram_addr;
reg     [6:0]   char_ram [0:4799];                   // 80 * 60 = 4800
wire    [6:0]   ascii = char_ram[char_ram_addr];
always @(posedge sys_clk) begin
    if (wvram) char_ram[char_ram_addr] <= d_t_mem[6:0];
end
// font_table 128 x 8 x 8 x 1
wire [12:0] ft_a = {ascii,font_row,font_col};        // ascii,row,col
font_table ft (ft_a,font_dot);
// i2c master controller
wire        csn = io_rdn & io_wrn;                   // chip select
wire    [1:0]   addr = cpu_mem_a[3:2];               // i2c address
wire    [7:0]   d_in = d_t_mem[7:0];                 // data to be sent
wire    [7:0]   d_out;                               // data out
i2c eeprom (sys_clk,clrn,csn,addr,io_wrn,d_in,io_rdn,d_out,i2c_sda,
            i2c_scl);
// cpu
single_cycle_cpu_io cpu (sys_clk,clrn,pc,inst,cpu_mem_a,d_f_mem,
                        d_t_mem,write,io_rdn,io_wrn,rvram,wvram);
// instruction memory
scinstmem_i2c_eeprom imem (pc,inst);
// data memory
wire [31:0] dm_out;                                  // data mem data out
```

Figure 14.46   I2C EEPROM write

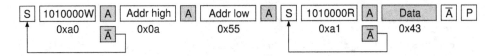

Figure 14.47   I2C EEPROM read

```
scdatamem_i2c_eeprom dmem (sys_clk,cpu_mem_a,d_t_mem,write,dm_out);
    assign d_f_mem = io_rdn ? dm_out : {24'h0,d_out};
endmodule
```

The string is "Computer Principles and Design in Verilog HDL", 45 characters. In our implementation, we write one character in each data transfer procedure. 24LC32A takes about 5 ms after stop condition for each EEPROM write operation. If the program starts a next write operation while the current character was not yet written to EEPROM, the I2C master will receive a not-acknowledge ($\overline{A}$). If so, our program will wait until it gets an acknowledge (A), as shown in Figure 14.46.

The slave addresses of the 24LC32A are 0xa0 for write and 0xa1 for read. The CPU can get the acknowledge or not-acknowledge with the I/O status read operation. Similarly, in the I2C EEPROM read operation, the CPU waits for an acknowledge, as shown in Figure 14.47.

We have given the Verilog HDL code of the CPU in the previous section. Below is the assembly program that writes and reads the I2C EEPROM. We let the starting byte address of the EEPROM to be 0x0a55. Note that, in the real implementation, we set the character count to 45 but for the simulation we set it to 1 for simplicity (only write/read the character "C"). In our implementation, the data memory has 256 bytes: the lower part is used to store the string, and the upper part is used as the stack to save the return address and some registers for the subroutine calls.

```
.text    0                       # 0: starting address of instruction memory
main:
    ori   $sp, $0, 0x0100        # stack pointer, 256-byte data memory
    lui   $8,  0xc000            # vram space: c0000000 - dfffffff
    lui   $9,  0xa001            # i/o  space: a0000000 - bfffffff
    la    $1,  msg               # address of data
    ori   $4,  $0, 0x0a55        # eeprom address
    ori   $7,  $0, 1 #45             # number of chars = 1 for simulation
write_data:                      # write chars to i2c eeprom
    lw    $6,  0($1)             # 4 bytes (chars)
    srl   $5,  $6, 24            # 1st byte
    jal   write_eeprom           # write the byte to eeprom
    addi  $7,  $7, -1            # counter--
    beq   $7,  $0, w_finished    # if counter is 0, go to w_finished
    addi  $4,  $4, 1             # eeprom addr + 1
    sll   $5,  $6, 8             # 2nd byte
```

```
        srl    $5,  $5, 24            # 2nd byte
        jal    write_eeprom          # write the byte to eeprom
        addi   $7,  $7, -1            # counter--
        beq    $7,  $0, w_finished    # if counter is 0, go to w_finished
        addi   $4,  $4,  1            # eeprom addr + 1
        sll    $5,  $6, 16            # 3rd byte
        srl    $5,  $5, 24            # 3rd byte
        jal    write_eeprom          # write the byte to eeprom
        addi   $7,  $7, -1            # counter--
        beq    $7,  $0, w_finished    # if counter is 0, go to w_finished
        addi   $4,  $4,  1            # eeprom addr + 1
        sll    $5,  $6, 24            # 4th byte
        srl    $5,  $5, 24            # 4th byte
        jal    write_eeprom          # write the byte to eeprom
        addi   $7,  $7, -1            # counter--
        beq    $7,  $0, w_finished    # if counter is 0, go to w_finished
        addi   $4,  $4,  1            # eeprom addr + 1
        addi   $1,  $1,  4            # next word
        j      write_data
w_finished:
        ori    $4,  $0, 0x0a55        # eeprom address
        ori    $7,  $0, 1 #45          # number of chars = 1 for simulation
read_data:
        jal    read_eeprom           # read a byte from i2c eeprom
        sw     $2,  0($8)            # to display on vga
        addi   $8,  $8,  4            # vram addr + 4
        addi   $4,  $4,  1            # eeprom addr + 1
        addi   $7,  $7, -1            # counter--
        beq    $7,  $0, r_finished    # if counter is 0, go to r_finished
        j      read_data             # continue to read
r_finished:
        j      r_finished            # should return to os
write_eeprom:                        # $4: addr, $5: data
        addi   $sp, $sp, -16         # reserve stack space
        sw     $ra, 12($sp)          # save return address
        sw     $fp, 08($sp)          # save frame pointer
        move   $fp, $sp              # new frame pointer
        jal    write_eeprom_addr     # write eeprom addr
        jal    wait_ready            # wait for ready
        sw     $5,  4($9)            # send byte to store
        jal    wait_ready            # wait for ready
        sw     $0,  12($9)           # send stop bit
        move   $sp, $fp              # restore stack pointer
        lw     $fp, 08($sp)          # restore frame pointer
        lw     $ra, 12($sp)          # restore return address
        addi   $sp, $sp, 16          # release stack space
        jr     $ra                   # return
```

```
read_eeprom:                        # $4, addr, $2: data (return value)
    addi  $sp, $sp, -16             # reserve stack space
    sw    $ra, 12($sp)              # save return address
    sw    $fp, 08($sp)              # save frame pointer
    move  $fp, $sp                  # new frame pointer
    jal   write_eeprom_addr         # write eeprom addr
    jal   wait_ready                # wait for ready
    sw    $0,  0($9)                # send repeat start bit
    ori   $2,  $0, 0xa1             # i2c eeprom slave addr, read
    jal   wait_ready                # wait for ready
    sw    $2,  4($9)                # send slave address, read
    ori   $2,  $0,  1               # not-ack
    jal   wait_ready                # wait for ready
    sw    $2,  8($9)                # go to RX state, not-ack
    jal   wait_ready                # wait for ready
    lw    $2,  0($9)                # get data of eeprom
    sw    $0, 12($9)                # send stop bit
    move  $sp, $fp                  # restore stack pointer
    lw    $fp, 08($sp)              # restore frame pointer
    lw    $ra, 12($sp)              # restore return address
    addi  $sp, $sp, 16              # release stack space
    jr    $ra                       # return
write_eeprom_addr:                  # $4: addr, $9: i/o addr
    addi  $sp, $sp, -20             # reserve stack space
    sw    $ra, 16($sp)              # save return address
    sw    $fp, 12($sp)              # save frame pointer
    sw    $6,  08($sp)              # save $6
    sw    $2,  04($sp)              # save $2
    move  $fp, $sp                  # new frame pointer
    jal   wait_ready                # wait for ready
send_start:
    sw    $0,  0($9)                # send start bit
    ori   $2,  $0, 0xa0             # i2c eeprom slave addr, w
    jal   wait_ready                # wait for ready
    sw    $2,  4($9)                # send slave address, w
    jal   wait_ready                # wait for ready
    lw    $6,  4($9)                # get i2c status, check ack
    andi  $6,  $6, 0x08             # check rx_ack {0,0,x,0,rx_ack,xxx}
    bne   $6,  $0, send_start       # if received not-ack
    srl   $2,  $4, 8                # high 8-bit eeprom addr
    andi  $2,  $2, 0x0f             # high 4-bit eeprom addr = 0
    sw    $2,  4($9)                # send eeprom high 8-bit addr
    andi  $2,  $4, 0xff             # low 8-bit eeprom addr
    jal   wait_ready                # wait for ready
    sw    $2,  4($9)                # send eeprom low 8-bit addr
    move  $sp, $fp                  # restore stack pointer
    lw    $2,  04($sp)              # restore $2
```

```
     lw    $6,   08($sp)         # restore $6
     lw    $fp,  12($sp)         # restore frame pointer
     lw    $ra,  16($sp)         # restore return address
     addi  $sp,  $sp, 20         # release stack space
     jr    $ra                   # return
wait_ready:                      # wait for ready
     addi  $sp,  $sp, -20        # reserve stack space
     sw    $ra,  16($sp)         # save return address
     sw    $fp,  12($sp)         # save frame pointer
     sw    $6,   08($sp)         # save $6
     move  $fp,  $sp             # new frame pointer
check_status:
     lw    $6,   4($9)           # get i2c status
     andi  $6,   $6, 0x20        # check ready {0,0,ready,0,x,xxx}
     beq   $6,   $0, check_status # if not ready, continue to check
     move  $sp,  $fp             # restore stack pointer
     lw    $6,   08($sp)         # restore $6
     lw    $fp,  12($sp)         # restore frame pointer
     lw    $ra,  16($sp)         # restore return address
     addi  $sp,  $sp, 20         # release stack space
     jr    $ra                   # return
.data     0                      # 0: starting address of data memory
msg:                             # address of ascii characters
     .ascii "Computer Principles and Design in Verilog HDL"
.end
```

Figure 14.48 shows the simulation waveforms when the CPU writes the character "C" (ASCII code 0x43) to the location 0x0a55 of the I2C EEPROM.

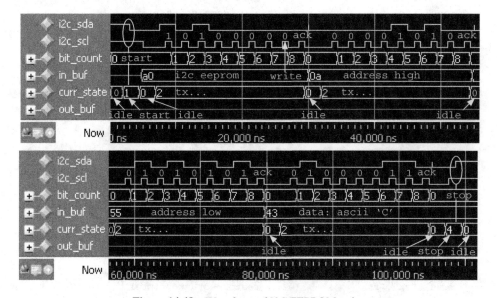

Figure 14.48   Waveform of I2C EEPROM write

At the beginning, the I2C bus is free (the current state of the I2C controller is IDLE and both `i2c_sda` and `i2c_scl` are float high). When the CPU executes the instruction `sw $0, 0($9)`, the I2C controller enters the START state (`curr_state = 1`) and sends a start condition. After that, the controller returns to the IDLE state. Then the CPU executes `sw $2, 4($9)` to send 0xa0, and the I2C controller enters TX state (`curr_state = 2`) and transmits the 8-bit 0xa0 to I2C bus. Then the controller receives an acknowledge from SDA (a 0 in the SDA line at the position where the `bit_count` is 8). The controller enters the IDLE state again. After the CPU sends 0x0a, 0x55, and 0x43 in a similar manner, the CPU executes `sw $0, 12($9)` to let the controller enter the STOP state. After transmitting a stop condition, the I2C bus becomes free.

Figure 14.49 shows the simulation waveforms when the CPU reads the character "C" from the location 0x0a55 of the I2C EEPROM.

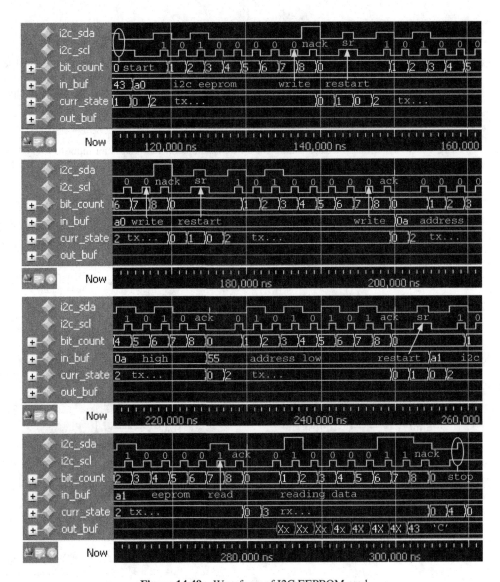

Figure 14.49 Waveform of I2C EEPROM read

Same as in the write operation, the CPU first sends 0xa0 to the I2C controller but the controller receives a not-acknowledge from SDA (a 1 in the SDA line at the position where the `bit_count` is 8). The CPU knows this matter through the execution of the `lw $6, 4($9)` instruction (reads I2C status). Then the CPU sends the 0xa0 with a prefixed repeated start condition to the controller and reads the I2C status again and again until the CPU knows that the I2C controller has got an acknowledge. Then, the CPU sends the EEPROM address of 0x0a55 and 0xa1 to the I2C controller with a prefixed repeated start condition for reading from the EEPROM. After getting the data byte 0x43, the controller transmits a not-acknowledge and a stop condition to terminate the data transfer. Then the I2C bus becomes free. Note that in the simulation we let the I2C controller receive only two not-acknowledges, but in the real situation this number will be much larger because the EEPROM takes about 5 ms to store a data byte to an EEPROM location before it can be read from the same location.

14.6.2   PCI Parallel Bus

PCI is a parallel bus proposed by Intel for attaching or inserting peripheral devices into a computer. The PCI bus uses 47 pins (49 pins for a master) to connect PCI cards. In the most common implementations, it is 32-bit wide with a frequency of 33 or 66 MHz. Figure 14.50 shows the pins in functional groups. We describe only the required pins on the left side of the figure. If a pin name is suffixed with a #, that signal is active-low.

1. CLK: Clock signal that provides timing for all transactions on PCI and is an input to every PCI device. Except RST#, all other input signals are sampled on the rising edge of CLK and all the output signals are transited at this edge.

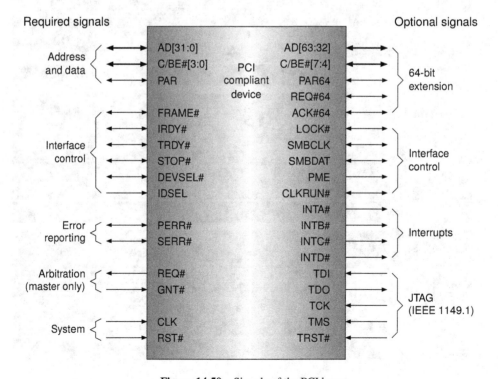

Figure 14.50   Signals of the PCI bus

Table 14.6   C/BE bus command definitions

C/BE#	Command type	C/BE#	Command type
0000	Interrupt acknowledge	1000	Reserved
0001	Special cycle	1001	Reserved
0010	I/O read	1010	Configuration read
0011	I/O write	1011	Configuration write
0100	Reserved	1100	Memory read multiple
0101	Reserved	1101	Dual address cycle
0110	Memory read	1110	Memory read line
0111	Memory write	1111	Memory write and invalidate

2. RST#: Reset signal that is used to place PCI registers, sequencers, and signals into a consistent state. When RST# is asserted, all PCI output signals will be tri-stated.

3. AD[31:0]: Address and Data signals that are multiplexed on the same PCI pins. A bus transaction consists of an address phase followed by one or more data phases.

4. C/BE#[3:0]: Bus Command and Byte Enable signals that are multiplexed on the same PCI pins. During the address phase of a transaction, C/BE#[3:0] define the bus command, as shown in Table 14.6. During the data phase, C/BE#[3:0] are used as byte enables. The byte enables are valid for the entire data phase and determine which byte lanes carry meaningful data. C/BE#[0] applies to byte 0 (LSB) and C/BE#[3] applies to byte 3 (MSB).

5. PAR: Parity bit that is even parity across AD[31:0] and C/BE#[3:0]. PAR has the same timing as AD[31:0], but it is delayed by one clock cycle. The master drives PAR for address and write data phases; the target drives PAR for read data phases.

6. IRDY#: Initiator Ready. During a write, IRDY# indicates that valid data is present on AD[31:0]. During a read, it indicates that the master is prepared to accept data. Wait cycles are inserted until both IRDY# and TRDY# are asserted together.

7. TRDY#: Target Ready. During a read, TRDY# indicates that valid data is present on AD[31:0]. During a write, it indicates that the target is prepared to accept data. Wait cycles are inserted until both IRDY# and TRDY# are asserted together.

8. FRAME#: Frame cycle that is driven by the current master to indicate the beginning and duration of an access. FRAME# is asserted to indicate that a bus transaction is beginning. While FRAME# is asserted, data transfers continue. When FRAME# is deasserted, the transaction is in the final data phase or has completed.

9. DEVSEL#: Device Select. When actively driven, it indicates that the driving device has decoded its address as the target of the current access. As an input, DEVSEL# indicates whether any device on the bus has been selected.

10. STOP#: Stop that indicates the current target is requesting the master to stop the current transaction.

11. IDSEL: Initialization Device Select that is used as a chip select during configuration read and write transactions.

12. REQ#: Request that indicates to the arbiter that this agent desires use of the bus. This is a point-to-point signal. Every master has its own REQ#, which must be tri-stated while RST# is asserted.

13. GNT#: Grant that indicates to the agent that access to the bus has been granted. This is a point-to-point signal. Every master has its own GNT#, which must be ignored while RST# is asserted.

14. PERR#: Parity Error that is only for reporting data parity errors during all PCI transactions except a Special Cycle.

15. SERR#: System Error that is for reporting address parity errors, data parity errors on the Special Cycle command, or any other system error where the result will be catastrophic.

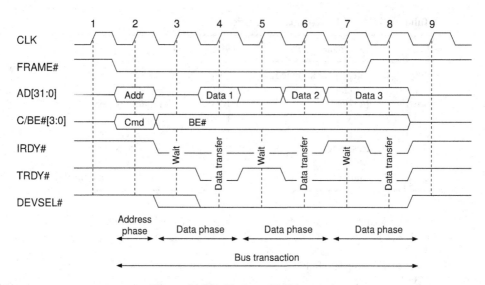

Figure 14.51   Timing of PCI bus read

Figure 14.51 illustrates a read transaction which starts when FRAME# is asserted. Data transfer occurs when both IRDY# and TRDY# are asserted on the same rising clock cycle. We explain the actions cycle by cycle next.

1. Cycle 1: The bus is idle.
2. Cycle 2: FRAME#, Addr (address), and Cmd (command) are asserted by the initiator to start a read transaction (the Cmd is a read command). This is an address phase.
3. Cycle 3: The initiator releases the address line and asserts BE# (Byte Enable) and IRADY#. The target device may assert DEVSEL# in this cycle. Because TRDY# is not yet asserted by the target device, the Cycle 3 is a wait cycle.
4. Cycle 4: The target device puts Data1 on the AD bus and asserts TRDY#. The initiator can get the data from the AD bus. This is the first data phase.
5. Cycle 5: The target device deasserts TRDY# to tell the initiator that the next data is not yet ready. This is a wait cycle.
6. Cycle 6: The target device puts Data2 on AD bus and asserts TRDY#. The initiator can get the data from the AD bus. This is the second data phase.
7. Cycle 7: The target device provides Data3 and keeps the TRDY# active. But the initiator deasserts IRDY# to tell the target device that it is not ready for receiving the data. This is a wait cycle.
8. Cycle 8: This is the third data phase. The initiator asserts IRDY# and gets Data3. Meanwhile, it deasserts FRAME# to tell the target device that this is the last data phase. The target device deasserts DEVSEL#.
9. Cycle 9: The bus becomes idle again.

Figure 14.52 illustrates a write transaction in which the initiator sends three data words to the target device.

1. Cycle 1: The bus is idle.
2. Cycle 2: FRAME#, Addr, and Cmd are asserted by the initiator to start a write transaction (the Cmd is a write command). This is an address phase.
3. Cycle 3: This is the first data phase. The initiator puts Data1 on the AD bus and asserts BE#1 and IRADY#. The target device asserts DEVSEL# and TRDY# and gets Data1 from the AD bus.

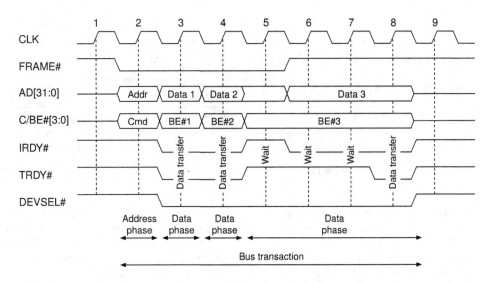

Figure 14.52 Timing of PCI bus write

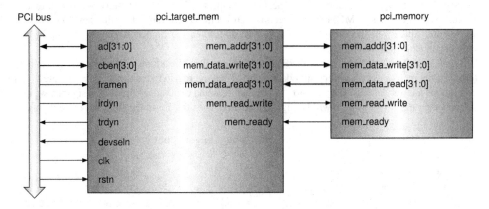

Figure 14.53 PCI memory example

4. Cycle 4: This is the second data phase. Both IRDY# and TRDY# are asserted. The initiator puts Data2 on the AD bus and asserts BE#2. The target device gets Data2 from the AD bus.
5. Cycle 5: The initiator deasserts IRDY# to tell the target device that it is not ready for sending the data. This is a wait cycle.
6. Cycle 6: The initiator asserts IRDY# and puts Data3 on the AD bus. But the target device deasserts TRDY# to tell the initiator that it is not ready for receiving the data. This is a wait cycle. The initiator deasserts FRAME# to tell the target device that this is the last data phase.
7. Cycle 7: This is also a wait cycle, caused by the target device.
8. Cycle 8: The target device asserts TRDY# and gets Data3.
9. Cycle 9: All signals are deasserted and the bus becomes idle again.

Next, we show a simple design example of a PCI target controller. Referring to Figure 14.53, a memory module can be accessed by a PCI master through the PCI target controller. The design shows how to translate the PCI signals to the signals for the memory access.

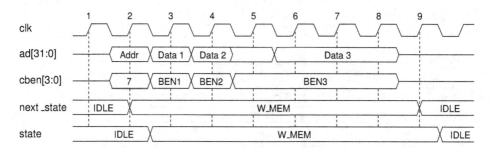

Figure 14.54   Next state signal for PCI controller

The PCI signals are shown on the left side of the controller. Instead of "#", we used the character "n" (negative) in the signal names. We only implemented the commands of "Memory Read" and "Memory Write" (Table 14.6). Therefore, we defined three states (`state`) of the controller: IDLE, R_MEM (memory read), and W_MEM (memory write). The `state` is transited on the falling edge of the clock, based on the next state signal (`next_state`). The `next_state` is transited on the raising edge of the clock, based on the command type in the address phase, as shown in Figure 14.54.

If the command type in the address phase is a Memory Write command, the `next_state` becomes W_MEM. Similarly, if the command type in the address phase is a Memory Read command, the `next_state` becomes R_MEM. It will become IDLE after a bus transaction finishes. In this example, we let the capacity of the memory to be 64 KB, and when the higher 16-bit address is 0xffff, this memory module is selected (`devseln` = 0).

Below is the Verilog HDL code of the PCI target controller. The address in the address phase is registered for the memory access in the first data phase. Then the registered address is incremented by 4 for continuous memory access. Because `ad[31:0]` (address and data) are bidirectional signals, we defined them as the `inout` type, which will be high-impedance when the controller does not drive them.

```
`define IDLE    0                    // pci target controller for memory access
`define R_MEM   1                    // define states
`define W_MEM   2
module pci_target_mem (clk,rstn,framen,cben,ad,irdyn,trdyn,devseln,
                       mem_read_write,mem_ready,mem_addr,mem_data_write,
                       mem_data_read,state);
    input   [31:0] mem_data_read;              // data from memory
    input   [3:0]  cben;                       // command/byte enable
    input          clk;                        // clock
    input          rstn;                       // reset
    input          framen;                     // frame
    input          irdyn;                      // initiator ready
    input          mem_ready;                  // memory ready
    inout   [31:0] ad;                         // bi-directional addr/data
    output  [31:0] mem_addr;                   // memory address
    output  [31:0] mem_data_write;             // data to memory
    output  [1:0]  state;                      // state of controller
    output         trdyn;                      // target ready
    output         devseln;                    // device select
    output         mem_read_write;             // memory read(1) / write(0)
    reg     [31:0] mem_addr;                    // memory address
```

```verilog
reg     [1:0] state;                      // state for memory access
reg           pre_framen;                 // for detecting falling edge
always @ (posedge clk) begin
    pre_framen <= framen;
end
// state transition
reg     [1:0] next_state = `IDLE;         // next state
reg     [31:0] auto_addr  = 0;            // address for burst mode
always @(posedge clk or negedge rstn) begin
    if (!rstn) begin
        next_state <= `IDLE;
        auto_addr  <= 0;
    end else begin
        if (!framen && pre_framen) begin   // start transaction
            if (ad[31:16] == 16'hffff) begin
                case (cben)
                    4'b0110: begin next_state <= `R_MEM;
                                   auto_addr  <= ad;    end
                    4'b0111: begin next_state <= `W_MEM;
                                   auto_addr  <= ad;    end
                    default: begin next_state <= `IDLE;
                                   auto_addr  <=  0;    end
                endcase
            end
        end else begin
            case (next_state)
                `R_MEM: begin
                    if (!irdyn && !trdyn) begin
                        auto_addr <= auto_addr + 4;
                    end else begin
                        if (framen && irdyn) begin
                            next_state <= `IDLE;
                        end
                    end
                end
                `W_MEM: begin
                    if (!irdyn && !trdyn) begin
                        auto_addr <= auto_addr + 4;
                    end else begin
                        if (framen && irdyn) begin
                            next_state <= `IDLE;
                        end
                    end
                end
            endcase
        end
    end
end
```

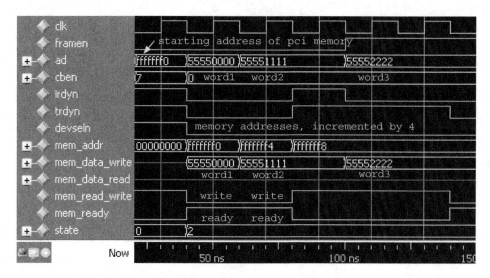

Figure 14.55   Waveform of PCI bus write

```
// memory signals
wire write = (state == 'W_MEM);
assign mem_read_write = ~(write & ~irdyn & ~trdyn);
assign mem_data_write = write ? ad : 32'hzzzzzzzz;
always @(negedge clk) begin
    state    <= next_state;
    mem_addr <= auto_addr;
end
// pci output signals
wire enable = (state == 'R_MEM) & !irdyn;
assign ad = enable ? mem_data_read : 32'hzzzzzzzz;
assign trdyn = ~mem_ready;
assign devseln = ~((state != 'IDLE) & ~(framen & irdyn));
endmodule
```

Figure 14.55 shows the waveforms when a PCI master writes three data words into the continuous locations starting from 0xfffffff0 of the memory module. The three words are 0x55550000, 0x55551111, and 0x55552222. We can see that in the W_MEM state (`state` = 2), the address is automatically incremented by 4 and the three words are written into memory sequentially. The write operation happens only when both `irdyn` and `irdyn` are asserted.

Figure 14.56 shows the waveforms when a PCI master reads three data words from the continuous locations starting from 0xfffffff0 of the memory module. The three words are 0x55550000, 0x55551111, and 0x55552222. We can see that in the R_MEM state (`state` = 1), the address is automatically incremented by 4 and the three words are read from memory sequentially. The read operation happens only when both `irdyn` and `irdyn` are asserted.

The parallel PCI bus starts and stops a bus transaction by asserting and deasserting the FRAME# signal, respectively. In contrast, the serial I2C bus starts and stops a bus transaction by issuing a start condition and a stop condition, respectively.

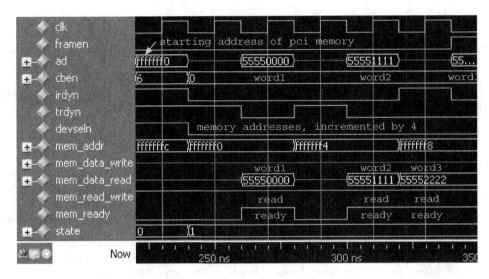

Figure 14.56 Waveform of PCI bus read

Exercises

14.1 What will the 16-bit extended Hamming code say if d_3, d_6, d_9, and d_{12} are errors? Explain the reason.

14.2 Referring to the Verilog HDL code of the PS/2 keyboard, redesign the UART by using two FIFOs for the receiver and transmitter.

14.3 Redesign PS/2 keyboard interface so that it can detect a frame error.

14.4 Compare PS/2 keyboard/mouse protocol with the UART protocol.

14.5 Develop the Verilog HDL code to implement a PS/2 mouse interface.

14.6 Display colorful characters on VGA: add a 24-bit color data to each character stored in character RAM.

14.7 Implement the keyboard's key reading with the interrupt mechanism.

14.8 Write a program to translate scan code to ASCII code and display the ASCII code on VGA. Also implement it on an FPGA board.

14.9 Add a cursor to the design described in the previous exercise.

14.10 Write a test bench i2c_tb.v to verify the i2c.v.

14.11 Design a computer system that contains a CPU, instruction memory, data memory, character RAM (text mode), font table, PS/2 keyboard interface, VGA interface, and an I2C controller with an I2C EEPROM, and develop a simple editor that deals with not only the standard keys but also the extended keys by referring to Figure 14.35.

14.12   Design a computer system that contains a CPU, instruction memory, data memory, video RAM (graphics mode), PS/2 keyboard interface, VGA interface, and an I2C controller with an I2C EEPROM. The video RAM stores true color pixels that will be displayed on VGA, so that we can display computer graphics images or photo images on display. If you also want to display the ASCII characters, you can store the ASCII font table in the data memory. The ASCII fonts can be read from the font table with load instruction and then written to the video RAM with store instruction.

14.13   Extension to the last exercise: implement a font table to show (a small part of) your native characters on VGA if your native language is not English.

14.14   A DMAC performs the data transfer between memory and I/O devices. Design a simple DMAC to perform the data transfer between two different regions of the same memory module.

14.15   This chapter gave a simple PCI target controller. Design a PCI master controller that contains all the required signals.

14.16   If your FPGA board has an SD card slot, try to develop an SD card controller so the your computer system can use it as an external storage for storing files, such as text files, executable binary files, and bitmap image files.

14.17   Develop a computer system that contains an SD card, OS kernel, and some applications. The OS kernel can accept commands from the keyboard and invoke the application. The PS/2 mouse is optional.

15

High-Performance Computers and Interconnection Networks

A high-performance computer system delivers much higher performance than can be got out of a typical desktop computer or workstation. It consists of a set of compute nodes as well as an interconnection network that connects the nodes in the systems. This chapter describes the architectures of high-performance computer systems and the interconnection networks.

15.1   Category of High-Performance Computers

In general, a high-performance computer system can be a shared-memory parallel multiprocessor system or a message-passing distributed multicomputer system.

15.1.1   Shared-Memory Parallel Multiprocessor Systems

A shared memory parallel multiprocessor system consists of a set of CPUs (central processing units), a set of memory modules, and an interconnection network. All CPUs share the memory. Communication between processes running on different CPUs is performed through writing to and reading from the global memory, as shown in Figure 15.1(a).

Generally, a small-scale, parallel multiprocessor system uses a centralized shared memory, and the interconnection network is commonly a high-performance parallel bus or a crossbar. Commercial examples of such systems are the Sun Microsystems SMP (symmetric multiprocessor) clusters and the SGI (Silicon Graphics Inc.) SMP clusters.

On the other hand, large-scale parallel multiprocessor systems use distributed shared memory (DSM). Each CPU has a local memory (in the same CPU/memory board), and the memory is shared by all CPUs. The interconnection networks used in these systems are scalable static interconnection networks, such as 3D torus, fat-tree, and hypercubes. Commercial examples are the supercomputers made by IBM and Cray Inc. In addition to the writing to and reading from the shared memory, some multiprocessor systems also provide the message-passing interface (MPI) for data communication among the CPUs.

15.1.2   Message-Passing Distributed Multicomputer Systems

A message-passing distributed multicomputer system consists of a set of computers connected with a high-speed interconnect. Each computer in the system has its own CPU(s) and storages. Unlike that of

---

Computer Principles and Design in Verilog HDL, First Edition. Yamin Li.
© 2015 Tsinghua University Press. All rights reserved. Published 2015 by John Wiley & Sons Singapore Pte Ltd.
Companion Website: www.wiley.com/go/li/verilog

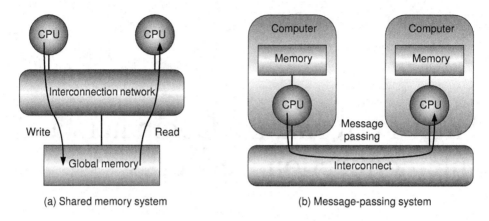

(a) Shared memory system   (b) Message-passing system

Figure 15.1   Communications in high-performance computers

the shared-memory parallel system, the CPU of a computer in the distributed system cannot directly access the memory of other computers. Communication between tasks running on different computers is performed through message passing, as shown in Figure 15.1(b).

Today message passing is supported by a wide variety of APIs (application programming interfaces), including runtime libraries such as **MPI**, network interfaces such as **TCP/IP** (transmission control protocol/internet protocol) or **UDP/IP** (user datagram protocol/internet protocol), and internet protocols such as **HTTP** (hypertext transfer protocol).

The interconnect that connects all the nodes (computers) in a distributed system may be an interconnection network, a LAN (local area network), a WAN (wide area network), or an internetwork (internet). Distributed systems are used in cloud computing, whose infrastructure may contain supercomputers that are shared-memory parallel systems.

15.2   Shared-Memory Parallel Multiprocessor Systems

Depending on the interconnection networks, shared-memory parallel multiprocessor systems can be classified as centralized shared-memory systems and DSM systems.

15.2.1   Centralized Shared-Memory (SMP) Systems

In a centralized shared memory system, all CPUs have equal opportunity and equal time to access all parts of the shared memory. The interconnection network can be a single bus, multiple buses, or a crossbar switch. Such a system is also called an SMP. It has a uniform memory access (UMA) architecture, meaning that all the CPUs have the same pattern to access the shared memory. The scale of SMP systems cannot be very large because of the limited communication capability of the interconnection network.

Figure 15.2 shows the typical architecture of an SMP system. The main memory is equally accessible to all the CPUs via a common bus. A bus arbiter is used to determine which CPU gets to use the bus.

In addition to the shared main memory, each CPU contains a local cache or multilevel caches in order to speed up the memory access and, more importantly, to solve the bus contention problem in the shared-memory system. The copies in the caches are coherent if they are equal to the same value in the memory. However, if one of the CPUs writes over the value of one of the copies, then the copy becomes inconsistent. Therefore, the cache coherency protocols are needed to maintain a level of consistency throughout the multiprocessor system.

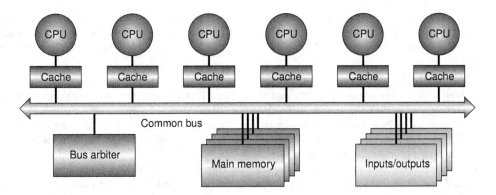

Figure 15.2 Architecture of a symmetric multiprocessor system

Since all CPUs and their caches are connected to a common bus, the cache controllers can snoop on the bus for maintaining coherent data. Each cache block is marked with a state, and the cache controller will modify the states to track changes to cache blocks made either locally or remotely. We give a four-state MESI bus snooping cache coherency protocol below. The four states are as follows:

1. Modified (M): Indicates that the cache block is present only in the current cache and has been modified from the value in the main memory.
2. Exclusive (E): Indicates that the cache block is present only in the current cache and matches the main memory.
3. Shared (S): Indicates that the cache block may be stored in other caches and matches the main memory.
4. Invalid (I): Indicates that the cache block is invalid.

A hit on a read implies that the cache data is usable (the state of the block is not Invalid). The state of the block is left unchanged. A read miss leads to a request for the data. This request is satisfied in one of the following four cases:

1. If no other cache has a copy of the data, the block data are provided by the main memory and the state of the cache block is set to Exclusive.
2. If other caches have a copy of the data in Shared state, the block data are provided by the main memory or by a cache that has a copy of the Shared data. The state of the cache block is also set to Shared.
3. If another cache has a copy of the data with the Exclusive state, the block data are provided by the main memory or by that cache. The block states of both caches are set to Shared.
4. If another cache has a copy of the data with the Modified state, the block data are provided by that cache. The block states of both caches are set to Shared. The main memory is also updated with the modified data.

Consider what happens when a CPU attempts to write to a cache block. On a write hit, there are three cases, corresponding to the states of the block:

1. If the state of the cache block is Shared, an invalidation signal is broadcast on the common bus, so that all other caches will set their cache blocks to the Invalid state. Meanwhile, the write can be completed in the cache, and the state is changed to Modified.
2. If the state of the block is Exclusive, the write can proceed without any delay, and the state is changed to Modified.
3. If the state is Modified, the write can proceed without any delay, and the state is left unchanged.

On a write miss occurs (the state of the cache block is Invalid), a write request is placed on the common bus. The block must be loaded to the cache either from the main memory or from another cache that has the Modified data. This is because the data to be written is a small part of the whole block. There are four cases:

1. If no other cache contains a copy, the data block is read from the main memory, then the cache write is completed by the CPU, and the cache block is set to the Modified state.
2. If other caches have the data in the Shared state, the copies of the caches are invalidated, the data block is read from the main memory, the cache write is completed by the CPU, and the cache block is set to the Modified state.
3. If another cache has data in the Exclusive state, the copy of the cache is invalidated, the data block is read from the main memory, the cache write is completed by the CPU, and the cache block is set to the Modified state.
4. If another cache has the data in the Modified state, the data of that cache is written to both the requesting cache and the main memory. The block state of that cache is changed from Modified to Invalid. Then the cache write can be completed by the CPU, and the cache block is set to the Modified state.

There are many other bus snooping cache coherency protocols that have different number of states and different state transitions. The protocol described above uses the copy-back and write invalidate policies. Other choices include write through and write update. Instead of broadcasting an invalidation signal, the write update policy lets the CPU that is writing the data to broadcast the new data over the bus so that all caches that contain copies of the data are then updated. In general, a good bus snooping cache coherency protocol should reduce the main memory access traffic and speed up the cache accesses.

15.2.2   Distributed Shared-Memory (DSM) Systems

The centralized shared memory architecture is not suitable for constructing a large-scale parallel system because the common bus will become a bottleneck as the number of CPUs increases. In a DSM system, every CPU has its local memory and those memories together create the shared memory. A high-performance interconnection network, such as a hypercube, an n-dimensional torus, or a fat-tree, is used to connect all the compute nodes together. Figure 15.3 shows a typical architecture of a DSM parallel multiprocessor system.

The local memory can be accessed directly by the CPU without disturbing the interconnection network. Access to shared memory parts of other CPUs (remote memory) must go through the interconnection network. Therefore, access to the local memory is faster than that to the remote memory. This kind of systems is said to have a nonuniform memory access (NUMA) architecture.

Maintaining cache coherence in a DSM system is a little bit complex. In an SMP system, we can use snoopy protocols that are able to listen to and broadcast invalidations on a common bus. However, listening to the actions on a non-bus interconnection network in DSM is impossible, and broadcast is at high cost. Commonly, directory-based cache coherency protocols are used in DSM systems to ensure that invalidation messages are received by all caches with copies of the shared data.

In a directory-based cache coherency protocol, a directory entry corresponding to each memory block is needed that keeps track of shared copies or the identification of the CPU that contains modified data. This information is used to perform the necessary point-to-point communications, instead of broadcast operation. Bit-map and linked-list schemes are two categories of directory protocols.

The full-map directory contains p presence bits with each memory block, 1 bit per node (Figure 15.4). If a copy of the shared block is contained in the cache of a node, the corresponding presence bit is set. The directory also has a dirty (modified) bit. If the dirty bit is set, only one node in the system has a copy of the corresponding shared block. This is sometimes known as the $p + 1$ directory. The advantage of this scheme is that only the nodes caching the block receive the invalidation messages. The disadvantage

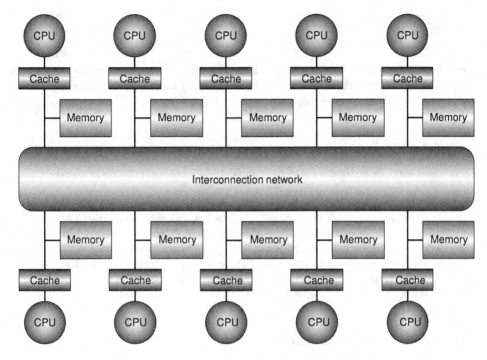

Figure 15.3   Architecture of a distributed shared memory system

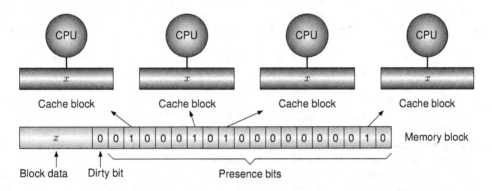

Figure 15.4   Full bitmap in directory-based cache coherency protocols

is the unscalable directory memory requirement. Another solution is a limited directory protocol whose directory contains pointers to each cache that has a copy of the memory block.

The full map or limited directory maintains a centralized directory. The linked lists use a distributed way to keep track of multiple copies of a block. The directory at the memory home node stores only a pointer to the first cached copy. Each cache stores a pointer to the next cached copy (a singly linked list) or pointers to the next and previous cached copies (a doubly linked list, see Figure 15.5). The advantage of this scheme is the scalable and potentially lower directory memory requirement. The disadvantages are that the invalidates must be serialized, and updating the linked list is more complex than that of full map or limited directory.

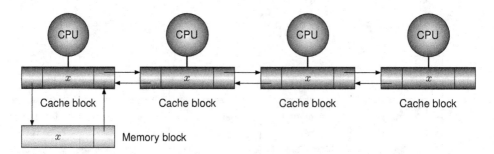

Figure 15.5   Doubly linked list in directory-based cache coherency protocols

The IEEE Scalable Coherence Interface (SCI) standard uses a doubly linked list of cache blocks to keep track of cached copies. Numerous variations have been proposed to improve the performance of directory-based protocols. Tree-based linked lists try to reduce the invalidation time from $O(n)$ to $O(\log n)$, where n is the average number of shared cache copies of a memory block. Hybrid techniques that combine snoopy protocols with directory-based protocols can be used in hybrid systems in which multiple bus-based clusters are connected by a non-bus interconnection network.

15.3   Inside of Interconnection Networks

This section describes the basic concepts of interconnection networks and gives the structure of a network switch that uses a crossbar to provide communication paths for each input–output port pair.

The purpose of an interconnection network is to provide a communication path between the CPU/memory boards in a parallel system. The simplest example is a common bus to which the CPU and memory boards are plugged in. It is most inexpensive but has lowest bandwidth since only one CPU can send data at a time.

A non-bus interconnection network is usually constructed from switches (or routers) and links, as shown in Figure 15.6. A switch is a device with input ports and output ports that can route data selectively from the former to the latter. Links are communication cables used to connect switch output ports to the input ports of other switches by following a certain topology. The combination of a switch and a CPU/memory board is called a compute node or simply a node.

A switch has multiple communication ports, which can be serial (low cost) or parallel (high perfor- mance). It must pass the data through several ports at the same time as long as none of them is sending data to the same output port. This can be done by a crossbar.

An $n \times n$ crossbar switch is organized as an $n \times n$ matrix to connect n input ports to n output ports. Crossbar switches can transfer data from multiple input ports to multiple outputs simultaneously. Each pair of input and output ports has a dedicated path through the switch. Figure 15.7 shows an 8×8 crossbar switch that is implemented with eight independent 8-to-1 multiplexers to allow each output port to be independently connected to any input port, and any input port can be connected to any output port (point-to-point communication), to some output ports (multicast), or to all output ports (broadcast).

If data at two input ports are routed to the same output port at the same time, the switch must make a choice to transfer the data from one of the input ports and block the other. In order to reduce the conflicts for output ports, a single FIFO (first-in first-out) buffer can be added at each input port for temporarily holding incoming data, as shown in Figure 15.8(a).

If the data at the head of the buffer cannot be forwarded because its destination output port is busy, more recent data in the buffer destined for other output ports also cannot be forwarded. This is known as head-of-line (HOL) blocking. HOL blocking can be reduced in the crossbar architecture by maintaining

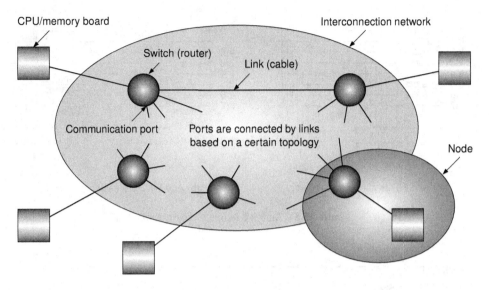

Figure 15.6   Switches and cables in interconnection networks

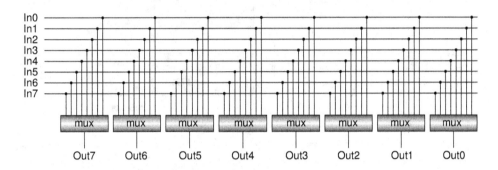

Figure 15.7   An 8 × 8 crossbar switch

separate FIFO buffers for each input–output port pair. That is, a single FIFO buffer is placed at each crosspoint, as shown in Figure 15.8(b).

The control circuit of the switch has the responsibility to generate the selection signals of the multiplexers. Figure 15.9 shows a detailed structure of a switch that uses thirty-two 32-to-1 multiplexers as the 32 × 32 switch matrix. The multiplexer select signals are configured using 32 configuration registers. The load registers are loaded with Data for each port individually by asserting the Load and CS (Chip Select) signals. The Address lines are decoded to select an output port's register, and the Data lines are latched to the selected register. After the load registers have been configured, the Config and CS signals are asserted, configuring all 32 output ports simultaneously. The Reset signal clears all the registers so that all the output ports select the input port 0, entering a broadcast state.

Figure 15.10 shows a possible structure of a switch that uses information from the incoming data to make the port selection (self-routing). The incoming data contain the destination address information that can be used by the switch controller to select an output port. There are 32 input ports and 32 output ports. Through the configuration registers, each output port can be selected individually. The FIFO buffers located at input ports are needed for storing at least the address information.

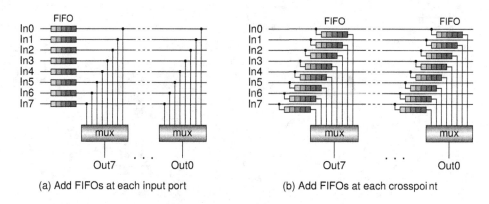

(a) Add FIFOs at each input port   (b) Add FIFOs at each crosspoint

Figure 15.8   An 8 × 8 buffered crossbar switch

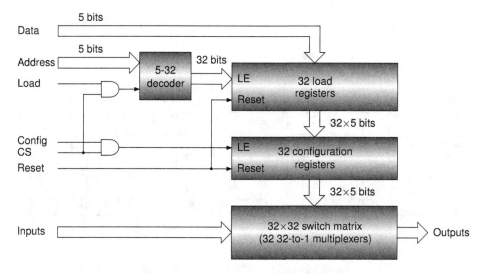

Figure 15.9   Implementation of crossbar

15.4   Topological Properties of Interconnection Networks

We take a ring and a complete graph network as examples to describe the topological properties of interconnection networks, including the node degree, network diameter, average distance, and bisection bandwidth.

Figure 15.11 shows two networks, a ring and a complete graph, both consisting of eight nodes. In a ring network, each node connects to exactly two other nodes. In a complete graph, every pair of distinct nodes is connected by a unique link. The complete graph is also known as a fully connected network.

15.4.1   Node Degree

Two nodes are neighbors if there is a link connecting them. The degree of a node is defined as the number of its neighbors. It can be also measured in terms of number of links of a node. The node degree of a ring is 2, irrespective of the number of nodes. The node degree of a complete graph with N nodes is $N - 1$.

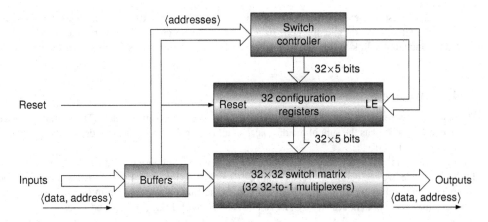

Figure 15.10   Crossbar with self-routing function

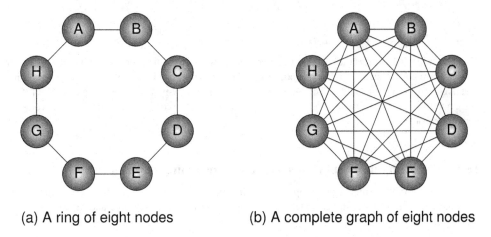

(a) A ring of eight nodes   (b) A complete graph of eight nodes

Figure 15.11   Ring and complete graph topologies

The node degree affects the cost of the network switch. As we discussed in the previous section, the switch of a node is usually implemented with a crossbar. Therefore, the cost is proportional to the square of the node degree.

15.4.2   Network Diameter

The diameter of a network is defined as the maximum of the shortest distances between any two nodes. The distance is measured in terms of the number of links in a path connecting two nodes. The diameter of the eight-node ring in Figure 15.11(a) is 4. Generally, the diameter of an N-node ring is $\lfloor N/2 \rfloor$. The diameter of a complete graph is always 1, irrespective of the number of nodes.

The network diameter affects the communication time between two nodes. The larger the diameter, the longer the communication time. We want to have a network that has a low node degree (low cost) and short diameter (fast communication). The ring has a low node degree, but its diameter is long. Imagine a system consisting of one million nodes; its diameter will be half a million. On the other hand, a complete graph has a shortest diameter of 1 because there is a link between any two distinct nodes, but the node degree is too large. For example, the node degree of a complete graph with one million nodes will be 1,000,000 (including a link to the CPU/memory board). It is almost impossible to design such a switch.

15.4.3   Average Distance

The distance between any two nodes is the length of the shortest path between them. The average distance of a network is defined as the sum of the distances of all node pairs (including a node and itself) divided by the number of pairs.

If the network is symmetric, the calculation can be simplified. Suppose there are N nodes in a system. Taking any node, we can get the average distance by calculating the sum of the distances between it and all nodes, and dividing the sum by N. For example, in the ring network shown in Figure 15.11(a), the distances from node A to nodes A, B, C, D, E, F, G, and H are 0, 1, 2, 3, 4, 3, 2, and 1, respectively. The average distance is $(0 + 1 + 2 + 3 + 4 + 3 + 2 + 1)/8 = 2$. Generally, the average distance of an N-node ring is $N/4$ if N is even, and $(N + 1)(N - 1)/(4N)$ or $N/4 - 1/(4N)$ otherwise. The average distance of the complete graph is $(N - 1)/N$.

15.4.4   Bisection Bandwidth

The bisection bandwidth is the smallest number of links we have to cut in order to separate the network into two parts of the same number of nodes (±1). The bisection bandwidth of a ring is always 2, irrespective of the number of nodes. The bisection bandwidth of an N-node complete graph is $N^2/4$ if N is even, and $(N^2 - 1)/4$ otherwise. The bisection bandwidth can be thought of as a measure of worst case network capacity.

15.5   Some Popular Topologies of Interconnection Networks

This section introduces some popular topologies of interconnection networks, including two-dimensional (2D) mesh and torus, three-dimensional (3D) mesh and torus, hypercube, tree and fat tree, and k-ary n-cube.

15.5.1   2D/3D Mesh

Figure 15.12 shows a 4-ary 2D mesh and a 3-ary 3D mesh.

An m-ary n-dimensional mesh has m^n nodes. The degree of interior nodes is $2n$, and the diameter is $n(m - 1)$. Each node u has an address $\{u_1, u_2, \ldots, u_n\}$, where $0 \leq u_i \leq m - 1$ for $1 \leq i \leq n$. Two nodes $\{u_1, u_2, \ldots, u_n\}$ and $\{v_1, v_2, \ldots, v_n\}$ are connected if and only if their addresses differ in one and only one dimension, say dimension i, and $|u_i - v_i| = 1$. Nodes along each dimension are connected as an m-node linear array.

15.5.2   2D/3D Torus

Mesh networks are not symmetric because the degrees of the boundary and corner nodes are different from those of interior nodes. Figure 15.13 shows two symmetric networks: a 4-ary 2D torus and a 3-ary 3D torus.

An m-ary n-dimensional torus has m^n nodes. The node degree is $2n$, and the diameter is $n\lfloor m/2 \rfloor$. Each node u has an address $\{u_1, u_2, \ldots, u_n\}$, where $0 \leq u_i \leq m - 1$. Two nodes $\{u_1, \ldots, u_i, \ldots, u_n\}$

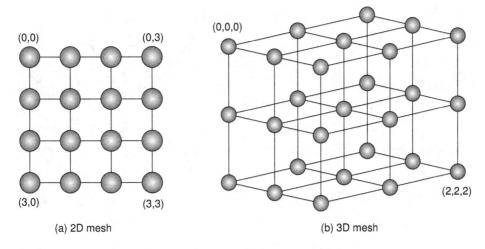

(a) 2D mesh (b) 3D mesh

Figure 15.12 Mesh topology

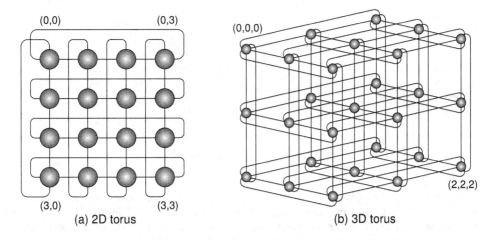

(a) 2D torus (b) 3D torus

Figure 15.13 Torus topology

and $\{u_1, \ldots, (u_i + 1) \bmod m, \ldots, u_n\}$ are connected for $1 \leq i \leq n$. Nodes along each dimension are connected as an m-node ring. Because of its features of symmetry and low node degree, torus topology is used by IBM and Cray Inc. for constructing their supercomputers. The drawback of the torus is that the diameter becomes large as the number of nodes increases. For example, a 1,000,000-node 3D torus has a diameter of 150.

15.5.3 Hypercubes

An n-dimensional hypercube, also called an n-cube, has the same topology of an m-ary n-dimensional mesh with a constant $m = 2$. An n-cube has 2^n nodes. Both the node degree and diameter are n. Figure 15.14 shows n-cubes with $n = 0, 1, 2, 3, 4, 5$.

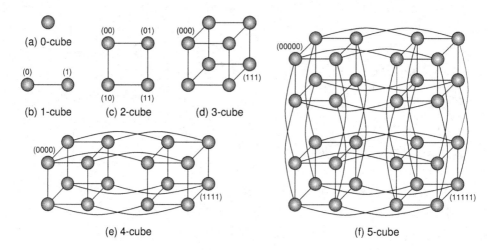

Figure 15.14   Hypercube topology

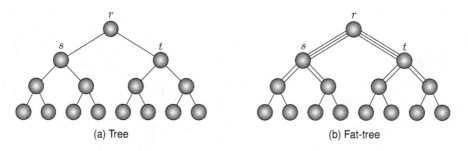

Figure 15.15   Tree and fat-tree topologies

Each node in an n-cube has an n-bit binary address. Two nodes are connected if and only if their addresses differ in exactly 1 bit. An n-cube can be constructed recursively: duplicate the $(n-1)$-cube including node addresses, add links between nodes with the same address in the original and duplicate, and prefix a 0 to each node address in the original and a 1 to each node address in the duplicate.

Hypercubes were widely used as interconnection networks for constructing supercomputers and servers. The drawback of the hypercubes is that the node degree increases logarithmically as the number of nodes increases. Because of the limitation on the number of switch ports, the scale of pure hypercube parallel systems is not large.

15.5.4   Tree and Fat Tree

A tree is defined as a connected graph without cycles. There is one and only one path joining any two nodes. An N-node tree has $N-1$ links. Figure 15.15(a) shows a complete binary tree with $N = 2^4 - 1 = 15$.

In a binary tree, each node has at most two child nodes and a parent node. A node without a parent node is called the root, which is the ancestor of all nodes. A node without children is called a leaf. A complete binary tree is a binary tree in which every node has two children except for the leaves, and all the leaves are at the same depth (same distance from the root).

Table 15.1 Topological properties of popular interconnection networks

Topology	Nodes	Degree	Diameter	Ave. distance	Bisection
1D Array	N	2	$N-1$	$(N-1)/2$	1
2D Mesh	N	4	$2(\sqrt{N}-1)$	$\sqrt{N}-1$	$\sqrt{N}$
3D Mesh	N	6	$3(\sqrt[3]{N}-1)$	$3(\sqrt[3]{N}-1)/2$	$\sqrt[3/2]{N}$
1D Ring	N	2	$N/2$	$N/4$	2
2D Torus	N	4	$\sqrt{N}$	$\sqrt{N}/2$	$2\sqrt{N}$
3D Torus	N	6	$3\sqrt[3]{N}/2$	$3\sqrt[3]{N}/4$	$2\sqrt[3/2]{N}$
Hypercube	2^n	n	n	$n/2$	2^{n-1}
k-ary n-cube	k^n	$2n$	$nk/2$	$nk/4$	$nk/4$
Binary tree	2^k-1	3	$2(k-1)$	$k-1$	1

Table 15.2 Types of collective communications

Pattern	Broadcast	Personalized communication
One-to-all	One-to-all broadcast	One-to-all personalized communication
All-to-all	All-to-all broadcast	All-to-all personalized communication

The root node becomes a bottleneck for communication. To solve this problem, we can use the fat tree, as shown in Figure 15.15(b). A fat tree is a tree in which the number of links between nodes increases closer to the root, that is, the links get "fatter" toward the top of the tree. Tree and fat tree were also used for building supercomputers.

15.5.5 Summary

The interconnection network that connects all the nodes together plays an important role in the parallel and distributed high-performance systems. An interconnection network consists of switches and links. The switch provides multiple communication ports and is usually designed with a crossbar. The links connect the switch ports with a certain topology.

Ring, mesh, torus, and hypercube are topologically isomorphic to a family of k-ary n-cube networks, where n is the dimension of the cube and k is the radix, or number of nodes in each dimension. Table 15.1 summarizes the topological properties of some popular interconnection networks.

The 3D torus, hypercube, and fat-tree topologies are widely used in real, high-performance supercomputers. The node degree and diameter of an interconnection network are two important measures for achieving high performance of supercomputers at low cost. What we want is an interconnection network that has a low node degree and a short diameter.

15.6 Collective Communications

This section describes collective communications on a hypercube. The types of the collective communications are listed in Table 15.2.

We assume the CPU-bounded model (one-port model) in which each CPU can access the network through a single input port and a single output port at a time. We use the store-and-forward switching

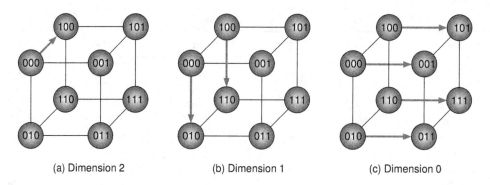

(a) Dimension 2   (b) Dimension 1   (c) Dimension 0

Figure 15.16   One-to-all broadcast in a hypercube

method: a switch will wait to forward a message until it has received the entire message. The communication time for a message of length m words to be sent to a neighbor node is evaluated with $t_s + mt_w$, where t_s is the start-up latency and t_w is the per-word transfer time. Another switching method is cut-through: switches begin forwarding the message as soon as they have read the destination address.

15.6.1   One-to-All Broadcast

Parallel algorithms often require a single CPU to send identical data to all other CPUs or to a subset of them. This operation is known as one-to-all broadcast or one-to-many multicast. Figure 15.16 shows the case where node 000 performs one-to-all broadcast in a 3-cube.

At the first step, node 000 sends the data to node 100 along with the dimension 2 of the node address. This takes $t_s + mt_w$ time. At the second step, nodes 000 and 100 send the data to nodes 010 and 110, respectively, along with the dimension 1. This also takes $t_s + mt_w$ time. At the third step, nodes 000, 010, 100, and 110 send the data to nodes 001, 011, 101, and 111, respectively, along with the dimension 0. This also takes $t_s + mt_w$ time. After this step, all the nodes have the data. Therefore, the total time is $3(t_s + mt_w)$.

Generally, the total time for a hypercube node performing one-to-all broadcast is $t_{oab} = (\log_2 N)(t_s + mt_w)$, where N is the number of nodes in the hypercube.

15.6.2   All-to-All Broadcast

All-to-all broadcast is a generalization of one-to-all broadcast in which all nodes simultaneously initiate a broadcast. A node broadcasts the same m-word message to every other node, but different nodes may broadcast different messages. The communication pattern of all-to-all broadcast can be used to perform some other operations, such as reduction and prefix sums.

Figure 15.17 shows the case where every node performs broadcast in a 3-cube. Initially, each node i has a message (i), which will be sent to other nodes. In the first step, each node sends its data to its neighbor along with the dimension 2 of the node address. This takes $t_s + mt_w$ time. After this step, each node holds the data doubled in size: the data of itself and the data received. In the second step, each node sends the double-sized data to its neighbor along with the dimension 1. This takes $t_s + 2mt_w$ time. Similarly, in the third step, each node sends the data to its neighbor along with the dimension 0. This takes $t_s + 4mt_w$ time. After this step, all the nodes have the data from all others. Therefore, the total time is $3t_s + (1 + 2 + 4)mt_w = 3t_s + 7mt_w$.

Generally, the total time for performing the all-to-all broadcast in a hypercube is $t_{aab} = (\log_2 N)t_s + (N - 1)mt_w$, where N is the number of nodes in the hypercube.

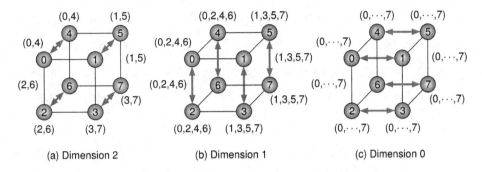

Figure 15.17 All-to-all broadcast in a hypercube

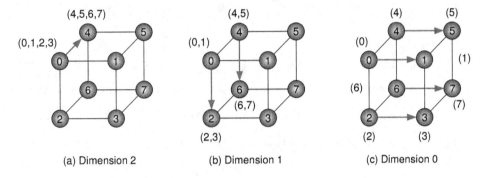

Figure 15.18 One-to-all personalized communication in a hypercube

15.6.3 One-to-All Personalized Communication

In one-to-all personalized communication, a single node sends a unique message of size m to every other node. Figure 15.18 shows the case where node 000 performs one-to-all personalized communication in a 3-cube.

Initially, node 0 has eight messages (assuming it also has a message to itself). In the first step, node 0 sends four messages to node 4 along with the dimension 2 of the node address. This takes $t_s + 4mt_w$ time. In the second step, nodes 0 and 4 send two messages to nodes 2 and 6, respectively, along with the dimension 1. It takes $t_s + 2mt_w$ time. In the third step, nodes 0, 2, 4, and 6 send one message to nodes 1, 3, 5, and 7, respectively, along with the dimension 0. This takes $t_s + mt_w$ time. After this step, all nodes have a unique message of node 0. Therefore, the total time is $3t_s + (4 + 2 + 1)mt_w = 3t_s + 7mt_w$.

Generally, the total time for a hypercube node performing one-to-all personalized communication is $t_{oap} = (\log_2 N)t_s + (N - 1)mt_w$, which is the same as the all-to-all broadcast time, where N is the number of nodes in the hypercube.

15.6.4 All-to-All Personalized Communication

In all-to-all personalized communication, each node sends a distinct message of size m to every other node. The total number of messages is N^2 (assuming each node also has a message to itself), where N is the total number of nodes. All-to-all personalized communication is the most complex pattern in

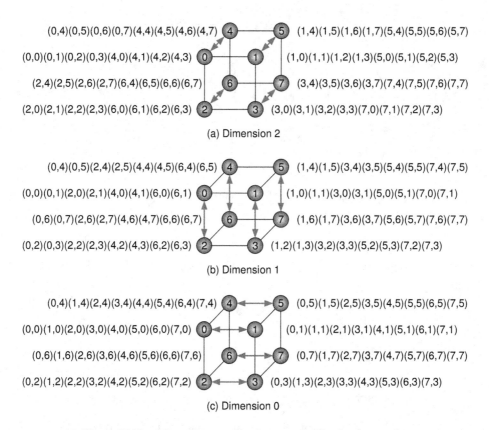

(a) Dimension 2

(b) Dimension 1

(c) Dimension 0

Figure 15.19   All-to-all personalized communication in a hypercube

collective communications. It is also known as total exchange. Figure 15.19 shows the case where every node performs personalized communication in a 3-cube.

Initially, each node has eight messages. The message (i,j) will be sent from node i to node j, for $0 \le i,j \le 7$. In the first step, each node sends four messages to its neighbor along with the dimension 2 of the node address. This takes $t_s + 4mt_w$ time. In the second step, each node sends the repacked four messages to its neighbor along with the dimension 1. It also takes $t_s + 4mt_w$ time. Similarly, in the third step, each node sends the repacked four messages to its neighbor along with the dimension 0. This also takes $t_s + 4mt_w$ time. After this step, all the nodes have a distinct message from every other node. Therefore, the total time is $3(t_s + 4mt_w)$.

Generally, the total time for performing all-to-all personalized communication in a hypercube is $t_{aap} = (\log_2 N)(t_s + Nmt_w/2)$, where N is the number of nodes in the hypercube.

15.7   Low-Node-Degree Short-Diameter Interconnection Networks

Because of the advances in computer and networking technologies, supercomputers containing millions of nodes have been built. The interconnection network plays an important role for achieving high performance in such ultrascale parallel systems. The node degree and the diameter will be the critical measures for the efficiency of the interconnection networks. The node degree is limited by the hardware technologies, and the diameter affects all kinds of communication schemes directly.

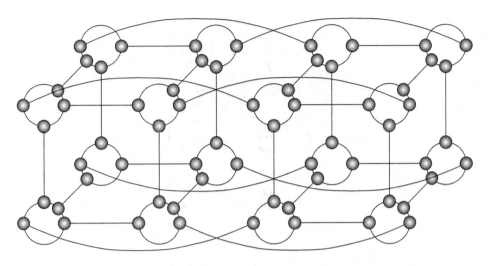

Figure 15.20   Cube-connected cycles

This section introduces some interconnection networks that have low node degree, or short diameter, or both.

15.7.1   Cube-Connected Cycle (CCC)

In an n-cube, each node has a degree n, and the total number of nodes is $N = 2^n$. As the system becomes large, the node degree n increases logarithmically to N. For example, a 21-port switch is required by each CPU in a 20-cube system which has 1,048,576 nodes. One port is used for connecting the CPU/memory board.

An n-dimensional cube-connected cycle CCC (n) has a low and fixed node degree. It replaces an n-cube node with an n-node ring and adds a link to each node of the ring for the hypercube connection; thus each node has only three links. A CCC(4) is shown in Figure 15.20.

The total number of nodes of a CCC(n) is $n2^n$. The node degree is 3, irrespective of n. The diameter is $2n + \lfloor n/2 \rfloor - 2$ for any $n \geq 4$.

15.7.2   Hierarchical Cubic Networks (HCNs)

The hierarchical cubic network HCN(n) has 2^n clusters and a cluster is an n-cube. A node address is denoted as (x, y), where x and y are n-bit numbers. Each node (x, y) is adjacent to (i) $(x, y^{(k)})$ for all $0 \leq k \leq n - 1$, where $y^{(k)}$ differs from y at the kth bit position; (ii) (y, x) if $x \neq y$; and (iii) $(\bar{x}, \bar{y})$ if $x = y$, where $\bar{x}$ and $\bar{y}$ are the bitwise complements of x and y, respectively.

Figure 15.21 shows an HCN(2). The total number of nodes of an HCN(n) is n^{2n}; the node degree is $n + 1$; and the diameter is $n + \lfloor (n + 1)/3 \rfloor + 1$.

15.7.3   Dual-Cube

The dual-cube is motivated from Origin 2000—the multiprocessor systems produced by SGI. An Origin 2000 is constructed with the hypercube or folded cube when the number of processors is not too large (≤ 64 CPUs). Figure 15.22 shows a 3D Origin 2000 and a 4D Origin 2000.

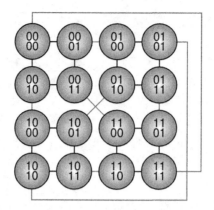

Figure 15.21 Hierarchical cubic network

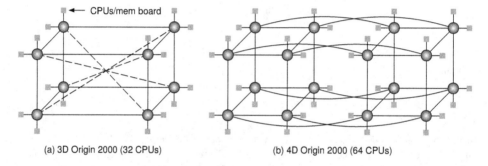

(a) 3D Origin 2000 (32 CPUs)   (b) 4D Origin 2000 (64 CPUs)

Figure 15.22 Origin 2000 of three and four dimensions

A switch (router) in Origin 2000 has six ports (links). As shown in Figure 15.22(b), two links are used to connect to compute boards, and four links are used to build a 4-cube. Each compute board contains two CPUs and the main memory. Therefore, the 4D Origin 2000 has $2^4 \times 2 \times 2$, or 64 CPUs. Figure 15.22(a) shows a folded cube configuration that contains 32 CPUs.

As shown in Figure 15.23, Origin 2000 reduces the number of links required when the system size increases by introducing a Cray Router to connect the clusters. A Cray Router is a high-level switch that does not connect CPUs directly. The CPUs are attached to regular switches within the clusters. Three links of the regular switch are used to build a 3-cube, and one link is connected to the Cray Router.

There are eight switches in a Cray Router. There are no connections among these switches. Each switch has four links connecting to four regular switches, forming a star topology. To build a hypercube would require routers with seven links. However, this configuration is made up of six-link switches. There are four clusters, and each cluster has 32 CPUs. Therefore, the total number of CPUs is 128.

The dual-cube we will introduce next can connect $32 \times 16 = 512$ CPUs with six-link regular switches and without using a Cray Router.

The node address of a dual-cube DC(m) has $2m + 1$ bits, as shown in Figure 15.24. The most significant bit is the Class ID. For the Class 0, the m least significant bits are the Node ID, and the remaining m bits are the Cluster ID. For the Class 1, the m least significant bits are the Cluster ID, and the remaining m bits are Node ID.

There are 2^m clusters in each class. A cluster is an m-cube: if the addresses of two nodes inside a cluster differ in 1 bit in Node ID, there is a link connecting the two nodes. We call these links the cube edges.

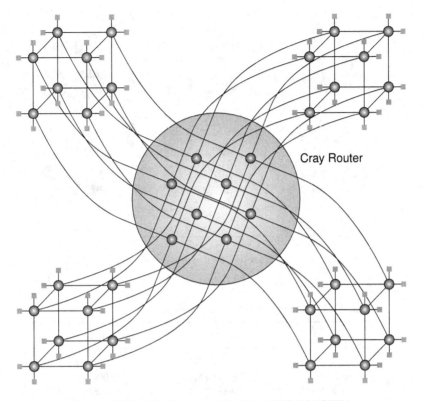

Figure 15.23 Cray Router in 5D Origin 2000 (128 CPUs)

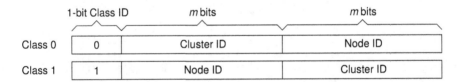

Figure 15.24 Node address format of a dual-cube

There is also a cross edge if the addresses of two nodes differ only in the Class ID. For example, the node 00000 in DC(2) is connected to nodes 00001 and 00010 with cube edges and to node 10000 with cross edge. Another example is where the node 10000 in a DC(2) is connected to nodes 10100 and 11000 with cube edges and to node 00000 with cross edge.

The number of nodes in a DC(m) is 2^{2m+1}, and the node degree is $m + 1$. Figure 15.25 depicts a DC(2). In each node, the Class ID is shown at the top position. For the nodes of class 0 (class 1), the Node ID (Cluster ID) is shown at the bottom and the Cluster ID (Node ID) is shown at the middle. There are four clusters in each class. Each cluster is a 2-cube. The total 32 nodes are connected with three-port switches. For building a hypercube of 32 nodes, five-port switches are required.

Table 15.3 shows the cases of routing from node s to node t in a DC(4). The routing in a dual-cube is almost the same as that in a hypercube. The last case has the longest distance. From the table, we know that the diameter of a DC(m) is $2m + 2$, which is only one longer than that of the n-cube where $n = 2m + 1$, but the number of links in the dual-cube is about half the links in the hypercube.

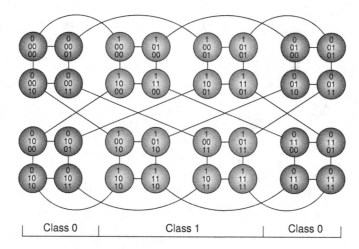

Figure 15.25   Dual-cube DC(2)

Table 15.3   Three cases of dual-cube routing

Same cluster	Different classes	Same class, different clusters
s = 0 0000 0000	s = 0 0000 0000	s = 0 0000 0000
0 0000 0001	0 0000 0001	0 0000 0001
0 0000 0011	0 0000 0011	0 0000 0011
0 0000 0111	0 0000 0111	0 0000 0111
t = 0 0000 1111	0 0000 1111	0 0000 1111
	1 0000 1111	1 0000 1111
	1 0001 1111	1 0001 1111
	1 0011 1111	1 0011 1111
	1 0111 1111	1 0111 1111
	t = 1 1111 1111	1 1111 1111
		t = 0 1111 1111

By using the dual-cube DC(3), we can build a system like Origin 2000. Referring to Figure 15.26, a cluster is a 3-cube which contains 32 CPUs. There are 16 clusters. The total number of CPUs is 512. For simplicity, the figure shows only those cross-edges connecting to a cluster. This configuration is also made up of six-link switches and does not use the Cray Router.

15.7.4   Metacube

A metacube (MC) is an extension of the dual-cube. In a dual-cube DC(m), the Class ID has only 1 bit, and two m-bit fields are used to denote the Cluster ID and the Node ID. In a metacube MC(k, m), we extend the Class ID to k bits. Correspondingly, there are 2^k fields, and each field has m bits.

Figure 15.27 shows the node address format of the metacube. The address of MC(k, m) has $k + m2^k$ bits: $(c, m_{2^k-1}, \ldots, m_c, \ldots, m_0)$. The k-bit c is the Class ID, the m-bit m_c is the Node ID, and the $(m(2^k - 1))$-bit $(m_{2^k-1}, \ldots, m_{c+1}, m_{c-1}, \ldots, m_0)$ is the Cluster ID.

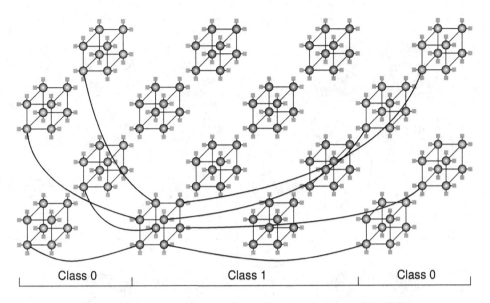

Figure 15.26 Implementing Origin 2000 with a dual-cube (512 CPUs)

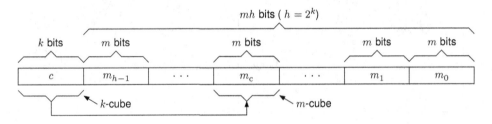

Figure 15.27 Node address format of an m-cube oriented metacube

The node degree of an MC(k, m) is $k + m$. Inside a cluster, the m-bit Node ID forms an m-cube. Two nodes in distinct clusters are linked if they differ in one and only one bit in the Class ID (k-cube). For example, the neighbors in the cluster of the node with address (01,111,101,110,000) in an MC(2, 3) have addresses (01,111,101,11$\underline{1}$,000), (01,111,101,1$\underline{0}$0,000), and (01,111,101,$\underline{0}$10,000). The underlined bits are those that differ from the corresponding bits in the address of the referenced node. The two neighbors in the high-level cube are (0$\underline{0}$,111,101,110,000) and ($\underline{1}$1,111,101,110,000).

Figure 15.28 shows an MC(2,2). It has $2^2 = 4$ classes, each class has $2^{2(2^2-1)} = 64$ clusters, and each cluster is a 2-cube. The total number of node is 1024, and the node degree is 4. The figure shows only the cross-edges in the four high-level cubes, each of which contains a node in the cluster 0 of the class 0.

With the node degree $k + m$, an MC(k, m) can connect 2^{k+m2^k} nodes. The value of k affects strongly the growth rate of the size of the network. An MC(1, m) containing 2^{2m+1} nodes is a dual-cube, which was already described. Similarly, an MC(2, m), an MC(3, m), and an MC(4, m) containing 2^{4m+2} nodes, 2^{8m+3} nodes, and 2^{16m+4} nodes, respectively, are called the quad-cube, the oct-cube, and the hex-cube, respectively. Since an MC(3, 3) contains $2^{27} = 134, 217, 728$ nodes, the oct-cube is sufficient to construct practically parallel computers of very large size. The hex-cube is of theoretical interest only. Note that MC(0, m) is a traditional hypercube (m-cube).

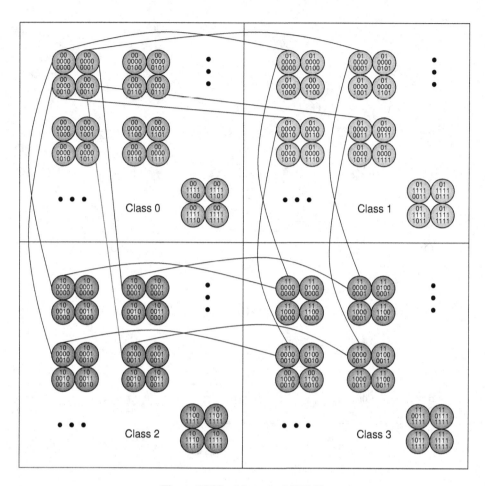

Figure 15.28   Metacube MC(2,2)

Table 15.4   Number of nodes in metacubes

Node degree	3	4	5	6	7	8
Hypercube	8	16	32	64	128	256
$MC(1, m)$	32	128	512	2048	8192	32768
$MC(2, m)$	64	1024	16384	2^{18}	2^{22}	2^{26}
$MC(3, m)$	–	2048	2^{19}	2^{27}	2^{35}	2^{43}
$MC(4, m)$	–	–	2^{20}	2^{36}	2^{52}	2^{68}

Table 15.4 lists the number of nodes a metacube can connect. We can see that an MC(2, 3), which requires five links per node, can have $2^{2+3\times2^2}$, or 16,384 nodes. To build a system of the same size with a hypercube, each node requires 14 links. The diameter of an MC(2, 3) is 16, only 2 more than that of the hypercube.

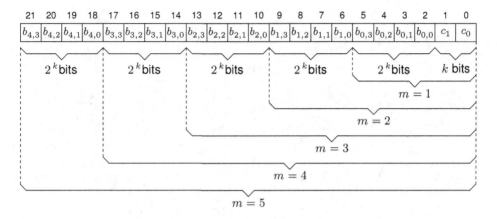

Figure 15.29 Node address format of a k-cube-oriented metacube

The metacube described above is m-cube-oriented. It can be also constructed in a k-cube-oriented manner. The address format of a k-cube-oriented metacube MC(k, m) is shown in Figure 15.29.

The k-bit Class ID $c = c_{k-1} \ldots c_0$ is placed in the rightmost position. These k bits form a k-cube. The rest (Cluster ID and Node ID) are divided into m parts, and each part contains 2^k bits. We use $b_{i,j}$ to denote a bit in these IDs, where $i = 0, 1, \ldots, m - 1$ and $j = 0, 1, \ldots, 2^k - 1$. Each part has only one link at $b_{i,c}$, where $c = c_{k-1} \ldots c_0$ is the Class ID for each $i = 0, 1, \ldots, m - 1$. Thus, m parts contribute m links, which form an m-cube. For example, the three neighbors in an m-cube of the node with address $(1100, 0110, 1011, 10)$ in an MC(2, 3) have addresses $(1100, 0110, 1\underline{1}11, 10)$, $(1100, 0\underline{0}10, 1011, 10)$, and $(1\underline{0}00, 0110, 1011, 10)$. The underlined bits are those that differ from the corresponding bits in the address of the referenced node. The two neighbors in the k-cube are $(1100, 0110, 1011, 1\underline{1})$ and $(1100, 0110, 1011, \underline{0}0)$.

Figures 15.30 and 15.31 show two k-cube-oriented metacubes, an MC(1,2) and an MC(2,1), respectively. The number in parentheses in the figures is the node address in decimal format.

The MC(1,2) shown in Figure 15.30 is a dual-cube that has the same topology as that shown in Figure 15.25. From the figures we can see that the k-cube-oriented metacubes are easier to layout the links when we design the networks on a chip.

15.7.5 Recursive Dual-Net (RDN)

A recursive dual-net (RDN) can connect a very large number of nodes with low node degree and short diameter. An RDN(m, k) can be recursively defined as follows: RDN($m, 0$) is a symmetric graph with m nodes, called the base network. Any symmetric network, such as a hypercube or a torus, can be used as the base network. For $k > 0$, RDN(m, k) is constructed from RDN($m, k - 1$) by a dual construction as explained below.

Referring to Figure 15.32, let RDN($m, k - 1$) be a cluster of level k and $n = |$ RDN($m, k - 1$)$|$. An RDN(m, k) is a graph that contains $2n$ clusters of level k as subgraphs. These clusters are divided into two sets, with each set containing n clusters. A cluster in one set is said to be of Type 0, denoted as C_i^0, where i is the cluster ID for $0 \leq i \leq n - 1$. A cluster in the other set is of Type 1, denoted as C_j^1, where j is the cluster ID for $0 \leq j \leq n - 1$. At level k, each node in every cluster has a new link to a node in a distinct cluster of the other type. We call this link the cross-edge of level k. By following this rule, for each pair of clusters C_i^0 and C_j^1, there is a unique cross-edge connecting a node in C_i^0 and a node in C_j^1, for $0 \leq i, j \leq n - 1$. Then, we get an RDN(m, k) with $|$ RDN(m, k)$| = 2n^2$ nodes.

Figure 15.33 depicts an RDN(4, 1) network with a 2-cube as its base network.

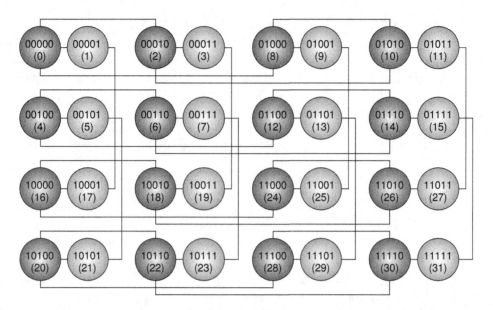

Figure 15.30   A k-cube-oriented metacube MC(1,2)

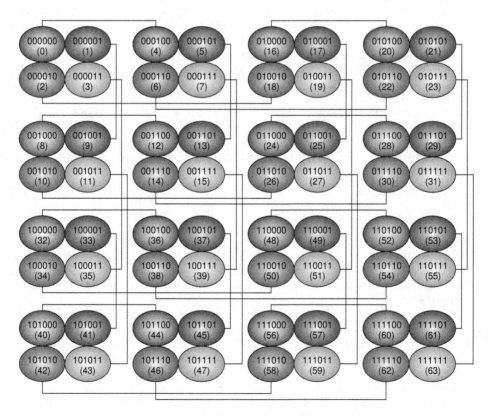

Figure 15.31   A k-cube-oriented metacube MC(2,1)

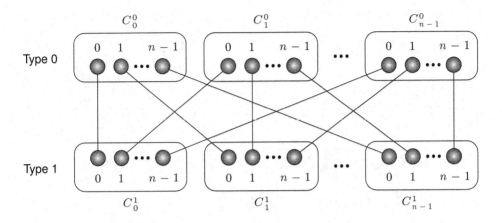

Figure 15.32 Recursive construction of RDN

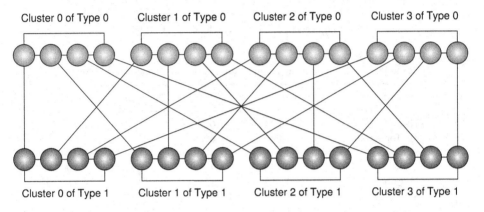

Figure 15.33 Recursive dual-net RDN(4,1)

There are four nodes in the base network. For $k = 1$, there are four clusters in each type, and each cluster is a 2-cube. The total number of nodes is $2 \times 4 \times 4 = 32$. The node degree is $2 + 1 = 3$. The diameter is $1 + 2 + 1 + 2 = 6$.

Figure 15.34 shows the RDN(4, 2), constructed from the RDN(4, 1). There are 32 clusters in each type, and each cluster is an RDN(4, 1). The total number of nodes is $2 \times 32 \times 32 = 2048$. The node degree is $2 + 2 = 4$. The diameter is $1 + 6 + 1 + 6 = 14$.

A k-level RDN(m, k) contains $(2m)^{2^k}/2$ nodes with node degree $d_0 + k$, where m and d_0 are the number of nodes and the node degree, respectively, of the base network. Concerning the diameter D_k of an RDN(m, k), we know that the worst case for the shortest path connecting any two nodes u and v is that u and v are in different clusters of the same type. The path must go through a cluster of the other type. Therefore, the diameter of RDN(m, k) satisfies the recurrence condition $D_k = (1 + D_{k-1}) + (1 + D_{k-1})$ for $k > 0$. Solving the recurrence, we get $D_k = 2^k D_0 + 2^{k+1} - 2$, where D_0 is the diameter of the base network.

The RDN is node- and edge-symmetric and can contain huge number of nodes with small node-degree and short diameter. For example, we can construct a symmetric RDN$(5^2, 2)$ connecting more than 3 million nodes with only six links per node and a diameter of 22, where the base network is a 5×5 torus.

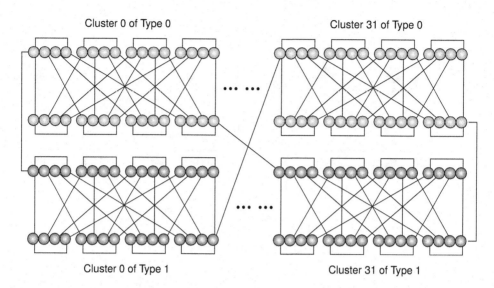

Cluster 0 of Type 0   Cluster 31 of Type 0

Cluster 0 of Type 1   Cluster 31 of Type 1

Figure 15.34   Recursive dual-net RDN(4,2)

Table 15.5   Cost ratio comparison of symmetric interconnection networks

Network	Nodes	Degree	Diameter	Cost ratio
10-Cube	1,024	10	10	1.00
3D Torus (10)	1,000	6	15	1.05
RDN(5^2, 1)	1,250	5	10	0.73
RDN(3^3, 1)	1,458	7	8	0.71
21-Cube	2,097,152	21	21	1.00
3D Torus (128)	2,097,152	6	192	4.72
DC(10)	2,097,152	11	22	0.74
RDN(5^2, 2)	3,125,000	6	22	0.65
RDN(3^3, 2)	4,251,528	8	18	0.59
RDN(5, 3)	50,000,000	5	30	0.68

We use the cost ratio $CR(G)$ to measure the cost of an interconnection network. Let $|(G)|$, $d(G)$, and $D(G)$ be the number of nodes, the node degree, and the diameter of G, respectively. We define $CR(G)$ as

$$CR(G) = \frac{d(G) \times w_1 + D(G) \times w_2}{\log_2 |(G)|}$$

where w_1 and w_2 are the weights of the degree and diameter, respectively, with $w_1 + w_2 = 100\%$. It is clear from the definition of $CR(G)$ that the smaller the value of $CR(G)$, the better the graph G. The cost ratio of an n-cube is

$$CR(n\text{-cube}) = \frac{n \times w_1 + n \times w_2}{n} = w_1 + w_2 = 1$$

irrespective of n, w_1, and w_2. It can be considered as a base performance to be compared to other networks. Table 15.5 summarizes the number of nodes, the node degree, the diameter, and the cost ratio for a 3D torus, hypercube, dual-cube, and RDNs, with $w_1 = w_2 = 50\%$.

Table 15.6 Topological properties of interconnection networks

Network	Nodes	Degree	Diameter
3D Torus	n^3	6	$3\lfloor n/2 \rfloor$
n-cube	2^n	n	n
CCC(n)	$n2^n$	3	$2n + \lfloor n/2 \rfloor - 2$
HCN(n)	2^{2n}	$n + 1$	$n + \lfloor (n+1)/3 \rfloor + 1$
Dual-cube(n)	2^{2n-1}	n	$2n$
Metacube(k, m)	$2^{m2^k + k}$	$m + k$	$(m+1)2^k$
RDN(m, k)	$(2m)^{2^k}/2$	$d_0 + k$	$2^k D_0 + 2^{k+1} - 2$

The base network of RDN(5^2, 1) and RDN(5^2, 2) is a 5×5 two-dimensional torus; the base network of RDN(3^3, 1) and RDN(3^3, 2) is a $3 \times 3 \times 3$ three-dimensional torus; and the base network of RDN(5, 3) is a five-node ring. From the table we know that RDN(3^3, 2) has the lowest cost ratio. It has 4,251,528 nodes with a degree of 8, and the diameter is just 18.

The benefit of RDN is that we can select any interesting interconnection network as the base network. Through adjusting the base network size and the recursion level k, we can construct a supercomputer of a desirable size.

As the summary of this section, we list the topological properties of the introduced interconnection networks in Table 15.6.

This is the final chapter. Thanks for choosing and reading this book. Good luck!

Exercises

15.1 What are the differences between a multiprocessor system and a multicomputer system?

15.2 Compare the singly linked list and doubly linked list in the directory-based cache coherency protocols.

15.3 Prove that the average distance of the n-cube is $n/2$.

15.4 Derive the diameter equation of CCC(n).

15.5 Derive the average distance equation of the m-ary n-dimensional mesh.

15.6 Derive the average distance equation of the m-ary n-dimensional torus.

15.7 Develop algorithms for performing collective communications (one-to-all broadcast, all-to-all broadcast, one-to-all personalized communication, and total exchange) on CCC, HCN, dual-cube, metacube, and RDN.

15.8 Investigate the star graph and pancake networks.

15.9 Investigate the multistage switch-based interconnection networks Omega, Benes, and Clos.

15.10 Design a switch circuit with eight ports that can perform self-routing based on the destination information in the coming data in Verilog HDL (hardware description language).

Bibliography

[1] J. Hennessy and D. Patterson. *Computer Architecture: A Quantitative Approach*. Morgan Kaufmann Publishers, Inc., San Francisco, CA, 2006.

[2] IDT. IDT Interprise IDT79RC32355, Integrated Communications Processor RISCore 32300 Family, User Reference Manual. 2003.

[3] IEEE. IEEE 754 *Standard for Binary Floating-Point Arithmetic*. IEEE, 1985.

[4] IEEE. IEEE Standard Verilog Hardware Description Language, IEEE Std 1364-2001 (Revision of IEEE Std 1364-1995). 2001.

[5] V. Kumar, A. Grama, A. Gupta, and G. Karypis. *Introduction to Parallel Computing: Design and Analysis of Algorithms*. Benjamin/Cummings Press, Redwood City, CA, 1994.

[6] Y. Li. *Computer Organization and Architecture*. The Press of Tsinghua University, 2000 (in Chinese).

[7] Y. Li. *Computer Principles and Design in Verilog HDL*. The Press of Tsinghua University, 2011 (in Chinese).

[8] Y. Li. AsmSim: a web-based graphical education support tool for teaching the course of assembly language programming. In *Proceedings of the IET International Conference on Frontier Computing—Theory, Technologies and Applications*, pages 18–24, Xining, China, August 2012.

[9] Y. Li and W. Chu. A performance prediction model for a parallel multithreaded RISC processor architecture. In *Proceedings of the Sixth IASTED International Conference on Parallel and Distributed Computing and System*, pages 162–166, Washington, DC, October 1994.

[10] Y. Li and W. Chu. The effects of STEF in finely parallel multithreaded processors. In *Proceedings of the First IEEE Symposium on High-Performance Computer Architecture (HPCA-1)*, pages 318–325, North Carolina, January 1995.

[11] Y. Li and W. Chu. Aizup: a pipelined processor design and implementation on Xilinx FPGA chip. In *Proceedings of the IEEE Symposium on FPGAs for Custom Computing Machines*, pages 98–106, Napa, CA, April 1996.

[12] Y. Li and W. Chu. A new non-restoring square root algorithm and its VLSI implementations. In *Proceedings of the International Conference on Computer Design—VLSI in Computers and Processors*, pages 538–544, Austin, TX, October 1996.

[13] Y. Li and W. Chu. Using computer architecture/organization at the University of Aizu. In *Workshop on Computer Architecture Education (WCAE-2)*, San Jose, CA, February 1996.

[14] Y. Li and W. Chu. Implementation of single precision floating-point square root on FPGAs. In *Proceedings of the IEEE Symposium on FPGAs for Custom Computing Machines*, pages 226–232, Napa, CA, April 1997.

[15] Y. Li and W. Chu. Parallel-array implementations of a non-restoring square root algorithm. In *Proceedings of the International Conference on Computer Design—VLSI in Computers and Processors*, pages 690–695, Austin, TX, October 1997.

[16] S. Li and Y. Li. *RISC Architecture: Single Issue and Multiple Issues*. The Press of Tsinghua University, 1993 (in Chinese).

[17] Y. Li, S. Peng, and W. Chu. Efficient collective communications in dual-cube. *The Journal of Supercomputing*, 28(1):71–90, 2004.

[18] Y. Li, S. Peng, and W. Chu. Metacube: a versatile family of interconnection networks for extremely large-scale supercomputers. *The Journal of Supercomputing*, 53(2):329–351, 2010.

[19] Y. Li, S. Peng, and W. Chu. Recursive dual-net: a new versatile network for supercomputers of the next generation. *Journal of Chinese Institute of Engineer*, 32(7):931–938, 2009.

[20] MIPS. *MIPS32 Architecture For Programmers Volume I, II, III*. MIPS Technologies, Inc., 2001.

[21] K. Pagiamtzis and A. Sheikholeslami. Content-addressable memory (CAM) circuits and architectures: a tutorial and survey. *IEEE Journal of Solid-State Circuits*, 41(3):712–727, 2006.

[22] Y. Patt and S. Patel. *Introduction to Computing Systems: From Bits and Gates to C and Beyond* (2nd Edition). McGraw-Hill Science Engineering, 2003.

[23] D. Patterson and J. Hennessy. *Computer Organization and Design: The Hardware/Software Interface*. Morgan Kaufmann Publishers, Inc., San Francisco, CA, 2008.

[24] PCISIG. *PCI Local Bus Specification, Revision 2.3*. PCI Special Interest Group, March 2002.

[25] Philips Semiconductors. *The I2C-Bus Specification*. NXP, January 2000.

[26] E. M. Schwarz, L. Sigal, and T. J. McPherson. CMOS floating-point unit for the S/390 parallel enterprise server G4. *IBM Journal of Research and Development*, 41(4/5):475–488, 1997.

[27] S. Suzuki. *The Digital Circuit Design*. CQ Press, 2006 (in Japanese).

[28] Texas Instrument. FIFO Architecture, Functions, and Applications. November 1999.

[29] Z. Zhu and Y. Li. *CPU Chip Logic Design*. The Press of Tsinghua University, January 2005 (in Chinese).

Index